Handbook of
Plant Growth

BOOKS IN SOILS, PLANTS, AND THE ENVIRONMENT

Soil Biochemistry, Volume 1, edited by A. D. McLaren and G. H. Peterson
Soil Biochemistry, Volume 2, edited by A. D. McLaren and J. Skujiņš
Soil Biochemistry, Volume 3, edited by E. A. Paul and A. D. McLaren
Soil Biochemistry, Volume 4, edited by E. A. Paul and A. D. McLaren
Soil Biochemistry, Volume 5, edited by E. A. Paul and J. N. Ladd
Soil Biochemistry, Volume 6, edited by Jean-Marc Bollag and G. Stotzky
Soil Biochemistry, Volume 7, edited by G. Stotzky and Jean-Marc Bollag
Soil Biochemistry, Volume 8, edited by Jean-Marc Bollag and G. Stotzky
Soil Biochemistry, Volume 9, edited by G. Stotzky and Jean-Marc Bollag
Soil Biochemistry, Volume 10, edited by Jean-Marc Bollag and G. Stotzky

Organic Chemicals in the Soil Environment, Volumes 1 and 2, edited by C. A. I. Goring and J. W. Hamaker
Humic Substances in the Environment, M. Schnitzer and S. U. Khan
Microbial Life in the Soil: An Introduction, T. Hattori
Principles of Soil Chemistry, Kim H. Tan
Soil Analysis: Instrumental Techniques and Related Procedures, edited by Keith A. Smith
Soil Reclamation Processes: Microbiological Analyses and Applications, edited by Robert L. Tate III and Donald A. Klein
Symbiotic Nitrogen Fixation Technology, edited by Gerald H. Elkan
Soil–Water Interactions: Mechanisms and Applications, Shingo Iwata and Toshio Tabuchi with Benno P. Warkentin

Soil Analysis: Modern Instrumental Techniques, Second Edition, edited by Keith A. Smith

Soil Analysis: Physical Methods, edited by Keith A. Smith and Chris E. Mullins

Growth and Mineral Nutrition of Field Crops, N. K. Fageria, V. C. Baligar, and Charles Allan Jones

Semiarid Lands and Deserts: Soil Resource and Reclamation, edited by J. Skujiņš

Plant Roots: The Hidden Half, edited by Yoav Waisel, Amram Eshel, and Uzi Kafkafi

Plant Biochemical Regulators, edited by Harold W. Gausman

Maximizing Crop Yields, N. K. Fageria

Transgenic Plants: Fundamentals and Applications, edited by Andrew Hiatt

Soil Microbial Ecology: Applications in Agricultural and Environmental Management, edited by F. Blaine Metting, Jr.

Principles of Soil Chemistry: Second Edition, Kim H. Tan

Water Flow in Soils, edited by Tsuyoshi Miyazaki

Handbook of Plant and Crop Stress, edited by Mohammad Pessarakli

Genetic Improvement of Field Crops, edited by Gustavo A. Slafer

Agricultural Field Experiments: Design and Analysis, Roger G. Petersen

Environmental Soil Science, Kim H. Tan

Mechanisms of Plant Growth and Improved Productivity: Modern Approaches, edited by Amarjit S. Basra

Selenium in the Environment, edited by W. T. Frankenberger, Jr., and Sally Benson

Plant–Environment Interactions, edited by Robert E. Wilkinson

Handbook of Plant and Crop Physiology, edited by Mohammad Pessarakli

Handbook of Phytoalexin Metabolism and Action, edited by M. Daniel and R. P. Purkayastha

Soil–Water Interactions: Mechanisms and Applications, Second Edition, Revised and Expanded, Shingo Iwata, Toshio Tabuchi, and Benno P. Warkentin

Stored-Grain Ecosystems, edited by Digvir S. Jayas, Noel D. G. White, and William E. Muir

Agrochemicals from Natural Products, edited by C. R. A. Godfrey

Seed Development and Germination, edited by Jaime Kigel and Gad Galili

Nitrogen Fertilization in the Environment, edited by Peter Edward Bacon

Phytohormones in Soils: Microbial Production and Function, William T. Frankenberger, Jr., and Muhammad Arshad

Handbook of Weed Management Systems, edited by Albert E. Smith

Soil Sampling, Preparation, and Analysis, Kim H. Tan

Soil Erosion, Conservation, and Rehabilitation, edited by Menachem Agassi

Plant Roots: The Hidden Half, Second Edition, Revised and Expanded, edited by Yoav Waisel, Amram Eshel, and Uzi Kafkafi

Photoassimilate Distribution in Plants and Crops: Source–Sink Relationships, edited by Eli Zamski and Arthur A. Schaffer

Mass Spectrometry of Soils, edited by Thomas W. Boutton and Shinichi Yamasaki

Handbook of Photosynthesis, edited by Mohammad Pessarakli

Chemical and Isotopic Groundwater Hydrology: The Applied Approach, Second Edition, Revised and Expanded, Emanuel Mazor

Fauna in Soil Ecosystems: Recycling Processes, Nutrient Fluxes, and Agricultural Production, edited by Gero Benckiser

Soil and Plant Analysis in Sustainable Agriculture and Environment, edited by Teresa Hood and J. Benton Jones, Jr.

Seeds Handbook: Biology, Production, Processing, and Storage, B. B. Desai, P. M. Kotecha, and D. K. Salunkhe

Modern Soil Microbiology, edited by J. D. van Elsas, J. T. Trevors, and E. M. H. Wellington

Growth and Mineral Nutrition of Field Crops: Second Edition, N. K. Fageria, V. C. Baligar, and Charles Allan Jones

Fungal Pathogenesis in Plants and Crops: Molecular Biology and Host Defense Mechanisms, P. Vidhyasekaran

Plant Pathogen Detection and Disease Diagnosis, P. Narayanasamy

Agricultural Systems Modeling and Simulation, edited by Robert M. Peart and R. Bruce Curry

Agricultural Biotechnology, edited by Arie Altman

Plant–Microbe Interactions and Biological Control, edited by Greg J. Boland and L. David Kuykendall

Handbook of Soil Conditioners: Substances That Enhance the Physical Properties of Soil, edited by Arthur Wallace and Richard E. Terry

Environmental Chemistry of Selenium, edited by William T. Frankenberger, Jr., and Richard A. Engberg

Principles of Soil Chemistry: Third Edition, Revised and Expanded, Kim H. Tan

Sulfur in the Environment, edited by Douglas G. Maynard

Soil–Machine Interactions: A Finite Element Perspective, edited by Jie Shen and Radhey Lal Kushwaha

Mycotoxins in Agriculture and Food Safety, edited by Kaushal K. Sinha and Deepak Bhatnagar

Plant Amino Acids: Biochemistry and Biotechnology, edited by Bijay K. Singh

Handbook of Functional Plant Ecology, edited by Francisco I. Pugnaire and Fernando Valladares

Handbook of Plant and Crop Stress: Second Edition, Revised and Expanded, edited by Mohammad Pessarakli

Plant Responses to Environmental Stresses: From Phytohormones to Genome Reorganization, edited by H. R. Lerner

Handbook of Pest Management, edited by John R. Ruberson

Environmental Soil Science: Second Edition, Revised and Expanded, Kim H. Tan

Microbial Endophytes, edited by Charles W. Bacon and James F. White, Jr.

Plant–Environment Interactions: Second Edition, edited by Robert E. Wilkinson

Microbial Pest Control, Sushil K. Khetan

Soil and Environmental Analysis: Physical Methods, Second Edition, Revised and Expanded, edited by Keith A. Smith and Chris E. Mullins

The Rhizosphere: Biochemistry and Organic Substances at the Soil–Plant Interface, Roberto Pinton, Zeno Varanini, and Paolo Nannipieri

Woody Plants and Woody Plant Management: Ecology, Safety, and Environmental Impact, Rodney W. Bovey

Metals in the Environment: Analysis by Biodiversity, M. N. V. Prasad

Plant Pathogen Detection and Disease Diagnosis: Second Edition, Revised and Expanded, P. Narayanasamy

Handbook of Plant and Crop Physiology: Second Edition, Revised and Expanded, edited by Mohammad Pessarakli

Environmental Chemistry of Arsenic, edited by William T. Frankenberger, Jr.

Enzymes in the Environment: Activity, Ecology, and Applications, edited by Richard G. Burns and Richard P. Dick

Plant Roots: The Hidden Half, Third Edition, Revised and Expanded, edited by Yoav Waisel, Amram Eshel, and Uzi Kafkafi

Handbook of Plant Growth: pH as the Master Variable, edited by Zdenko Rengel

Biological Control of Crop Diseases, edited by Samuel S. Gnanamanickam

Additional Volumes in Preparation

Handbook of Postharvest Technology, edited by A. Chakraverty, Arun S. Mujumdar, G. S. V. Raghavan, and H. S. Ramaswamy

Plant Biotechnology and Transgenic Plants, edited by Kirsi-Marja Oksman-Caldentey and Wolfgang Barz

Pesticides in Agriculture and the Environment, edited by Willis B. Wheeler

Mathematical Models of Crop Growth and Yield, Allen R. Overman and Richard Scholtz

Handbook of Plant Growth

pH as the Master Variable

edited by

Zdenko Rengel
University of Western Australia
Perth, Western Australia, Australia

CRC Press
Taylor & Francis Group
Boca Raton   London   New York

CRC Press is an imprint of the
Taylor & Francis Group, an **informa** business

First published 2002 by Marcel Dekker, Inc.

Published 2019 by CRC Press
Taylor & Francis Group
6000 Broken Sound Parkway NW, Suite 300
Boca Raton, FL 33487-2742

First issued in paperback 2019

No claim to original U.S. Government works

ISBN 13: 978-0-367-44707-6 (pbk)
ISBN 13: 978-0-8247-0761-3 (hbk)

Visit the Taylor & Francis Web site at
http://www.taylorandfrancis.com

and the CRC Press Web site at
http://www.crcpress.com

Preface

Few would argue that pH is truly a master variable that permeates just about any area of study of living organisms and extends into the physical and chemical world in which living organisms come about, grow, develop, reproduce, and die. There is arguably no other master variable that captures the complexities of the interactions between the biological, physical, and chemical aspects of the world to a similar extent.

This book aims to provide a unifying view of the role of pH in plant growth, taking into account molecular, biochemical, functional, structural, and developmental factors in such growth, as well as environmental processes involved in plant interaction with the biotic and abiotic environment. The book tries to capture the multitude of roles played by H^+ ions in the processes that sustain life on this planet. It deals with pH in plant symplasm, plant apoplasm, the rhizosphere, the ecosystem, and soil biotic and abiotic components, thus covering plant life from the general environment all the way down to cell organelles and molecules. The book covers four main subjects: (1) dynamics of H^+ fluxes across membranes (plasma membrane, tonoplast, chloroplast thylakoids and mitochondria), (2) the role of H^+ activity (pH) in cellular, subcellular, and whole plant processes, (3) the role of pH and H^+ fluxes in soil biotic processes involving microorganisms as well as in soil–plant–microbe interactions, and (4) the interdependence of pH changes and soil abiotic processes (ion availability). The book covers a wide range of topics spanning many scientific disciplines, for example, plant biology, cell physiology, botany, microbiology, ecology, soil science, agronomy, and forestry.

All chapters have been reviewed according to the standards of high-impact international journals. I would like to thank the authors, who patiently went with me through a number of revisions of their chapters. I would also like to thank the Marcel Dekker, Inc., staff for capable handling of numerous issues and for their dedication to producing a high-quality multidisciplinary book.

Zdenko (Zed) Rengel

Contents

Preface   *iii*

Contributors   *vii*

1. H$^+$-ATPases in the Plasma Membrane: Physiology
   and Molecular Biology   1
   Thomas Jahn and Michael Gjedde Palmgren

2. H$^+$-ATPase and H$^+$-PPase in the Vacuolar Membrane:
   Physiology and Molecular Biology   23
   Masayoshi Maeshima and Yoichi Nakanishi

3. The Cytoplasmic pH Stat   49
   Robert J. Reid and F. Andrew Smith

4. Confocal pH Topography in Plant Cells: Shifts of Proton
   Distribution Involved in Plant Signaling   73
   Werner Roos

5. pH as a Signal and Regulator of Membrane Transport   107
   Hubert H. Felle

6. The Role of the Apoplastic pH in Cell Wall Extension
   and Cell Enlargement   131
   Robert E. Cleland

7. Mechanisms and Physiological Roles of Proton Movements
   in Plant Thylakoid Membranes   149
   W. S. Chow and Alexander B. Hope

8.   Dynamics of H^+ Fluxes in Mitochondrial Membrane   173
     Francis E. Sluse and Wiesława Jarmuszkiewicz

9. H^+ Fluxes in Nitrogen Assimilation by Plants   211
     Fernando Gallardo and Francisco M. Cánovas

10. Crassulacean Acid Metabolism: A Special Case of pH
     Regulation and H^+ Fluxes   227
     Karl-Josef Dietz and Dortje Golldack

11. Dynamics of H^+ Fluxes in the Plant Apoplast   255
     Jóska Gerendás and Burkhard Sattelmacher

12. H^+ Currents around Plant Roots   299
     Miguel A. Piñeros and Leon V. Kochian

13. Role of pH in Availability of Ions in Soil   323
     Zdenko Rengel

14. Regulation of Microbial Processes by Soil pH   351
     David E. Crowley and Samuel A. Alvey

15. The Role of Acid pH in Symbiosis between Plants
     and Soil Organisms   383
     Karen G. Ballen and Peter H. Graham

16. Distribution of Plant Species in Relation to pH of Soil
     and Water   405
     Jacqueline Baar and Jan G. M. Roelofs

Index   427

Contributors

Samuel A. Alvey Department of Environmental Sciences, University of California, Riverside, Riverside, California

Jacqueline Baar* Department of Aquatic Ecology and Environmental Biology, University of Nijmegen, Nijmegen, The Netherlands

Karen G. Ballen Biology Department, Augsburg College, Minneapolis, Minnesota

Francisco M. Cánovas Department of Molecular Biology and Biochemistry, Andalusian Institute of Biotechnology, University of Málaga, Málaga, Spain

W. S. Chow Research School of Biological Sciences, Australian National University, Canberra, Australia

Robert E. Cleland Department of Botany, University of Washington, Seattle, Washington

David E. Crowley Department of Environmental Sciences, University of California, Riverside, Riverside, California

Karl-Josef Dietz Department of Physiology and Biochemistry of Plants, University of Bielefeld, Bielefeld, Germany

Hubert H. Felle Botanisches Institut I, Justus Liebig University, Giessen, Germany

---

* Current affiliation: Department of Applied Plant Research, Wageningen University and Research Center, Horst, The Netherlands

Fernando Gallardo   Department of Molecular Biology and Biochemistry, Andalusian Institute of Biotechnology, University of Málaga, Málaga, Spain

Jóska Gerendás   Institute for Plant Nutrition and Soil Science, University of Kiel, Kiel, Germany

Dortje Golldack   Department of Physiology and Biochemistry of Plants, University of Bielefeld, Bielefeld, Germany

Peter H. Graham   Department of Soil, Water, and Climate, University of Minnesota, St. Paul, Minnesota

Alexander B. Hope   School of Biological Sciences, Flinders University, Adelaide, South Australia, Australia

Thomas Jahn   Department of Agricultural Sciences, The Royal Veterinary and Agricultural University, Copenhagen, Denmark

Wiesława Jarmuszkiewicz   Department of Bioenergetics, Adam Mickiewicz University, Poznan, Poland

Leon V. Kochian   U.S. Plant, Soil and Nutritional Laboratory, USDA-ARS, Cornell University, Ithaca, New York

Masayoshi Maeshima   Graduate School of Bioagricultural Sciences, Nagoya University, Nagoya, Japan

Yoichi Nakanishi   Graduate School of Bioagricultural Sciences, Nagoya University, Nagoya, Japan

Michael Gjedde Palmgren   Department of Agricultural Sciences, The Royal Veterinary and Agricultural University, Copenhagen, Denmark

Miguel A. Piñeros   U.S. Plant, Soil and Nutrition Laboratory, USDA-ARS, Cornell University, Ithaca, New York

Robert J. Reid   Department of Environmental Biology, Adelaide University, Adelaide, South Australia, Australia

Zdenko Rengel   Department of Soil Science and Plant Nutrition, The University of Western Australia, Perth, Western Australia, Australia

Jan G. M. Roelofs   Department of Aquatic Ecology and Environmental Biology, University of Nijmegen, Nijmegen, The Netherlands

Werner Roos   Institute of Pharmaceutical Biology, Martin Luther University, Halle (Saale), Germany

Burkhard Sattelmacher   Institute for Plant Nutrition and Soil Science, University of Kiel, Kiel, Germany

Francis E. Sluse   Laboratory of Bioenergetics, University of Liege, Liege, Belgium

F. Andrew Smith   Department of Soil and Water, Adelaide University, Adelaide, South Australia, Australia

Handbook of
Plant Growth

1

H^+-ATPases in the Plasma Membrane: Physiology and Molecular Biology

Thomas Jahn and Michael Gjedde Palmgren
The Royal Veterinary and Agricultural University, Copenhagen, Denmark

1 INTRODUCTION

A typical plant cell expresses four ATP-fueled proton pumps (H^+-ATPases), each one targeted to a specific cellular membrane. The F_oF_1 and CF_oCF_1 H^+-ATPases, present in the mitochondrial inner membrane and the thylakoid membrane, respectively, operate under physiological conditions to synthesize ATP at the expense of H^+ gradients. Vacuolar H^+-ATPase and plasma membrane H^+-ATPase, on the other hand, generate H^+ gradients at the expense of ATP. The plant plasma membrane H^+-ATPase has been extensively discussed in a number of recent reviews [1–5]. In the present chapter we give an overview of recent structural and functional aspects of the plasma membrane H^+-ATPase, including its regulation and its role in intra- and extracellular pH regulation.

2 EVOLUTION OF PROTON PUMPS

Plasma membrane H^+-ATPase is not evolutionarily related to any other plant proton pumps. This enzyme is composed of a single polypeptide of around 100 kDa (around 950 amino acid residues) and belongs to the superfamily of P-type ATPases, mainly cation pumps characterized by forming a phosphorylated reaction cycle intermediate and by being inhibited by vanadate [6,7].

Plasma membrane H^+-ATPases are ubiquitous in higher plants, algae, and fungi but have not been identified in animals and eubacteria. This would suggest a relatively late evolutionary origin of this class of proton pumps. A sequence related to plasma membrane H^+-ATPase has been identified in the genome of the archaebacterium *Methanococcus jannaschii* [8]. In the four other archaebacterial genomes available so far, no sequences with homology to H^+-ATPases have been observed. Also, a plasma membrane H^+-ATPase–like gene has been cloned from *Leishmania donovani*, a protozoan [9]. In present-day bacteria, P-type ATPases are involved in pumping K^+, Mg^{2+}, Ca^{2+}, Cu^{2+}, and Cd^{2+}, suggesting that, early in evolution, P-type ATPases mainly pumped divalent cations. Evolution of mechanisms for extrusion of metal ions from cells might have been necessary to prevent metal salts from precipitating in the cell or giving rise to toxic effects.

Other H^+ pumps such as F_oF_1 and CF_oCF_1 (both F-type) H^+-ATPases are multisubunit pumps that probably evolved early in evolution in order to extrude excess H^+ generated in anaerobic metabolism [10]. Later, other mechanisms for pH homeostasis evolved, and F_oF_1 ATPases acquired a new role in ATP synthesis by operating in the reverse direction.

3 MOLECULAR BIOLOGY OF THE PLASMA MEMBRANE H^+-ATPase

Every single plant species investigated has a large number of plasma membrane H^+-ATPase isoforms. Twelve different isoforms have been identified in *Arabidopsis* (according to the P-type ATPase database—PAT-base; http://biobase. dk/~axe/Patbase.html), and so far there is evidence for the expression of eight of these isoenzymes. At least nine isoforms are present in the genome of the tobacco *Nicotiana plumbaginifolia* [4]. The expression patterns of *Arabidopsis* and tobacco plasma membrane H^+-ATPase isoforms have been studied extensively using reporter genes [11,12] and immunohistochemical localization of epitope-tagged H^+-ATPase [13]. The picture that has emerged supports the view that each isoform is expressed in a tissue- and development-specific manner [11–13]. Furthermore, it has been established that in certain cell types more than one isoform might be expressed at the same time in the same cell [11].

It is not known why there are so many plasma membrane H^+-ATPases in a given plant. When expressed in the yeast *Saccharomyces cerevisiae*, different isoforms of *Arabidopsis* [14] and tobacco [15] plasma membrane H^+-ATPases exhibit quantitative differences with respect to a number of kinetic parameters. However, depending upon the isoform, heterologously expressed plant H^+-ATPase is phosphorylated in this host at the penultimate threonine residue, which might alter the properties of the enzyme [16,17]. Therefore, it is still unclear whether the functional differences observed between the H^+-ATPases produced in yeast are relevant in planta.

The current completion of eukaryotic genome sequencing programs has showed that multigene families are a normal feature of multicellular organisms. For example, more than 60% of the genes on chromosome 2 of *Arabidopsis thaliana* have a significant match with another *Arabidopsis* gene [18]. Gene duplication events might give rise to a large number of functionally more or less equivalent isoforms. Evolution of different promoter regions for such isoenzymes might be the simplest means for providing tissue- and development-specific expression of a given enzyme.

3.1 Structure and Function of Plasma Membrane H^+-ATPase

How a single polypeptide of rather limited size is able to execute the whole sequence of functions of ATP hydrolysis, energy coupling, energy transfer, as well as highly specific binding and release of the transported substrate is not well understood.

Hydrophobicity analyses in combination with biochemical studies have suggested the presence of 10 transmembrane helices, with most of the protein mass (about 70% of the H^+-ATPase) facing the cytoplasmic side of the plasma membrane [19]. The cytoplasmic part of the enzyme involves the N- and the C-termini, a relatively small loop of about 135 residues between transmembrane segments 2 and 3, a large loop of about 345 residues between transmembrane domains 4 and 5, and two very small loops of less than 20 residues connecting transmembrane helices 6–7 and 8–9. The large central cytoplasmic loop includes a conserved DKTGT sequence in which the aspartate becomes reversibly phosphorylated during catalysis. In addition to phosphorylation, the large central loop has been implicated in binding of ATP [7].

Our picture of the structure of the H^+-ATPase has been improved by data obtained by two-dimensional (2D) crystallization of the *Neurospora crassa* homologue (NcPMA1) [20]. This fungal H^+-ATPase forms well-ordered 2D crystals in which the protein appears in a hexameric structure. The polypeptide crosses the membrane 10 times and a big cytoplasmic domain is connected with the membrane-embedded part of the enzyme.

The transmembrane part of the H^+-ATPase most likely contains the H^+ binding site(s). Intensive mutagenesis analysis of various P-type ATPases has pointed to a role for transmembrane oxygen atoms in binding and coordination of the transported cations. In all plasma membrane H^+-ATPases, one aspartate residue is conserved in transmembrane helix 6 (D684 in *Arabidopsis* AHA2; D730 in *Saccharomyces cerevisiae* PMA1). Thus, this group is a candidate residue directly involved in H^+ coordination. In a recent study, all charged residues in the postulated transmembrane segments of *S. cerevisiae* PMA1 were mutagenized and the effects on proton pumping by the H^+-ATPase were tested [21]. Nearly half of the mutant ATPases (including D730) were misfolded and consequently not properly secreted to the plasma membrane. In this case, the role of D730 in binding of H^+ was difficult to ascertain. However, mutagenesis studies with PMA1 have demonstrated that H^+ pumping can take place in the absence of D730 provided that second-site mutations are introduced to remove the positive charge of R695 in neighboring transmembrane helix 5 [22]. Therefore, D730 is apparently not part of the H^+ translocation mechanism but rather is important for proper folding of the enzyme, probably by forming a salt bridge with R695. Two additional charged residues, E703 in transmembrane segment 5 and E803 in transmembrane segment 8, appear to be important for the coupling between ATP hydrolysis and H^+ pumping in PMA1 [21], but these residues are not conserved in plant plasma membrane H^+-ATPases.

The C-terminus of the plasma membrane H^+-ATPase functions as an autoinhibitor of enzyme activity [23,24]. Successive deletions [25] and single point mutations have revealed the presence of two inhibitory regions within the C-terminal domain [26–29] as well as a binding site for regulatory 14-3-3 protein (see later).

4 PHYSIOLOGY OF THE PLASMA MEMBRANE H^+-ATPase

The plant plasma membrane H^+-ATPase is thought to play a crucial role in a number of essential physiological processes [1,2]. Among them are energization of nutrient uptake, phloem loading, opening of stomata, as well as the regulation of extra- and intracellular pH (Fig. 1).

The plasma membrane H^+-ATPase is very abundant in cells that are specialized for nutrient acquisition. Here it plays a role in generation of the electrochemical gradient of H^+ that provides the driving force for uptake of solutes through channel proteins and H^+-coupled carriers. Cells specialized for nutrient uptake are, for example, those of the root epidermis, phloem companion cells, and the transfer cells of the xylem. The root is a specialized organ that functions in uptake of nutrients from the soil and translocation of those nutrients to other parts of the plant. In roots, plasma membrane H^+-ATPase has been shown to be highly abundant in the epidermis and vascular tissues [30–32]. In the corn root epidermis and outer cortical cells, the plasma membrane H^+-ATPase has been shown to be asymmetrically localized, with very high abundance in plasma membrane domains facing the root-soil interface [32]. Asymmetric localization of plasma membrane H^+-ATPase has also been shown in other cell types [33].

Phloem companion cells are rich in mitochondria and thus have the capacity to synthesize large amounts of ATP. The ATP produced is believed to fuel mainly

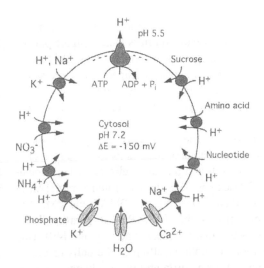

FIGURE 1 Plant plasma membrane H⁺-ATPase generates a membrane potential (negative on the inside of the cell) and a ΔpH gradient (acidic on the outside). This electrochemical gradient in turn can be used as an energy supply for transport proteins. Typical values for pH and membrane potential are indicated.

plasma membrane H⁺-ATPase, which is abundant in this cell type. In tobacco, promoter-GUS fusion studies have shown PMA4 to be expressed in companion cells but also in other tissues [12]. Based on epitope-tagging studies, *Arabidopsis thaliana* plasma membrane H⁺-ATPase AHA3 appears to be specifically expressed in phloem companion cells [13]. The H⁺ gradient established across the plasma membrane of companion cells is believed to energize sucrose uptake through H⁺-coupled sucrose transporters in the phloem [34]. Sucrose transporters were first localized in phloem companion cells in *Plantago major* [35] and *A. thaliana* [36] as the result of promoter-GUS fusion studies. Later, using immunolocalization and in situ hybridization techniques, sucrose transporters were localized in enucleated sieve elements of tobacco, potato, and tomato [37]. In the latter case, sucrose transporter *SUT1* messenger RNA (mRNA) and potentially SUT1 protein is targeted through plasmodesmata from the companion cells to the sieve elements. How energization of sucrose uptake through the plasma membrane of enucleate sieve elements takes place is still an open question. One possibility is that the mRNA of plasma membrane H⁺-ATPase is transported through plasmodesmata from companion cells to the sieve elements, where H⁺-ATPase is then synthesized. ATP would have to be transported in the same way as enucleate sieve elements lack mitochondria. Alternatively, the membrane potential generated by the H^+-ATPase in companion cells is transmitted to sieve elements, where it is used to drive uptake through sucrose transporters located in the plasma membrane of these cells. The possibility that plasma membrane H^+-ATPase is in fact present in the enucleate sieve elements but has escaped detection by the methods employed so far cannot be discarded. For example, modification of H^+-ATPase mRNA by insertion of a tag or a reporter gene might interfere with targeting to sieve elements.

An additional number of cell types in the plant body are specialized for intense active transport and contain high amounts of plasma membrane H^+-ATPase. Most prominent are stomatal guard cells and pulvinar cells. Guard cell localization of plasma membrane H^+-ATPases has been demonstrated for MHA2 in *Zea mays* [38], VHA1 and VHA2 in *Vicia faba* [39], and PMA2 and PMA4 in *Nicotiana plumbaginifolia* [12]. Activation of the H^+-ATPase in guard cells results in an increased uptake of potassium that precedes water uptake [40]. Due to the special wall anatomy of the cells, osmotic swelling of the cells results in opening of the stomatal pore, allowing transpiration and gas exchange with the environment of the plant. In accordance with this model, cosuppression of PMA4 gene expression in *N. plumbaginifolia* results in failure of stomatal opening [41]. Pulvinar cells are specialized cells functioning as osmotic motors driving leaf movements. Increased expression of plasma membrane H^+-ATPase in these cells [42] suggests a role for this enzyme in energizing the massive ion fluxes taking place here.

4.1 Role in Regulation of Apoplastic pH

Activation of proton pump activity in plant tissues results in an increase in acidification of external solutions, whose pH can be lowered by at least one unit concomitant with this hyperpolarization of the plasma membrane [43–45]. Although accurate determination of apoplastic pH is difficult, it has been accomplished by the use of pH-dependent fluorescent dyes [46,47] or sharp double-barreled microelectrodes [48]. Although the cell wall is buffered and able to maintain a steep pH gradient toward the external solution, compounds affecting plasma membrane H^+-ATPase positively such as fusicoccin or negatively such as vanadate provoke acidification and alkalization of the apoplast, respectively. However, the deviation from normal values (typically between pH 5.1 and 5.9) is not very dramatic and ranges from 0.1 to 0.6 pH units [46–48]. A more extensive discussion of factors contributing to regulation of apoplastic pH can be found in Chapter 11. Here the specific roles of plasma membrane H^+-ATPase are discussed.

Acidification at the root surface is important for increasing nutrient accessibility. This is due to the fact that most nutrients in the soil are not readily available for uptake by the plant. Soil particles can be of either organic or inorganic origin, but in both cases they are characterized by having negative charges on their surface. The negative charge of soil particles causes cationic mineral nutrients such as K^+, Ca^{2+}, Mg^{2+}, Mn^{2+}, Fe^{3+}, and Al^{3+} to be absorbed to their surface, whereas anions in principle do not bind to soil particles and remain dissolved. However, sulfate (SO_4^{2-}) is typically bound strongly to Ca^{2+}, whereas phosphate (PO_4^{3-}) forms very strong complexes with soil particles by replacing hydroxyl ions (OH^-) that are complexed to Fe^{2+}, Fe^{3+}, and Al^{3+}. Solubilization of nutrients bound to soil particles occurs by cation exchange with H^+, which is very efficient in exchanging bound cations. Factors affecting soil pH are the decomposition of organic matter, acidic rain, and active extrusion of organic acids and H^+ from the roots. Because the plasma membrane H^+-ATPase is responsible for the extrusion of H^+ from roots to the soil, its role in nutrient mobilization cannot be overestimated.

Acidification of the apoplast is believed to play an important role in plant physiology, namely in cell elongation growth. The plant cell wall is a complex cross-linked network of carbohydrate polymers limiting cell expansion. Therefore, for the cell to expand, this rigid structure has to be softened by breaking cross-linking bonds. This process appears to be strictly pH dependent. Growth-promoting substances such as the plant hormone auxin and the fungal toxin fusicoccin have promoted H^+ extrusion from plant tissues [49–51]. This increased H^+ secretion has been ascribed to increased activity of the plasma membrane H^+-ATPase. Rayle and Cleland [52] and Hager et al. [53] independently formulated the acid growth theory of auxin action that adopts elements of the growth theory proposed by Ruge [54]. According to this model, auxin-induced growth is triggered by acidification of the cell wall, resulting in disruption of chemical or physical bonds in the cell wall matrix. Loosening of cell wall bonds would then allow turgor-driven cell expansion in a well-ordered manner. Initiation of root hairs has been shown to be accompanied by, and strictly dependent on, the formation of local changes in apoplastic and cytoplasmic pH around the initiation zone [55]. This strongly suggests a role of the plasma membrane H^+-ATPase in the regulation of cell expansion.

A molecular mechanism for the acid-induced softening of cell walls is beginning to emerge. A number of tissues, such as hypocotyl segments, respond with an increase in plastic extensibility when incubated in buffers with a pH of 4–4.5. When such segments are treated with boiling to denature proteins, the cell walls refuse to expand. However, Cosgrove and coworkers [56–58] could show that addition of extracts of native cell wall proteins to the incubation solution restored the pH-sensitive response. This bioassay has been used to identify a class of cell wall proteins, so-called expansins. The ability of expansins to promote cell wall relaxation is strictly regulated by the pH of the apoplast. Expansins comprise a big family with a high degree of functional conservation and, although their biochemical function has not been clarified, they show sequence similarity to a family of endoglucanases [59]. Endoglucanases, xyloglucan endotransglycosylase, and other enzymes also modify cell wall structure during cell elongation but appear to act secondarily to expansin action [59].

In corn coleoptiles, the growth hormone auxin induces expression of an isoform of plasma membrane H^+-ATPase [38] concomitant with increased turnover of the enzyme [60], indicating that H^+-ATPase is an element of auxin-induced growth in this tissue. However, auxin-induced growth stimulation of tobacco leaf strips does not appear to involve plasma membrane H^+-ATPase, although cell wall loosening in some form does occur [61]. This would suggest that cell wall loosening induced by auxin does not require H^+-ATPase per se but may involve other proteins. Thus, a connection between auxin and the expression of cell wall–modifying proteins is beginning to emerge. The expression of tomato genes encoding xyloglucan endotransglycosylase (*LeEXT1*) and an endo-1,4-beta-glucanase (*Cel7*) is auxin regulated in etiolated hypocotyls [62]. Similarly, expression of expansin genes from tomato [63] and *Pinus taeda* [64] is upregulated in hypocotyls during incubation with auxin.

4.2 Role in Regulation of Cytoplasmic pH

Several factors contribute to the formation of a pH-stat that keeps cytoplasmic pH more or less constant. These factors are discussed in detail by Reid and Smith (Chapter 3). However, the specific role of plasma membrane H^+-ATPase will be discussed here.

As the result of each catalytic cycle, the plasma membrane H^+-ATPase transfers at least one proton out of the cytosol and into the apoplast [65]. Therefore, it would be expected that activation of proton pumping results in alkalization of the cytoplasm. Assuming a cellular volume of 20 pL, a typical cell contains about 10^8 H^+ at pH 7. A number of 1 million plasma membrane H^+-ATPase molecules per cell is probably not unreasonable; for example, much smaller kidney cells each contain more than a million Na^+/K^+-ATPase molecules in the plasma membrane [66]. Extrapolating from the maximal turnover rate of 6000 per minute for the purified enzyme (T. Jahn et al., unpublished data), 10^8 H^+ would be pumped out of the cell per second. Obviously, this should have dramatic implications for intracellular pH.

The pH dependence of the activity of the plasma membrane H^+-ATPase does not allow it to operate under alkaline conditions. The curve describing the pH versus activity profile of the enzyme is bell shaped with an optimum that is typically pH 6.5 (Fig. 2). The activity drops sharply when the pH becomes either more alkaline or acidic than the optimal pH. This pH dependence is highly dependent upon the ATP levels in the cell. Similarly, the ATP affinity of the pump is strictly linked to pH. The closer the pump is at its pH optimum, the higher its affinity for ATP is [25].

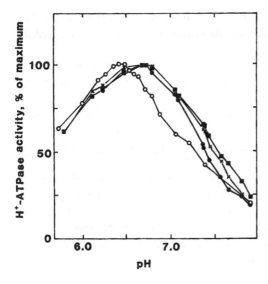

Figure 2 The pH optimum of plasma membrane H⁺-ATPase from oat roots is slightly acidic but moves toward physiological pH as the ATPase is activated, e.g., by lysophospholipids or free fatty acids. (O) No additions; (●) 24 µg/mL lysophosphatidylcholine; (×) 6 µM (18:3) linolenic acid; (■) 6 µM (20: 4) arachidonic acid. (Adapted from Ref. 82.)

The shape of the pH dependence profile implies that as the pH in the cell becomes alkaline, the activity of the pump diminishes. Similarly, if the pH in the cytosol acidifies from a typical value of pH 7.2, the pump increases in activity severalfold, resulting in adjustment of the pH back to normal. In this way the plasma membrane H⁺-ATPase by itself serves as an efficient pH-stat of the cell. This is illustrated by the difficulty in changing cytoplasmic pH, even by treating the plant cell with a compound known to activate H⁺-ATPase activity. Fusicoccin is a powerful agent causing activation of plasma membrane H⁺-ATPase (see later). An addition of 1 µM fusicoccin to root hairs of *Medicago sativa* resulted in an increase in plasma membrane potential of 30 mV, which is probably due to activation of H⁺-ATPase, but the pH of the cytoplasm remained constant at around pH 7 [67]. Other factors that might lead to a decrease in cytoplasmic pH are low oxygen and acidic soil, both of which are factors that might result in cytoplasmic acidosis. The plasma membrane H⁺-ATPase operates during anoxia and might therefore play a role under these conditions [68]. Overexpression of a modified form of AHA3 H⁺-ATPase in transgenic *Arabidopsis thaliana* led to increased growth at low pH compared with wild-type plants [69]. These data suggest that proton extrusion via the plasma membrane H^+-ATPase can contribute to stabilization of cytoplasmic pH.

A role for plasma membrane H^+-ATPases in intracellular pH regulation is supported by the data obtained from heterologous expression of regulatory mutants of plant plasma membrane H^+-ATPase in the yeast *Saccharomyces cerevisiae*. The more activated the plant H^+-ATPase, the higher its capability is to confer acid tolerance to yeast cells lacking the endogenous H^+-ATPase PMA1 [4,26]. In addition, plant isoforms with different pH optima when expressed in yeast support growth at different external pH values [15].

Sudden changes in the cytoplasmic pH are often seen in plant cells attacked by certain pathogens. One example is the rapid acidification of the cytosol of tobacco cells [70] and *Chenopodium rubrum* [71] after treatment with various fungal elicitors. The role of plasma membrane H^+-ATPase under these conditions is unclear. It is also questionable whether these changes in pH are related to elicitor-induced gene expression [71].

5 REGULATION OF THE PLASMA MEMBRANE H^+-ATPase

Because of the fundamental roles plasma membrane H^+-ATPase plays in plant physiology, the enzyme is likely to be subject to tight regulation. Indeed, regulation of the H^+-ATPase occurs at different levels. Salt stress [72,73], hydroponic growth conditions [74], and addition of exogenous glucose [75] or auxin [38] are factors that have been found to lead to increased expression of plasma membrane H^+-ATPase in a number of species. Increased H^+-ATPase activity concomitant with an increase in immunodetectable protein occurs in response to wounding [76] and aging [77]. The number of fusicoccin binding sites can be taken as a measure of the number of H^+-ATPases in complex with the activator 14-3-3 protein (see later). Auxin application to corn coleoptiles increased the number of fusicoccin receptors [78]. Under cold stress, an increase in fusicoccin binding activity in plasma membranes of suspension cultures of sugar beet cells was accompanied by an increase in H^+-ATPase activity [79].

Some plasma membrane H^+-ATPases with short open reading frames in the 5'-untranslated region appear to be regulated at the translational level [4,80]. The physiological significance of this phenomenon is at present unknown.

At least fusicoccin (see later), blue light [81], lysophosphatidylcholine and free fatty acids [82,83], and products of phospholipase A_2 action [84,85] activate proton pumping and ATPase activity by the plasma membrane H^+-ATPase via a mechanism not involving increased gene expression. Considerable attention has been paid to the study of posttranslational regulation of H^+-ATPase activity; this work is summarized next.

5.1 Posttranslational Regulation

The C-terminus of the H^+-ATPase functions as an autoinhibitory domain. Deletion of this domain either by trypsin treatment [24] or at the gene level [25] results in activation of the pump. The activated state is characterized by an increase in V_{max}, an increased affinity for ATP, a changed pH dependence with increased activity toward physiological pH (Fig. 2), and an increase in the ratio of H^+ pumped to ATP hydrolyzed. Based on these results, it has been hypothesized that in vivo activation of the H^+-ATPase involves displacement of the C-terminal from the catalytic site. In principle, this could occur (1) irreversibly via alternative splicing of mRNA transcripts or by proteolysis or (2) reversibly by displacement of the C-terminus by phosphorylation or by binding of a regulatory polypeptide.

5.1.1 The Role of the 14-3-3 Protein

The fungus *Fusicoccum amygdali* is a parasite that invades peach and almond trees by growing with its hyphae in the leaf parenchyma and participating in the primary metabolism of the host plants [86]. To gain access to the leaf parenchyma, the fungus spreads a toxin, fusicoccin, on the lower leaf surface, which promotes irreversible opening of stomata irrespective of the plants' endogenous regulation. When applied to different plant tissues, fusicoccin results in increased H^+ secretion, increased K^+ uptake, and an increase in turgor. Marrè [87] suggested early that fusicoccin was a direct activator of plasma membrane H^+-ATPase. Fifteen years later, 14-3-3 proteins were identified as being part of a fusicoccin receptor in plant cells [88–90]. The 14-3-3 proteins are ubiquitous in eukaryotes and are known as regulatory players in the cell cycle and the activity of various enzymes [91,92]. It is now known that the fusicoccin receptor is a complex of two proteins: 14-3-3 protein and plasma membrane H^+-ATPase [93–95]. Neither of the two proteins alone binds fusicoccin.

In plants, 14-3-3 proteins bind directly to the C-terminal regulatory domain of the H^+-ATPase [96,97]. Fusicoccin stabilizes the association between 14-3-3 protein and the C-terminal regulatory domain of the plant H^+-ATPase [96–98], resulting in an almost irreversible complex (K_d = 7 nM) [16] between the two proteins in which H^+-ATPase is stabilized in an active comformation [27]. The complex can also be formed in the absence of fusicoccin, but in such a case binding is phosphatase sensitive [99]. Similarly, heterologous expression studies have shown that activation of the plant H^+-ATPase in yeast involves binding of 14-3-3 protein to the H^+-ATPase in a phosphatase-sensitive manner [16].

Recently, direct evidence for a physiological role of 14-3-3 protein in activation of plasma membrane H^+-ATPase has been obtained. Blue light activates the plasma membrane H^+-ATPase in stomatal guard cells [81]. A pulse of blue light results in phosphorylation and subsequent binding of 14-3-3 protein to the

C-terminus of the H^+-ATPase in guard cells. The time course of phosphorylation and binding of 14-3-3 protein parallels an increase in H^+ pumping and ATP-hydrolytic activity in isolated microsomes.

5.1.2 Regulation by Protein Kinases and Phosphatases

Several lines of evidence indicate that plasma membrane H^+-ATPase is a target for protein kinase action in vivo, and apparently both serine and threonine residues in the pump molecule are phosphorylated [81,100,101]. Plasma membrane H^+-ATPases from the algae *Dunaliella acidophila* [102] and the higher plants *Vicia faba* and spinach are phosphorylated at their C-terminal end. Sequencing of peptides derived from spinach H^+-ATPase phosphorylated in vivo has led to the identification of an in vivo phosphorylation site at the penultimate C-terminal residue, a threonine [101].

The effect of phosphorylation on the activity of plasma membrane H^+-ATPase is highly disputed. Thus, it has been reported that H^+-ATPase activity is increased concomitantly with either phosphorylation [101,103] or dephosphorylation [104–107].

Binding of 14-3-3 protein to the H^+-ATPase involves its three C-terminal residues (YTV in almost all H^+-ATPase sequences so far) and requires phosphorylation of the penultimate threonine residue [16,17,108]. Based on these results, it is hypothezised that phosphorylation-dependent activation of the H^+-ATPase involves binding of 14-3-3 to the extreme end of the C-terminus and subsequent displacement of the autoinhibitory domain from the catalytic site (Fig. 3) [16,26,108].

In corn roots, a protein kinase with characteristics of a calcium-dependent protein kinase (CDPK) was partially purified and shown to phosphorylate the last 103 amino acids of the corn H^+-ATPase expressed as a recombinant protein in *Escherichia coli* [109]. However, phosphorylation did not induce binding of 14-3-3 protein, suggesting that the site is different from the penultimate C-terminal residue (see earlier). Thus, it appears that, apart from the threonine in the penultimate position [16,17,108], there is more than one regulatory phosphorylation site within the C-terminal domain. In addition, a sequence with homology to a common 14-3-3 binding site involving a phosphoserine has been identified in the small loop connecting transmembrane segments 8 and 9 of the H^+-ATPase [99]. However, it is not known whether the corresponding serine residue in the H^+-ATPase is phosphorylated in vivo. When a peptide corresponding to this sequence is phosphorylated at the serine residue, the peptide inhibits 14-3-3–dependent activation of the H^+-ATPase [110]. Similar results have been obtained using a phosphorylated peptide derived from the 14-3-3 protein binding site in Raf-1 kinase [26].

Phosphorylation of a second site in the C-terminal domain could in principle disrupt binding of 14-3-3 protein to the C-terminal end, thus leading to deacti-

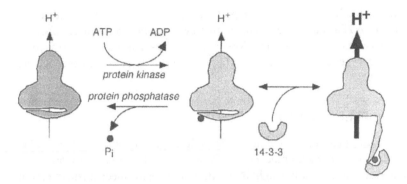

FIGURE 3   Model for regulation of plant plasma membrane H⁺-ATPase by a protein kinase, a protein phosphatase, and 14-3-3 protein. A protein kinase is likely to promote binding of activator 14-3-3 protein to the pump, whereas a specific protein phosphatase inhibits this association. Further phosphorylation of the H⁺-ATPase by a second protein kinase (not shown) might impair binding of 14-3-3 protein.

vation of the H⁺-ATPase. A phosphorylation-dependent inactivation mechanism for plasma membrane H⁺-ATPase has been suggested on the basis of elicitor studies. Thus, elicitor induced dephosphorylation of tomato H⁺-ATPase leads to increased H⁺ pumping [104]. In this case, two kinases, one with characteristics of a protein kinase C (PKC) and the second one with characteristics of a Ca²⁺/calmodulin-dependent kinase II, may be involved in rephosphorylation and concomitant inhibition of the H⁺-ATPase [106]. To draw further conclusions, it will be important to pinpoint the respective phosphorylation sites, as the involvement of different kinases does not per se exclude the possibility that the same modification occurred at the H⁺-ATPase.

The complex between plasma membrane H⁺-ATPase and 14-3-3 protein is labile in plant materials in the absence of fusicoccin. This is probably due to the presence of endogenous phosphatases in the preparation [99]. One such phosphatase having the plasma membrane H⁺-ATPase as a target has been partially purified from maize plasma membranes and was found to have characteristics resembling those of protein phosphatase 2A [99]. The activity of this enzyme disrupts the phosphorylation-dependent interaction between 14-3-3 and H⁺-ATPase and might function in vivo as a natural regulator of plasma membrane H⁺-ATPase activity.

Phosphorylation of the H⁺-ATPase might have additional roles, such as being part of maturation of the enzyme or its passage through the secretory pathway to the plasma membrane. In yeast, it has been shown that the H⁺-ATPase PMA1 becomes phosphorylated during intracellular trafficking [111]. The degree of phosphorylation increased along the path of intracellular trafficking and secretion of the H^+-ATPase to the plasma membrane. The molecular mechanism of this phenomenon is not understood. However, it seems that the phosphorylation of PMA1 is a necessary modification for a regulated targeting to the plasma membrane [111].

6 FUTURE PERSPECTIVES

In the near future, site-directed mutagenesis studies in combination with crystallization of plant plasma membrane H^+-ATPase and related P-type ATPases are expected to result in structural insight that will illuminate the mechanism of cation translocation by this class of ion pumps [112]. Phenotypic description of mutant plants carrying single or multiple knockouts in H^+-ATPase genes [113], or plants that by other means have altered expression of H^+-ATPase genes [41], will provide useful tools for analyzing the physiological roles of individual H^+-ATPase isoforms. Finally, bioimaging techniques, which allow detection of protein-protein interactions [114,115], movement of protein domains [116,117], and the activity state of proteins [118], should be applicable to plants [119] and would be useful in analyzing the physiological factors regulating plasma membrane H^+-ATPase activity and cellular trafficking in vivo.

REFERENCES

1. B Michelet, M Boutry. The plasma membrane H^+-ATPase. A highly regulated enzyme with multiple physiological functions. Plant Physiol 108:1–6, 1995.
2. MG Palmgren. Proton gradients and plant growth: role of the plasma membrane H^+-ATPase. Adv Bot Res 28:2–70, 1998.
3. H Sze, X Li, MG Palmgren. Energization of endomembranes and the plasma membrane by H^+ pumps: roles, isoforms, and regulation. Plant Cell 11:677–690, 1999.
4. P Morsomme, M Boutry. The plant plasma membrane H^+-ATPase: structure, function and regulation. Biochim Biophys Acta 1465:1–16, 2000.
5. F Portillo. Regulation of plasma membrane H^+-ATPase in fungi and plants. Biochim Biophys Acta 1469:31–42, 2000.
6. KB Axelsen, MG Palmgren. Evolution of substrate specificities in the P-type ATPase superfamily. J Mol Evol 46:84–101, 1998.
7. JV Møller, B Juul, M le Maire. Structural organization, ion transport, and energy transduction of P-type ATPases. Biochim Biophys Acta 1286:1–51, 1996.
8. CJ Bult, O White, GJ Olsen, L Zhou, RD Fleischmann, GG Sutton, JA Blake, LM Fitzgerald, RA Clayton, JD Gocayne, AR Kerlavage, BA Dougherty, JF Tomb, MD Adams, CI Reich, R Overbeek, EF Kirkness, KG Weinstock, JM Merrick, A Glodek, JL Scott, NSM Geoghagen, JC Venter. Complete genome sequence of the methanogenic archaeon, *Methanococcus jannaschii*. Science 273:1058–1073, 1996.

9. JC Meade, J Shaw, S Lemaster, G Gallagher, JR Stringer. Structure and expression of a tandem gene pair in *Leishmania donovani* that encodes a protein structurally homologous to eucaryotic cation-transporting ATPase. Mol Cell Biol 7:3937–3946, 1987.

10. VP Skulachev. Bioenergetics: the evolution of molecular mechanisms and the development of bioenergetic concepts. Antonie Leeuwenhoek 65:271–284, 1994.

11. JF Harper, L Manney, MR Sussman. The plasma membrane H⁺-ATPase gene family in *Arabidopsis*: genomic sequence of AHA10 which is expressed primarily in developing seeds. Mol Gen Genet 244:572–587, 1994.

12. L Moriau, B Michelet, P Bogaerts, L Lambert, A Michel, M Oufattole, M Boutry. Expression analysis of two gene subfamilies encoding the plasma membrane H⁺-ATPase in *Nicotiana plumbaginifolia* reveals the major transport functions of this enzyme. Plant J 19:31–41, 1999.

13. ND DeWitt, MR Sussman. Immunocytological localization of an epitope-tagged plasma membrane proton pump (H⁺-ATPase) in companion cells. Plant Cell 7: 2053–2067, 1995.

14. MG Palmgren, G Christensen. Functional comparisons between plant plasma membrane H⁺-ATPase isoforms expressed in yeast. J Biol Chem 269:3027–3033, 1994.

15. H Luo, P Morsomme, M Boutry. The two major types of plant plasma membrane H⁺-ATPases show different enzymatic properties and confer differential pH sensitivity of yeast growth. Plant Physiol 119:627–634, 1999.

16. AT Fuglsang, S Visconti, K Drumm, T Jahn, A Stensballe, B Mattei, ON Jensen, P Aducci, MG Palmgren. Binding of 14-3-3 protein to the plasma membrane H⁺-ATPase AHA2 involves the three C-terminal residues Tyr[946]-Thr-Val and requires phosphorylation of Thr[947]. J Biol Chem 274:36774–36780, 1999.

17. O Maudoux, H Batoko, C Oecking, K Gevaert, J Vandekerckhove, M Boutry, P Morsomme. A plant plasma membrane H⁺-ATPase expressed in yeast is activated by phosphorylation at its penultimate residue and binding of 14-3-3 regulatory proteins in the absence of fusicoccin. J Biol Chem 275:17762–17770, 2000.

18. X Lin, S Kaul, S Rounsley, TP Shea, MI Benito, CD Town, CY Fujii, T Mason, CL Bowman, M Barnstead, TV Feldblyum, CR Buell, KA Ketchum, J Lee, CM Ronning, HL Koo, KS Moffat, LA Cronin, M Shen, G Pai, S Van Aken, L Umayam, LJ Tallon, JE Gill, JC Venter, et al. Sequence and analysis of chromosome 2 of the plant *Arabidopsis thaliana*. Nature 402:761–768, 1999.

19. A Wach, A Schlesser, A Goffeau. An alignment of 17 deduced protein sequences from plant, fungi, and ciliate H⁺-ATPase genes. J Bioenerg Biomembr 24:309–317, 1992.

20. M Auer, GA Scarborough, W Kühlbrandt. Three-dimensional map of the plasma membrane H⁺-ATPase in the open conformation. Nature 392:840–843, 1998.

21. VV Petrov, KP Padmanabha, RK Nakamoto, KE Allen, CW Slayman. Functional role of charged residues in the transmembrane segments of the yeast plasma-membrane H⁺-ATPase. J Biol Chem 275:15709–15716, 2000.

22. SS Gupta, ND DeWitt, KE Allen, CW Slayman. Evidence for a salt bridge between transmembrane segments 5 and 6 of the yeast plasma-membrane H⁺-ATPase. J Biol Chem 273:34328–34334, 1998.

23. MG Palmgren, C Larsson, M Sommarin. Proteolytic activation of the plant plasma membrane H$^+$-ATPase by removal of a terminal segment. J Biol Chem 265:13423–13426, 1990.

24. MG Palmgren, M Sommarin, R Serrano, C Larsson. Identification of an autoinhibitory domain in the C-terminal region of the plant plasma membrane H$^+$-ATPase. J Biol Chem 267:20470–20475, 1991.

25. B Regenberg, JM Villalba, FC Lanfermeijer, MG Palmgren. C-terminal deletion analysis of plant plasma membrane H$^+$-ATPase: yeast as a model system for solute transport across the plant plasma membrane. Plant Cell 7:1655–1666, 1995.

26. KB Axelsen, K Venema, T Jahn, L Baunsgaard, MG Palmgren. Molecular dissection of the C-terminal regulatory domain of the plant plasma membrane H$^+$-ATPase AHA2: mapping of residues that when altered give rise to an activated enzyme. Biochemistry 38:7227–7234, 1999.

27. L Baunsgaard, K Venema, KB Axelsen, JM Villalba, A Welling, B Wollenweber, MG Palmgren. Modified plant plasma membrane H$^+$-ATPase with improved transport coupling efficiency identified by mutant selection in yeast. Plant J 10:451–458, 1996.

28. P Morsomme, A De Kerchove d'Exaerde, S De Meester, D Thines, A Goffeau, M Boutry. Single point mutations in various domains of a plant plasma membrane H$^+$-ATPase expressed in *Saccharomyces cerevisiae* increase H$^+$-pumping and permit yeast growth at low pH. EMBO J 15:5513–5526, 1996.

29. P Morsomme, S Dambly, O Maudoux, M Boutry. Single point mutations distributed in 10 soluble and membrane regions of the *Nicotiana plumbaginifolia* plasma membrane PMA2 H$^+$-ATPase activate the enzyme and modify the structure of the C-terminal region. J Biol Chem 273:34837–34842, 1998.

30. A Parets-Soler, JM Pardo, R Serrano. Immunocytolocalization of plasma membrane H$^+$-ATPase. Plant Physiol 93:1654–1658, 1990.

31. AL Samuels, L Fernando, ADM Glass. Immunofluorescent localization of plasma membrane H$^+$-ATPase in barley roots and effects of K nutrition. Plant Physiol 99:1509–1514, 1992.

32. T Jahn, F Baluska, W Michalke, JF Harper, D Volkmann. Plasma membrane H$^+$-ATPase in the maize root apex: evidence for strong expression in xylem parenchyma and asymmetric localization within cortical and epidermis cells. Physiol Plant 104:311–316, 1998.

33. S Bouchè-Pillon, P Fleurat-Lessard, R Serrano, JL Bonnemain. Asymmetric distribution of plasma membrane H$^+$-ATPase in embryos of *Vicia faba* L. with special reference to transfer cells. Planta 193:392–397, 1994.

34. C Kühn, L Barker, L Burkle, WB Frommer. Update on sucrose transport in higher plants. J Exp Bot 50:935–953, 1999.

35. R Stadler, J Brandner, A Schulz, M Gahrtz, N Sauer. Phloem loading by the PmSUC2 sucrose carrier from *Plantago major* occurs into companion cells. Plant Cell 7:1545–1554, 1995.

36. R Stadler, N Sauer. The *Arabidopsis thaliana AtSUC2* gene is specifically expressed in companion cells. Bot Acta 109:299–306, 1996.

37. C Kühn, VR Franceschi, A Schulz, R Lemoine, WB Frommer. Macromolecular

trafficking indicated by localization and turnover of sucrose transporters in enucleate sieve elements. Science 275:1298–1300, 1997.

38. I Frias, MT Caldeira, JR Perez-Castineira, JP Navarro-Avino, FA Culianez-Macia, O Kuppinger, H Stransky, M Pages, A Hager, R Serrano. A major isoform of the maize plasma membrane H⁺-ATPase: characterization and induction by auxin in coleoptiles. Plant Cell 8:1533–1544, 1996.

39. AE Hentzen, LB Smart, LE Wimmers, HH Fang, JI Schroeder, AB Bennett. Two plasma membrane H⁺-ATPase genes expressed in guard cells of *Vicia faba* are also expressed throughout the plant. Plant Cell Physiol 37:650–659, 1996.

40. EV Kearns, SM Assmann. The guard cell–environment connection. Plant Physiol 101:711–715, 1993.

41. R Zhao, V Dielen, JM Kinet, M Boutry. Cosuppression of a plasma membrane H⁺-ATPase isoform impairs sucrose translocation, stomatal opening, plant growth, and male fertility. Plant Cell 12:535–554, 2000.

42. P Fleurat-Lessard, N Frangne, M Maeshima, R Ratajczak, JL Bonnemain, E Martinoia. Increased expression of vacuolar aquaporin and H⁺-ATPase related to motor cell function in *Mimosa pudica* L. Plant Physiol 114:827–834, 1997.

43. RE Cleland, HBA Prins, JR Harper, N Higinbotam. Rapid hormone-induced hyperpolarization of the oat coleoptile transmembrane potential. Plant Physiol 59:395–397, 1977.

44. WG Bates, MHM Goldsmith. Rapid response of the plasma-membrane potential in oat coleoptiles to auxin and other weak acids. Planta 159:231–237, 1983.

45. H Barbier-Brygoo, G Ephritikine, D Klämbt, C Maurel, K Palme, J Schell, J Guern. Perception of the auxin signal at the plasma membrane of tobacco mesophyll protoplasts. Plant J 1:83–93, 1991.

46. H Kosegarten, F Grolig, A Esch, KH Glusenkamp, K Mengel. Effects of NH₄⁺, NO₃⁻ and HCO₃⁻ on apoplast pH in the outer cortex of root zones of maize, as measured by the fluorescence ratio of fluorescein boronic acid. Planta 209:444–452, 1999.

47. A Amtmann, TC Jelitto, D Sanders. K⁺-selective inward-rectifying channels and apoplastic pH in barley. Plant Physiol 120:331–338, 1999.

48. HH Felle. The apoplastic pH of the *Zea mays* root cortex as measured with pH-sensitive microelectrodes: aspects of regulation. J Exp Bot 49:987–995, 1999.

49. H Lüthen, M Bigdon, M Böttger. Reexamination of the acid growth theory of auxin action. Plant Physiol 93:931–939, 1990.

50. P Schopfer. Determination of auxin-dependent pH changes in coleoptile cell walls by a null-point method. Plant Physiol 103:351–357, 1993.

51. AP Senn, MHM Goldsmith. Regulation of electrogenic proton pumping by auxin and fusicoccin as related to the growth of *Avena* coleoptiles. Plant Physiol 88:131–138, 1988.

52. DL Rayle, RE Cleland. Enhancement of wall loosening and elongation by acid solutions. Plant Physiol 46:250–253, 1970.

53. A Hager, H Menzel, A Krauss. Versuche und Hypothese zur Primärwirkung des Auxins beim Streckungswachstum. Planta 100:47–75, 1971.

54. U Ruge. Untersuchungen über den Einfluss des Heteroauxins auf das Streckungswachstums des Hypokotyls von *Helianthus annuus*. Z Bot 31:1–56, 1937.

55. TN Bibikova, T Jacob, I Dahse, S Gilroy. Localized changes in apoplastic and cytoplastic pH are associated with root hair development in *Arabidopsis thaliana*. Development 125:2925–2934, 2000.

56. DJ Cosgrove, ZC Li. Role of expansins in cell enlargement of oat coleoptiles: analysis of developmental gradients and photocontrol. Plant Physiol 103:1321–1328, 1993.

57. ZC Li, DM Durachko, DJ Cosgrove. An oat coleoptile wall protein that induces wall extension in vitro and that is antigenetically related to a similar protein from cucumber hypocotyls. Planta 191:349–356, 1993.

58. S McQueen-Mason, DL Cosgrove. Disruption of hydrogen bonding between wall polymers by proteins that induce plant wall extension. Proc Natl Acad Sci USA 91:6574–6578, 1994.

59. DJ Cosgrove. Enzymes and other agents that enhance cell wall extensibility. Annu Rev Plant Physiol Plant Mol Biol 50:391–417, 1999.

60. A Hager, G Debus, HG Edel, H Stransky, R Serrano. Auxin induces exocytosis and the rapid high-turnover pool of plasma-membrane H^+-ATPase. Planta 185: 527–537, 1991.

61. CP Keller, E Van Volkenburgh. Evidence that auxin-induced growth of tobacco leaf tissues does not involve cell wall acidification. Plant Physiol 117:557–564, 1998.

62. C Catala, JK Rose, AB Bennett. Auxin regulation and spatial localization of an endo-1,4-beta-D-glucanase and a xyloglucan endotransglycosylase in expanding tomato hypocotyls. Plant J 12:417–426, 1997.

63. C Catala, JK Rose, AB Bennett. Auxin-regulated genes encoding cell wall–modifying proteins are expressed during early tomato fruit growth. Plant Physiol 122: 527–534, 2000.

64. KW Hutchison, PB Singer, S McInnis, C Diaz-Sala, MS Greenwood. Expansins are conserved in conifers and expressed in hypocotyls in response to exogenous auxin. Plant Physiol 120:827–832, 1999.

65. DP Briskin, S Basu, SM Assmann. Characterization of the red beet plasma membrane H^+-ATPase reconstituted in a planar bilayer system. Plant Physiol 108:393–398, 1995.

66. JL Shaver, C Stirling. Ouabain binding to renal tubules of the rabbit. J Cell Biol 76:278–292, 1978.

67. HH Felle. Control of cytoplasmic pH under anoxia conditions and its implications for plasma membrane proton transport in *Medicago sativa* root hairs. J Exp Bot 47:967–973, 1996.

68. JH Xia, JKM Roberts. Regulation of H^+ extrusion and cytoplasmic pH in maize root tips acclimated to a low-oxygen environment. Plant Physiol 111:227–233, 1996.

69. JC Young, ND DeWitt, MR Sussman. A transgene encoding a plasma membrane H^+-ATPase that confers acid resistance in *Arabidopsis thaliana* seedlings. Genetics 149:501–507, 1998.

70. Y Mathieu, D Lapous, S Thomine, C Lauriere, J Guern. Cytoplasmic acidification as an early phosphorylation-dependent response of tobacco cells to elicitors. Planta 199:416–424, 1996.

71. M Hofmann, R Ehness, TK Lee, T Roitsch. Intracellular protons are not involved in elicitor dependent regulation of mRNAs for defence related enzymes in *Chenopodium rubrum*. J Plant Physiol 155:527–532, 1999.

72. XM Niu, B Damsz, AK Kononowicz, RA Bressan, PM Hasegawa. NaCl-induced alterations in both cell structure and tissue-specific plasma membrane H⁺-ATPase gene expression. Plant Physiol 111:679–686, 1996.

73. E Perez-Prats, ML Narasimhan, X Niu, MA Botella, RA Bressan, V Valpuesta, PM Hasegawa, ML Binzel. Growth-cycle stage-dependent NaCl induction of plasma membrane H⁺-ATPase messenger-RNA accumulation in de-adapted tobacco cells. Plant Cell Environ 17:327–333, 1994.

74. B Michelet, M Lukaszewicz, V Dupriez, M Boutry. A plant plasma membrane proton-ATPase gene is regulated by development and environment and shows signs of a translational regulation. Plant Cell 6:1375–1389, 1994.

75. N Mito, LE Wimmers, AB Bennett. Sugar regulates mRNA abundance of H⁺-ATPase gene family members in tomato. Plant Physiol 112:1229–1236, 1996.

76. AM Noubhani, S Sakr, MH Denis, S Delrot. Transcriptional and post-translational control of the plant plasma membrane H⁺-ATPase by mechanical treatments. Biochim Biophys Acta 1281:213–219, 1996.

77. R Papini, MI De Michelis. Changes in the level and activation state of the plasma membrane H⁺-ATPase during aging of red beet slices. Plant Physiol 114:857–862, 1997.

78. P Aducci, A Ballio, M Marra. Incubation of corn coleoptiles with auxin enhances in vitro fusicoccin binding. Planta 167:129–132, 1986.

79. VV Chelysheva, IN Smolenskaya, MC Trofimova, AV Babakov, GS Muromtsev. Role of the 14-3-3 proteins in the regulation of H⁺-ATPase activity in the plasma membrane of suspension-cultured sugar beet cells under cold stress. FEBS Lett 456:22–26, 1999.

80. M Lukaszewicz, B Jerouville, M Boutry. Signs of translational regulation within the transcript leader of a plant plasma membrane H⁺-ATPase gene. Plant J 14:413–423, 1998.

81. T Kinoshita, K Shimazaki. Blue light activates the plasma membrane H⁺-ATPase by phosphorylation of the C-terminus in stomatal guard cells. EMBO J 18:5548–5558, 1999.

82. MG Palmgren, M Sommarin, P Ulvskov, PL Jørgensen. Modulation of plasma membrane H⁺-ATPase by lysophosphatidylcholine, free fatty acids and phospholipase A₂. Physiol Plant 74:11–19, 1988.

83. MG Palmgren, M Sommarin. Lysophosphatidylcholine stimulates ATP dependent proton accumulation in isolated oat root plasma membrane vesicles. Plant Physiol 90:1009–1014, 1989.

84. U Stahl, B Ek, S Stymne. Purification and characterization of a low-molecular-weight phospholipase A₂ from developing seeds of elm. Plant Physiol 117:197–205, 1998.

85. U Stahl, M Lee, S Sjodahl, D Archer, F Cellini, B Ek, R Iannacone, D MacKenzie, L Semeraro, E Tramontano, S Stymme. Plant low-molecular-weight phospholipase A₂s (PLA2s) are structurally related to the animal secretory PLA2s and are present as a family of isoforms in rice (*Oryza sativa*). Plant Mol Biol 41:481–490, 1999.

86. A Graniti. Azione fitotossica di *Fusicoccum amygdali* Del. su mandorlo *Prunus amygdali* St. Phytopathol Mediterr 1:182–185, 1962.

87. E Marrè. Fusicoccin: a tool in plant physiology. Annu Rev Plant Physiol 30:273–288, 1979.

88. C Oecking, C Eckerkorn, EW Weiler. The fusicoccin receptor of plants is a member of the 14-3-3 superfamily of eucaryotic regulatory proteins. FEBS Lett 352:163–166, 1994.

89. HAAJ Korthout, AH De Boer. A fusicoccin binding protein belongs to the family of 14-3-3 brain protein homologs. Plant Cell 6:1681–1692, 1994.

90. M Marra, MR Fullone, V Fogliano, S Masi, M Mattei, J Pen, P Aducci. The 30 kD protein present in purified fusicoccin receptor preparations is a 14-3-3–like protein. Plant Physiol 106:1497–1501, 1994.

91. A Aitken. 14-3-3 and its possible role in co-ordinating multiple signalling pathways. Trends Cell Biol 6:341–347, 1996.

92. HJ Chung, PC Sehnke, RJ Ferl. The 14-3-3 proteins: cellular regulators of plant metabolism. Trends Plant Sci 4:367–371, 1999.

93. L Baunsgaard, AT Fuglsang, T Jahn, HA Korthout, AH De Boer, MG Palmgren. The 14-3-3 proteins associate with the plant plasma membrane H$^+$-ATPase to generate a fusicoccin binding complex and a fusicoccin responsive system. Plant J 13: 661–671, 1998.

94. M Piotrowski, P Morsomme, M Boutry, C Oecking. Complementation of the *Saccharomyces cerevisiae* plasma membrane H$^+$-ATPase by a plant H$^+$-ATPase generates a highly abundant fusicoccin binding site. J Biol Chem 273:30018–30023, 1998.

95. C Oecking, K Hagemann. Association of 14-3-3 proteins with the C-terminal autoinhibitory domain of the plant plasma-membrane H$^+$-ATPase generates a fusicoccin-binding complex. Planta 207:480–482, 1999.

96. T Jahn, AT Fuglsang, A Olsson, IM Brüntrup, DB Collinge, D Volkmann, M Sommarin, MG Palmgren, C Larsson. The 14-3-3 protein interacts directly with the C-terminal region of the plant plasma membrane H$^+$-ATPase. Plant Cell 9:1805–1814, 1997.

97. C Oecking, M Piotrowski, J Hagemeier, K Hagemann. Topology and target interaction of the fusicoccin-binding 14-3-3 homologs in *Commelina communis*. Plant J 12:441–453, 1997.

98. C Olivari, C Meanti, MI De Michelis, F Rasi-Caldogno. Fusicoccin binding to its plasma membrane receptor and the activation of the plasma membrane H$^+$-ATPase. IV. Fusicoccin induces the association between the plasma membrane H$^+$-ATPase and the fusicoccin receptor. Plant Physiol 116:529–537, 1998.

99. L Camoni, V Iori, M Marra, P Aducci. Phosphorylation-dependent interaction between plant plasma membrane H$^+$-ATPase and 14-3-3 proteins. J Biol Chem 275: 9919–9923, 2000.

100. GE Schaller, MR Sussman. Phosphorylation of the plasma membrane H$^+$-ATPase of oat roots by calcium-stimulated protein kinase. Planta 173:509–518, 1988.

101. A Olsson, F Svennelid, B Ek, M Sommarin, C Larsson. A phosphothreonine residue at the C-terminal end of the plasma membrane H$^+$-ATPase is protected by fusicoccin-induced 14-3-3 binding. Plant Physiol 117:551–555, 1998.

102. I Sekler, M Weiss, U Pick. Activation of the *Dunaliella acidophila* plasma mem-

brane H$^+$-ATPase by trypsin cleavage of a fragment that contains a phosphorylation site. Plant Physiol 105:1125–1132, 1994.

103. PC Van der Hoeven, M Siderius, HA Korthout, AV Drabkin, AH De Boer. A calcium and free fatty acid–modulated protein kinase as putative effector of the fusicoccin 14-3-3 receptor. Plant Physiol 111:857–865, 1996.

104. R Vera-Estrella, BJ Barkla, VJ Higgins, E Blumwald. Plant defense response to fungal pathogen: activation of host–plasma membrane H$^+$-ATPase by elicitor-induced enzyme dephosphorylation. Plant Physiol 104:209–215, 1994.

105. G Desbrosses, J Stelling, JP Renaudin. Dephosphorylation activates the purified plant plasma membrane H$^+$-ATPase. Possible function of phosphothreonine residues in a mechanism not involving the C-terminal domain of the enzyme. Eur J Biochem 251:496–503, 1998.

106. T Xing, VJ Higgins, E Blumwald. Regulation of plant defense to fungal pathogens: two types of protein kinases in the reversible phosphorylation of the host plasma membrane H$^+$-ATPase. Plant Cell 8:555–564, 1996.

107. B Lino, VM Baizabal-Aguirre, LEG De la Varra. The plasma membrane H$^+$-ATPase from beet root is inhibited by a calcium-dependent phosphorylation. Planta 204:352–359, 1998.

108. F Svennelid, A Olsson, M Piotrowski, M Rosenquist, C Ottman, C Larsson, C Oecking, M Sommarin. Phosphorylation of Thr-948 at the C-terminus of the plasma membrane H$^+$-ATPase creates a binding site for the regulatory 14-3-3 protein. Plant Cell 11:2379–2391, 1999.

109. L Camoni, MR Fullone, M Marra, P Aducci. The plasma membrane H$^+$-ATPase from maize roots is phosphorylated in the C-terminal domain by a calcium-dependent protein kinase. Physiol Plant 104:549–555, 1998.

110. M Marra, C Olivari, S Visconti, C Albumi, P Aducci, MI De Michelis. A phosphopeptide corresponding to the cytosolic stretch connecting transmembrane segments 8 and 9 of the plasma membrane H$^+$-ATPase binds 14-3-3 proteins and inhibits fusicoccin-induced activation of the H$^+$-ATPase. Plant Biol 2:11–16, 2000.

111. A Chang, CW Slayman. Maturation of the yeast plasma membrane [H$^+$]ATPase involves phosphorylation during intracellular transport. J Cell Biol 115:289–295, 1990.

112. DL Stokes, M Auer, P Zhang, W Kühlbrandt. Comparison of H$^+$-ATPase and Ca^{2+}-ATPase suggests that a large conformational change initiates P-type ion pump reaction cycles. Curr Biol 9:672–679, 1999.

113. PJ Krysan, JC Young, MR Sussman. T-DNA as an insertional mutagen in *Arabidopsis*. Plant Cell 11:2283–2290, 1999.

114. NP Mahajan, K Linder, G Berry, GW Gordon, R Heim, B Herman. Bcl-2 and Bax interactions in mitochondria probed with green fluorescent protein and fluorescence resonance energy transfer. Nat Biotechnol 16:547–552, 1998.

115. Y Nagai, M Miyazaki, R Aoki, T Zama, S Inouye, K Hirose, M Iino, M Hagiwara. A fluorescent indicator for visualizing cAMP-induced phosphorylation in vivo. Nat Biotechnol 18:313–316, 2000.

116. A Cha, GE Synder, PR Selvin, F Bezanilla. Atomic scale movement of the voltage-sensing region in a potassium channel measured via spectroscopy. Nature 16; 403: 809–813, 1999.

117. Y Suzuki, T Yasunaga, R Ohkura, T Wakabayashi, K Sutoh. Swing of the lever arm of a myosin motor at the isomerization and phosphate-release steps. Nature 396:380–383, 1999.

118. T Ng, A Squire, G Hansra, F Bornancin, C Prevostel, A Hanby, W Harris, D Barnes, S Schmidt, H Mellor, PI Bastiaens, PJ Parker. Imaging protein kinase Cα activation in cells. Science 283:2085–2089, 1999.

119. TWJ Gadella, GNM Van der Krogt, T Bisseling. GFP-based FRET microscopy in living plant cells. Trends Plant Sci 4:287–291, 1999.

2

H+-ATPase and H+-PPase in the Vacuolar Membrane: Physiology and Molecular Biology

Masayoshi Maeshima and Yoichi Nakanishi

Nagoya University, Nagoya, Japan

1 INTRODUCTION

Vacuoles are ubiquitous, multifaceted, indispensable, and acidic organelles. The vacuoles occupy a large part of the plant cells. Typical plant vacuoles are 50 to 100 μm in diameter and accumulate many soluble substances, such as inorganic ions, organic acids, enzymes, and secondary metabolites. The acidic environment in the vacuole is essential not only for the hydrolytic enzymes but also for the active transport and accumulation of substances. Plant vacuoles have two distinct proton-transporting enzymes, namely H+-translocating inorganic pyrophosphatase (V-PPase) and H+-ATPase (V-ATPase, vacuolar-type H+-ATPase). V-ATPase is known as the third ion-translocating ATPase in addition to the F-type and P-type ATPases (see Chapter 1). V-PPase is a unique proton pump different from these three types of ATPases. V-PPase uses inorganic pyrophosphate (PPi, diphosphate) as an energy source comparable to ATP.

Most plant cells contain both V-ATPase and V-PPase in the same vacuolar membrane. The proton pumps acidify vacuoles and generate an electrochemical potential difference of protons across the vacuolar membrane. In most plants, a vacuolar membrane potential is $+20$ to $+30$ mV relative to the cytosol, and the lumenal pH ranges from 3 to 6. Certain active transport systems in the vacuolar membrane work by a mechanism of H^+/substrate antiport, and some transporters and channels utilize a membrane potential in an electrophoretic manner (Fig. 1). The vacuolar proton pumps also contribute to keeping the pH in the cytosol constant at about 7.2.

Detailed information on the molecular aspects of the vacuolar proton pumps obtained during the past decade is described in this chapter along with the biochemical and physiological properties of vacuolar proton pumps. The relationship between proton pumps and various types of vacuoles in higher plants, such as

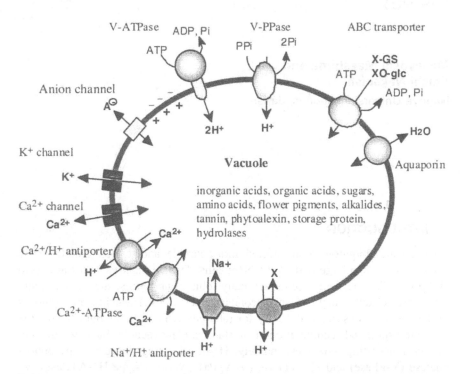

FIGURE 1 Transporters and channels in the plant vacuolar membrane. H^+-ATPase (V-ATPase) and H^+-pyrophosphatase (V-PPase) actively translocate H^+ ions into vacuoles. Several antiporters are driven by a pH gradient generated by proton pumps. ABC transporters use the free energy from ATP hydrolysis [4]. Aquaporin facilitates the water transport in an osmotic pressure–dependent manner [108].

lytic, protein storage, pigment, volume-regulating, and tannin vacuoles, is also described in the latter part.

2 PHYSIOLOGICAL ROLES OF PLANT VACUOLES

Vacuoles are functionally related to the lysosomes of animal cells, containing a variety of hydrolytic enzymes, but their functions are remarkably diverse. Some of the physiological roles of plant vacuoles can be summarized as follows.

  1.   Space-filling function. The space-filling role of the vacuole is essential to the growth of plant cells. The presence of a large vacuole is peculiar to plant cells. In most of the nonphotosynthetic tissues, such as the shoot and the root, a considerable part of each cell is occupied by a large central vacuole. Many plant cells maintain an almost constant turgor pressure by changing the osmotic pressure of the cytosol and vacuole. Osmotic pressure is regulated in part by changes in rates of transport of sugars, ions, and other metabolites across the vacuolar membrane and the plasma membrane and in part by the controlled breakdown and resynthesis of polymers in the lytic or the protein storage vacuoles.

  2.   Accumulation and storage of metabolites. The vacuole is the site of a temporary and/or long-term pool of nutrients for its own cell and for new, growing organs. Vacuoles accumulate large quantities of carboxylic acids (malate, citrate, and oxalate) and sugars in addition to inorganic ions [1]. Seed storage proteins are actively synthesized and transported into the vacuoles during seed maturation. These protein storage vacuoles are called protein bodies. The secondary metabolites, such as terpenoids, berverine, and betaine, are also concentrated in the vacuole [2]. Furthermore, plant vacuole compartmentalizes flower pigments, such as anthocyanins and flavonols. The anthocyanins require special attention because these compounds can exist in various structural forms depending on the pH of the vacuolar lumen. For example, "heavenly blue" anthocyanin of the flower petal of morning glory (*Ipomoea tricolor*) exhibits red color at pH 5–5.8, purple at pH 6–7, and blue at pH 7.7 [3]. The color of flowers changes from purple in the buds to clear blue in the open flowers. This means that the vacuolar pH increases during the flower opening.

  3.   Compartmentalization of toxic substances. Because of their inability to move to avoid abiotic stresses and environmental invaders, such as bacteria, fungi, and animals, plants have evolved unique adaptations, some of which involve production of toxic compounds. For example, certain plant tissues contain cytotoxic substances, such as cyanogenic glycosides and alkaloids [2]. These noxious molecules, which are released from vacuoles when a plant is eaten or damaged by animals, provide a defense against predators. Furthermore, Ca^{2+}, Na^+, and Cd^{2+} ions inhibit many enzymes in the cytoplasm; these ions are compartmentalized and concentrated in the vacuole. Some metal ions in vacuoles are

chelated with organic acids, phytic acid, and metallothionein (a cysteine-rich small protein). Vacuoles participate in compartmentalization of detoxicated compounds. When particular phytotoxic foreign compounds, such as microbial toxins and xenobiotics, are incorporated into plant cells, these compounds are modified to glutathione S-conjugates and then transported into vacuoles to avoid toxicity.

4.   Regulation of cytosolic levels of inorganic ions. The vacuole is important as a homeostatic device. Plant vacuoles actively accumulate inorganic ions, such as nitrate and Ca^{2+}, and resupply these ions to the cytosol as the need arises. For example, Ca^{2+}-ATPase and Ca^{2+}/H^+ antiporter transport Ca^{2+} from the cytosol into the vacuole, thus keeping the cytosolic concentration of Ca^{2+} at the extremely low level of about 100 nM. These Ca^{2+} uptake systems allow Ca^{2+} ions to act as second messengers in the cytosol [5]. The Ca^{2+}/H^+ antiporter utilizes a pH gradient across the vacuolar membrane. Recently, this antiporter has been identified and characterized [6–8].

5.   Hydrolysis and recycling of cellular components. Seed proteins in protein storage vacuoles are hydrolyzed by proteases during germination. The vacuole, as the site of such degradation, corresponds to the lysosome of animal cells. The plant vacuole contains a variety of hydrolytic enzymes, such as proteases, ribonucleases, phosphodiesterases, glycosidases, phytase, and acid phosphatase. In the case of germinating castor bean endosperm, the total activity of autolysis of storage proteins is maximal at pH 5.0 [9]. The optimal pH of aspartic endopeptidase in protein storage vacuoles of castor bean endosperm is 3.0 and the optimal pH of vacuolar processing enzyme is 5.5 [10]. It is not clear how the cytoplasmic components, including organelles, are incorporated into vacuoles during their recycling. In yeast, during nutrient starvation, intracellular organelles and the cytosol are sequestered in autophagosomes that subsequently fuse with vacuoles [11,12]. Autophagy is thought to be a common mechanism of protein degradation among all eukaryotes, including plants.

3   MOLECULAR STRUCTURE AND ENZYMATIC PROPERTIES OF VACUOLAR PROTON PUMPS

3.1   Structure of V-ATPase

Vacuolar-type H^+-ATPase (V-ATPase) is a universal enzyme distributed in the membrane of a variety of acidic organelles in eukaryotic cells and in the plasma membrane of certain animals and bacteria. Several fine reviews are available on V-ATPase in various organisms [13–17]. V-ATPase is mainly localized in the vacuoles in plant cells. However, it can be associated with the endoplasmic reticulum (ER), Golgi bodies, coated vesicles, and provacuoles [18–20]. Furthermore, V-ATPase could be detected in the plasma membrane of the seed storage tissue [21].

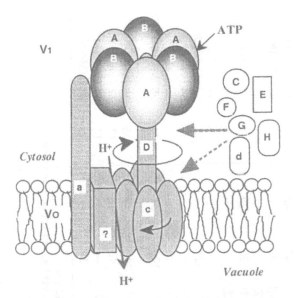

FIGURE 2 Schematic model of V-ATPase. V-ATPase consists of the V_1 and V_0 sectors. The shaft subunit (probably D subunit) and a c-subunit hexamer rotate against the A_3B_3 complex in a counterclockwise direction.

The overall structure of V-ATPase is similar to that of the F-type ATPase [17,22]. V-ATPase is composed of two domains, a peripheral catalytic sector (V_1) and a transmembranous sector (V_0) (Fig. 2). The subunit compositions of the V_1 and V_0 sectors are summarized in Table 1. The catalytic domain consists of A and B subunits, and the other subunits (C–H) in the V_1 sector form a central stalk part between V_1 and V_0. The proton channel in V_0 is composed of six copies of the c subunit that has four membrane-spanning domains, and the subunits a and d are also localized in the membrane.

The rotational mechanism of F-ATPase during ATP hydrolysis has been clearly demonstrated as counterclockwise rotation of the γ subunit in the F_1 sector [23–25]. It has been shown that the c subunit in F_0 also rotates together with the γ subunit during ATP hydrolysis [26]. The rotation of the γ subunit of chloroplast ATP synthase has also been demonstrated [27]. Thus, F-ATPase is called a molecular rotary motor. V-ATPase may also function via the rotational mechanism, judging from a similarity between the V-ATPase and the F-ATPase in the primary sequences of subunits and the tertiary structure. The rotation of V_1 has been demonstrated recently by two Japanese groups (unpublished). In this respect, a question arises which subunit of V_1 corresponds to the γ subunit of F_1. It has been suggested from the high α-helical content of D and E subunits that either

TABLE 1 Comparison of the Subunits of V-ATPases in Plants, Yeast, and Bovine[a]

Subunit	Subunit function	Molecular mass of subunit (kDa)					
		Mung bean	Pea	Oat	Kalanchoë	Bovine	Yeast (loci)
V_1 sector							
A	Catalytic ATP binding	68	68	70	72	73	69 (VMA1)
B	Noncatalytic ATP binding	57	57	60	56	58	57 (VMA2)
?	Unknown	—	—	44	48	—	—
C	Activity, assembly	—	—	42	42	40	42 (VMA5)
D	Activity, central stalk (?)	32 (D1)	35 (D1)	32	—	34	32 (VMA8)
		29 (D2)	29 (D2)				
E	Activity, central stalk (?)	38 (E1)	40 (E1)	29	28	33	27 (VMA4)
		37 (E2)	37 (E2)				
F	Activity, assembly	13*,[b]	13*	13*	—	14	14 (VMA7)
G	Activity, assembly	12*	12*	12*	—	15	13 (VMA10)
H	Activity	51*	51*	—	—	50	54 (VMA13)
V_0 sector							
a	H+ translocation, assembly, targeting	100*	100*	100	97	100	95 (VPH1/STV1)
d	Activity, assembly	44*	45*	36*	—	38	36 (VMA6)
c	H+ translocation	16	16	16	16	17	17 (VMA3)
c′	H+ translocation	—	15.5	—	—	17	17 (VMA11)
c″	H+ translocation	—	—	24	—	19	23 (VMA16)

[a] Data for mung bean [34], pea [34], oat [14], Kalanchoë [75], bovine [15], and yeast [13].
[b] Asterisks indicate the subunits for which the correspondence to yeast subunits has not yet been examined.

of the subunits may form the central stalk as the V-ATPase homologue of the γ subunit [15]. The fine structural model of V-ATPase remains to be determined by crystallography.

It is noteworthy that certain subunits of V-ATPase in plants have isoforms (for a review, see Ref. 17). Two cDNAs encoding the c subunit have been identified in *Mesembryanthemum crystallinum* (ice plant) [28] and *Gossypium hirsutum* (cotton) [29] and four cDNAs in *Arabidopsis thaliana* [30]. Also, two isoforms have been found for the subunits A [28,31], B [28,32], and G [33] of several plant species. In addition to the previous reports, Kawamura et al. [34] demonstrated that V-ATPases from *Vigna radiata* (mung bean) and *Pisum sativum* (pea) have two isoforms of subunits D (D1 and D2) and E (E1 and E2). Characteristically, the molecular masses of the D and E subunits differ remarkably with the isoform. From the stoichiometry of the subunit composition, each V-ATPase complex is thought to contain one of the two isoforms of the subunits D and E. The isotypes of V-ATPase that contain either E1 or E2 have been separated by ion-exchange column chromatography [34]. Enzymatic properties such as V_{max} and K_m slightly differed between the V-ATPase isotypes.

The c subunit (proteolipid subunit) of V-ATPase is a critical element for H^+ transport and membrane anchoring. The membrane targeting of the enzyme also depends on this subunit. The c subunit is essential for early development of the animal body [35]. In the yeast V-ATPase, the additional V_0 subunits c′ (17 kDa) and c″ (23 kDa) have been demonstrated to play a role in the proton translocation [36]. Four genes for the c subunit of *Caenorhabditis elegans* (nematode) V-ATPase, *VHA-1* (16 kDa), *VHA-2* (16 kDa), *VHA-3* (16 kDa), and *VHA-4* (23 kDa), are expressed in a cell-specific manner [37]. The number of c subunits in the V_0 sector is six per V-ATPase complex. This value corresponds to 12 copies of the c subunit (8 kDa) in an F-ATPase complex, because the c subunit of V-ATPase is thought to be generated by duplication of the gene for F-ATPase c subunit during molecular evolution. However, this number of 6 should be reexamined because it has been reported that there are only 10 c subunits in the oligomeric ring of yeast F_0 [38]. Therefore, there is a possibility that the proton channel in V_0 is composed of 5 c subunits.

The largest subunit of 100 kDa (subunit a) has a transmembranous domain. This subunit is thought to be essential for V-ATPase assembly and proton translocation in yeast [39,40]. However, Li and Sze [41] reported that the 100-kDa subunit did not appear to be associated at stoichiometric levels with the fully assembled enzyme. They proposed that the main function of the 100-kDa subunit is for assembly of the enzyme in the ER membrane [41]. F-ATPases do not contain any subunit corresponding to the 100-kDa subunit of V-ATPase. In the case of F-ATPase, the subunit a (23 to 30 kDa) has been estimated to form the a_1b_2 complex and function as a stator for the rotary motor [42]. Their model indicates that the subunit a is essential for rotation of the central cylinder composed of c subunits and for the H^+ translocation. The stator is also essential for the linkage between the F_1 and F_0 as well as the V_1 and V_0. Thus, the subunit a (100 kDa) of V-ATPase is the most probable candidate for the stator component. The V-ATPases of *C. elegans* and mouse have four and three isoforms of the subunit a, respectively [16]; the isoform of the subunit has not been identified in plants.

3.2 Molecular Properties of V-PPase

V-PPase is one of the simplest proton pumps, and its substrate PPi also has a simple structure [43–46]. The cDNAs for V-PPases have been cloned from various land plants (references in Ref. 46), algae (*Acetabularia acetabulum* [47], *Chara corallina* [48]), protozoans (*Trypanosoma cruzi* [49]), photosynthetic bacteria (*Rhodospirillum rubrum* [50]), and archaebacteria (*Pyrobaculum aerophilum* [51]). However, it should be noted that V-PPase does not exist in any membranes of animal or yeast cells. The distribution of V-PPases among living organisms, sequence similarity, and structure-function relationship of the enzyme have been discussed in recent reviews [45,46]. Thus, we describe here only V-PPases in land plants. Plant V-PPases have been reported to consist of 761 to 771 amino acid residues (Fig. 3). Their calculated molecular masses range from 80 to 81 kDa. The plant V-PPase migrates on sodium dodecyl sulfate (SDS)-polyacrylamide gel as a polypeptide of about 73 kDa [52]. The amino acid sequences are highly conserved among land plants, with 86 to 91% identity. With respect to the tertiary structure of V-PPase, experiments using radiation inactivation [53,54] and gel permeation [55] showed that the enzyme exists as a homodimer of 80-kDa subunits.

By comparing all H^+-PPases of various organisms, some highly conserved segments have been found [43,45,46]. Seven conserved segments are shown in Fig. 3. One of the segments (segment i) contains a common motif DVGADLVG-KVE that is the essential sequence for the catalytic site. The immunochemical analysis confirmed that this sequence is exposed to the cytosol [56]. Furthermore, using the mutagenesis analysis of the plant enzyme with a heterologous expres-

---

FIGURE 3 Multiple amino acid sequence alignment of V-PPases. Deduced amino acid sequences of the enzymes in *Chara corallina* [48], mung bean [78], *Acetabularia acetabulum* [47], and *Rhodospirillum rubrum* [50] were aligned by the Clustal method. Identical and conserved residues are marked by asterisks (*); gaps introduced to maximize alignment scores are denoted by hyphens. The 14 putative transmembrane domains (1 to 14) are shaded. Thick lines (i to vii) mark conserved segments. The total number of amino acid residues in each sequence is shown in brackets.

Chara
Mung bean
Acetabularia
R. rubrum

```
                                                          1
MAAVAEGLNAMTTIPTTTAPADLALAAVTGNTIVSETAVLIFIPAACVIGILFAVLQWSVVGKISVRPS----------G
M---------------------------GAAILPDLGTELLIPVCAVIGIAFAIFQWLLVSKVKLSAVRDASPNAAAKN
M------------------GQTIT------IATPVFGVLALLYIFWRSSKVSKQEVG----
```

```
                                      2
GGMNYPLMGDEGLEDSSVVTRCAEIQEAISEGAVSFLMTEYKELSYFMVGFFIVIFAFLGATEDFGTDRKPCEWDATKLC
GYNDYLIEEEGINDHNVVVKCAEIQNAISEGATSFLFTEYKRVGIFMVAFAILIFLFLSVEGFSTSPQACSYDKTKTC
------------------TERMGRIAKNITDGAMAFLKAEYRVLAIFVIAVAILLGISG-------N-RE---------G
----------------MQEGASAFLNRQYKIIAVVGAVVFVILTALLGSVGFG--------------
```

```
    3                                                    4
GSGVMNALLSAVAFALGAITSTLCGFLGMKIATFANARTSRRGGGVGPAFKAAFRSGAVMGFLLTSLGLIVLYFTILIP
KPALATAIFSTVSFLGGVTSLVSGFLGMKIATYANARTTLEARKGVGKAFITAFRSGAVMGFLLAANGLLVLYIAINLF
-----TSFLIAVSFILGAICSALAGYIIGMIVATKANVRTTNARRSSLGRALEVAFAGGSVMGLGVVGLGVLGLGTLFLAY
-------FLIGAVCSGIAGYVGMYISVRANVRVAAGAQQGLARGLELAFQSGAVTGNLVAGLALLSVAFYY-IL
```

```
     5
QRYYGDDWIGLYESIAGYGLGQSSVALFGRVGGGIYTKAADVGADLVGKVERNIPEDDPRNPAVIADNVGDNVGDIAGMG
KIYYGDDWGGLFEAITGYGLGGSSNALFGRVGGGIYTKAADVGADLVGKVERNIPEDDPRNPAVIADNVGDNVGDIAGMG
SN-IGWDINRVITVITGFSFGASBIALFARVGGGIYTKAADVAADLVGKVEAGIPEVHPLNPATIADNVGDNVGDVAGMG
LVGIGATGRALIDPLVALGFGASLISIFARLGGGIFTKCADVGADLVGKVEAGIPEDDPRNPAVIADNVGDCAGMA
                    i                      ii
```

```
                                   6
ADLFGSLAESTCAALVVSS------LSDFGKEMNYVAMSFPLLYTGAGILVCLIFTLVADLTSGVSNIKGIEPALKQQ
SDLFGSYAESSCAALVVAS------ISSFGLNHELTAMLYPLIVSSVGILVCLLTTLFADFFEIKA-VKEIEPALKKQ
ADLFESIVGSIIGTMVLGATFIGVAGFQETNAFNGLNAVLLPLVLAGTGIITSIVGTFFVKVKEGGNP-----QKALNTG
ADLFETYAVTVVATMVLAS------IFFAGVPAMTSMMAYPLAI-GG--VCILASILGTKFVKLGP-KNNIMGALYRG
     _***_____*_____**__*_____*_____*_*____
```

```
7                            8
LVISTVLMTPVIALLAWGCLPDTFEIINGAETKVVKKWYNFFCVACGLWAGLLIGYTIEYFTSHQFTPVRDVADSCRTGA
LVISTVLMTIGVAVVSFVAIPTSFTIFNFGVQKDVKSWQLFLCVAVGLWAGLLIGFVTEYYTSNAYSPVQDVADSCRTGA
EFLASSIMLVLTYLIVDWMIPSTWTSATGVS---YSSFGVFMAVIFGLVWAGLLIGMVTEFYTGTGTRPVKGIVSQSLTGS
FLVSAGASFVQIILATADVPGFGDIQGANGVLYSGGFDLFLCAVIGLVLVTGLLINVIEYYTTGTNFRPVRSVAKASTTGH
```

```
        9
ATNVIFGLALGYKSVIIPILAIAPTVFVSHTLAAMYGIACAALGMLSTLSTCLAIDAYGPISDNAGGIAEMAEMGPAIRE
ATNVIFGLALGYKSVIIPIFAIAISIFVSFTLAAMYGIAVAALGMLSTIATGLAIDAYGPISDNAGGIAEMAGMSERIRE
ATNIIAGLGVGMQSTAIPIVIILAAIIGAHEFAGLYGIAIAAVGMLSNTGIQLAVDAYGPVTDNAGGIAEMGELPKEVRG
GTNVIQGLAISMEATALPALIICAAIITTVQLSQIFGIAITVTSMLALAGMVVALDAYGPVTDNAGGIAEMANLPEDVRK
                                              iii
```

```
             10                                          11
KTDALDAAGNTTAAIGKAFAIGSAALVSLALFGAYINRAGITSVDVIL----------PKEFVGLIVGAMLPYWF
RTDALDAAGNTTAAIGKGFAIGSAALVSLALFGAFVSRASITTVDVLT----------PKVFIGLIVGAMLPYWF
RTDKLDAVGNTTVAIGKGFAIGSALTALALFAAFMETAKITEINVAD----------PLVMAGLFLGGMLPSLF
TTDALDAVGNTTKAVTKGYAIGSSGLGALVLFAAYTSDLAFFKANVDAYPAFAGVDVNFSLSSPYVVVGLFIGGLLPYLF
            iv                                              
```

```
                                                          12
SAMTMKSVGKAALAMVEEVRRQFNTIAGLMQGTVKP--------------------DYKRCVEISTDASLREMI
SAMTMKSVGKAALKMVEEVRRQFNTIPGLMEGTAKP--------------------DYATCVKISTDASIKEMI
SSLAMNAVGRAAMDMIQEVRRQFKTIPELKAALDTMRKNDGKEFAEWSEADQTIFNAADGKAETSKCVEISTKASIREMV
GSMGMTAVGRAAGSVVEEVRRQFREIPGIMEGTAKP--------------------EYGRCVDMLTKAAIKEMI
        v
```

```
             13
PPGCLVMLTPLVVGG---------LLGKETLAG-ILRGALVSGVQIAISASNTGGAWDNAKKYIEAGGNDEARTLGPKGSDC
PPGALVMLTPLVVGI--------LFGVETLSG-VLAGSLVSGVQIAISASNTGGAWDNAKKYIEAGASEHARSLGPKGSDC
LPGLIAVLTPVVIGF--------AGGAEMLGG-LLAGVTVSGVQIAISASNSGGAWDNAKKYIEAGNSEHARSLGPKGSDC
IPSLLFPVLAPIVLYFVILGIADKSAAFSALGAMLLGVIVTGLFVAISMTAGGGAWDNAKKYIEDG------HYGGKGSEA
                                        vi
```

```
             14
HKAAVIGDTVGDPLKDTSGPSLNILIKLMAVESLVFAPFFKTYGGVLFVLWDKFVAGKY   (793)
HKAAVIGDTIGDPLKDTSGPSLNILIKLMAVESLVFAPFFATHGGLLFKIF   (766)
HKAAVIGDTIGDPLKDTSGPSLNILIKLMAVESLVFAPFFATYGGLLFKYI   (721)
HKAAVTGDTVGDPYKDTAGPAVNPMIKITNIVALLLLAVLAH   (660)
*****__***_***_**___*_*___**_____
                    vii
```

sion system in yeast, the acidic residues in this motif have been demonstrated to be critical for the enzymatic activity (Y. Nakanishi and M. Maeshima, unpublished data). In addition to the catalytic site, the binding sites for Mg^{2+}, K^+, and some reagents, such as N,N'-dicyclohexylcarbodiimide (DCCD), 7-chloro-4-nitrobenzo-2-oxa-1,3-diazole (NBDCl), and N-ethylmaleimide (NEM), have been investigated (references in [44] and [46]). The detailed structure-function relationship and an accurate tertiary structure should be examined by crystallography. Fortunately, V-PPase is an abundant component of the vacuolar membrane of growing organs, such as seedling hypocotyls. About 10 mg of the highly purified V-PPase can be obtained from 10 kg of mung bean hypocotyls. The crystallographic, high-resolution structure is expected to be determined within several years, and we will then be able to discuss the coupling mechanism of PPi hydrolysis and H^+ translocation at the molecular level as has been done for F-ATPase and Ca^{2+}-ATPase [57].

3.3 Enzymatic Properties of V-ATPase

The activities and physiological roles of the enzymes in cells cannot be accurately understood without the detailed enzymatic characteristics. Here, some properties of V-ATPase and V-PPase are briefly described.

Usually, the V-ATPase activity has been assayed in the presence of 2 to 3 mM ATP and 3 mM $MgCl_2$. The Pi released during the enzyme reaction is assayed by the colorimetric method. In some cases, $\alpha[^{32}P]$-ATP is used. V-ATPase also hydrolyzes GTP and ITP at relatively low rates. Magnesium is essential for formation of the Mg-ATP complex and for enzyme activity, as in the case of other ATP-utilizing enzymes. The K_m value of V-ATPase for Mg-ATP is about 0.1–0.2 mM. The other essential factor is Cl^- ion because the enzyme interacts directly with that ion [58]. Thus, the assay medium contains KCl or choline chloride at a final concentration of 50 mM. The chloride ion can be substituted by Br^- [59]. The well-known inhibitors of V-ATPase are nitrate, bafilomycin A_1, and concanamycin A. Nitrate inhibits the enzyme at high concentrations as a chaotropic anion that causes the destruction of a precise higher order structure of V-ATPase. Concanamycin A is more effective than bafilomycin A_1 [60], but bafilomycin A_1 has been widely used to define the V-ATPase. Bafilomycin A_1 is a macrolide antibiotic with a 16-member lactone ring, isolated from *Streptomyces*; it specifically inhibits V-ATPase among the three ATPase classes (P-, F-, and V-ATPases) [61], although at high concentrations bafilomycin A_1 inhibits Na^+,K^+-ATPases from *E. coli* and ox brain at concentrations higher than 20 μM. Bafilomycin A_1 completely inhibits V-ATPase in the vacuolar membrane of *Neurospora crassa* at 100 nM [61], the purified V-ATPases from lemon cotyledons at 1 nM [62], and the purified enzyme from mung bean hypocotyls at 500 nM.

There are a few exceptions, though. Müller et al. [63] reported that V-ATPase of lemon fruit is insensitive to bafilomycin A_1 at 1 nM.

3.4 Enzymatic Properties of V-PPase

V-PPase hydrolyzes PPi but not ATP or other nucleotides. The actual substrate is an Mg^{2+}-PPi complex, Mg_2PPi or $MgPPi^{2-}$ [43,44,64]. Magnesium ion is essential for the formation of the Mg^{2+}-PPi complex and the stabilization and activation of the enzyme [65]. Figure 4 shows the dependence of the proton pumping activity of V-PPase on Mg^{2+} concentration. For the maximal activity, Mg^{2+} should be added at more than 0.5 mM. Most of Mg^{2+} added is incorporated in the Mg^{2+}-PPi complex (K_d for MgPPi, 55 μM) and the remaining free Mg^{2+} activates V-PPase. The K_m value of V-PPase for free Mg^{2+} is 20 to 42 μM [46]. Under physiological conditions, Mg^{2+} and PPi have been reported to have cytosolic concentrations of more than 0.4 and 0.2 mM, respectively. Thus, the V-PPase can express almost the maximal activity [43,46].

The V-PPase activity is stimulated more than three fold by K^+. The K_m for K^+ stimulation is 1.27 mM [65], except for the V-PPase of *Vicia faba* guard cells (51 mM) [66]. Under the standard assay conditions, the full activity of V-PPase

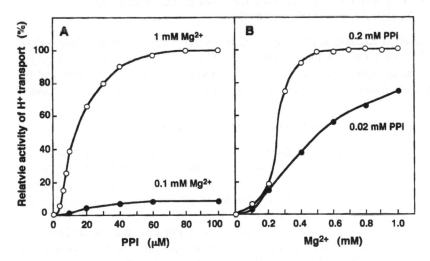

Figure 4 Dependence of proton pump activity by V-PPase on the concentrations of pyrophosphate (PPi) and Mg^{2+}. The vacuolar membrane was prepared from mung bean hypocotyls and assayed for the activity of PPi-dependent H^+ transport at various concentrations of PPi (A) and Mg^{2+} (B) by using the fluorescent dye acridine orange.

can be obtained in the presence of more than 30 mM K^+, which is the physiological level. Interestingly, K^+ has been substituted with Rb^+, NH^+_4, and partially with Cs^+ and Na^+ [67,68]. On the other hand, Tris has been demonstrated to be a competitive inhibitor of K^+ stimulation [67]. Tris base increases the K_m value of V-PPase for K^+ and decreases V_{max}. The question is whether or not V-PPase transports K^+ into the vacuole [44]. From electrochemical and patch-clamp experiments [69,70], V-PPase has been proposed to transport K^+ actively into vacuoles. In contrast, biochemical experiments using radioactive $^{42}K^+$ [71] and fluorescent probes for K^+ and H^+ [72] revealed that the enzyme did not transport K^+. The proposal of K^+ transport by V-PPase is attractive, but its physiological meaning may be restricted to special cells such as guard cells of stoma.

The V-PPase described in this chapter is encoded by *AVP1* of *Arabidopsis* (previous name, *AVP3*) [68,73]. A new type of V-PPase of *Arabidopsis*, composed of 800 amino acid residues, has been identified and named AVP2p [68] or AVPL1p [74]. Interestingly, AVP2p does not require K^+ or NH^+_4 [68]. The amino acid sequence identity between AVP1p and AVP2p (AVPL1p) has been reported to be 36% (51% in similarity). The intracellular localization and physiological role of AVP2p (AVPL1p) remain to be determined.

4   EXPRESSION LEVELS AND ORGAN SPECIFICITY OF VACUOLAR PROTON PUMPS

Information concerning the changes in the activity, protein amount, and mRNA levels of the proton pumps has accumulated [14,43,75]. In this section, the physiological changes of proton pumps under normal and stress conditions and the organ specificity of V-ATPase are described with reference to recent reports.

4.1   Cell Growth and Vacuolar Proton Pumps

Young plant tissues, such as growing hypocotyls, show relatively high activities of V-ATPase and V-PPase compared with the mature, differentiated tissues [14,17,46,75]. The greater general expression of V-ATPase in young growing cells than in differentiated cells has been thought to reflect a more active endomembrane system in young cells [14]. It should be noted that V-PPase is a predominant vacuolar proton pump in young cells. For example, the activity ratio of V-PPase to V-ATPase was 4.0 for the elongating region of 3.5-day-old mung bean hypocotyl [76], 5.2 for 1-month-old pear fruit, and 1.3 for developing cotton locule [77]. The V-PPase/V-ATPase ratio markedly decreased during the cell maturation in mung bean hypocotyl [76,78], radish taproot [79], and pear fruit [80]. The decrease in the V-PPase/V-ATPase ratio is mainly due to the decrease in the protein amount and mRNA level of V-PPase in growing hypocotyls and leaves of mung bean [78].

The germinating seeds can be considered as the developing tissues. During transformation of protein storage vacuoles to lytic vacuoles, both V-ATPase and V-PPase are actively synthesized and accumulated in the seed vacuoles. However, neither V-ATPase nor V-PPase could be detected in the dry seed. In pumpkin, the specific activities of V-PPase (0.35 units mg^{-1}) and V-ATPase (0.13 units mg^{-1}) in the vacuolar membrane from 6-day-old cotyledons were relatively high compared with those in vegetative tissue (6-day-old hypocotyls) (1.25 and 0.27 units mg^{-1} for V-PPase and V-ATPase, respectively) [81].

The relatively high level of V-PPase in developing tissues may be related to the supply of PPi that is a by-product of biosyntheses of RNA, protein, starch, and cellulose. The β-oxidation of fatty acids also produces PPi, although it is not clear whether PPi generated in this process is released to the cytosol from microbodies and whether PPi is hydrolyzed by V-PPase or cytosolic soluble PPase.

The usage of PPi for acidification of a large central vacuole in the plant cell is reasonable from the perspective of cell energetics. Scavenging of PPi is the second role of V-PPase because high concentrations of cytosolic PPi inhibit the polymerization reactions of DNA, RNA, protein, starch, and cellulose. However, the cytosolic PPi concentration must be kept at a constant level because some enzymes, such as PPi-phosophofructokinase of the glycolytic pathway, require PPi. Thus, in the mature cells, the rate of PPi production decreases together with the V-PPase level [78].

It is clear that two or more different types of vacuoles exist in the same cell. For example, both lytic and protein storage vacuoles exist in cells of barley aleurone [82], barley roots [83], and pea cotyledons [21]. A motor cell of *Mimosa pudica* (mimosa) contains a tannin vacuole in addition to an aqueous central vacuole [84]. The pulvini motor cell of *M. pudica* has two types of vacuoles, a large lytic vacuole and a small tannin vacuole. There are marked differences in the distribution density among V-ATPase, V-PPase, and aquaporin on the membrane surface [84]. Aquaporin is a water-specific channel that facilitates the water transport across biomembranes in an osmotic pressure–dependent manner. The membrane of lytic vacuole in the motor cells contains these components at an extremely high density and exhibits quick shrinking and swelling in response to mechanical stimuli. One explanation is that rapid water efflux requires a high density of aquaporin and that active incorporation of H$^+$ and solutes during reswelling of the vacuole requires a large population of proton pumps. The vacuolar type specificity of proton pumps should be examined with reference to recent reports on the vacuolar type-specific aquaporins [82,85].

V-ATPase has been suggested to play a role in the control of cell elongation and photomorphogenesis of plants. The gene *DET3* (*de-etiolated 3*) has been identified as the C subunit of V-ATPase. The *Arabidopsis det3* mutant develops morphologically as a light-grown plant even when it is grown in the dark [86].

This mutant showed organ-specific defects in cell elongation and a dysfunction in response to brassinosteroids. On the other hand, carrot plants transformed with the antisense DNA encoding the A subunit of V-ATPase exhibited altered leaf morphologies and reduced cell expansion [87]. This antisense experiment also indicates the role of V-ATPase in plant growth and development.

4.2 Organ Specificity of V-ATPase

Taiz and his colleagues [62,63,88] did intensive work on lemon V-ATPase with respect to the tissue specificity. They compared the enzymatic and molecular properties of V-ATPases in the juice sac of lemon fruit with the enzyme in the epicotyl to investigate how the juice sac vacuole acidifies its lumen to pH 2.2. The value is 3 pH units lower than in typical plant vacuoles. V-PPase does not function in lemon fruit vacuoles, although it is not clear whether V-PPase is inactive or absent. Thus, this steep pH gradient in lemon fruit is close to the calculated maximum for a V-ATPase operating at thermodynamic equilibrium assuming an H^+/ATP stoichiometry of 2 [89]. The calculation was done with the following equation [88]:

$$\Delta G_{ATP}/n \geq F \, \Delta\psi -2.303 \, RT \, \Delta pH$$

in which ΔG_{ATP} is the Gibbs free energy for ATP ($-50,000$ J mol^{-1}), n is the H^+/ATP stoichiometry (2 under standard conditions), F is the Faraday constant (96.5 J mV^{-1} mol^{-1}), $\Delta\psi$ is the transmembrane electrical potential (inside positive, $+20$ to $+30$ mV), R is the gas constant (8.31 J K^{-1} mol^{-1}), and T is the absolute temperature (300 K).

The lemon fruit V-ATPase in the vacuolar membrane has been demonstrated to express unusual properties, namely insensitivity to bafilomycin A$_1$, nitrate, and N-ethylmaleimide; sensitivity to vanadate; and no requirement for Cl$^-$ [88]. Interestingly, the solubilized, purified enzyme from juice sac vacuolar membranes exhibited the same inhibitor profiles as the typical V-ATPase [62]. However, the H^+/ATP coupling ratio has been determined to be still about one half that of the typical epicotyl V-ATPase. The loose coupling enables juice sac V-ATPase to make a steep pH gradient across the membrane.

Müller et al. [63] proposed that the lemon fruit V-ATPase exists in two states, a nitrate-sensitive (normal) and a nitrate-insensitive (altered) state, with a low H^+/ATP coupling ratio, and designated this concept the "one pump/two states model." The relative amount of each state varied with the season [63]. The altered state of V-ATPase could be induced in vitro by treatment with KNO$_3$. It is assumed that the tightness of the linkage between V$_1$ and V$_0$ may be essential to keep the normal state. The biochemical mechanism of the transition between the normal and altered states in vivo remains to be resolved. Patch-clamp experiments revealed that the H^+/ATP coupling ratio ranged from 1.75 to 3.28 depending on the cytoplasmic and vacuolar lumenal pH in red beet vacuoles [90]. However, a recent paper reported that the proton transport by V-ATPase or V-PPase is not strictly regulated by the proton chemical gradient (ΔpH) [91]. The authors pointed out that the vacuolar membrane system keeps a constant membrane potential ($\Delta\psi$). The loose coupling (or slip) mechanism of the altered state should be considered with respect to the rotational model of V-ATPase.

The other topic is the presence of organ-specific isoforms of V-ATPase subunits. Developmental and environmental regulation of the c subunit of V-ATPase has been demonstrated (for reviews see Refs. 14 and 17). Among the four genes of subunit c, two genes are constitutively expressed in all organs of *Arabidopsis* and the two genes (*AVA-P3* and *AVA-P4*) are expressed in a tissue-specific manner. Also, Kawamura et al. [34] showed that the distribution of E subunit isoforms differs with the organs of pea, mung bean, and tomato. In pea, the E2 subunit, but not E1, was present in the leaf and cotyledon, whereas both E1 and E2 subunits were present in the petal, epicotyl, and root. Also, the pea leaf lacked the D2 subunit. The steady-state pH gradient across the vacuolar membrane from pea epicotyls was higher than that of the vesicles from leaves. They suggested that the presence of the isoforms of D and E subunits is characteristic of plants and that the isoforms are closely related to the enzymatic properties and physiological demand in each organ [34].

4.3 Expression of V-ATPase and V-PPase under Stress Conditions

The amount of V-ATPase is altered in response to environmental conditions. In particular, the effect of salt stress on the amounts of V-ATPase subunits has been well examined with respect to Na⁺ accumulation in vacuoles and the salt-induced transition from the C_3 state of photosynthesis to the CAM state of CAM plants, such as *Mesembryanthemum crystallinum*. The mRNA level of subunit A has been reported to be increased two- to three-fold by treatment of plants or the cultured cells with 100 mM NaCl. The expression of V-ATPase subunit genes in response to salt stress has been reviewed in detail [17,75]. The increase in V-ATPase activity under salt stress is thought to be an adaptive response to support the energy requirement of an Na⁺/H⁺ antiporter.

With reference to chilling injury of plants, the cold sensitivity of V-ATPase is critical for cold-sensitive plant species such as mung bean [92,93]. Cold inactivation of V-ATPase is a well-known phenomenon for plant and animal enzymes [94]. The activity of V-ATPase on the basis of vacuolar membrane protein decreased to half of the original activity after chilling of mung bean seedlings at 0°C for 3 days [95,96]. Of the several subunits of V-ATPase, the specific contents of the V_1 subunits (A, B, D, and E subunits) in vacuolar membranes decreased after chilling. This might be a result of the release of the V_1 sector from the membrane at 0°C. Yoshida et al. [93,97] proposed that the dysfunction of V-ATPase in chilling-sensitive plants caused by low temperature results in acidosis of the cytosol and alkalization of the vacuoles and that this is a critical point in the chilling injury of plant cells. From these observations we can infer the importance of the physiological role of V-ATPase in plant cells. It must be noted that the total protein amount recovered in the vacuolar membrane from chilling-treated seedlings was about half of that from nontreated seedlings [96]. The chilling treatment has been thought to cause degradation of the vacuolar membrane and dissociation of the V-ATPase complex in hypocotyl cells.

On the other hand, V-PPase activity in mung bean seedlings decreased slightly after chilling treatment at 0°C for 3 days [92] but increased two-fold at 4°C [98]. It has been reported that the V-PPase activity increased more than 50-fold under anoxia and chilling at 10°C in seedlings of rice, an anoxia-tolerant species [99]. This is a reversible change because the increased level shifted down to the original level after return to air. These physiological changes are explained by the induced V-PPase replacing V-ATPase under energy stress, low ATP production, to maintain the vacuolar acidity.

In the marine green algae *Chlorococcum littorale*, the number of vacuoles per cell increased from 6 to 12 and the volume of vacuoles increased markedly under extremely high CO_2 conditions (aerated with air enriched with 40% CO_2) [100]. In this case, both the V-ATPase activity and the level of the B subunit on the basis of membrane protein increased two-fold after transfer to high-CO_2 conditions. The previous observation that ferric reductase was induced under high-CO_2 conditions [101] implied that the increase in size and number of vacuoles under these conditions is related to increased accumulation of iron in vacuoles. Cytosolic pH regulation has also been proposed as a possible function of the vacuole developed under high-CO_2 conditions [100,102].

5   REGULATION OF LUMENAL pH IN VACUOLE

The vacuolar lumenal pH may be regulated through the vacuolar proton pumps themselves (number and activity) and the other transporters and channels. The membrane integrity, namely H^+ leakage of the vacuolar membrane, is also a critical factor in the regulation of the vacuolar pH. However, the mechanism underlying the differential pH regulation in the vacuole is not fully understood. Some possible mechanisms are described and briefly discussed in this section.

5.1   Regulation by Proton Pumps

The number of proton pumps per unit surface area of the vacuolar membrane (density) is a critical factor for the vacuolar pH. The density of proton pumps is regulated through their biosynthesis and degradation rates. Most investigations revealed that the amount of proton pumps is regulated at the transcriptional level. In A subunit of V-ATPase in mung bean hypocotyls, however, the mRNA level was not consistent with the protein amount accumulated in the vacuolar membrane [78]. Such a result may be due to the difference in the life span between mRNAs and proteins. In general, the life span of mRNA (~20 min) in plant cells is shorter than that of protein (at least several hours). Also, we consider the posttranscriptional regulation, such as turnover of mRNA and the efficiency of translation, and the posttranslational regulation, such as chemical modification and proteolytic degradation of the enzyme. Several mechanisms have been proposed for the regulation of the V-ATPase activity: oxidation-reduction of cysteine residues in the catalytic subunit [103], cytosolic inhibitors [104], activators [105], and change in the H⁺/ATP coupling ratio ("slip") [63]. However, neither an activator nor an inhibitor has been identified in plants so far. Furthermore, a reversible dissociation of the V_1 sector from the membrane has been reported as a regulatory mechanism for V-ATPase activity [106].

The volume of vacuoles is also an important factor in defining the vacuolar pH. This "size effect" should be considered in the cell growth process. In mung bean hypocotyls, the small vacuole in the elongating cell expands more than 20-fold during cell elongation [76]. The vacuole expansion itself causes a dilution of H⁺ in the vacuole. Furthermore, several active transporters may consume protons concentrated in the growing vacuoles in elongating cells to generate a high osmotic pressure to expand the vacuole. A small vacuole acidifies the lumen more easily than a large one if they have V-ATPase and V-PPase at an equal density. In mung bean hypocotyls, the membrane of the small vacuole had V-ATPase and V-PPase at the same densities as the expanded vacuole in the mature cell [76]. This even distribution of vacuolar proton pumps in seedlings may be effective in maintaining the lumenal pH.

5.2 Regulation through Other Transport Systems

There may be other mechanisms for regulating the lumenal pH in the vacuole. The vacuole incorporates many ions (e.g., K⁺, Na⁺, Ca²⁺, Mg²⁺, Cl⁻, HCO₃⁻, malate, citrate, amino acids, nitrate, and oxalate) in addition to H⁺. The concentrations of these ions may change the lumenal pH in the vacuole. Only several transporters and channels have been identified in the plant vacuolar membrane. Figure 5 shows possible pH regulatory mechanisms in vacuoles. It is well known that Na⁺-driven antiporters, such as an Na⁺/H⁺ antiporter, in the plasma membrane of mammalian cells regulate not only the Na⁺ concentration but also the cytosolic pH. With the vacuolar Na⁺/H⁺ antiporter, the active incorporation of Na⁺ in vacuoles increases the vacuolar pH. The vacuolar membrane vesicles from

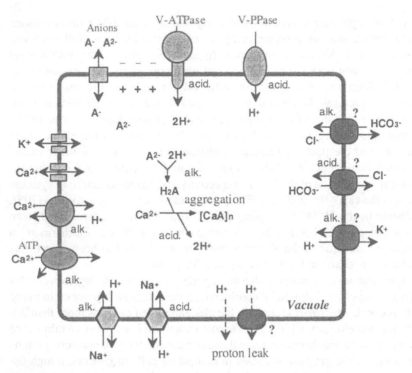

FIGURE 5 Schematic diagram showing pH regulatory mechanisms in the vacuole. Components that have not been identified are marked by question marks. The transport processes resulting in the acidification (acid.) and neutralization (alk.) of the vacuolar lumen are marked.

various plant species exhibit high activity of the Ca^{2+}/H^+ antiporter [5–8]. The antiporter also causes an increase in the vacuolar pH. A K^+/H^+ antiporter has been identified in the plasma membrane of insect epithelial cells [107]. A Cl^-/HCO_3^- antiporter (exchanger), the band 3 protein in red blood cells, has also been characterized at the molecular level. If a Cl^-/HCO_3^- antiporter exists in the vacuolar membrane, Cl^- is transported out of the vacuole and HCO_3^- is brought into the vacuole. This process neutralizes H^+ in the vacuole.

As already mentioned, the V-ATPase has the potential to decrease the vacuolar pH to about 3. However, the actual vacuolar pH value is approximately at 5.5 in most plants. Furthermore, the vacuolar pH of most cells in certain tissues may be regulated within a narrow range because the flower pigments, such as anthocyanin, show the same color in all cells. Anthocyanin is a sensitive natural pH indicator.

6 CONCLUDING REMARKS

The plant vacuole is an alternative external environment that is under the control of the cell. Plant vacuoles accumulate various components from the cytoplasm, process and resupply them, and support the cell development. Multifaceted functions of the vacuole depend on the various membrane transport systems. Vacuolar H⁺-ATPase and H⁺-PPase are key elements in the regulation of the vacuolar pH and many transport systems. Now it is possible to obtain a plant mutant that lacks the gene for the proton pump and to express the gene for plant proton pumps in yeast. Thus, further studies of the proton pumps at physiological and molecular levels may provide new information on how plant cells control the vacuolar functions. Also, the proton pumps attract considerable attention from the viewpoint of the structure-function relationship, especially the coupling mechanism between the substrate hydrolysis and H⁺ transport across the membrane. A detailed study of V-PPase, a unique proton pump, can provide useful information on general bioenergetics.

REFERENCES

1. M Martinoia, R Ratajczak. Transport of organic molecules across the tonoplast. Adv Bot Res 25:365–400, 1997.
2. M Wink. Compartmentation of secondary metabolites and xenobiotics in plant vacuoles. Adv Bot Res 25:141–169, 1997.
3. K Yoshida, T Kondo, Y Okazaki, K Katou. Cause of blue petal colour. Nature 373: 191, 1995.
4. PA Rea, ZS Li, YP Lu, YM Drozdowicz, E Martinoia. From vacuolar GS-X pumps to multispecific ABC transporters. Annu Rev Plant Physiol Plant Mol Biol 49:727–760, 1998.
5. D Sanders, C Brownlee, JF Harper. Communicating with calcium. Plant Cell 11: 691–706, 1999.
6. KD Hirschi, RG Zhen, KW Cunningham, PA Rea, GR Fink. CAX1, an H⁺/Ca²⁺ antiporter from *Arabidopsis*. Proc Natl Acad Sci USA 93:8782–8786, 1996.
7. H Ueoka-Nakanishi, Y Nakanishi, Y Tanaka, M Maeshima. Properties and molecular cloning of Ca²⁺/H⁺ antiporter in the vacuolar membrane of mung bean. Eur J Biochem 262:417–425, 1999.
8. H Ueoka-Nakanishi, T Tsuchiya, M Sasaki, Y Nakanishi, KW Cunningham, M Maeshima. Functional expression of mung bean Ca²⁺/H⁺ antiporter in yeast and its intracellular localization in the hypocotyl and tobacco cells. Eur J Biochem 267: 3090–3098, 2000.
9. M Nishimura, H Beevers. Hydrolysis of protein in vacuoles isolated from higher plant tissue. Nature 277:412–413, 1979.
10. N Hiraiwa, M Kondo, M Nishimura, I Hara-Nishimura. An aspartic proteinase is involved in the maturation of storage proteins in concert with the vacuolar processing enzyme. Eur J Biochem 246:133–141, 1997.

11. N Mizushima, T Noda, T Yoshimori Y Tanaka, T Ishii, MD George, DJ Klionsky, M Ohsumi, Y Ohsumi. A protein conjugation system essential for autophagy. Nature 395:395–398, 1998.

12. T Kirisako, M Baba, N Ishihara, K Miyazawa, M Ohsumi, T Yoshimori, T Noda, Y Ohsumi. Formation process of autophagosome is traced with Apg8/Aut7p in yeast. J Cell Biol 147:435–446, 1999.

13. TH Stevens, M Forgac. Structure, function and regulation of the vacuolar H^+-ATPase. Annu Rev Cell Dev Biol 13:779–808, 1997.

14. H Sze, X Li, MG Palmgren. Energization of plant cell membranes by H^+-pumping ATPases: regulation and biosynthesis. Plant Cell 11:677–689, 1999.

15. M Forgac. Structure and properties of the vacuolar H^+-ATPases. J Biol Chem 274: 12951–12954, 1999.

16. M Futai, T Oka, G Sun-Wada, Y Moriyama, H Kanazawa, Y Wada. Luminal acidification of diverse organelles by V-ATPase in animal cells. J Exp Biol 203:107–116, 2000.

17. R Ratajczak. Structure, function and regulation of the plant vacuolar H^+-translocating ATPase. Biochim Biophys Acta 1465:17–36, 2000.

18. X Li, HT Hsu, RT Su, H Sze. The molecular chaperone calnexin associates with the vacuolar H^+-ATPase from oat seedling. Plant Cell 10:119–130, 1998.

19. EM Herman, X Li, RT Su, P Larsen, HT Hsu, H Sze. Vacuolar-type H^+-ATPases are associated with the endoplasmic reticulum and provacuoles of root tip cells. Plant Physiol 106:1313–1324, 1994.

20. K Matsuoka, T Higuchi, M Maeshima, K Nakamura. A vacuolar type H^+-ATPase in a nonvacuolar organelle is required for the sorting of soluble vacuolar protein precursors in tobacco cells. Plant Cell 9:533–546, 1997.

21. DG Robinson, HP Haschke, G Hiz, B Hoh, M Maeshima, F Marty. Immunological detection of tonoplast polypeptides in the plasma membrane of pea cotyledons. Planta 198:95–103, 1996.

22. WJ Dschida, BJ Bowman. Structure of the vacuolar ATPase from *Neurospora crassa* as determined by electron microscopy. J Biol Chem 267:18783–18789, 1992.

23. JP Abrahams, AGW Leslie, R Lutter, JE Walker. Structure at 2.8 Å resolution of F_1-ATPase from bovine heart mitochondria. Nature 370:621–628, 1994.

24. H Noji, R Yasuda, M Yoshida, K Kinoshita. Direct observation of the rotation of F_1-ATPase. Nature 386:299–302, 1997.

25. PD Boyer. The ATP synthase: a splendid molecular machine. Annu Rev Biochem 66:717–749, 1997.

26. Y Sambongi, Y Iko, M Tanabe, H Omote, A Iwamoto-Kihara, I Ueda, T Yanagida, Y Wada, M Futai. Mechanical rotation of the c subunit oligomer in ATP synthase (F_0F_1): direct observation. Science 286:1722–1724, 1999.

27. T Hisabori, A Kondoh, M Yoshida. The γ subunit in chloroplast F_1-ATPase can rotate in a unidirectional and counter-clockwise manner. FEBS Lett 463:35–38, 1999.

28. R Löw, B Rockel, M Kirsch, R Ratajczak, U Lüttge, S Hörtensteiner, E Martinoia, T Rausch. Early salt stress effects on the differential expression of vacuolar H^+-ATPase genes in roots and leaves of *Mesembryanthemum crystallinum*. Plant Physiol 110:259–265, 1996.

29. M Hasenfratz, CL Tsou, TA Wilkins. Expression of two related vacuolar H^+-ATPase 16-kilodalton proteolipid genes is differentially regulated in a tissue specific manner. Plant Physiol 108:1395–1404, 1995.

30. IY Perera, X Li, H Sze. Several genes encode nearly identical 16 kDa proteolipids of the vacuolar H^+-ATPase from *Arabidopsis thaliana*. Plant Mol Biol 29:227–244, 1995.

31. JP Gogarten, L Taiz. Evolution of proton pumping ATPases: rooting the tree of life. Photosynth Res 33:137–146, 1992.

32. T Berkelman, KA Houtchens, FM DuPont. Two cDNA clones encoding isoforms of the B subunit of the vacuolar ATPase from barley roots. Plant Physiol 104:287–288, 1994.

33. D Rouquie, C Tournaire-Roux, W Szponarski, M Rossignol, P Doumas. Cloning of the V-ATPase subunit G in plant: functional expression and sub-cellular localization. FEBS Lett 437:287–292, 1998.

34. Y Kawamura, K Arakawa, M Maeshima, S Yoshida. Tissue specificity of E subunit isoforms of plant vacuolar H^+-ATPase and existence of isotype enzymes. J Biol Chem 275:6515–6522, 2000.

35. H Inoue, T Noumi, M Nagata, H Murakami, H Kanazawa. Targeted disruption of the gene encoding the proteolipid subunit of mouse vacuolar H^+-ATPase leads to early embryonic lethality. Biochim Biophys Acta 1413:130–138, 1999.

36. R Hirata, LA Graham, A Takatsuki, TH Stevens, Y Anraku. *VMA11* and *VMA16* encode second and third proteolipid subunits of *Saccharomyces cerevisiae* vacuolar membrane H^+-ATPase. J Biol Chem 272: 4795–4803, 1997.

37. T Oka, R Yamamoto, M Futai. Multiple genes for vacuolar type ATPase proteolipids in *Caenorhabditis elegans*. A new gene, *vha-3*, has a distinct cell-specific distribution. J Biol Chem 273:22570–22576, 1998.

38. D Stock, AGW Leslie, JE Walker. Molecular architecture of the rotary motor in ATP synthase. Science 286:1700–1705, 1999.

39. XH Leng, MF Manolson, Q Liu, M Forgac. Site-directed mutagenesis of the 100 kD subunit (Vph1p) of the yeast vacuolar H^+-ATPase. J Biol Chem 271:22487–22493, 1996.

40. XH Leng, MF Manolson, M Forgac. Function of the COOH-terminal domain of VpH1 in activity and assembly of the yeast V-ATPase. J Biol Chem 273:6717–6723, 1998.

41. X Li, H Sze. A 100 kD polypeptide associates with the V_0 membrane sector but not with the active oat vacuolar H^+-ATPase suggesting a role in assembly. Plant J 17:19–30, 1999.

42. D Stock, AGW Leslie, JE Walker. Molecular architecture of the rotary motor in ATP synthase. Science 286:1700–1705, 1999.

43. PA Rea, RJ Poole. Vacuolar H^+-translocating pyrophosphatase. Annu Rev Plant Physiol Plant Mol Biol 44:157–180, 1993.

44. RG Zhen, EJ Kim, PA Rea. The molecular and biochemical basis of pyrophosphate-energized proton translocation at the vacuolar membrane. Adv Bot Res 25:297–337, 1997.

45. M Baltscheffsky, A Schultz, H Baltscheffsky. H^+-PPases: a tightly membrane-bound family. FEBS Lett 457:527–533, 1999.

46. M Maeshima. Vacuolar H^+-pyrophosphatase. Biochim Biophys Acta 1465:37–51, 2000.

47. M Ikeda, E Tanabe, MH Rahman, H Kadowaki, C Moritani, R Akagi, Y Tanaka, M Maeshima, Y Watanabe. A vacuolar inorganic H^+-pyrophosphatase in *Acetabularia acetabulum*: partial purification, characterization and molecular cloning. J Exp Bot 50: 139–140, 1999.

48. Y Nakanishi, N Matsuda, K Aizawa, T Kashiyama, K Yamamoto, T Mimura, M Ikeda, M Maeshima. Molecular cloning of the cDNA for vacuolar H^+-pyrophosphatase from *Chara corallina*. Biochim Biophys Acta 1418:245–250, 1999.

49. DA Scott, W De Souza, M Benchimol, L Ahong, HG Lu, SNJ Moreno, R Docampo. Presence of a plant-like proton-pumping pyrophosphatase in acidocalcisomes of *Trypanosoma cruzi*. J Biol Chem 273:22151–22158, 1998.

50. M Baltscheffsky, S Nadanaciva, A Schultz. A pyrophosphate synthase gene: molecular cloning and sequencing of the cDNA encoding the inorganic pyrophosphate synthase from *Rhodospirillum rubrum*. Biochim Biophys Acta 1364:301–306, 1998.

51. YM Drozdowicz, YP Lu, V Patel, S Fitz-Gibbon, JH Miller, PA Rea. A thermostable vacuolar-type membrane pyrophosphatase from the archaeon *Pyrobaculum aerophilum*: implications for the origins of pyrophosphate-energized pumps. FEBS Lett 460:505–512, 1999.

52. M Maeshima, S Yoshida. Purification and properties of vacuolar membrane proton-translocating inorganic pyrophosphatase from mung bean. J Biol Chem 264:20068–20073, 1989.

53. A Chanson, PE Pilet. Target molecular size and sodium dodecylsulfate polyacrylamide gel electrophoresis analysis of the ATP-dependent and pyrophosphate-dependent proton pumps from maize root tonoplast. Plant Physiol 90:934–938, 1989.

54. CM Tzeng, CY Yang, SJ Yang, SS Jiang, SY Kuo, SH Hung, JT Ma, RL Pan. Subunit structure of vacuolar proton-pyrophosphatase as determined by radiation inactivation. Biochem J 316:143–147, 1996.

55. MH Sato, M Maeshima, Y Ohsumi, M Yoshida. Dimeric structure of H^+-translocating pyrophosphatase from pumpkin vacuolar membranes. FEBS Lett 290:177–180, 1991.

56. A Takasu, Y Nakanishi, T Yamauchi, M Maeshima. Analysis of the substrate binding site and carboxyl terminal region of vacuolar H^+-pyrophosphatase of mung bean with peptide antibodies. J Biochem 122:883–889, 1997.

57. C Toyoshima, M Nakasato, H Nomura, H Ogawa. Crystal structure of the calcium pump of sarcoplasmic reticulum at 2.6 Å resolution. Nature 405:647–655, 2000.

58. JM Ward, H Sze. Proton transport activity of the purified vacuolar H^+-ATPase from oats: direct stimulation by Cl^-. Plant Physiol 99:925–931, 1992.

59. Y Wang, H Sze. Similarities and differences between the tonoplast-type and the mitochondrial H^+-ATPase of oat roots. J Biol Chem 260:10434–10443, 1985.

60. S Dröse, KU Bidseil, EJ Bowman, A Siebers, A Zeeck, K Altendorf. Inhibitory effects of modified bafilomycins and concanamycins on P- and V-type adenosinetriphosphatases. Biochemistry 32:3902–3906, 1993.

61. EJ Bowman, A Siebers, K Altendorf. Bafilomycins: a class of inhibitors of mem-

brane ATPases from microorganisms, animal cells, and plant cells. Proc Natl Acad Sci USA 85:7972–7976, 1988.

62. ML Müller, U Irkens-Kiesecker, D Kramer, L Taiz. Purification and reconstitution of the vacuolar H⁺-ATPases from lemon fruits and epicotyls. J Biol Chem 272: 12762–12770, 1997.

63. ML Müller, M Jensen, L Taiz. The vacuolar H⁺-ATPase of lemon fruits is regulated by variable H⁺/ATP coupling and slip. J Biol Chem 274:10706–10716, 1999.

64. PJ White, J Marshall, JAC Smith. Substrate kinetics of the tonoplast H⁺-translocating inorganic pyrophosphatase and its activation by free Mg^{2+}. Plant Physiol 93: 1063–1070, 1990.

65. M Maeshima. H⁺-translocating inorganic pyrophosphatase of plant vacuoles: inhibition by Ca^{2+}, stabilization by Mg^{2+} and immunological comparison with other inorganic pyrophosphatases. Eur J Biochem 196:11–17, 1991.

66. CP Darley, LA Skiera, FD Northrop, D Sanders, JM Davies. Tonoplast inorganic pyrophosphatase in Vicia faba guard cells. Planta 206:272–277, 1998.

67. R Gordon-Weeks, VD Koren'kov, SH Steele, RA Leigh. Tris is a competitive inhibitor of K⁺ activation of the vacuolar H⁺-pumping pyrophosphatase. Plant Physiol 114:901–905, 1997.

68. YM Drozdowicz, JC Kissinger, PA Rea. AVP2, a sequence-divergent, K⁺-insensitive H⁺-translocating inorganic pyrophosphatase from Arabidopsis. Plant Physiol 123:353–362, 2000.

69. JM Davies, RJ Poole, PA Rea, D Sanders. Potassium transport into plant vacuoles energized directly by a proton-pumping inorganic pyrophosphatase. Proc Natl Acad Sci USA 89:11701–11705, 1992.

70. G Obermeyer, A Somer, FW Bentrup. Potassium and voltage dependence of the inorganic pyrophosphatase of intact vacuoles from Chenopodium rubrum. Biochim Biophys Acta 1284:203–212, 1996.

71. MH Sato, M Kasahara, N Ishii, H Homareda, H Matsui, M Yoshida. Purified vacuolar inorganic pyrophosphatase consisting of a 75-kDa polypeptide can pump H⁺ into reconstituted proteoliposomes. J Biol Chem 269:6725–6758, 1994.

72. R Ros, C Romieu, R Gibrat, C Grignon. The plant inorganic pyrophosphatase does not transport K⁺ in vacuole membrane vesicles multilabeled with fluorescent probes for H⁺, K⁺, and membrane potential. J Biol Chem 270:4368–4374, 1995.

73. V Sarafian, Y Kim, RJ Poole, PA Rea. Molecular cloning and sequence of cDNA encoding the pyrophosphate-energized vacuolar membrane proton pump (H⁺-PPase) of Arabidopsis thaliana. Proc Natl Acad Sci USA 89:1775–1779, 1992.

74. Y Nakanishi, M Maeshima. Isolation of a cDNA for a H⁺-pyrophosphatase-like protein from Arabidopsis thaliana (accession no. AB034696) and its functional expression in yeast (PGR00-026). Plant Physiol 122:619, 2000.

75. U Lüttge, R Ratajczak. The physiology, biochemistry and molecular biology of the plant vacuolar ATPase. Adv Bot Res 25:253–296, 1997.

76. M Maeshima. Development of vacuolar membranes during elongation of cells in mung bean hypocotyls. Plant Cell Physiol 31:311–317, 1990.

77. LB Smart, F Vojdani, M Maeshima, TA Wilkins. Genes involved in osmoregulation during turgor driven cell expansion of developing cotton fibers are differentially regulated. Plant Physiol 116:1539–1549, 1998.

78. Y Nakanishi, M Maeshima. Molecular cloning of vacuolar H^+-pyrophosphatase and its developmental expression in growing hypocotyl of mung bean. Plant Physiol 116:589–597, 1998.

79. M Maeshima, Y Nakanishi, C Matsuura-Endo, Y Tanaka. Proton pumps of the vacuolar membrane in growing plant cells. J Plant Res 109:119–125, 1996.

80. K Shiratake, Y Kanayama, M Maeshima, S Yamaki. Changes in H^+-pumps and a tonoplast intrinsic protein of vacuolar membranes during the development of pear fruit. Plant Cell Physiol 38:1039–1045, 1997.

81. M Maeshima, I Hara-Nishimura, Y Takeuchi, M Nishimura. Accumulation of vacuolar H^+-pyrophosphatase and H^+-ATPase during reformation of the central vacuole in germinating pumpkin seeds. Plant Physiol 106:61–69, 1994.

82. SJ Swanson, PC Bethke, RL Jones. Barley aleurone cells contain two types of vacuoles: characterization of lytic organelles by use of fluorescent probes. Plant Cell 10:685–698, 1998.

83. N Paris, CM Stanley, RL Jones, JC Rogers. Plant cells contain two functionally distinct vacuolar compartments. Cell 85:563–572, 1996.

84. P Fleurat-Lessard, N Frangne, M Maeshima, R Ratajczak, JL Bonnemain, E Martinoia. Increased expression of vacuolar aquaporin and H^+-ATPase related to motor cell function in *Mimosa pudica* L. Plant Physiol 114:827–834, 1997.

85. GY Jauh, TE Phillips, JC Rogers. Tonoplast intrinsic protein isoforms as markers for vacuolar functions. Plant Cell 11:1867–1882, 1999.

86. K Schumacher, D Vafeados, M McCarthy, H Sze, T Wilkins, J Chory. The *Arabidopsis det3* mutant reveals a central role for the vacuolar H^+-ATPase in plant growth and development. Genes Dev 13:3259–3270, 1999.

87. JP Gogarten, J Fichmann, Y Braun, L Morgan, P Styles, SL Taiz, K Delapp, L Taiz. The use of antisense mRNA to inhibit the tonoplast proton ATPase in carrot. Plant Cell 4:851–864, 1992.

88. ML Müller, U Irkens-Kiesecker, B Rubinstein, L Taiz. On the mechanism of hyperacidification in lemon. J Biol Chem 271:1916–1924, 1996.

89. AL Schmidt, DP Briskin. Energy transduction in tonoplast vesicles from red beet (*Beta vulgaris* L.) storage tissue: H^+/substrate stoichiometries for the H^+-ATPase and H^+-PPase. Arch Biochem Biophys 301:165–173, 1993.

90. JM Davies, I Hunt, D Sanders. Vacuolar H^+-pumping ATPase variable transport coupling ratio controlled by pH. Proc Natl Acad Sci USA 91:8547–8551, 1994.

91. T Hirata, N Nakamura, H Omote, Y Wada, M Futai. Regulation and reversibility of vacuolar H^+-ATPase. J Biol Chem 275:386–389, 2000.

92. S Yoshida, C Matsuura, S Etani. Impairment of tonoplast H^+-ATPase as an initial physiological response of cells to chilling in mung bean (*Vigna radiata* L. Wilczek). Plant Physiol 89:634–642, 1989.

93. S Yoshida, K Hotsubo, Y Kawamura, M Murai, K Arakawa, D Takezawa. Alterations of intracellular pH in response to low temperature stresses. J Plant Res 112:225–236, 1999.

94. Y Moriyama, N Nelson. Cold inactivation of vacuolar proton-ATPases. J Biol Chem 264:3577–3582, 1989.

95. C Matsuura-Endo, M Maeshima, S Yoshida. Subunit composition of vacuolar membrane H^+-ATPase from mung bean. Eur J Biochem 187:745–751, 1990.

96. C Matsuura-Endo, M Maeshima, S Yoshida. Mechanism of the decline in vacuolar H⁺-ATPase activity in mung bean hypocotyls during chilling. Plant Physiol 100: 718–722, 1992.

97. S Yoshida. Low temperature–induced cytoplasmic acidosis in cultured mung bean (*Vigna radiata* L. Wilczek) cells. Plant Physiol 104:1131–1138, 1994.

98. CP Darley, JM Davies, D Sanders. Chill-induced changes in the activity and abundance of the vacuolar proton-pumping pyrophosphatase from mung bean hypocotyls. Plant Physiol 109:659–665, 1995.

99. GD Carystinos, HR MacDonald, AF Monroy, RS Dhindsa, RJ Poole. Vacuolar H⁺-translocating pyrophosphatase is induced by anoxia or chilling in seedlings of rice. Plant Physiol 108:641–649, 1995.

100. T Sasaki, NA Pronina, M Maeshima, I Iwasaki, N Kurano, S Miyachi. Development of vacuoles and vacuolar H⁺-ATPase activity under extremely high CO_2 conditions in *Chlorococcum littorale* cells. Plant Biol 1:68–75, 1999.

101. T Sasaki, N Kurano, S Miyachi. Induction of ferric reductase activity and of iron uptake capacity in *Chlorococcum littorale* cells under extremely high-CO_2 and iron-deficient conditions. Plant Cell Physiol 39:405–410, 1998.

102. I Iwasaki, N Kurano, S Miyachi. Effects of high-CO_2 stress on photosystem II in a green alga, *Chlorococcum littorale*, which has a tolerance to high CO_2. J Photochem Photobiol 36:327–332, 1996.

103. Y Feng, M Forgac. Inhibition of vacuolar H⁺-ATPase by disulfide bond formation between cysteine 254 and cysteine 532 in subunit A. J Biol Chem 269:13224–13230, 1994.

104. K Zhang, ZQ Wang, S Gluck. A cytosolic inhibitor of vacuolar H⁺-ATPases from mammalian kidney. J Biol Chem 267:14539–14542, 1992.

105. XS Xie, BP Crider, DK Stone. Isolation of a protein activator of the clathrin-coated vesicle proton pump. J Biol Chem 268:25063–25067, 1993.

106. PM Kane. Disassembly and reassembly of the yeast vacuolar H⁺-ATPase in vivo. J Biol Chem 270:17025–17032, 1995.

107. H Wieczorek, M Putzenlechner, W Zeiske, U Klein. A vacuolar-type proton pump energizes K⁺/H⁺ antiport in an animal plasma membrane. J Biol Chem 266:15340–15347, 1991.

108. C Maurel. Aquaporins and water permeability of plant membranes. Annu Rev Plant Physiol Plant Mol Biol 48:399–429, 1997.

3

The Cytoplasmic pH Stat

Robert J. Reid and F. Andrew Smith
Adelaide University, Adelaide, South Australia, Australia

1 INTRODUCTION

Many plant species are able to grow well under extremely alkaline or acidic conditions. For example *Catharanthus* suspension-cultured cells are able to grow well at pH 2.3 (but not at pH 2) [1], and many algae are able to live in lakes with pH above 10 [2]. Furthermore, plant cells produce and consume large amounts of protons in normal metabolism and in membrane transport of nutrients. Despite these potential "threats" to plants in terms of proton concentrations and production, the pH of the cytoplasm of plant cells is maintained within a narrow range. In the absence of such pH regulation, the ionization of acidic and basic groups, especially on proteins and other polyelectrolytes, would cause considerable changes in structure that would ultimately affect their function and the viability of the cell. However, there remains sufficient variability in cytoplasmic pH to enable pH to act as a controller of many cellular functions, both in the bulk cytoplasm and in membranes. This chapter considers the processes that contribute to the maintenance of a stable cytoplasmic pH. Regulation of vacuolar pH is not considered here, and the effect of cytoplasmic pH on cellular processes is addressed only in relation to pH regulation itself. Useful previous reviews of this topic include those by Smith and Raven [3], Kurkdjian and Guern [4], Guern et al. [5], and Sakano [6].

2   MEASUREMENT OF CYTOPLASMIC pH

A good understanding of cellular pH homeostasis is founded upon the need to measure accurately the pH of the cytoplasm. This presents a range of technical challenges because, unlike animal cells, the cytoplasm of plant cells shares its intracellular space with a vacuole that characteristically occupies 80–90% of the cell volume. Whole cell measurements, by whatever method, are therefore a rough average of the two compartments, and more direct approaches are required to obtain the cytoplasmic pH alone. In view of the technical difficulties, it is perhaps not surprising that few researchers have attempted to measure cytoplasmic pH in plant cells (or, more likely, have succeeded). Much of what is known about cytoplasmic pH in plants is due to a handful of laboratories that have been able to master one of three main techniques: weak acid distribution [7–10], pH microelectrodes [10,11] and ^{31}P nuclear magnetic resonance (NMR) [12,13].

The weak acid distribution method is based on the fact that weak acids accumulate in compartments with pH higher than the pK of the weak acid and as long as only the neutral form of the acid can cross the membrane, the degree of accumulation can be used to calculate the intracellular pH. The most widely used weak acid is 5,5-dimethyl-oxazolidine dione (DMO), the concentration of which can be easily measured if it is radioactively labeled. The disadvantage of this technique when applied to the measurement of cytoplasmic pH of plant cells is that it is necessary to know the volume of the cytoplasm, which is often difficult. Inaccuracies are increased (1) where there is a significant amount of the acid in the vacuole and (2) through leakage of the ionized form of the weak acid across membranes. These problems tend to render the technique unsuitable for small cells, but Smith and his coworkers [7–9,14,15] have had considerable success with giant algal cells, where the concentration of the dye in the vacuole can be accurately measured and the extent of membrane leakage estimated and corrected for. However, time resolution was poor (up to tens of minutes, depending on the method used).

Liquid membrane pH-sensitive microelectrodes have been widely used in animal cells, but the sensor liquid is immediately displaced on insertion into plant cells because of the high turgor pressure. This problem was overcome by stabilizing the sensor in the tip with polyvinyl chloride [10] or nitrocellulose [11], and the development of double-barreled microelectrodes overcame problems associated with the reference and pH-sensing electrodes being in different compartments.

^{31}P-NMR has been widely used to investigate phosphorus metabolism in cells [e.g., 16,17], but the chemical shifts associated with changes in protonation of P compounds can also be used to measure the pH of cellular compartments.

The main advantage of this technique is that it is noninvasive and gives good resolution between vacuolar pH and cytoplasmic pH, even in small cells. The drawbacks are that it has poor time resolution (several minutes), poor spatial resolution in complex tissues, and requires large amounts of material, especially when dealing with mature cells, where the proportion of cytoplasm is small relative to the total cell volume.

The use of these three techniques provided an understanding of the extent to which the cytoplasmic pH changes when challenged with a variety of natural and experimentally imposed treatments such as metabolic inhibition, application of weak acids and bases, and light-dark transitions. Examples are given in Table 1. In most cases these treatments alter the cytoplasmic pH only slightly. Likewise, large fluctuations in external pH produce relatively small changes in cytoplasmic pH. In *Chara*, measurements on a large number of cells bathed in solutions over the pH range 4.5–10.5 showed that the cytoplasmic pH varied by only about 1 unit (Fig. 1a). Similar measurements were obtained with both the DMO method [7] and pH-sensitive microelectrodes [11], which gives some confidence in the two techniques. Measurements of the cytoplasmic pH in suspension-cultured cells [25] (Fig. 1b) and in barley root tips [13] using ^{31}P-NMR indicated an even lower sensitivity to external pH.

The more recent technique of ratiometric imaging of pH-sensitive fluorescent dyes using confocal laser scanning microscopy is being increasingly widely employed, despite problems of introducing the dye into plant cells and also possible artifacts caused by buffering by the dyes [26]. Nevertheless, the method is proving useful, for example, in demonstrating the presence of different domains of cytoplasmic pH in individual cells such as pollen tubes [26]. This raises questions about the usefulness of "point" pH measurements with microelectrodes or "averaging" measurements, e.g., with DMO (see also Chapter 4).

3 THE ORIGIN OF CHANGES IN CYTOPLASMIC pH

3.1 Sources of Protons and Hydroxyl Ions

It soon becomes obvious that it is fairly difficult experimentally to induce large changes in cytoplasmic pH (e.g., Table 1 and Fig. 1). Cell death ultimately results in the loss of control of intracellular pH, which then approaches the pH of the environment, but while the cell remains viable, even if its metabolism is severely restricted, it retains the ability to limit any shift in cytoplasmic pH. Before considering the processes that contribute to this stability, it is useful to ask the question, why does the pH change at all? The simple answer is that the change in pH is caused by increases in the concentration of protons or hydroxyl ions in the cytoplasm. But where do these extra protons or hydroxyl ions come from? The answer

TABLE 1 Effects of a Range of Experimental Treatments on the Cytoplasmic pH of Different Plant Cells[a]

Treatment	Method	Plant material	pH_o	ΔpH_{cyt}	Reference
NaN$_3$ (30 μM)	DMO	*Chara* internodes	5.0	−0.35	9
CCCP (4 μM)	DMO	*Chara* internodes	5.1	−0.57	9
DNP (10 μM)	DMO	*Chara* internodes	5.0	−0.85	9
DCMU (4 μM)	DMO	*Chara* internodes	5.0	−0.24	9
Ca^{2+} starvation	DMO	*Chara* internodes	7.5	−0.30	14
Cl$^-$ starvation	DMO	*Chara* internodes	6.2	+0.05	18
NH$_4^+$ (0.2 mM)	DMO	*Chara* internodes	7.1	+0.13	7
Darkness	DMO	*Chara* internodes	5.2	−0.24	15
Darkness	Microelectrodes	*Riccia* thallus	7.3	−0.30[b]	19
CN$^-$ (0.1 mM)	Microelectrodes	*Riccia* rhizoids	7.3	−0.30	10
Darkness	Microelectrodes	*Chara* internodes	6.5	−0.40	11
Acetate (1 mM)	Microelectrodes	*Riccia* thallus	5.1	−0.18	19
Acetate (2 mM)	Microelectrodes	Maize root hairs	5.2	−0.70	20
Butyrate (1 mM)	Microelectrodes	*Chara* internodes	5.0	−0.82	21
Butyrate (10 mM)	^{31}P-NMR	Barley root tips	6.0	−0.40	13
Fusicoccin (10 μM)	^{31}P-NMR	Maize root tips	6.0	+0.09	13
NaN$_3$ (0.5 mM)	^{31}P-NMR	Barley root tips	6.0	−0.30	13
Anoxia	^{31}P-NMR	Maize root tips	6.0	−0.30	22
NH$_4^+$ (10 mM)	^{31}P-NMR	Maize root tips	9.0	+0.26	23
NH$_4^+$ (2 mM)	Fluorescent dye	Rice root hair	7.0	+0.35[b]	24

[a] The responses are shown as the change in pH (ΔpH_{cyt}) from the normal control value, recorded by one of four different techniques used for measuring intracellular pH. In most cases the pH changes were recorded within 1 h of application of the treatment. pH_o = external pH during the measurement.
[b] Transient.

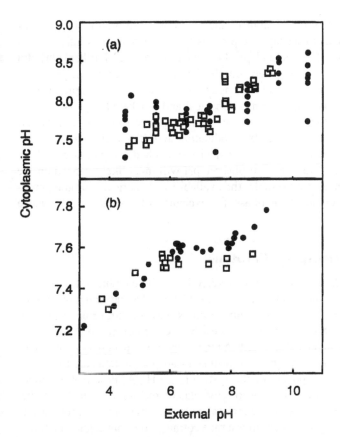

FIGURE 1 Effect of external pH on cytoplasmic pH in (a) *Chara corallina* measured by either weak acid distribution (open squares) or double-barreled pH microelectrodes (data from Ref. 11) and (b) suspension-cultured cells of oil palm (open squares) and carrot measured by ^{31}P-NMR (data from Ref. 25).

to this question is more difficult because protons and hydroxyl ions are central to almost all activities occurring within cells and so alteration of any activity could potentially perturb the local pH.

The sources of protons or hydroxyl ions can be separated into those originating from fluxes across the membranes that bound the cytoplasm and those that are produced or consumed by reactions occurring within the cytoplasm. The former are usually referred to as biophysical and the latter as biochemical proton or hydroxyl loads. However, under any given set of conditions in which the cytoplasmic pH changes, it is often difficult to ascribe the change to just one of these categories. For example, if metabolism is inhibited by anoxia, is the reduction in cytoplasmic pH of 0.3 units (Table 1) due to inhibition of proton efflux across the plasma membrane, to leakage of protons from the outside solution, or to metabolic production of acidic end products such as lactate? Further, why did the pH fall by only 0.3—were there biophysical and/or biochemical processes that were activated to limit the pH change? Again, there is no simple answer, but it seems certain that just as both biophysical and biochemical processes can threaten the stability of intracellular pH by producing excess protons or hydroxyl ions, both can counteract changes by modulating activities that lead to the excesses. The relative importance of biophysical and biochemical processes in both perturbation and maintenance of cytoplasmic pH will most likely depend on the particular circumstances, for example, the availability of energy for proton pumping or access to an extracellular phase for exchange of (usually) acidic waste products.

3.2 Membrane Transport of Protons

Protons are involved, either directly or indirectly, in the membrane transport of almost all solutes, with the exception of nonelectrolytes that have high membrane permeability (e.g., gases, alcohols). Thus, protons play a central role in nutrient acquisition through the use of proton electrochemical gradients to drive uptake and efflux. These gradients, along with ATP, are the principal energy currencies of plant cells. The main classes of proton-linked transport systems are illustrated in Fig. 2. The central component of the system is the H^+-ATPase that generates both pH and electrical differences across the plasma membrane, which together provide an inward-directed electrochemical potential gradient for protons that can be used as an energy source for membrane cotransport of nutrients. The fluxes of many of these nutrients are small and therefore the flux of associated protons will have little effect on pH on either side of the membrane. However, the fluxes of major nutrients, in particular NO_3^- and $H_2PO_4^-$ (Pi), involve significant proton influx that needs to be countered by active extrusion of protons through the H^+-ATPase.

3.3 Metabolic Production of H^+ and OH^-

Many steps in metabolic sequences involve chemical conversions in which protons are either produced or consumed. Commonly, production of protons at one step is balanced by consumption at a later step. For example, synthesis of ATP consumes a proton that is released again during hydrolysis to ADP and Pi. Oxidative breakdown of glucose to CO_2 and H_2O involves production and consumption of protons at many stages of the pathway, but the overall process is virtually neutral (Table 2), although CO_2 itself forms a weakly acidic solution. Under anaerobic conditions, energy cycling depends mainly on the breakdown of carbohy-

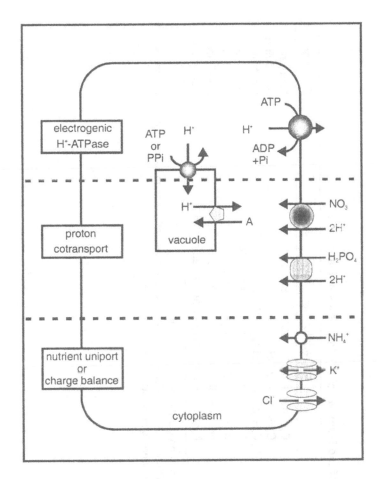

FIGURE 2 Principal transport processes involved in biophysical pH regulation in plants.

drates to ethanol and lactic acid. Production of ethanol does not in itself alter pH and metabolism can continue as long as ethanol can be released from the cell because its accumulation can be toxic. Breakdown of carbohydrates to lactic acid via decarboxylation of pyruvate does produce protons, and this causes the pH to fall. Thus, unless lactic acid can readily leave the plant, lactic fermentation can be sustained only over relatively short periods because of the negative impact on pH homeostasis. Fox et al. [27] found that the switch from lactic to alcoholic fermentation could be induced by treatment with weak acids, which suggested that the shift to alcoholic fermentation was activated by cytoplasmic acidification.

TABLE 2 Effect on Cytoplasmic pH of the Main Metabolic Processes in Plant Cells

Reaction	Transformation	Effect on pH_{cyt}
Oxidation of glucose	$C_6H_{12}O_6 + 6O_2 \rightarrow 6CO_2 + 6H_2O$	Neutral
Lactic fermentation [a]	$C_6H_{12}O_6 \rightarrow 2C_3H_4O_3^- + 2H^+$	Acidifying
Ethanolic fermentation[a]	$C_6H_{12}O_6 \rightarrow 2C_2H_6O + 2CO_2$	Neutral
Dehydrogenase[b]	$R{-}H_2 + NAD^+ \rightarrow R + NADH + H^+$	Acidifying
ATP hydrolysis[b]	$ATP^{4-} \rightarrow ADP^{3-} + H_2PO_4^{2-} + H^+$	Acidifying
Nitrate assimilation	$NO_3^- + 8H \rightarrow NH_4^+ + 2H_2O + 2OH^-$	Alkalizing
	$NH_4^+ + $ organic acid $\rightarrow$ amino acid $+ H^+$	Acidifying

[a] In most plants these reactions are important only under anaerobic conditions.
[b] Note that under normal metabolic conditions these reactions are cyclic (oxidation-reduction, dephosphorylation-phosphorylation) so that H^+ is produced and consumed and there is no effect on cytoplasmic pH.

4 THE pH-STAT CONCEPT

The idea that there are key processes in plant cells that are responsible for regulating cytoplasmic pH has been widely discussed in the literature. According to this view, changes in cytoplasmic pH would activate specific biophysical (e.g., membrane H^+-ATPase) or biochemical (e.g., carboxylation-decarboxylation) reactions that would counter the pH change. The alternative proposition is that pH regulation is the overall outcome of the autoregulation, by cytoplasmic pH, of the rates of the many proton-consuming and proton-producing activities within the cell, including transport of protons across membranes.

4.1 The Biochemical pH-Stat

Biochemical control of cytoplasmic pH relies on (1) metabolites acting as pH buffers and (2) enzymic rearrangements of metabolites to convert weak acids into strong acids and vice versa. It should be noted that cytoplasmic pH buffers are not capable of long-term net consumption or production of protons or hydroxyl ions and therefore should be considered as short-term mechanisms for preventing large fluctuations in pH that would otherwise have adverse consequences for metabolism. Likewise, enzymic rearrangement of metabolites cannot cope with long-term acidification, although it can with alkalizing reactions in assimilation of NO_3^-, as we shall show. It is important to emphasize that both mechanisms operate in a cytoplasmic environment that has already been "set up" so that cytoplasmic pH is near neutrality (see Ref. 3).

4.1.1 Chemical Buffering

The cell is full of natural pH buffers—amino acids, organic acids, phosphates, etc.—that potentially can resist quite large short-term proton or hydroxyl loads. Estimates of the buffering capacity vary widely, from less than 20 mM pH^{-1} [21] up to 100 mM pH^{-1} of cytoplasm [4]. The pK_a values of organic acids (generally < pH 5) and the carboxyl groups of amino acids (< pH 3) are usually too low to be useful for maintaining cytoplasmic pH near neutrality. Likewise, the pK values of amino groups of free amino acids are mostly above 9, the exception being an amino group of cystine with a pK_a of 8.0. In peptides and proteins, the alpha-amino and alpha-carboxyl groups that are combined in peptide linkages cannot ionize in the pH range 0 to 14, but the NH_2-terminal and COOH-terminal groups are free and display pK_a values different from those of their free amino acids. Due to the greater distance between the terminal $-NH_2$ and $-COOH$ groups in proteins and peptides compared with single amino acids, electrostatic interactions between them are diminished, so that the pK_a values for the terminal carboxyl groups are higher and those for the terminal amino groups lower than in the free amino acid. The weakening of the acid nature of the terminal carboxyl groups in proteins can reduce the pK_a of many of the terminal amino groups to around 8, which provides significant resistance to alkalization of the cytoplasm. The other amino acid groups that have useful pK_a values are the imidazolyl group of histidine ($pK_a = 6.0$) and the $-SH$ group of cysteine ($pK_a = 8.3$).

At acid pH, the CO_2/HCO_3^- couple ($pK_a = 6.3$), along with the buffering by histidine, provides a degree of protection against cytoplasmic acidification. Perhaps the most useful chemical buffer at neutral pH is phosphate (Pi), which has a pK_a around 7.2 and occurs in the cytoplasm at a free concentration of approximately 10 mM [28]. Thus, there are three main chemical buffering systems in plant cells, one that is useful at alkaline pH, one at neutral pH, and one at acid pH. It must be remembered that a buffer is a mixture of undissociated and dissociated solutes, with the dissociated forms balanced by other ions, both inorganic and organic. As noted before, the setting up of the buffer mixture cannot be solely an intracellular event—it requires transport of ions into or out of the cells. An obvious example is exchange of cations such as K^+ for protons that are produced during synthesis of carboxylic acids. Also, formation of amino groups involves uptake of NH_4^+ or NO_3^-, as discussed in the following.

4.1.2 Metabolic Interconversions

The concept of control of cytoplasmic pH through interconversions of metabolites was first enunciated by Davies [29,30], who asserted that cytoplasmic pH could be controlled by regulating the degree of carboxylation of certain organic acids within the cytoplasm. The key elements of the Davies pH-stat are shown in Fig. 3. It relies on the different pH optima of the carboxylating (OH^--consuming) enzyme PEP carboxylase that is stimulated by high pH and the decarboxylating (OH^--producing) malic enzyme that is stimulated by more acidic conditions. The carboxylating and decarboxylating steps as well as the intermediate dehydrogenase reaction can be written in simple form as shown at the top of Fig. 3. This system gained considerable acceptance because it explained changes in malate concentrations that occurred under conditions in which the cytoplasm would be expected to alkalinize (e.g., excess cation uptake, i.e., K^+/proton exchange [31]). Empirical measurements of pH under such conditions have not shown significant increases [32], and so the signal for the increased carboxylation has not been satisfactorily identified. Current models for the regulation of PEP carboxylase in CAM plants focus on phosphorylation states of protein kinases and pay little attention to regulation by pH, e.g., as a second messenger [33]. Increase in cytoplasmic pH is part of the signaling cascade involved in light stimulation of PEP carboxylase kinase in C_4 plants [34]. However, this involves increased gene expression rather than "kinetic" control of enzyme activity as envisaged in the Davies model.

The Davies model has been challenged by Sakano [6], who maintained that when the fate of the H^+ involved in the dehydrogenase reactions is taken into account, proton consumption or production does not occur as predicted by the

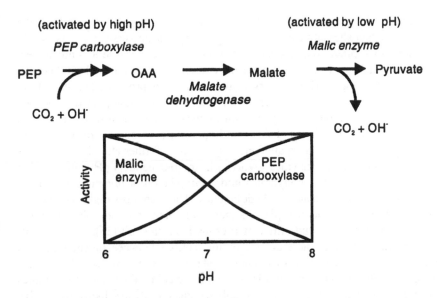

FIGURE 3 Key features of the biochemical pH-stat as proposed by Davies [29,30] involving a strong pH sensitivity of carboxylating and decarboxylating reactions. PEP, phosphoenolpyruvate; OAA, oxaloacetate.

model. Sakano suggested that the following is a more accurate representation of the carboxylating and decarboxylating reactions.

$$PEP^{3-} + HCO_3^- \rightarrow OAA^{2-} + Pi^{2-}$$
$$OAA^{2-} + NADH + H^+ \rightarrow malate^{2-} + NAD^+$$
$$Malate^{2-} + NAD^+ \rightarrow pyruvate^- + NADH + CO_2$$

However, if NADH is recycled by other metabolic reactions, the proton exchanges of the dehydrogenase reactions can be ignored, at least in terms of ongoing synthesis or breakdown of malate. Nevertheless, there may be other steps within the NADH-cycling pathways in which there is net production or consumption of protons, and this needs to be more closely scrutinized. Certainly, the need to maintain balance between NADH and NAD$^+$ introduces new possible control mechanisms for biochemical pH regulation. Sakano [6] also pointed out that the biochemical pH-stat is complicated by the consumption of protons associated with hydrolysis of ATP during the lead up to malate formation (see Table 2). Again, ADP is recycled within the cell in order to maintain a "functional" ATP/ADP ratio: there can be no long-term accumulation of ADP (or ATP). Accordingly, the concept that synthesis or breakdown of carboxylic acids is important in biochemical regulation of pH remains valid. Sakano [6] also proposed that the cyanide-resistant alternative oxidase pathway, which is universal in plants, would be a more appropriate route for the oxidation of malate because it is not subject to the same allosteric control as the normal cytochrome pathway and would therefore allow more rapid pH regulatory adjustment. It was proposed that short-term pH control is the primary function of the alternative oxidase pathway and that its dependence on alpha-keto acids such as pyruvic acid is strong evidence in support of the hypothesis [6]. This is an important development of the concept of biochemical pH regulation and certainly requires further attention.

Activation of key enzymes involved in the biochemical pH-stat, resulting in a decrease in stored carbohydrates and accumulation of organic acid anions, occurs as a response of plants to Fe deficiency. There are also slight increases in both cytoplasmic and vacuolar pH, changes in distribution of phosphate between vacuole and cytoplasm, and efflux of protons from the cell [35]. These events neatly demonstrate that there are major interactions between biochemical and biophysical components of cytoplasmic pH regulation. In fact, large-scale changes in intracellular concentrations of organic acid anions and other solutes that are associated with the biochemical pH-stat are always likely to involve membrane transport across the tonoplast.

5   THE BIOPHYSICAL pH-STAT

5.1   Proton-Pumping ATPases

The major limitation to biochemical pH regulation is that, although it can cope with metabolic reactions that generate OH^-, by synthesizing carboxylic acid, such reactions are much fewer than reactions that generate H^+ (especially but not exclusively from carboxylic acids). The biochemical pH-stat can cope with the latter only by breaking down carboxylates synthesized earlier in the life of the cell, but this synthesis would also have generated H^+ that the cell would have had to lose. In terms of intracellular metabolism alone, this becomes a reductio ad absurdum. In fact, the cytoplasm "loses" H^+ by means of transport across the boundary membranes. The key H^+ transporters are plasma membrane H^+-ATPases with a pH optimum around 6.6 [36–38] (see also Chapter 1). H^+ extrusion should therefore be stimulated (other factors permitting) by decreasing cytoplasmic pH, either allosterically or more directly through an increase in substrate (H^+) concentration. There is strong evidence that cytoplasmic acidification stimulates pump activity in vivo [39,40], whereas the observed depolarization of the plasma membrane in *Riccia* rhizoid cells following alkalization by procaine [10] is consistent with inhibition of the proton pump at high pH. A simple mechanism therefore exists to regulate cytoplasmic pH by altering the rate of proton pumping from the cell. Regulation of proton extrusion can also be caused by changes in expression of genes for the H^+-ATPases, of which there are approximately 10[41]. The existence of families of genes allows different properties in different types of plant cells.

Increased H^+-ATPase activity, causing acidification in the rhizosphere, is an important strategy used by plants to take up Fe from Fe-deficient soils. This results in changes in cytoplasmic pH that are moderated by the concomitant changes in the enzymes involved in the biochemical pH-stat and in distribution of Pi [35] as summarized earlier. The main complication is that the unidirectional transport of protons also transfers positive charge from the inside of the cell to the outside and in so doing causes membrane hyperpolarization. The absolute difference in charge between the inside and outside of a plant cell is minuscule [42,43], so proton pumping, without compensating fluxes of other ions, would rapidly hyperpolarize the membrane beyond the point at which ATP could drive the reaction in the forward direction. In other words, the H^+-ATPase would stall, and removal of protons from the cytoplasm would cease. To prevent this occurring, it is necessary for there to be either influx of other positive charge, usually K^+, or efflux of negative charge, usually Cl^-.

The large vacuoles present in most mature plant cells can form a significant sink for H^+ produced in metabolism, although the amounts disposed of in this way depend on the overall vacuolar composition, buffer capacity, and indeed on the plant's need to regulate its vacuolar pH. Crassulacean acid metabolism is a special case (see Chapter 10). Protons are actively transported across the tonoplast by both the H^+-ATPase and the H^+-pyrophosphatase (see Chapter 2).

Movement of H^+ into intracytoplasmic compartments, such as endoplasmic reticulum, is likely to be quantitatively very small in terms of the plant's capacity to shift H^+. However, short-term redistribution of H^+ in this way, often in exchange for intracellular Ca^{2+}, has important roles with respect to intracellular signaling and metabolic control. The coupling and membrane transport processes that are involved in cytoplasmic Ca^{2+}-H^+ exchange buffers in green algae have been elegantly measured by fluorescence ratio imaging [44].

5.2 Plasma Membrane Redox Systems

The possibility that a redox system on the plasma membrane of plant cells is a redox-driven H^+ pump (outward oriented) that can increase external acidification has implications with respect to cytoplasmic pH regulation [5]. However, the evidence suggests that increased extrusion of H^+, as well as transferring electrons outward (e.g., to reduce Fe^{3+} to Fe^{2+}), also increases H^+-ATPase activity [35]. In the absence of H^+ extrusion, oxidation of the electron donor (NADPH) can result in cytoplasmic acidification, an effect that occurs in early events of the hypersensitive response caused when the fungal elicitor cryptogein is applied to tobacco [45]. Once again, there is a cascade of biochemical events that include activation of the pentose phosphate pathway and changes in membrane transport that probably include inhibition of the H^+-ATPase.

5.3 Proton Cotransport

The other complication is that H^+-ATPase has a primary function in driving uptake of nutrients into the cell via cotransport, which may conflict with its role as a regulator of cytoplasmic pH. The activities of the various proton-linked transport systems have the capacity to perturb cytoplasmic pH, and it seems reasonable to suggests that the rates of these uptake systems might also be controlled by cytoplasmic pH. However, such a control system would undermine the regulation of nutrient uptake by internal nutrient status. It is uncertain how most membrane transporters are affected by intracellular pH or whether they might be regulated by cytoplasmic pH. This lack of understanding of the role of pH in regulating membrane transport is largely due to the difficulties in varying internal pH without altering a range of other variables that might affect transport activity. Some of these problems can be overcome experimentally by taking advantage of the large size of internodal cells of the giant alga *Chara*, in which it is possible to manipulate cytoplasmic conditions directly by intracellular perfusion. Sanders [46] and Reid and Walker [47] used this technique to study the $2H^+/Cl^-$ transport system. Although Cl^- influx was relatively insensitive to external pH [46,47], influx was acutely sensitive to changes in internal pH, increasing 10-fold as the cytoplasmic pH was raised from 7.0 to 7.8 [46] (Fig. 4). In intact cells, effects on Cl^- transport of other treatments that alter cytoplasmic pH, such as darkness [47] and pretreatment at high or low external pH [48], are consistent with high sensitivity of Cl^- transport to internal pH. What is not clear is why Cl^- transport should be so sensitive to changes in cytoplasmic pH. A lateral explanation would be that the main purpose is not to regulate the transport of Cl^- but to contribute to the biophysical regulation of cytoplasmic pH by providing a mechanism for increasing H^+ influx as the internal pH rises and to switch it off as the internal pH falls. More experiments are needed to determine whether other H^+ cotransport reactions such as NO_3^- and $H_2PO_4^-$ show a similar sensitivity to cytoplasmic pH. It should be remembered, though, that Cl^- is a relatively nontoxic ion that plays only a minor role in metabolism, so manipulation of Cl^- fluxes for the purposes of pH regulation is unlikely to have major consequences. This may not be true for N and P.

It can be seen from Fig. 2 that cotransport reactions are sometimes unbalanced with respect to charge. For example, uptake of NO_3^- or $H_2PO_4^-$ involves influx with two (possibly more) protons, and as a result the overall reaction becomes electrogenic. Under these circumstances, as with proton uniport through the H^+-ATPase, charge balance is maintained by passive counterfluxes of other ions, principally K^+ and Cl^-, through membrane ion channels. However, uptake

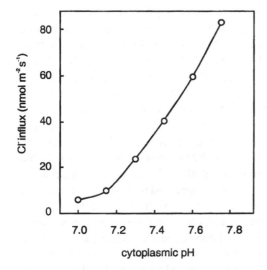

FIGURE 4 Acute sensitivity of proton-driven Cl⁻ influx in *Chara*. The cytoplasmic pH was controlled by intracellular perfusion. (Data from Ref. 46.)

of NO_3^- can be balanced in part by uptake of H^+, i.e., the balance between the "excess" H^+ taken up by cotransport and the H^+ effluxed through the ATPase. (This is equivalent to the efflux of OH^-, as described in the older literature). The uptake of "strong acid" is feasible because reduction and assimilation of NO_3^- generate OH^- (see later).

To a large extent, the biophysical pH-stat is self-regulating because inhibition of the H^+-ATPase (e.g., by low energy status) will automatically turn off many proton cotransport processes by reducing the electrical and/or the delta pH component of the driving force for these reactions, thereby protecting the cytoplasm from excessive acidification.

6 BIOPHYSICAL VERSUS BIOCHEMICAL pH REGULATION

In a clever series of experiments, Sanders and Slayman [49] were able to compare the relative contributions of biophysical and biochemical processes to the regulation of cytoplasmic pH. The experiments were conducted on the fungus *Neurospora*, but the results are probably equally applicable to plant cells to varying extents. What they discovered was that the fall in pH of 0.7 units caused by metabolic inhibition by cyanide was not associated with any significant uptake of protons across the plasma membrane; i.e., the internal acidification was not due to influx of external protons but to the accumulation of acidic metabolites.

By carefully controlling the degree of metabolic inhibition, they were able to vary the ATP concentration and found that there was a strong correlation between ATP level and cytoplasmic pH. Because they could find no correlation between ATP concentration and the activity of the H^+-ATPase, they proposed that the variation in the pH of the cytoplasm under these conditions was due directly to metabolic activity. Further, when they inhibited the H^+-ATPase with vanadate, the cytoplasmic pH was still maintained above neutrality, indicating that this stability must therefore be due to the regulation of metabolic acid-base production. The activity of the H^+-ATPase appeared not to be limited by thermodynamic constraints such as the availability of ATP or by the proton gradient against which it pumped because large changes in external pH had only a small effect on the pump activity. The limitation to pump activity appeared to be substrate concentration because pump activity was acutely sensitive to changes in the cytoplasmic pH, increasing by more than 20-fold for an increase in internal proton concentration of 8-fold. Sanders and Slayman [49] concluded that either the H^+-ATPase alone or metabolism alone could control cytoplasmic pH. It should be stressed that these were short-term experiments; there should be no implication that when the H^+-ATPase was inhibited normal growth processes (including nutrient uptake) could proceed.

7 INTRACELLULAR pH CHANGES ASSOCIATED WITH EXCESS CATION UPTAKE

In most cases, uptake of cations occurs as uniport, i.e., not associated with the flux of another ion. The consequence of this is that the inward transfer of positive charge needs to be compensated for either by uptake of an anion or by efflux of another cation, which in plants is nearly always H^+. Therefore, where cation uptake exceeds anion uptake, the net efflux of H^+ will cause the cytoplasmic pH to rise unless other processes prevent this. In the short term the protons would originate from intracellular buffers, but in the longer term the protons originate from newly synthesized organic acids. The end result is accumulation of inorganic cations and carboxylate anions (e.g., malate), most of which pass into the vacuole (where present).

It was suggested by Davies [29,30] as part of his biochemical pH-stat hypothesis that the small change in cytoplasmic pH that was expected to occur with excess cation uptake would be sufficient to explain the observed increase in PEP carboxylase activity, but this has been challenged [4,50] on the basis that (1) in vivo the enzyme may not have the necessary sensitivity to pH and (2) there is a lack of measurable pH increases in roots during excess cation uptake [32]. The earlier proposal by Jacoby and Laties [51] that the stimulation of PEP carboxylase with increasing pH was due to an increase in the formation of HCO^-_3 as the substrate for the enzymes was discounted by Chang and Roberts [52], who demonstrated that the HCO^-_3 concentration in maize root tips was normally around two orders of magnitude higher than the K_m of PEP carboxylase and would therefore not limit its activity.

Excess cation uptake and accumulation of carboxylate anions are also a feature of stomatal opening. Despite much speculation in the past, the involvement, if any, of cytoplasmic pH as a direct signal remains unresolved (see also Chapter 5).

8 NITROGEN ASSIMILATION AND LONG-DISTANCE pH REGULATION

For plants that are bathed in solution or those having roots surrounded by damp soil, pH regulation is relatively simple because excess H^+ or OH^- can be dumped into the external medium (see Chapter 9). This is not the case for bulky tissues such as tubers, where access to soil is limited because of the long pathways for diffusion, or for aerial tissues such as leaves and stems, where the external medium consists only of a thin cell wall and limited apoplast and the soil can be reached only by long-distance transport in the phloem. In such situations, pH regulation involves different strategies [53,54]. Because nitrogen assimilation is one of the major producers of excess H^+ or OH^-, it is worthwhile to consider briefly how plants cope with N metabolism in leaves (see Ref. 55 for a detailed review).

The equations for NO_3^- assimilation given in Table 2 show that the reduction of NO_3^- to NH_4^+ generates two OH^-, while the incorporation of the NH_4^+ into amino acids produces one H^+, which leaves an excess of one OH^-. In roots this can be neutralized by increased H^+ influx, resulting in increased pH within the rhizosphere, or by synthesis of organic acids. In shoots, the options are more limited and organic acid synthesis predominates. A comparison of the anion and cation concentrations in roots and leaves of plants fed with either NH_4^+-N or NO_3^--N is shown in Table 3. In NH_4^+-fed plants, cations were mostly balanced by inorganic anions in both the roots and the shoots [56]. In NO_3^--fed plants, inorganic anions accounted for a large proportion of the negative charge in roots, pointing to biophysical disposal of excess OH^- ions, but there was also a significant accumulation of carboxylate anions. In the leaves of NO_3^--fed plants, carboxylates made up most of the anions, no doubt due to the ineffectiveness of biophysical processes in coping with excess OH^- ions in situations in which the external sink for disposal is limited. An interesting feature of this study was the high proportion of oxalate in both roots and leaves of NO_3^--fed plants. This was interpreted by Raven and Smith [55] as a means of avoiding osmotic problems that might arise through having to store high concentrations of inorganic cations–organic anions in the vacuole, by forming insoluble Ca-oxalate salts. In simple terms, the levels of carboxylates in shoots are one measure of how much of its

TABLE 3   Cation-Anion Balance in Roots and Leaves of Tomato Plants Fed with Either NO_3^- or NH_4^+ [a]

Tissue	N source	Free cations	Anions		
			Inorganic	Carboxylate[b]	Oxalate
Leaves	NO_3	1820	510	1170	410
	NH_4	790	640	110	80
Roots	NO_3	1340	850	450	220
	NH_4	740	530	50	10

[a] Concentrations are in mmol (charge) kg^{-1} dwt.
[b] Excluding oxalate.
Source: Adapted from Ref. 56.

photosynthate the plant is devoting to metabolic reactions that at least relate to biochemical pH regulation during NO_3^- reduction and assimilation. However, high levels of carboxylates (e.g., 10% of total photosynthate in chenopods) are accompanied by high levels of inorganic cations, again implying high energy expenditure on biophysical processes.

Plants have two further strategies for dealing with nitrate nutrition of tissues with limited access to the external environment: (1) export of carboxylates plus K^+ in the phloem back to the root, where decarboxylation releases OH^- that can then be disposed of by biophysical means, and (2) assimilation of NO_3^- in the roots and transfer of organic N to the shoots or other remote parts. Analysis of the composition of xylem fluids shows that amino acids can occur in quite high concentrations in the xylem, although usually much lower than in the phloem [57,58]. In this way, much of the pH balancing has already occurred in the roots before export to the shoots. Sometimes xylem sap also contains NH_4^+, apparently balanced by carboxylate anions. This is again a mixture that minimizes pH stress when the NH_4^+ is assimilated in the shoots. In reality, it is rare for plants to be exposed to only NO_3^- or only NH_4^+ and the balance between uptake of the two N sources may be dictated, among other considerations, by the need for pH homeostasis [59].

9   CONCLUSIONS

Clearly, the cytoplasmic pH-stat in plants has both biochemical and biophysical components, and their relative importance will be different in different types of plants, in a single plant under different environmental conditions, and in different parts of the same plant under a given set of environmental conditions. Time-dependent changes in metabolism associated with developmental changes will add another layer of complication. Spatial differences between proton-consuming and proton-generating reactions that are associated in linked metabolic pathways also complicate the issue. A good example is the spatially separated carboxylation and decarboxylation reactions in leaf cells of plants with the various classes of C_4 photosynthesis. This would result in acidification in mesophyll cells (sites of PEP carboxylase) and alkalization in bundle sheath cells (sites of ribulose-1,5-diphosphate carboxylase) in the absence of other measures to prevent such changes [60]. An example at the other end of the plant, seemingly specialized but in fact very widespread, is the supply of phosphate via hyphae of arbuscular mycorrhizal fungi to the roots of their host. If P is translocated mainly as polyphosphate (polyP), as sometimes proposed (see Ref. 61), there is intracellular alkalization as the polyP is synthesized in fungal hyphae in soil and acidification as the polyP is hydrolyzed within the fungus at the intraradical interface (arbuscule). This is analogous to generation of H^+ when ATP is hydrolyzed (Table 2) and might cause acid stress within the limited confines of the arbuscule or, if transferred across the interface, in the root cells. If P is instead translocated and transferred as orthophosphate, there would be no such stress.

Finally, much remains unknown about the ways in which the limits of changes in cytoplasmic pH are defined, in other words, the nature of the signals and receptors. The possible involvement of H^+ as a second messenger in developmental processes has been obscured by the importance of Ca^{2+} in this respect. Concomitant changes in cytoplasmic Ca^{2+} concentrations and pH occur in many developmental processes, including early responses of legume root hairs to Nod factors [62] and cellular recognition events in some plant diseases [45]. They also occur in stomatal opening-closing cycles [63]. It is not yet clear whether the pH changes are merely responses to changed Ca^{2+} fluxes and intracellular distribution or whether they have a more active role. In simple terms, we know much more about regulation of pH than about regulation by pH. Nevertheless, there is no doubt that cytoplasmic pH is a master variable that is tightly regulated during plant growth.

ABBREVIATIONS

PEP   phosphoenolpyruvate
OAA   oxaloacetate
NAD   nicotinamine adenine dinucleotide

REFERENCES

1.   T Mimura, C Shindo, M Kato, E Yokota, K Sakano, H Asihara, T Shimmen. Regulation of cytoplasmic pH under extreme acid condition in suspension cultured cells

of *Catharanthus roseus*: a possible role of inorganic phosphate. Plant Cell Physiol 41:424–431, 2000.

2. GE Hutchinson. A Treatise on Limnology. Vol III. Limnological Botany. New York: Wiley, 1975.

3. FA Smith, JA Raven. Intracellular pH and its regulation. Annu Rev Plant Physiol 30:289–311, 1979.

4. A Kurkdjian, J Guern. Intracellular pH: measurement and importance in cell activity. Annu Rev Plant Physiol Plant Mol Biol 40:271–303, 1989.

5. J Guern, H Felle, Y Mathieu, A Kurkdjian. Regulation of intracellular pH in plant cells. Int Rev Cytol 127:111–173, 1991.

6. K Sakano. Revision of biochemical pH-stat: involvement of alternative pathway metabolisms. Plant Cell Physiol 39:467–473, 1998.

7. FA Smith. Comparison of the effects of ammonia and methylamine on chloride transport and intracellular pH in *Chara corallina*. J Exp Bot 31:597–606, 1980.

8. FA Smith. Regulation of the cytoplasmic pH of *Chara corallina*: response to changes in external pH. J Exp Bot 35:43–56, 1984.

9. FA Smith. Short-term measurements of the cytoplasmic pH of *Chara corallina* derived from the intracellular equilibration of 5,5-dimethyloxazolidine-2,4-dione (DMO). J Exp Bot 37:1733–1745, 1986.

10. H Felle, A Bertl. The fabrication of H^+-selective liquid membrane microelectrodes for use in plant cells. J Exp Bot 37:1416–1428, 1986.

11. RJ Reid, FA Smith. Measurements of the cytoplasmic pH of *Chara corallina* using double-barrelled pH microelectrodes. J Exp Bot 39:1421–1433, 1988.

12. JKM Roberts. Study of plant metabolism in vivo using NMR spectroscopy. Annu Rev Plant Physiol 33:375–86, 1984.

13. RJ Reid, LD Field, MG Pitman. Effects of external pH, fusicoccin and butyrate on the cytoplasmic pH in barley root tips measured by ^{31}P-nuclear magnetic resonance spectroscopy. Planta 166:341–347, 1985.

14. FA Smith, JL Gibson. Effect of cations on the cytoplasmic pH of *Chara corallina*. J Exp Bot 36:1331–1340, 1985.

15. NA Walker, FA Smith. Intracellular pH in *Chara corallina* measured by DMO distribution. Plant Sci Lett 4:125–132, 1975.

16. MJ Kime, BC Loughman, RG Ratcliffe. The application of ^{31}P-nuclear magnetic resonance to higher plant tissue. II. Detection of intracellular changes. J Exp Bot 33:670–681, 1982.

17. BC Loughman, RG Ratcliffe. Nuclear magnetic resonance and the study of plants. In: PG Tinker, A Läuchli, eds. Advances in Plant Nutrition. Vol 1. New York: Praeger, 1984, pp 241–283.

18. FA Smith, EAC MacRobbie. Comparison of cytoplasmic pH and chloride influx in cells of *Chara corallina* following chloride starvation. J Exp Bot 32:827–836, 1981.

19. H Felle, A Bertl. Light-induced cytoplasmic pH changes and their interrelation to the activity of the electrogenic pump in *Riccia fluitans*. Biochim Biophys Acta 848: 176–182, 1986.

20. H Felle. Cytoplasmic free calcium in *Riccia fluitans* L. and *Zea mays* L.: interactions of Ca^{2+} and pH? Planta 176:248–255, 1988.

21. RJ Reid, FA Smith, J Whittington. Control of intracellular pH in *Chara corallina* during uptake of weak acid. J Exp Bot 40:883–891, 1989.

22. RJ Reid, BC Loughman, RG Ratcliffe. ^{31}P-NMR measurements of cytoplasmic pH changes in maize root tips. J Exp Bot 36:889–897, 1985.

23. J Gerendas, RG Ratcliffe. Intracellular pH regulation in maize root tips exposed to ammonium at high external pH. J Exp Bot 51:207–219, 2000.

24. H Kosegarten, F Grolig, J Wieneke, G Wilson, B Hoffman. Differential ammonia-elicited changes of cytosolic pH in root hair cells of rice and maize as monitored by 2′,7′-bis-(s-carboxyethyl)-5 (and-6)-carboxyfluorescein fluorescence ratio. Plant Physiol 113:451–461, 1997.

25. GG Fox, RG Ratcliffe. ^{31}P NMR observations on the effect of the external pH on the intracellular pH of plant cell suspension cultures. Plant Physiol 93:512–521, 1990.

26. JA Feijó, J Sainhas, GR Hackett, JG Kunkel, PK Hepler. Growing pollen tubes possess a constitutive alkaline band in the clear zone and a growth dependent acidic tip. J Cell Biol 144:483–496, 1999.

27. GG Fox, NR McCallan, RG Ratcliffe. Manipulating cytoplasmic pH under anoxia: a critical test for the role of pH in the switch from aerobic to anaerobic metabolism. Planta 195:324–330, 1995.

28. T Mimura, RJ Reid, FA Smith. Control of phosphate transport across the plasma membrane of *Chara corallina*. J Exp Bot 49:13–19, 1998.

29. DD Davies. Control of and by pH. Symp Soc Exp Biol 27:513–529, 1973.

30. DD Davies. The fine control of cytosolic pH. Physiol Plant 67:702–706, 1986.

31. AJ Hiatt. Relationship of cell sap pH to organic acid change during ion uptake. Plant Physiol 42:294–298, 1967.

32. JKM Roberts, D Wemmer, PM Ray, O Jardetzky. Regulation of cytoplasmic and vacuolar pH in maize (*Zea mays*) root tips under different experimental conditions. Plant Physiol 69:1344–1347, 1982.

33. HG Nimmo. The regulation of phosphoenolpyruvate carboxylase in CAM plants. Trends Plant Sci 5:75–80, 2000.

34. J Vidal, R Chollet. Regulatory phosphorylation of C4 PEP carboxylase. Trends Plant Sci 2:230–237, 1996.

35. L Espen, M Dell'Orto, P De Nisi, G Zocchi. Metabolic responses in cucumber (*Cucumis sativus* L.) roots under Fe-deficiency: a ^{31}P-nuclear magnetic resonance in vivo study. Planta 210:985–992, 2000.

36. E Marré, A Ballarin-Denti. The proton pumps at the plasmalemma and the tonoplast of higher plants. J Bioenerg Biomembr 17:1–21, 1985.

37. H Sze. H^+-translocating ATPases of the plasma-membrane and tonoplast of plant cells. Physiol Plant 61:683–691, 1984.

38. B Michelet, M Boutry. The plasma membrane H^+-ATPase. Plant Physiol 108:1–6, 1995.

39. MT Marré, G Romani, E Marré. Transmembrane hyperpolarisation and increase of K^+ uptake in maize roots treated with permeant weak acids. Plant Cell Environ 6:617–623, 1983.

40. GW Bates, MHM Goldsmith. Rapid response of the plasma-membrane potential in oat coleoptiles to auxin and other weak acids. Planta 159:231–237, 1983.

41. P Morsonne, M Boutry. The plant plasma membrane H⁺-ATPase: structure, function and regulation. Biochim Biophys Acta 1465:104–126, 2000.
42. PS Nobel. Physicochemical and Environmental Plant Physiology. San Diego: Academic Press, 1991.
43. J Gerendas, U Schurr. Physicochemical aspects of ion relations and pH regulation in plants—a quantitative approach. J Exp Bot 50:1101–1114, 1999.
44. C Plieth, B Sattelmacher, U-P Hansen. Cytoplasmic Ca^{2+}-H⁺-exchange buffers in green algae. Protoplasma 198:107–124, 1997.
45. A Pugin, J-M Frachisse, E Tavernier, R Bligny, E Gout, R Douce, J Guern. Early events induced by the elicitor cryptogein in tobacco cells: involvement of a plasma membrane NADPH oxidase and activation of glycolysis and the pentose phosphate pathway. Plant Cell 9:2077–2091, 1997.
46. D Sanders. The mechanism of Cl⁻ transport at the plasma membrane of *Chara corallina*. I. Cotransport with H⁺. J Membr Biol 53:129–141, 1980.
47. RJ Reid, NA Walker. The energetics of Cl⁻ active transport in *Chara*. J Membr Biol 78:35–41, 1984.
48. D Sanders, FA Smith, NA Walker. Proton/chloride cotransport in *Chara*: mechanism of enhanced influx after rapid external acidification. Planta 163:411–418, 1985.
49. D Sanders, CL Slayman. Control of intracellular pH. Predominant role of oxidative metabolism, not proton transport, in the eukaryotic microorganism *Neurospora*. J Gen Physiol 80:377–402, 1982.
50. J Guern, Y Mathieu, A Kurkdjian. Phosphoenolpyruvate carboxylase activity and the regulation of intracellular pH in plant cells. Physiol Veg 21:855–856, 1983.
51. B Jacoby, GG Laties. Bicarbonate fixation and malate synthesis in relation to salt-induced stoichiometric synthesis of organic acid. Plant Physiol 47:525–531, 1971.
52. K Chang, JKM Roberts. Quantitation of rates of transport, metabolic fluxes and cytoplasmic levels of inorganic carbon in maize root tips during potassium uptake. Plant Physiol 99:291–297, 1992.
53. JA Raven. Biochemical disposal of excess H⁺ in growing plants. New Phytol 104:175–206, 1986.
54. JA Raven. Acquisition of nitrogen by the shoots of land plants: its occurrence and implications for acid-base regulation. New Phytol 109:1–20, 1988.
55. JA Raven, FA Smith. Nitrogen assimilation and transport in vascular land plants in relation to intracellular pH regulation. New Phytol 76:415–431, 1976.
56. EA Kirkby, K Mengel. Ionic balance in different tissues of the tomato plant in relation to nitrate, urea, or ammonium nutrition. Plant Physiol 42:6–14, 1967.
57. PJ Hocking. The composition of phloem exudate and xylem sap from tree tobacco (*Nicotiana glauca* Groh). Ann Bot 45:633–643, 1980.
58. J Pate, E Shedley, D Arthur, M Adams. Spatial and temporal variations in phloem sap composition of plantation-grown *Eucalyptus globulus*. Oecologia 117:312–322, 1998.
59. ML Van Beusichem, EA Kirkby, R Baas. Influence of nitrate and ammonium nutrition on the uptake, assimilation and distribution of nutrients in *Ricinus communis*. Plant Physiol 86:914–921, 1988.
60. CB Osmond, FA Smith. Symplasmic transport of metabolites in C_4 photosynthesis.

In: BES Gunning, AW Robards, eds. Intercellular Communication in Plants: Studies on Plasmodesmata. New York: Springer, 1976, pp 229–240.

61. SE Smith, DJ Read. Mycorrhizal Symbiosis. 2nd ed. London: Academic Press, 1997.
62. L Cárdenas, TL Holdaway-Clarke, F Sánchez, C Quinto, JA Feijó, JG Kunkel, PK Hepler. Ion changes in legume root hairs responding to Nod factors. Plant Physiol 123:443–451, 2000.
63. A Grabov, M Blatt. Co-ordination of signalling elements in guard cell ion channel control. J Exp Bot 49:351–360, 1998.

4

Confocal pH Topography in Plant Cells: Shifts of Proton Distribution Involved in Plant Signaling

Werner Roos
Martin Luther University, Halle (Saale), Germany

1 INTRODUCTION

The homeostasis of pH in the living cytoplasm and the nucleus is a prerequisite for the efficiency of most genetic, metabolic, and transport processes and hence is subject to powerful control mechanisms. The mechanisms involved do, however, allow the generation of pH shifts within distinct limits and their functionalization as signal elements.

In plant cells, pH control is required to cope not only with influences of the external medium but also with large intracellular proton gradients: the cytosolic pH is usually kept at 7.6 ± 0.2, whereas the interior of the vacuole typically attains a pH around 5 ± 0.5 but may range down to near 2.5, e.g., in vacuoles of lemon fruits [1]. Illuminated, photosynthesizing chloroplasts likewise display a steep proton gradient with pH around 8 in the stroma and down to 5 in the thylakoid matrix. Homeostatic control under such conditions requires a combination of high buffering capacity and flexibility that is realized by the concerted activities of passive buffers (proteins including cytoskeleton, salts, amino acids, etc.), pH-dependent formation-degradation of carboxylic acids, and proton transport systems of the plasmalemma and organellar membranes. The interplay of these components that arose during the evolution of plant cells constitutes a dynamic mode of pH control that allows deflections from "midpoint" values within distinct physiological limits.

Several elements of cellular pH control—mainly those connected with H^+ transport—respond to environmental or developmental signals and convert them into transient pH shifts. In this way plant cells, like those of other organisms, establish an apparent paradox: the use of a highly regulated cellular parameter as a signal module.

2 MAPPING THE INTRACELLULAR PROTON DISTRIBUTION IN PLANT CELLS

A typical plant cell displays an unequal distribution of pH that reflects the ion composition of its organelles, different local buffering capacities, and vectorial transport of protons. pH shifts of regulatory significance probably occur at subcellular levels, i.e., in distinct organelle(s), or in the proximity of membrane surfaces. As such shifts do not necessarily change the total cellular H^+ concentration or are small compared with the huge proton reservoir of the vacuole, they may not be detected by measurements that average (totally or partially) the intracellular proton concentration. They are best assayed by a procedure that measures local pH differences and their dynamics in living cells with the spatial resolution of light microscopy. Among the methods currently available,* fluorescence-based pH imaging appears to come closest to this goal.

2.1 The Principles

pH imaging by fluorescence microscopy is based on indicator molecules that accumulate inside living cells and change their fluorescence properties in response to their degree of protonation. A similar principle can be used for the imaging of other ions, mainly Ca^{2+} and Na^+ (see Refs. 3 and 4 for recent reviews). Almost all pH indicators presently available are ratiometric; i.e., upon protonation they display shifts of their excitation and/or emission spectra. Therefore, cells loaded with these molecules can be examined at two different bands of excitation or emission wavelengths that detect (exclusively or preferentially) the fluorescence of either the protonated or anionic form of the probe. A pair of fluorescence images is thus obtained (via video camera or photomultiplier) and the intensities of each data point (pixel) are ratioed by image analysis. The resulting ratio image mirrors the fluorescence changes caused by proton binding-dissociation only and

---

* ^{31}P nuclear magnetic resonance (NMR) [2] and microelectrode measurements (see Chapter 5) yield data on cytosolic or vacuolar pH but with less defined catchment areas.

thus maps the distribution of H^+. An example is presented in Fig. 1 (see color insert). The ratioing principle eliminates any heterogeneity of fluorescence that occurs in either of the paired images as nonuniform dye accumulation, leakage of the probe, photobleaching, thickness of the specimen, or inequalities of the fluorescence detection.

The maximum dynamic range of the probe is exploited by dividing signals obtained at two wavelengths with opposite pH dependence of fluorescence intensity. Alternatively, if the spectrum contains an isosbestic point, i.e., a wavelength at which the fluorescence is independent of pH, this can be selected to obtain one of the paired images that mirrors the distribution of the probe.

2.2 The pH Probes

Table 1 provides an overview of the pH-sensitive fluoroprobes currently in use. Compared with their parent compound fluorescein, the modern probes bear additional carboxyls that increase their polarity and intracellular retention.

The most widely applied probes are cSNARF-1, a seminaphthorhodafluor dye that allows both excitation and emission ratioing [5–7], and BCECF, a carboxyfluorescein derivative used for excitation ratioing [8]. The SNAFL probes are seminaphthofluoresceins that, in contrast to cSNARF-1, display a stronger emission in their protonated than in their unprotonated form. In the more acidic range (e.g., in vacuolar organelles) halogenated fluoresceins such as CDCF or the rhodamine derivatives Cl-NERF and DM-NERF are suited because of their pK values of <6.

A new development is the LysoSensor dyes, different heterocycles with basic side chains that can be trapped in their protonated form in acidic organelles. In contrast to fluoresceins or cSNARF-1, they display increasing fluorescence intensity at decreasing pH and may thus prove useful for the ratiometric detection of low pH values. As these probes are likely to accumulate in organellar membranes, quantitative measurements require the determination of fluorescence spectra of the membrane-bound molecules [8].

Recently, pH-sensitive mutants of the green fluorescent protein (GFP) of *Aequorea victoria* were introduced into animal cells by gene transfer and could be successfully applied for pH measurements [9]. These photoproteins are now available with different pK values and display sufficient differences of either excitation or emission in the pH range between 5.5 and 8. Compartment-specific expression vectors allow the targeting of pH-reporting green fluorescent proteins (fused to organelle-specific proteins) to the cytosol and the nucleus, the trans-Golgi compartment, and the mitochondrial matrix. This approach has the potential to extend fluorescence pH topography to intracellular sites that are "invisible" to light microscopy (i.e., not resolved by visible light transmission, for instance, small ER or Golgi vesicles) but emit detectable fluorescence.

TABLE 1 Fluorescent Probes for the Imaging of Intracellular pH

| Name | pK_a | Spectral optima of protonated/anionic probe (nm)[a] | |
		Excitation[b]	Emission
CarboxySNAFL-1 (cSNAFL-1)	7.8	508/540 (490/540)	543/623 (540/630)
CarboxySNARF-1 (cSNARF-1)	7.5	548/576	587/635 (580/640)
SNARF calcein	7.2	552/574	590/629 (535/625)
2′,7′-Bis-(2-carboxyethyl)-5- (and 6-) carboxyfluorescein (BCECF)	7.0	482/503 (440i/505)	520/528 (535)
5- (and 6-) Carboxyfluorescein (CF)	6.4	475/492 (490/450)	517
5- (and 6-) Carboxy-2′,7′-Dichlorofluorescein (CDCF)	4.8	488/504 (440i/490)	529
DM-NERF	5.4	497/510 (488/514)	527/536 (535)
CL-NERF	3.8	504/514 (488/514)	540
Oregon Green	4.7	478/492 (440i/490)	518

[a] In parentheses: typical settings of filter or laser lines (nm) used for ratioing.
[b] Absorption optima are given; these are close to excitation optima but less sensitive to the environment of the molecule. i = isosbestic wavelength (fluorescence not sensitive to pH).
Source: Ref. 8 (http://www.probes.com).

2.3 Loading and Cellular Compartmentation of pH Probes

Several strategies have been invented to load the usually polar fluorescent indicator molecules into the cell type of interest (Table 2).

The least invasive way is to incubate cells with membrane-permeant esters of the desired probes (usually acetylated hydroxyls and/or acetoxymethylated carboxyl groups). The absorbed molecules are split by intracellular esterases and the anions formed are trapped inside the cell. Although this method works well with many cultivated cells, plant tissues often give insufficient loading results because of either the cutin barrier of intact tissues or ester cleavage by extracellular esterases. The latter obstacle may be overcome by reducing the temperature during the uptake phase followed by the intracellular cleavage of the preaccumulated ester at a raised temperature [10].

If wound reactions are not considered a problem, the cell may be impaled by microelectrodes followed by pressure injection or iontophoresis, i.e., current-assisted loading of the dye. Plant protoplasts can be loaded by methods used with animal cells (e.g., via a patch microelectrode or by fusion with dye-containing liposomes) but cannot generally be used instead of intact cells because of impairments caused during the removal of the cell wall.

Once accumulated, pH probes usually undergo intracellular sequestration

TABLE 2 Strategies for Loading Fluorescent Ion Probes into Living Cells

Loading method	Resulting probe distribution	Advantage (+), problems (−)
Permeant esters (acetate-, acetoxymethyl-) (e.g., [3,14,30,50])	Initially: cytoplasm, sequestration into vacuole/vesicles	+ Low invasive − No/low accumulation in mature plant tissues
Free acid at low external pH (e.g., [107])	Cytoplasm, vesicular sequestration	− Stress by low pH_{ext}
Dextran conjugates: pressure injection, iontophoresis (e.g., [6,18])	Cytoplasm (or vacuole)	+ No/little sequestration − Wound reactions (?) − Large cells only
Reactive probes: coupling to cellular proteins (e.g., [14])	Mainly cytoplasm	+ Long-term stability − Reaction uncontrolled
Targeted organellar expression: aequorins, green fluorescent proteins (e.g., [9])	Directed by expression vector (cytosol, organelles)	+ Selective localization − Emission insufficient for subcellular imaging

that may include transfer to most if not all the cell's organelles ranging from the vacuole to small vesicles that might even escape detection by the light microscope.

The compartmentation of probes that are liberated from internalized esters is dominated by the distribution patterns of the involved esterases, at least in the initial phase of loading. Substantial species differences exist in this respect. In corn root hair cells, acetoxymethyl esters of BCECF and cSNAFL1 are split and accumulated exclusively in the vacuole or the cytoplasm, respectively, indicating different locations of dye-selective acetoxymethyl esterases [11,12]. In other cells, BCECF-AM seems to be split in the cytoplasm (protoplasts of *Digitaria* [13]) followed by the transfer of the liberated fluorophore into vacuolar or other vesicles (protoplasts of *Arabidopsis* and pollen tubes of *Lilium longiflorum* [14]). In cultured cells of *Eschscholzia californica*, acetoxymethyl esters of carboxyfluorescein are cleaved only in the cytoplasm, whereas the acetylated hydroxyl groups are deesterified both in the cytoplasm and in the vacuole as shown by comparing cells with intact and permeabilized plasma membranes (Fig. 2; see color insert).

Similarly, cSNARF-1-AM which bears one acetoxymethyl and one acetyl group, is completely deesterified in the cytoplasm and the fluorescent molecule transferred to the vacuole [6,14,15]. The mechanisms and the extent of intracellular transfer of fluoroprobes have not yet been systematically investigated. Summarizing the dispersed data, it appears that probes requiring a high number of ester groups to be cleaved (e.g., five in BCECF, five in most calcium indicators) show a strong tendency to become sequestered in the vacuole or in related vesicles, whereas probes with a lower number of ester bonds (e.g., cSNARF-1) tend to be retained longer in the cytosol and to yield brighter fluorescence [3,15,16]. This argues that the specificity and/or activity of cytoplasmic esterases limits the degree of cleavage of internalized esters.

High organellar density, as often prevails in plant or fungal cytoplasms, may complicate the exact localization of fluorescence signals (and derived pH data) received from the cytoplasmic area. Whereas conventional epifluorescent illumination may indicate a relatively even distribution of some probes throughout the cytoplasm, subsequent confocal laser scanning microscopy of the same cells often revealed localization in vesicular structures [17]. Especially in less vacuolated plant cells (e.g., pollen tubes or coleoptiles, but not guard cells) the free-acid forms of ion probes tend to become sequestered into the endoplasmic reticulum, mitochondria, and other small vesicles.

Vesicular compartmentation in particular dominates the distribution of highly esterified probes in fungal hyphae: in *Neurospora crassa*, BCECF or SNARF-calcein (a derivative of cSNARF-1 containing two additional iminodiacetic ester groups) liberated from its AM esters accumulated both in the vacuolar network and in endomembranal vesicles of different size, the smallest of which might even range below the resolution of the light microscope [16].

Chapter 4

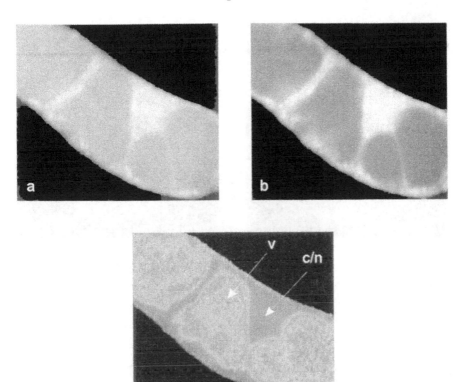

FIGURE 1 Confocal mapping of cellular pH via fluorescence intensity ratio imaging. Cultured cells of *Eschscholzia californica* were incubated with cSNARF-1-AM for 2 h and confocal fluorescence images were obtained with a Leica confocal laser scanning microscope at excitation (Ex) 514 nm. Emission (Em): >619 nm (image a, representing mainly the anionic form of the probe) and Em 583 nm (image b, representing mainly the acidic form of the probe). The ratio image a/b is color coded according to a self-defined lookup table and mirrors the actual pH distribution in the cell. The relation of color code to pH is shown in the pH scale given with Fig. 3. This relation was calibrated for the cytoplasm, the vacuole, and the extracellular dye as mentioned in the text (see Ref. 15 for details). Typical vacuolar (v) and cytoplasmic/nuclear (c/n) regions are marked by arrows. (From Ref. 15.)

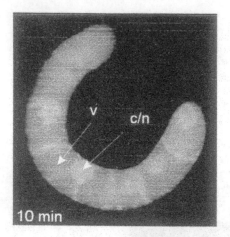

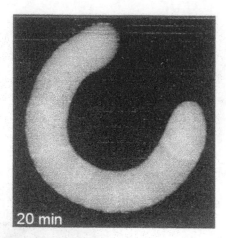

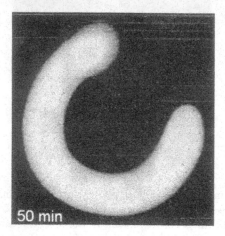

FIGURE 2 Loading of the pH-probe 5-carboxyfluorescein diacetate acetoxymethyl ester (CFDA-AM) into cultured cells of *Eschscholzia californica*. Cells were incubated with 5 µM CFDA-AM in culture liquid in a flow-through cell chamber. At the times indicated, conventional fluorescence images were obtained (Ex 440 nm, Em >520 nm) and digitized with a Sony 3CCD camera. The fluorophore first accumulates in the cytoplasmic/nuclear region before being transferred to the vacuole. Typical vacuolar (v) and cytoplasmic/nuclear (c/n) regions are marked by arrows. Similar experiments were performed after selective permeabilization of the plasma membrane leaving the vacuole intact within its macromolecular environment. Under such in situ conditions, accumulation of fluorescence was not observed with the acetoxymethyl ester but with the analogue bearing the nonesterified carboxyl group, i.e., 5 carboxyfluorescein diacetate (image not shown).

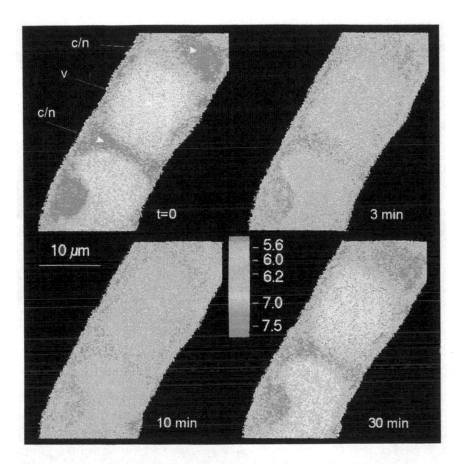

FIGURE 3 pH maps of *Eschscholzia* cells obtained after contact with a fungal elicitor. An elicitor preparation from baker's yeast (glycoproteins) was added to cells kept on agarose disks under the confocal microscope. At the times indicated, pH maps were obtained with the LEICA CLSM as described in the legend of Fig. 1. Some vacuolar (v) and cytoplasmic/nuclear (c/n) regions are marked by arrows. Within 7 min, the average pH in the cytoplasmic/nuclear region drops from 7.4 to 6.9 and the average pH of some vacuoles increases from 5.6 to 5.9 (see Ref. 15 for details). (From Ref. 15.)

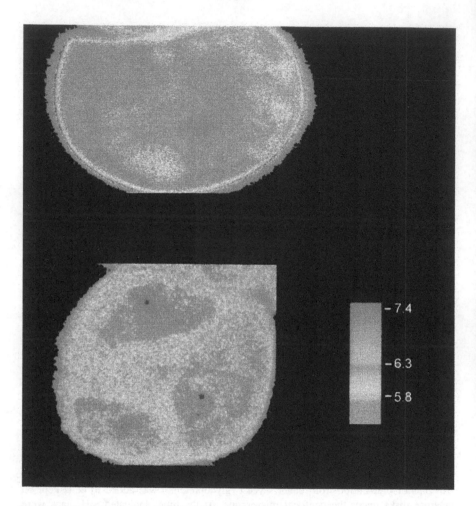

FIGURE 4 Localized areas of acidic pH in cultured cells of *Gossypium hirsutum*. Confocal pH maps (obtained with the BIORAD MRC 600 microscope) of cells loaded with cSNARF-1 show layers of acidic pH in the peripheral cytoplasm close to the plasma membrane (top image) and vesicles with acidic interior stacked in the areas of future cell wall formation in a dividing cell (bottom image). The pH scale used here is designed to emphasize pH differences between 6.3 and 5.8 (see Ref. 96 for details). (From Ref. 96.)

Hence, if a "cytoplasmic" location of fluoroprobes is assumed, the experimenter has to make clear whether the fluorescence emission used for imaging can really be attributed to the cytosolic phase (that may still include submicroscopic vesicles) or rather represents some average of all extravacuolar organelles. In such cases, the use of confocal fluorescence microscopy is almost indispensable. Confocal images not only allow identification of organelles at the cutting edge of spatial resolution but—equally important—yield maps of a defined optical section and with adjustable depth.

Ion probes with a prolonged retention time in the cytosol can be obtained by conjugating the ion-sensitive fluorophores to macromolecules. Dextran conjugates (molecular mass between 10 and 1000 kDa) are available for many of the actual pH probes. These can be loaded into cells by microinjection or by incorporation into liposomes that are then fused to protoplasts. Microinjected dextran conjugates are usually well retained in the cytoplasm of plant cells and are less likely to be sequestered into the vacuolar system than probes liberated from AM esters. For example, dextran conjugates of cSNARF-1 remained in the cytoplasm of *Fucus* zygotes [18], *Neurospora* hyphae, and *Agapanthus* pollen tubes [6]. However, conjugated dyes are not principally prevented from sequestration: injected dextran (10 kDa) conjugates of some Ca^{2+} probes (Calcium Green-1, Fura-2, or Indo-1) did not escape the high endocytotic activity of fungal cells and became sequestered in the hyphal vacuolar network of *Basidiobolus*, *Neurospora*, and *Uromyces* much as did their unconjugated analogues [19].

Some pH probes can be loaded as reactive molecules that allow conjugation to cellular constituents. The membrane-permeant compounds chloromethylfluorescein diacetate (CMFDA) and chloromethylSNARF-1 acetate react with cellular glutathione (catalyzed by glutathione-S-transferase) and with cysteine-containing proteins; the adducts formed show substantially longer retention times than the nonreactive molecules. The chloromethyl derivatives are nonfluorescent until the preceding conjugates are formed and the acetates are split off. Many glutathione-S-conjugates are transported into plant vacuoles by ATP-dependent pumps [20–22]. Expectedly, after incubation of epidermal cells of *Hordeum vulgare* with CMFDA, the fluorescence originally accumulated in the cytosol moved slowly into the vacuole [14]. Commercially available amine-reactive derivatives are useful for the preparation of defined dye conjugates [8] in order to optimize loading and intracellular behavior for the cell type under study.

2.4 Buffering Effects of Accumulated pH Probes

A pH probe that accumulates inside a cell adds to the cellular H^+ buffering capacity and thus influences the local concentration of free protons. In order to keep this interference at a negligible level, pH imaging should generally be conducted at the lowest concentration of the intracellular probe that yields an acceptable signal-to-noise ratio. A good compromise must take into account the fluorescence properties of the probe (e.g., quantum yield, dynamic range, bleaching), its distribution in the cell studied (rate of accumulation, compartmentation, quenching, relation to background fluorescence), initial ion concentrations as well as magnitude and rapidity of the expected pH shifts, and, last but not least, the sensitivity and accuracy of the available detection system(s).

Estimates of sufficient intracellular probe concentrations have been provided by comparing local fluorescence intensities in loaded cells with those of standard preparations of the active probe in known cell-sized volumes such as liquid droplets in oil emulsions [23] or defined three-dimensional (3D) elements (voxels) of confocal images. They range from around 50 μM for BCECF in guard cells [24] to 10–20 μM for cSNARF1 in cultured plant cells (W. Roos, unpublished).

The buffering effects of such concentrations may be considered negligible if they are compared with the high H^+-buffering capacity of the plant cytoplasm: relevant figures (in mM H^+/pH unit at pH 7–7.5) range from extreme values of 90 in *Vicia* guard cells [25] to the more common levels of 20–50 in root hairs of *Sinapis* [26], 36 in cultured tobacco cells [27], 17.9 in *Chara* [28], and 14.9 in *Asparagus* mesophyll cells [29]. However, the buffering capacity is most probably not distributed uniformly across the cytoplasm and its organelles (see Sec. 4.1.). Nonplasmatic organelles, such as the vacuole, have a much smaller buffering capacity than the cytosol, as evidenced, e.g., by the fast equilibration of vacuolar and extracellular pH in the presence of ammonia [15,30]. Thus, vacuoles require a careful adjustment of the level to which a pH probe is allowed to accumulate, at least if small pH changes are to be imaged. Furthermore, pH probes are weak acids and therefore might act to some degree as proton shuttles that dissipate pH gradients [31].

2.5 The Imaging Process

2.5.1 Conventional Fluorescence Microscopy

In conventional fluorescence microscopy, the excitation light forms a cone that penetrates both in-focus and out-of-focus areas of the specimen. Therefore, the emitted light seen by the detector is collected throughout the sample depth and the fluorescence images as well as the resulting ion maps display a less defined average of information that cannot be attributed to a distinct horizontal plane. With these limits accepted, conventional microscopy and ratio imaging nevertheless can yield 2D overviews of the pH distribution in cells or organelles, preferentially those with a simple structure, e.g., yeasts [32], protoplasts [33], guard cells [34,35], or plant vacuoles (W. Roos et al., unpublished).

The light from out-of-focus parts of the cell causes blurring of fluorescent images, thus obscuring details and complicating the alignment of intensity distributions and derived ion maps with small cellular structures. Optical methods as well as image processing algorithms have been developed to overcome this and other limitations of optical resolution and contrast.

2.5.2 Confocal Microscopy

At present, the most defined images to be used for ion mapping are provided by confocal fluorescence microscopy [36]. In the confocal microscope, the excitation light (a collimated laser beam) and even more so the emission light are confined by adjustable pinholes to small volumes that lie along the optical axis. Only the light emerging from focal regions of the specimen passes through the exit pinhole and reaches the photomultiplier detector. By simultaneous scanning of both the illuminating laser beam and the detection point over the specimen, a fluorescence image emerges pixel by pixel, thus generating a definable optical section through the cell that contains very little out-of-focus information [37]. Different scanning modes are in use that allow maximum scan speeds between 10 and >100 images per second (the latter, however, at lowered spatial resolution), and hence even fast pH shifts can be monitored [38]. Dual excitation as well as splitting of the emitted light into two or more channels with individual wavelength settings allows different modes of ratio imaging immediately after image acquisition.

The resolution in the x, y direction as determined by the resolution and numerical aperture (NA) of the objective can be further improved by the hardware zoom facility of confocal microscopes and the excellent image contrast. At a magnification of $100\times$ and a zoom factor of 8, the theoretical distance of two resolved pixels is 48 nm; in practice, organelles of 150 nm distance or diameter can be resolved well. The "depth of field," i.e., the thickness of one horizontal plane that appears to be sharply in focus, is, among other factors, strongly influenced by the diameter of the detection pinhole; for an objective of NA = 1.4 and at $\lambda = 488$ nm, it can reach minimally 0.7 μm [37]. By stepping the object through the vertical dimension, vertically aligned images can be obtained that may be combined to form 3D arrays.

2.5.3 Fluorescence Lifetime Imaging

An alternative way to discriminate fluorescence signals emitted by protonated and unprotonated forms of the ion probe is offered by their different lifetimes [39]. This fluorescence parameter that is independent of the signal intensity can be measured either by the time-domain method (excitation with pulsed light produced by picosecond lasers and ratioing of successively emitted signal intensities) or by the frequency-domain (phase-modulation) method. In the latter, more widespread technique, the decay of fluorescence intensity (lifetime) is represented by the phase angle and/or the modulation emitted in response to intensity-modulated light. The modulation frequency (typically in the range of 10–100 MHz) needs to be adjusted for a particular fluorescent probe. Coupled to a conventional or confocal fluorescence microscope, this method yields maps of either phase or modulation data, a series of which can be used to calculate a lifetime image that mirrors local ion concentrations. Successful examples have been reported for animal cells, e.g., mapping of pH with carboxySNAFL-1 [40] and of Ca^{2+} [41]. Lifetime images are not sensitive to differences in local fluorescence and thus do not require the ion probe to display ratioable shifts of the excitation or emission spectra. Furthermore, the phase-modulation methods allow the simultaneous acquisition of data at all pixels of the optical plane; i.e., no pixel-for-pixel scanning is necessary. Current improvements of optics, CCD cameras, and computerization [42] will considerably shorten the operation time, which is currently limiting application to dynamic processes, and open a way to gain broader experiences with this methodology.

2.5.4 Computational Approaches

Spherical and photometric aberrations are inherent in all fluorescence imaging procedures, although to different extents and proportions [43]. Many of them can be corrected by mathematical approaches that are based on the point-spread function of the microscope, i.e., the 3D shape that is formed by the light emitted by an ideal point source.

> Conventional epifluorescence images can be deblurred by subtracting out-of-focus information via deconvolution, a procedure that works with several digitized images taken from in-focus and the neighboring out-of-focus planes [44–46].
>
> Confocal images still contain spherical and chromatic aberrations that are due to imperfect alignment of illumination and detection volumes and the influence of highly refractive cellular structures, especially in thick specimens, and therefore benefit from modeling and eliminating errors by image processing as well [43]. Actual developments of both deconvolution software and computing power will soon shorten the processing time to a degree that allows viewing the corrected images in real time.

2.6 Calibrating Fluorescence Data to pH Values

The relationship of a measured fluorescence ratio to the actual proton concentration needs to be calibrated from a series of data obtained at known pH. Such calibration experiments should be preferentially performed in situ, i.e., with the probe present intracellularly in order to take into account all factors that might influence its binding equilibrium and fluorescence properties in the cell or organelle of interest (ionic and macromolecular composition, viscosity, degree of deesterification, binding to cell components, quenching, etc.). In situ calibrations make use of an ionophore to equilibrate the intracellular to the extracellular pH, the latter being controlled by appropriate buffers. At the beginning of calibration experiments, useful estimates of the pH dependence and the dynamic range of the probe may also be obtained by in vitro calibrations performed in buffers that mimic some important components of the cellular interior [3].

An established in situ calibration procedure rests on the equilibration of cytoplasmic and external pH with nigericin (an ionophore facilitating the exchange of K^+ for H^+) in the presence of 100–150 mmol/L KCl [47]. Equally high K^+ concentrations on both sides of the plasma membrane depolarize the membrane potential and eliminate effects of a strong transmembrane K^+ gradient on pH. In practice, besides some toxic effects of nigericin, systematic errors of about 0.2 units of the steady-state pH_{in} have been observed after such treatments, indicating slow or incomplete equilibration of cytoplasmic and extracellular pH [48,49].

A more elegant calibration procedure has been proposed by James-Kracke [48]. Cells loaded with BCECF were titrated in the presence of protonophores (CCCP or FCCP) with HCl and than back with KOH in order to drive pH_{in} to values that allow the determination of minimum and maximum fluorescence ratios that represent the fully protonated or fully deprotonated form of the intracellular indicator. By introducing these values in the Henderson-Hasselbalch equation, together with the increments of fluorescence measured over the pH range used, calibration graphs were obtained that were fairly coincident with those of the null method (see later). This approach requires neither complete depolarization of the membrane potential nor equilibration of proton concentrations across the plasma membrane and hence should be applicable to a broad spectrum of cell types, including plant cells [50]. It requires, however, that the dissociation constant of the intracellular probe is known, which is not generally the case in plant cells.

An alternative approach uses membrane-permeant weak acids or bases to facilitate pH equilibration. In animal cells, the so-called null method was successful in determining absolute values of pH_{in} by exposing cells to buffered mixtures of membrane-permeant acid (e.g., butyrate) plus base (e.g., trimethylamine) in concentrations that were previously determined to cause equal deflections of intracellular fluorescence quotients in the acidic and the basic direction. Hence, no ("null") resultant effect of the mixture on pH_{in} is observed if a null solution is found whose nominal pH matches pH_{in}. Using several null solutions, the pH_{in} as well as the cellular buffering power could be bracketed with high accuracy [49]. In intact plant cells, complete equilibration of cytoplasmic and external pH under the influence of permeant acids or bases is often prevented by the high cytoplasmic buffering capacities and the pH control exerted by the vacuole. In cultured cells of *Eschscholzia* treated with permeant acids (butyric or pivalic), fluorescence ratios of cSNARF-1 responded to changes in pH_{ext} with the same slope as did the external fluorescence ratio. However, the absolute values of cytoplasmic and external fluorescence ratios remained different and coincided only after combined administration of permeant acids, nigericin, and 100 mM KCl

[15]. In contrast, the vacuolar pH can easily be clamped to the pH of external buffers if membrane-permeant ammonia or methylamine is present, which indicates that the vacuole functions as a base trap. This has been demonstrated with BCECF in vacuoles of corn root cells [30] and with cSNARF-1 in vacuoles of *Eschscholzia* cells [15].

In situ calibration procedures generally face the problem that the identity of cellular and external ion concentrations after "equilibration" is hard to prove and hence deviations between nominated and true ion concentrations cannot totally be excluded. Therefore comparisons are helpful between different probes and equilibration procedures as well as between the increments of intracellular and extracellular fluorescence in response to the applied ion concentration. Fortunately, with some pH probes, the in situ approach indicated a pH dependence of fluorescence similar to that of the extracellular dye assayed in vitro. Calibration of BCECF in appropriate buffers (with no cells present) compared with the previous null method used with cultured rabbit aorta cells revealed only small differences (<0.1 pH unit) that were independent of pH_i [51]. The pH sensitivity of cSNARF-1 measured in cultured cells of *Eschscholzia* did not differ significantly irrespective of whether the fluorescence ratios were measured in the vacuole (after equilibration with methylamine), in the cytosol (after equilibration with butyrate plus nigericin and KCl), or in the external, buffered media of detergent-treated cell suspensions [15]. This seemed to be in contrast to reported changes of the pK_a of cSNARF-1 inside cells [52] that were attributed to the binding of the probe to cellular proteins [53]. However, more recent data revealed that a contaminant of the commercially distributed substance but not the molecule liberated from cSNARF-1 AM is able to bind to proteins and thereby changes its fluorescence [7].

In summary, the calibration of fluorescence data to local concentrations of H^+ may not always guarantee a perfect match. However, in most experiments the correct monitoring of experimental *changes* of pH is of higher importance than the determination of absolute proton concentrations. Most experimental conditions known to affect the relation of pH and fluorescence emission shift the fluorescence-versus-concentration curve in parallel rather than change its slope. Hence, changes of ion concentrations can often be correctly mirrored even if the absolute data contain some degree of uncertainty or error [3,6,15,30,50].

2.7 Problems with Chlorophyll Fluorescence

Chlorophyll-containing cells and tissues provide a specific problem because of the red fluorescence excited between the maxima of around 430/450 nm and 660/640 nm (chlorophyll *a*/chlorophyll *b*, respectively). Interference by chlorophyll fluorescence can be avoided or minimized by using probes with light emission well below 600 nm and high fluorescence yield, careful filtering of the emission light (modern detection systems allow emission spectra at distinct cellular areas), and/or subtraction of nonprobe fluorescence by digital image processing. In some cases, the anatomy of the cell may allow defining chloroplast-free areas for confocal imaging. The lifetime of chlorophyll fluorescence shows clear differences from that of other fluorophores, making lifetime imaging an elegant way to circumvent any interferences [42].

3 SHIFTS OF CELLULAR pH AS STEPS OF SIGNAL TRANSFER CHAINS

The following examples represent actual fields of plant cell physiology where pH shifts were found to have important functions in signal transduction. Most of them gained substantial input by pH mapping experiments.

3.1 Hormonal Control of Stomatal Movements in Guard Cells

Stomatal movements represent controlled changes of guard cell turgor via mass fluxes of K^+ and counteranions through ion channels of the plasma membrane and the tonoplast. These channels are targets of hormone-triggered signal chains [54,55] that include shifts of cytoplasmic H^+ and Ca^{2+} concentrations.

Abscisic acid (ABA) triggers closing of stomata preceded by cytoplasmic alkalization and an increase of cytoplasmic Ca^{2+} levels. These ionic transients were first detected by confocal imaging with BCECF (pH) or Fluo-3 (Ca^{2+}) of guard cells in epidermal strips of the orchid *Paphiopedilum tonsum* [56]. Similar changes are known from ion maps obtained in corn coleoptiles and parsley hypocotyls and roots [57]. On the contrary, in the same experimental systems auxin at low concentrations (<100 μmol/L) causes stomatal opening that is preceded by cytosolic acidification and likewise an increase of Ca^{2+}_{cyt} [56,57]. These imaging data are in agreement with microelectrode measurements of cytoplasmic pH [58].

The signal character of pH shifts is seriously considered since their relationship to stomatal movements (acidification precedes opening, and alkalization precedes closing) was shown to reflect a pH-dependent control over K^+ channels in the plasma membrane. Stomatal opening requires mass influx of K^+ via inwardly rectifying K^+ channels that are activated by increasing the cytosolic H^+ concentration. In a combined assay of pH_{cyt} via ratio photometry with BCECF and microelectrode measurements of K^+ current, the protonation site of this channel could even be titrated in vivo with a pK_a near 6.3 [25]. Conversely, stomatal closure involves K^+ efflux via the outwardly rectifying K^+ channel, which is activated by decreasing cytosolic H^+ concentrations [58], with an apparent pK_a of 7.4 [25]. The stimulating effect of increasing pH on K^+ efflux can be demonstrated by electrophysiological measurements with excised patches of the plasma membrane and probably reflects an increase in the number of channels available for activation rather than effects of pH on the single-channel conductance [59]. Hence, the alkalization caused by ABA can be expected to inhibit the K^+-influx channels and activate the K^+-efflux channels, thus promoting stomatal closure.

The link between pH_{cyt} and K^+ channel activity was further substantiated by combined pH mapping and voltage clamp experiments with a synthetic peptide homologue of the C-terminus of the auxin binding protein from *Zea mays* [60]. This peptide mimics the effect of high auxin concentrations (>100 µmol/L), i.e., stomatal closing, reversible alkalization in the cytoplasm (visualized by confocal pH mapping with BCECF), and a parallel, reversible inactivation of the inward rectifying K^+ channels. The latter inhibition was prevented by clamping pH_{cyt} to near 7.0 with butyric acid, irrespective whether it was evoked by the preceding peptide [60] or high auxin concentrations [61]. Taken together, the findings established that transient shifts of the cytoplasmic pH are essential intermediates of the signal pathways initiated by auxin or ABA, that the mutual antagonism of these hormones relies on their opposite effects on pH_{cyt}, and that both K^+-influx and K^+-efflux channels are regulated via pH_{cyt}.

In the ABA-dependent signal chain, pH regulation of channel activity might include serine/threonine protein phosphatases (type 2C), which in their mutant form (abil-1) provide ABA insensitivity to *Arabidopsis* guard cells by abolishing the response of K_{in}^+ and K_{out}^+ channels to ABA [62] and to experimental changes in pH_{cyt} [63]. These phosphatases are strongly activated by small pH increases, such as those evoked by ABA, and require millimolar concentrations of Mg^{2+} for activity [64]. They appear to be targets of the pH signal as well as of the Mg^{2+} status of the cytoplasm, which they convert into K^+ channel activity via a hitherto uncharacterized dephosphorylation step [65].

K^+ influx channels are also activated by extracellular (apoplastic) protons that are sensed by extracellular binding sites of the channel protein [66]. The well-known H^+ extrusion from auxin-treated cells might therefore be involved in the regulation of K^+ influx. Imaging of the apoplastic pH with dextran-coupled pH probes [67] might offer a way to verify this hypothesis.

The hormonal control of stomatal movements is a prominent example of how shifts of intracellular pH are coordinated or interact with Ca^{2+} transients in a complex signal cascade [68]. The signal character of Ca^{2+} in guard cell movements has long remained obscure as Ca_{cyt}^{2+} is raised by different hormones irrespective of whether they cause stomatal opening (low IAA, kinetin, fusicoccin) or closing (ABA, high IAA). The inwardly transporting K^+ channels in the plasma membrane, independent of their activation by cytosolic H^+, are inactivated by cytosolic Ca^{2+} (ratio photometry with Fura-2 and BCECF together with microelectrode measurements [25]). The outwardly transporting K^+ channel, i.e., the major path for K^+ efflux during stomatal closure, is activated by decreasing

H_{cyt}^+ (see earlier) but is also indirectly influenced by Ca_{cyt}^{2+} via the membrane potential [58]. Depolarization of the plasma membrane indeed occurs during ABA treatment [69] and can be explained by Ca^{2+}-dependent effects: anion efflux via Ca^{2+}-activated, slow-type anion channels causes a long-term depolarization [70,71] and Ca^{2+} influx through a voltage-gated channel immediately depolarizes the plasma membrane [72]. Actual imaging data indicate that oscillations of cellular Ca^{2+} provide the background for the tuning of ABA effects during stomatal closure. Ca^{2+} waves could be triggered by experimental oscillations of the membrane potential (in *Vicia* [73]) or by ABA treatment (in *Commelina* [74]) and are thought to start with either influx of the external ion or mobilization of endogenous Ca^{2+}, respectively.

The mobilization of intracellular Ca^{2+} stores is obviously influenced by the cytoplasmic pH and therefore appears as a major point of interaction of Ca^{2+}- and H^+-mediated signaling; in guard cells, an experimental decrease in pH_{cyt} below 7.0 led to an elevation of Ca^{2+} [25]. The background of this finding is most probably the H^+-dependent gating of vacuolar ion channels [75,76], but an influence of pH_i on Ca^{2+} fluxes across the plasma membrane cannot be excluded [68]. In addition to pH_{cyt}, further low-molecular-weight signal molecules are involved in the cellular control over endogenous Ca^{2+} stores, mainly cyclic adenosine diphosphate ribose (cADPR) [77,78] and inositol trisphosphate (IP_3) [74,79]. In animal cells, the binding of IP_3 to its receptor is pH dependent [80], which further argues for interlock of proton and Ca^{2+} transients in hormonal signal transduction. Finally, the cytoplasmic Ca^{2+} concentration itself may act on slow vacuolar channels and thus cause the phenomenon of Ca^{2+}-induced Ca^{2+} release (CICR) [75,81,82]. In the case of ABA signaling in guard cells, this situation has been pinpointed in a model that postulates a bifurcated signal path containing both a Ca^{2+}-dependent branch (that involves cADPR) and a Ca^{2+}-independent, pH-controlled branch [25,65,68,73]. Both branches might interact at later stages of signaling, thereby generating the ABA-specific Ca^{2+} signature.

3.2 Signaling by Gibberellic Acid

This hormone causes transient acidification of the cytoplasm, an effect that has been demonstrated most convincingly in corn coleoptile cells by confocal mapping with the probes BCECF and cSNARF-1. These images are reminiscent of the pH drop triggered by auxin [83].

A classical study object of gibberellic acid (GA)-dependent signaling is the cereal aleurone cells, which respond to gibberellins with the production and secretion of several hydrolases (mainly α-amylases) and vacuolation. Among the earliest events measurable after hormone contact are peaks of cytoplasmic Ca^{2+} and of H^+ (for a review see Ref. 84). Imaging data demonstrate that GA evokes an acidification of the vacuolar interior, which appears to be a prerequisite for the activation of vacuolar proteases that mobilize storage proteins. The increase in vacuolar protons is due to an enhanced activity (but not amount) of tonoplast H^+ pumps (measured with BCECF [85]).

A pH response in the cytoplasm of GA-treated aleurone cells has so far been demonstrated only by a null-point method that measures pH changes of weakly buffered external media caused after digitonin permeabilization of the plasma membrane [86]. pH mapping in these cells, i.e., visualization of the spatial dimension of the expected pH decrease by a less invasive method remains to be done.

Targets of the pH shift in the course of GA signaling have not yet been identified. An influence of artificial shifts of pH_{cyt} on the induction of the intact α-amylase promoter could not be established. Interestingly, a chimeric α-amylase promoter construct containing several GA-responsive elements was induced by lowering pH_{cyt}. On the other hand, induction of a promoter of the *rab* gene family in the presence of ABA was enhanced by a pH increase of a magnitude similar to that caused by ABA [87]. It seems therefore that shifts in pH_{cyt} are not a sufficient signal but a factor influencing the efficiency of ABA-controlled gene expression in aleurone cells.

3.3 pH and Gravitropism

Recent data suggest a role for pH shifts in the gravitropic response. Columella cells of *Arabidopsis* roots displayed rapid changes in cytoplasmic pH following gravistimulation (examined by ratio mapping of BCECF [88]). Magnitude and direction of pH change depended on the position of the cell in this tissue; i.e., cells in a lower tier of the columella became more alkaline, whereas cells in the next row up acidified. Acidification of the cytoplasm (by inhibiting the vacuolar ATPase with bafilomycin) enhanced gravitropic bending but alkalization disrupted it. The authors suggest that changes of pH_{cyt} in the gravisensing cells and the resultant pH gradients across the root cap are essential early events in the signaling toward the gravitropic response.

Complementary data on the apoplastic pH in corn roots support the need for a pH gradient across the plasma membrane for directed growth; during gravitropic stimulation the apoplastic pH on the convex side of the bending root decreased from 4.9 to 4.5. According to the acid growth theory, this pH shift might explain the enhanced growth on that side (confocal imaging with CL-NERF [67]).

3.4 Elicitation of Antimicrobial Defenses

Contact with elicitor molecules (distinct glycoproteins, oligosaccharides, or proteins of pathogenic microorganisms) evokes a complex variety of defense responses in susceptible plant cells. Among the earliest reactions are perturbations of the cellular ionic balance, i.e., external alkalization and cellular acidification, efflux of K^+ and Cl^-, and influx of Ca^{2+} [89] as well as an oxidative burst, i.e., the extracellular generation of reactive oxygen species by a plasma membrane NADPH oxidase (for a review see Ref. 90). They may trigger various downstream events, including oxidative cross-linking of cell wall constituents, overproduction of plant phenolics, hypersensitive cell death, and induction of several genes. Integrative signal systems obviously coordinate such responses at the cellular and the systemic level [91–94] but allow autonomous expression of distinct branches of signal chains, including the species-specific formation of phytoalexins, i.e., antimicrobial secondary metabolites.

Cytoplasmic acidification was frequently observed in elicitor-treated plant cells [15,27,89,95–97] but appears to be involved in different signal chains and to have different origins. Under conditions that finally led to the hypersensitive reaction, cytoplasmic acidification is tightly coupled to alkalization of the outer medium, indicating an influx of external protons [89,97]. Such a process has been studied by monitoring the distribution of [^{14}C] benzoic acid in tobacco cells that responded to oligogalacturonide elicitors [27]. Inhibitor experiments suggest that this proton influx is preceded and controlled by protein phosphorylation-dephosphorylation; cytoplasmic acidification and external alkalization were inhibited by protein kinase inhibitors and could be mimicked by protein phosphatase inhibitors [98]. The signal character of the observed pH shifts was supported by showing that artificial intracellular acidification (achieved by either permeant acids, phosphate uptake, or inhibition of the PM ATPase and optimized for duration and magnitude) led to an increase of mRNA levels of phenylalanine ammonia-lyase (PAL) and 3-hydroxy-3-methylglutaryl-coenzyme A reductase, the first enzymes of phenylpropanoid and isoprenoid phytoalexin pathways [99].

Similarly, in suspension-cultured rice cells artificial acidification by propionic acid induced accumulation of mRNAs of PAL and two other, as yet undefined genes in a manner similar to the natural elicitor N-acetyl-chitoheptaose but not of chitinase and beta-glucanase mRNAs [100]. The artificial pH shift did not trigger the production of reactive oxygen species, and its effect on mRNA induction was less sensitive to kinase inhibitors than to the elicitor.

An elicitation process that is not coupled to the hypersensitive reaction was studied in cultured cells of *Eschscholzia californica* (California poppy). These cells respond to fungal elicitor preparations with the induction of biosynthesis of benzophenanthridine alkaloids that are potent phytoalexins. This elicitation process neither required alkalization of the outer medium nor was linked to an oxidative burst. Cultured cells reacted to low concentrations of a yeast glycoprotein elicitor with a rapid, transient acidification of the cytoplasm and a concomitant increase in the vacuolar pH as shown by confocal pH mapping with cSNARF-1 [15] (Fig. 3; see color insert). Quantification of the pH maps revealed a nearly constant proportion between the concentration of protons disappearing from the vacuole and those arriving in the cytoplasm, suggesting an efflux of H^+ from the vacuole and a constant buffering capacity of the cytoplasm. pH mapping also allowed controlled gain- and loss-of-function experiments that corroborated the signal character of cytoplasmic pH shifts; artificial acidification of the cytoplasm with butyric or pivalic acid (e.g., from 7.3 to 6.8) triggered alkaloid biosynthesis, whereas treatment with strong buffers or methylamine at pH_{ext}=7.4 led to the reversible deprivation of vacuolar acidity and within this period prevented the effects of elicitors on alkaloid biosynthesis. Furthermore, during the recovery from methylamine treatment, cells generated significant amounts of new acidity that transiently decreased pH_{cyt} by approximately 0.3 units. Under these conditions, alkaloid biosynthesis was evoked in the absence of elicitor, and the increase of productivity correlated with the degree of acidification. According to other systems, only transient, short-time peaks of pH_{cyt} lead to enhanced alkaloid biosynthesis. It appears therefore that a transient cytoplasmic acidification provided by vacuolar protons is a necessary step of the signal path toward phytoalexin formation [15]. A vacuolar origin of cytoplasmic pH shifts has also been suggested in guard cell protoplasts where the auxin-triggered acidification of the cytoplasm was absent after evacuolation, although the characteristic stimulation of H^+ secretion was retained [101].

The plant polypeptide systemin is most probably a systemically acting, mobile wound signal released after herbivorous attacks. It enables, e.g., tomato plants to produce proteinase inhibitors in both wounded and unwounded leaves. Among the earliest responses of leaves to systemin are transient depolarization of the plasma membrane and H^+ efflux into mesophyll cells. Both effects are diminished in the presence of fusicoccin, which also prevents the systemin-induced synthesis of proteinase inhibitor, indicating a pH_{cyt}-dependent control over the systemin-triggered signal path [102].

3.5   Cytosolic pH and Cellular Development

A gradient in cytosolic pH has been implicated in the regulation of polarized growth in plant and fungal cells. However, evidence of such a gradient in tip-growing cells is not unequivocal and an actual debate reflects the subtlety needed to map fragile transients of pH.

Gibbon and Kropf [103] examined growing rhizoid cells in embryos of the brown alga *Pelvetia* by imaging with dextran-conjugated cSNARF-1 and microelectrode measurements. Both methods showed a longitudinal pH gradient with the apex more acidic than the base of the cell (0.3 to 0.5 units). The magnitude of the proton gradient correlated well with the rate of elongation growth, and both parameters were strongly reduced by treatment with the membrane-permeant propionic acid. In zygotes of this alga, the formation of a developmental axis was accompanied by a pH shift in the cortical region of 0.1 units, with the presumed rhizoid pole becoming significantly more acidic than the thallus pole. Small experimental perturbations of pH_{cyt} (0.2–0.3 pH units) inhibited the ability of the zygotes to form a developmental axis and to initiate rhizoid outgrowth [104].

In filamentous fungi, growing hyphal cells appear to maintain a relatively alkaline tip region. Robson et al. [105] visualized such a gradient in *Neurospora crassa*, with the tip region up to 1.4 units more alkaline than the more distal areas (confocal imaging of hyphal cells with BCECF). The magnitude and length of this gradient were strongly correlated with the rate of hyphal extension; dissipation of the gradient with either propionic acid or the uncoupler CCCP arrested growth. The authors therefore concluded that a pH gradient with an alkaline tip region is an essential regulatory element in hyphal extension. Alkalization at the hyphal tip was also observed during early stages of conidiation of the mold *Penicillium cyclopium* and proved to require the presence of Ca^{2+} (ratioing with cSNARF1 [106]).

In contrast, Bachewich and Heath [107] did not find a cytoplasmic pH gradient by imaging hyphal cells of the oomycete fungus *Saprolegnia ferax* containing acid-loaded cSNARF-1. However, experimental acidification of pH_{cyt} by permeant acids (e.g., from 7.2 to 6.8) caused reversible inhibition of tip growth together with relocations of mitochondria and nuclei, disruptions of peripheral actin, and even changes of the chromatin structure. As even smaller additions of permeant acids were found to induce similar changes of cytoplasmic organization (implying that small, hitherto undetected pH shifts are effective), the authors suggested a role for H^+ in regulating growth-correlated cell functions. In an extended imaging study with AM-ester- and dextran-conjugated cSNARF-1, Parton et al. [6] did not observe cytoplasmic pH gradients >0.1 units in tip growing cells of *Neurospora*, pollen tubes of *Agapanthus umbellatus*, and rhizoids of gametophytes of *Dryopteris affinis*. Again, artificial acidification of the cytoplasm to pH near 7 reduced (*Neurospora*) or completely inhibited (*Agapanthus* and *Dryopteris*) apical growth. In pollen tubes of *Lilium longiflorum*, Fricker et al. [50] could not visualize a consistent pH gradient in the growth zone (ratio imaging with BCECF), but, as in previous examples, growth stopped upon increasing pH_{cyt} by either external alkalization or vanadate inhibition of the plasma membrane H^+-ATPase (as well as by adding Ca^{2+} or La^{3+}). In root hairs of *Sinapis alba*, measurements with ion-selective microelectrodes revealed several Ca^{2+}- and pH-dependent fluxes and steady states that appeared essential for tip growth, i.e., Ca^{2+} influx and elevated Ca^{2+} levels in the tip area and H^+ extrusion via the PM-ATPase. Only a minor intracellular pH gradient (if any) was observed and was not considered important for growth; however, a tightly regulated pH_{cyt} appeared essential [108].

A very careful contribution to the imaging of fragile pH distribution patterns in tip-growing cells by Feijo et al. [31] was based on a critical assessment and precise adjustment of experimental imaging parameters and demonstrated the risk of artifactual dissipation of small pH gradients. In growing pollen tubes of *Lilium longiflorum*, a significant pH gradient was detected by wide-field fluorescence microscopy with pressure-injected BCECF-dextran. This gradient was, however, easily dissipated by concentrated external buffers if present at concentrations of 1 μM or higher.

Under optimized conditions (0.3–0.5 μM BCECF-dextran, exposure times below 1 s), the pollen tubes displayed an acidic domain at the extreme apex that appeared only during growth and a more distally located, constitutive alkaline band near the clear zone (a region of low organellar density). Buffer injection into the tip region completely destroyed the pH gradient but did not stop growth. The pH maps and the extracellular proton currents measured simultaneously by a vibrating electrode support a model that assumes recycling of protons across the plasma membrane, i.e., influx at the apex region and extrusion by PM-ATPases that are enriched at the more distal regions. The resulting current loop near the apex of the pollen tube may act as a sensor of extracellular signals and convert them into altered F-actin structures and/or altered vesicle transfer, thus adjusting the direction of growth [31].

4 GENERAL ASPECTS OF pH-DEPENDENT SIGNAL TRANSDUCTION IN PLANT CELLS

The preceding examples not only document typical features of pH-dependent signaling in plant cells but also illustrate the necessity for, and problems with, adjusting the imaging methodology to the peculiarities of cellular pH control. Actually, it appears that pH shifts with signaling functions are highly organized both in time and in space. In this respect they show some similarities to, but also important differences from the better documented Ca^{2+} transients.

4.1 Spatial Characteristics

Free protons are more mobile than Ca^{2+} as indicated by a nearly 10-fold higher diffusion coefficient. At the same time, the high efficiency of cytoplasmic pH homeostasis establishes a threshold that might prevent weak proton fluxes from generating detectable shifts of steady-state pH. Thus, pH gradients can be expected to last shorter than Ca^{2+} gradients, span lower concentration differences, and become detectable only in close proximity to the sites of proton generation or consumption. Indeed, reported H^+ gradients are generally smaller than those frequently found in Ca^{2+} signaling.

Although the molecular background of local pH control is not yet known in detail, it appears realistic to assume that the proton pumps of the plasma membrane as well as the cellular buffering capacity are distributed unevenly at least in some cell types or during some developmental stages. Support for this hypothesis comes from observations of spatial separation of cellular organelles.

For instance, in growing pollen tubes vacuoles and organelles with active metabolism and high buffering capacity are excluded from the clear zone that contains a band of alkaline pH [31] (see Sec. 4.4.1). This region also lacks cytoplasmic streaming and cable-forming actin microfilaments [109]. Vesicle transfer is another, variable source of localized pH gradients. Confocal imaging of dividing cells of *Gossypium hirsutum* revealed an accumulation of vesicles with acidic interior close to the sites of subsequent cell wall formation [96] (Fig. 4; see color insert). The cytoskeleton directs cytoplasmic streaming, vesicle transfer, and distribution of organelles and at the same time mediates regulatory impacts of pH shifts on these levels of subcellular organization. As already mentioned, in tips of fungal hyphae artificial acidification reversibly caused alterations of the cytoskeleton-dependent organization of tip morphology, including repositioning of organelles and disruptions of peripheral actin [107]. This pH sensitivity of cytoskeletal structures implicates influences of pH shifts on local buffering and pH homeostasis that in turn may condition the extent and duration of such proton transients.

Highly localized pH gradients are represented by the acidic layers of peripheral cytoplasm in close proximity to the plasma membrane as they have been imaged in some fungal and cultured plant cells and protoplasts [33,96] (see Fig. 4). The pH difference from the surrounding cytoplasm could be increased by inhibitors of P-ATPase (e.g., vanadate) but was not dissipated by uncouplers such as CCCP. They might thus reflect interplay between peripheral cytoplasmic buffers or Donnan-active macromolecules and the proton pump. Another cause of acidification of peripheral cytoplasm may be the activity of redox systems localized at the plasma membrane that catalyze the transfer of electrons from cellular donors to external acceptors (heavy metals, nitrate, oxygen, some toxins). The electron flow can be experimentally triggered by adding artificial, impermeant electron acceptors (e.g., ferricyanide or hexabromoiridate), and in its initial phase it is accompanied by a transient acidification of the peripheral cytoplasm followed by an excretion of protons. It is suggested that the trans–plasma membrane redox process activates the P-ATPase via membrane depolarization and cytosolic acidification [110].

Rapid shifts of the cytoplasmic pH usually reflect proton fluxes across either the plasma membrane or the tonoplast. It is well known that the H^+ pumps of both membranes are activated by decreasing pH_{cyt} and thus may recycle incoming protons before they reach the core cytoplasm [2]. There are various examples that indicate how effectively the vacuolar proton pool contributes to the pH homeostasis of the cytoplasm; pH changes connected with uptake of ammonium are almost fully compensated by vacuolar accumulation, thus resulting in relatively minor effects on pH_{cyt} [15,30]. Acid loads imposed by permeant acids have a much lower impact on pH_{cyt} if the vacuolar pumps are present and active [101]. Therefore, only relatively strong proton fluxes have a chance to establish a measurable shift of the cytoplasmic steady-state pH. On the other hand, the loss of protons from the (less buffered) vacuolar interior can be taken as another indication of proton fluxes into the cytoplasm [15].

4.2 Time-Dependent Variations

pH shifts that might act as second messengers were shown to be transient in nature when their time course was characterized (e.g., in hormonal signaling or pathogen defense; see earlier). This behavior is similar to that of other low-molecular-weight signal transmitters in animal and plant cells (Ca^{2+}, phosphoinositides, etc.) and appears a basic prerequisite of flexibly adjustable signal systems operating on the background of cellular homeostasis. It is therefore not surprising to find that artificial shifts of intracellular pH to a new, persistent level do not function as a signal, e.g., for the accumulation of transcripts of PAL or invertase after artificial acidification [111].

Single, reversible pH shifts of different length have been documented in plant cells treated with microbial elicitors. In *Eschscholzia californica* the cytoplasmic H^+ concentration attained a peak after nearly 15 min of contact with yeast elicitor [15]. Tobacco cells responded to elicitor-active oligogalacturonides with maximum acidification at around 30 min. In the same cells, cryptogein or an elicitor from *Phytophthora megasperma* evoked a continuous decrease in pH_{cyt} that proceeded over several hours [99]. It is therefore likely that the duration and amplitude of pH transients are an essential part of the messages to be transduced. In tobacco cells, artificial cytosolic acidification that was too long and/or too intense no longer allowed the increase in mRNA transcripts of PAL and HMGR to occur, although the viability of the cells was not significantly impaired [99]. Similarly, alkaloid biosynthesis in *Eschscholzia* cells was induced by short pulses of artificial acidification (<30 min) but not by prolonged treatment, even if the growth rates remained unimpaired (W. Roos et al., unpublished).

Detection of local oscillations of steady-state pH in plant cells is now within reach of imaging techniques. The pH gradients visualized in some tip-growing cells fall into this category. As an example, in growing lily pollen tubes, pH_{cyt} displayed periodic changes that were most prominent in the relatively alkaline band near the clear zone [31]. Moreover, pulses of acidic pH were observed to move from the apex to the tube shaft. These oscillations, which followed the growth pulses of the tube, were detectable even though no standing pH gradient was observed under the conditions tested [112].

Oscillating proton fluxes may be generated by both the proton pumps and passive components of the cellular pH-stat. As an example, proton fluxes with a pronounced periodicity were detected by microelectrode ion flux estimation in cells located in the elongation region of corn roots [113]. They could be separated into a fast-oscillating component (7-min period) that most probably represented rhythmic changes of the activity of the H^+-ATPase and a slow-oscillating component (90-min period) that was thought to mirror the oscillatory behavior of passive H^+ transporters. This idea was consistent with the finding that the short proton pulses responded to the actual cytosolic pH, whereas the slow oscillations were suppressed at low external pH.

Unraveling the molecular and cytological background behind the generation and maintenance of local domains and transients of cellular pH is at the borderline of actual ultrastructural research in plant cells. It might profit, e.g., from the opportunity for site-directed expression of pH-sensitive hybrids of the green fluorescent protein and similar reporter proteins as well as advanced immunolocalization techniques.

4.3 Links between pH and Ca^{2+} Signaling

There is ample evidence that plant cells—much like animal cells—use Ca^{2+} as an almost universal signal molecule. Various stimulus-specific distribution patterns of intracellular Ca^{2+} have been visualized and their time-dependent variations have been monitored. Such complex "signatures" are now increasingly identified as central intermediates of signal transfer cascades initiated by plant hormones (e.g., auxin or abscisic acid in the regulation of ion channels that govern stomatal movements) or environmental stimuli (e.g., cold, touch, osmotica, redox shifts) or governed by developmental programs such as the determination of cell polarity (for reviews see Refs. 114, 115).

In many of these areas, Ca^{2+} signatures are associated with pH shifts. It appears that H^+ and Ca^{2+} signaling can be related at different levels: they may exert concerted effects on their targets (for instance, in the ABA-triggered signal transduction in guard cells, where a Ca^{2+}- and a pH-dependent branch are distinguishable; see earlier), or both ions may interact directly in controlling their mutual ion activity changes. It has long been known that artificial acidification of the cytoplasm with permeant acids causes a concomitant increase in free Ca^{2+}_{cyt}, whereas alkalization with methylamine reduces this activity [25,116,117]. Although the details of the pH-dependent control of Ca^{2+}_{cyt} are not yet clear, it probably reflects pH-sensitive mechanisms of Ca^{2+} release from intracellular stores, such as the pH-dependent gating of vacuolar ion channels [76,118]. The Ca^{2+} reservoir of the cell wall may also be mobilized by H^+ extrusion as suggested by strongly synchronized oscillations between H^+ and Ca^{2+} fluxes that occur around the elongation region of corn roots [113]. In turn, cytoplasmic Ca^{2+} may well influence pH homeostasis via its inhibitory effect on the plasma membrane H^+-ATPase [119,120].

A spatial and functional coordination between gradients of Ca^{2+} and H^+

has been suggested by imaging data for tip-growing cells. In *Pelvetia* embryos, a pH gradient has been visualized [103] (see also earlier), spatially coinciding with the cytosolic Ca^{2+} gradient [18]. This pH gradient was dissipated not only by treatment with propionic acid but also by La^{3+}, an inhibitor of calcium influx that reduced Ca^{2+}_{cyt} but did not change the average cytoplasmic pH. Both treatments inhibited tip growth, which confirms the regulatory importance of the pH gradient but at the same time indicates a role of Ca^{2+} in the pH_{cyt} homeostasis [103,120]. It is likely that the cytoskeleton provides a major level of spatial coordination between H^+ and Ca^{2+} signaling as its structure is clearly influenced by pH_{cyt} [107,121] and at the same time controls the extent of Ca^{2+} shifts [122].

In summary, when regulatory functions of pH gradients are considered, the possibility of interacting with Ca^{2+} gradients should be investigated, for instance, by parallel imaging of Ca^{2+} and H^+. This method is not yet in routine use with plants but has been successfully applied to animal cells (for an example, see Ref. 123).

4.4 Targets of pH Control

Molecular targets of pH shifts with regulatory significance have rarely been identified in detail. Nevertheless, some signal modules are likely to operate downstream of Δ pH as judged by their pH dependence and the necessity of their functioning in a defined signal transduction process. In these cases, it needs to be distinguished whether the pH shift is a sufficient intermediary step within a signal sequence or a necessary condition that allows other events to occur (an example of the latter type might be the induction of the α-amylase promoter in ABA-treated aleurone cells; see Sec. 3.2). Furthermore, even a localized change of pH can affect more than one target protein or enzyme and thus diversify the signal process. Some prominent, pH-responding targets are now briefly summarized.

4.4.1 Ion Channels

The guard cell plasma membrane harbors at least three pH-responding ion channels: the outwardly rectifying K^+ channel is activated by decreasing cytosolic H^+ and the inwardly rectifying K^+ channels show an opposite pH dependence (see Sec. 3.1). Furthermore, the anion channel 1 (GCAC1) is activated by ATP binding in a pH-dependent manner; i.e., its activity increases at increasing H^+_{cyt} [124].

4.4.2 Protein Kinases and Phosphatases

It has long been argued that the degree of phosphorylation of some proteins might be controlled by the local pH [125]. A highly pH-dependent serine/threonine protein phosphatase (ABI-1) has now been shown to be involved in one of the ABA-triggered signal chains [65]. In tobacco cells, cytosolic acidification by permeant acids triggered transient activation of a MAP kinase that was insensitive to the general protein kinase inhibitor staurosporine [126]. Furthermore, a protein kinase that is activated by low pH (e.g., pH 4.3) is involved in the light-dependent redox control of some thylakoid proteins [127].

4.4.3 Cytoskeleton

As mentioned before, the structure of the cytoskeleton, especially the remodeling of actin during tip growth, is very likely to respond to cytoplasmic pH (4.1). Some of the pH-sensitive actin binding proteins of animal cells (e.g., the eukaryotic elongation factor 1 and members of the cofilin/ADF family) have now been identified in pollen and at the apex of root hairs, but their pH-dependent functioning in planta remains to be documented (for a more thorough discussion see Ref. 31).

5 CONCLUSIONS

During the last decade, suggestions about signal functions of intracellular pH [128,129] have been studied in detail and corroborated in a number of plant cell systems. Actually, it appears that shifts in the intracellular H^+ distribution play essential roles in several areas of plant signal transfer, be that as intermediate steps of defined signal chains originating from hormones, pathogens, etc. or by providing specific conditions for signal transduction, vesicle transfer, cytoskeletal organization, or other growth-related processes. In the near future, proton signaling might prove as subtle as the Ca^{2+}-mediated information but with different characteristics. It seems realistic to expect cellular H^+ signatures that provide and connect fast-signaling events at distinct microenvironments and thereby complement the longer lasting Ca^{2+} signature.

There is growing evidence that the final intracellular message initiated by a distinct environmental or developmental stimulus is not a one-way sequence but rather a specific combination of activated ion channels, protein kinases, protein phosphatases, phospholipases, or other stimulus-response couplers [115]. Such a signaling network is very likely to include a distinct signature and dynamics of pH, Ca^{2+}, and membrane potential and may include other ions such as Na^+ and Mg^{2+}. Unraveling the topography, generators, and addresses of peaks, oscillations, waves, and other patterns of ion distribution therefore becomes a central task in signal transduction research. It appears that approaches monitoring more than one signal parameter (e.g., simultaneous imaging of H^+ and Ca^{2+}) have the best chance of finding causal relationships and signal sequences within these complexities.

New methodical developments in confocal, lifetime, and two-photon imaging can be expected to improve the resolution and alignment of imaging information with subcellular structures, such as the vesicular and reticular organelles

of the plant cell that are often stacked at high densities. Fluorescence or luminescence signals of defined organellar origin are the goal of designing organelle-
specific fluorescent probes by either liposome-based targeting or site-directed
expression of aequorins and pH-sensitive green fluorescent proteins [9]. It may
further be anticipated that expression vectors with varying targeting sequences
and improved transgenic technologies will allow the simultaneous visualization
and quantification of pH and other ion concentrations in different organelles.

ABBREVIATIONS

ABA	abscisic acid
APB	auxin-binding protein
CCCP	carbonyl cyanide m-chlorophenylhydrazone
FCCP	carbonyl cyanide p-trifluoromethoxyphenylhydrazone
GA	gibberellic acid
HMGR	3-hydroxy-3-methylglutaryl-coenzyme A reductase
IAA	indole-3-acetic acid
NA	numerical aperture
PAL	phenylalanine ammonia-lyase

Abbreviated names of fluorescent probes are compiled in Table 1. Trademarks
are properties of Molecular Probes, Inc.

REFERENCES

1. ML Müller, U Irkens-Kiesecker, B Rubinstein, L Taiz. On the mechanism of hyper-
    acidification in lemon—comparison of the vacuolar H^+-ATPase activities of fruits
    and epicotyls. J Biol Chem 271:1916–1924, 1996.
2. E Gout, R Bligny, R Douce. Regulation of intracellular pH values in higher plant
    cells (carbon-13 and phosphorus-31 nuclear magnetic resonance studies). J Biol
    Chem 267:13903–13909, 1992.
3. MD Fricker, C Plieth, H Knight, E Blancaflor, MR Knight, NS White, S Gilroy.
    Fluorescence and luminescence techniques to probe ion activities in living plant
    cells. In: WT Mason, ed. Fluorescent and Luminescent Probes for Biological Activ-
    ity. 2nd ed. San Diego: Academic Press, 1999, pp 569–596.
4. W Roos. Ion mapping in plant cells—methods and applications in signal transduc-
    tion research. Planta 210:347–370, 2000.
5. SH Cody, PN Dubbin, AD Beischer, ND Duncan, JS Hill, AH Kaye, DA Williams.
    Intracellular pH mapping with SNARF-1 and confocal microscopy. I. A quantita-
    tive technique for living tissues and isolated cells. Micron 24:573–580, 1993.
6. RM Parton, S Fischer, R Malho, O Papasouliotis, TC Jelitto, T Leonard, ND Read.
    Pronounced cytoplasmic pH gradients are not required for tip growth in plant and
    fungal cells. J Cell Sci 10:1187–1198, 1997.
7. M Yassine, JM Salmon, J Vigo, P Viallet. cSNARF-1 as a pH_i fluoroprobe: discrep-

ancies between conventional and intracellular data do not result from protein interactions. J Photochem Photobiol 37:18–25, 1997.

8. RP Haugland. Handbook of Fluorescent probes and Research Chemicals. 6th ed. Molecular Probes Inc, 1996. 7th ed. (CD-ROM-version), 1999.

9. J Llopis, JM McCaffery, A Miyawaki, M Farquhar, RY Tsien. Measurement of cytosolic, mitochondrial and Golgi pH in single living cells with green fluorescent proteins. Proc Natl Acad Sci USA 95:6803–6808, 1998.

10. WH Zhang, Z Rengel, J Kuo. Determination of intracellular Ca^{2+} of intact wheat roots: loading of acetoxymethyl ester of Fluo-3 under low temperature. Plant J 15: 147–151, 1998.

11. D Brauer, J Otto, SI Tu. Selective accumulation of the fluorescent pH indicator, BCECF, in vacuoles of maize root-hair cells. J Plant Physiol 145:57–61, 1995.

12. D Brauer, J Uknalis, R Triana, SI Tu. Subcellular compartmentation of different lipophilic fluorescein derivatives in maize root epidermal cells. Protoplasma 192: 70–79, 1996.

13. N Giglioli-Guivarc'h, JN Pierre, J Vidal, S Brown. Flow cytometric analysis of cytosolic pH of mesophyll cell protoplasts from the crabgrass *Digitaria sanguinalis*. Cytometry 23:241–249, 1996.

14. MD Fricker, M Tlalka, J Ermantraut, G Obermeyer, M Dewey, S Gurr, J Patrick, NS White. Confocal fluorescence ratio imaging of ion activities in plant cells. Scanning Microsc 8:391–405, 1994.

15. W Roos, S Evers, M Hieke, M Tschöpe, B Schumann. Shifts of the intracellular pH distribution as a part of the signal mechanism leading to the elicitation of benzophenanthridine alkaloids in cultured cells of *Eschscholtzia californica*. Plant Physiol 118:349–364, 1998.

16. CL Slayman, VV Moussatos, WW Webb. Endosomal accumulation of pH indicator dyes delivered as acetoxymethyl esters. J Exp Biol 96:419–438, 1994.

17. DA Williams, SH Cody, PN Dubbin. Introducing and calibrating fluorescent probes in cells and organelles. In: WT Mason, ed. Fluorescent and Luminescent Probes for Biological Activity. London: Academic Press, 1993, pp 321–334.

18. F Berger, C Brownlee. Ratio confocal imaging of free cytoplasmic calcium gradients in polarizing and polarized *Fucus* zygotes. Zygote 1:9–15, 1993.

19. MR Knight, ND Read, AK Campbell, AJ Trewaves. Imaging calcium dynamics in living plants using semi-synthetic recombinant aequorins. J Cell Biol 121:83–90, 1993.

20. R Tommasini, E Martinoia, E Grill, KJ Dietz, N Amrhein. Transport of oxidized glutathione into barley vacuoles: evidence for the involvement of the glutathione-S-conjugate ATPase. Z Naturforsch 48C:867–871, 1993.

21. JOD Coleman, R Randall, MMA Blakekalff. Detoxification of xenobiotics in plant cells by glutathione conjugation and vacuolar compartmentalization: a fluorescent assay using monochlorobimane. Plant Cell Environ 20:449–460, 1997.

22. PA Rea, ZS Li, YP Lu, YM Drozdowicz, E Martinoia. From vacuolar GS-X pumps to multispecific abc transporters. Annu Rev Plant Physiol Plant Mol Biol 49:727–760, 1998.

23. SG Gilroy, MD Fricker, ND Read, AJ Trewavas. Role of calcium in signal transduction of *Commelina* guard cells. Plant Cell 3:333–344, 1991.

24. MD Fricker, M Tester, S Gilroy. Fluorescence and luminescence techniques to probe ion activities in living plant cells. In: WT Mason, ed. Fluorescent and Luminescent Probes for Biological Activity. London: Academic Press, 1993, pp 360–377.
25. A Grabov, MR Blatt. Parallel control of the inward rectifier K^+ channel by cytosolic free Ca^{2+} and pH in *Vicia* guard cells. Planta 201:84–95, 1997.
26. H Felle. Proton transport and pH control in *Sinapis alba* root hairs: a study carried out with double-barrelled pH microelectrodes. J Exp Bot 38:340–354, 1987.
27. Y Mathieu, D Lapous, S Thomine, C Lauriere, J Guern. Cytoplasmic acidification as an early phosphorylation-dependent response of tobacco cells to elicitors. Planta 199:416–424, 1996.
28. RJ Reid, FA Smith, J Whittington. Control of intracellular pH in *Chara corallina* during uptake of weak acid. J Exp Bot 40:883–891, 1989.
29. LA Crawford, AW Brown, KE Breitkreuz, FC Guinel. The synthesis of gamma-aminobutyric acid in response to treatments reducing cytosolic pH. Plant Physiol 104:865–871, 1994.
30. D Brauer, J Uknalis, R Triana, SI Tu. Effects of external pH and ammonium on vacuolar pH in maize root hair cells. Plant Physiol Biochem 35:31–39, 1997.
31. JA Feijo, J Sainhas, GR Hackett, JG Kunkel, PK Hepler. Growing pollen tubes possess a constitutive alkaline band in the clear zone and a growth-dependent acidic tip. J Cell Biol 144:483–496, 1999.
32. J Slavik, A Kotyk. Intracellular pH distribution and transmembrane pH profile of yeast cells. Biochim Biophys Acta 766:679–684, 1984.
33. W Roos, J Slavlík. Intracellular pH topography of *Penicillium cyclopium* protoplasts. Maintenance of pH by both passive and active mechanisms. Biochim Biophys Acta 899:67–75, 1987.
34. MR McAinsh, C Brownlee, AM Hetherington. Visualizing changes in cytosolic-free Ca^{2+} during the response of stomatal guard cells to abscisic acid. Plant Cell 4:1113–1122, 1992.
35. MR McAinsh, AAR Webb, JE Taylor, AM Hetherington. Stimulus-induced oscillations in guard cell cytosolic free calcium. Plant Cell 7:1207–1219, 1995.
36. M Opas. Measurement of intracellular pH and pCa with a confocal microscope. Trends Cell Biol 7:75–80, 1997.
37. S Inoué. Foundations of confocal scanned imaging in light microscopy. In: JB Pawley, ed. Handbook of Biological Confocal Microscopy. 2nd ed. New York: Plenum, 1990, pp 1–14.
38. WT Mason, J Dempster, J Hoyland, TJ McCann, B Somasundaram, W O'Brien. Quantitative digital imaging of biological activity in living cells with ion-sensitive fluorescent probes. In: WT Mason, ed. Fluorescent and Luminescent Probes for Biological Activity. 2nd ed. San Diego: Academic Press, 1999, pp 175–195.
39. H Szmacinski, JR Lakowicz. Fluorescence lifetime-based sensing and imaging. Sensors Actuators 29:16–24, 1995.
40. R Sanders, A Draaijer, HC Gerritsen, PM Houpt, YK Levine. Quantitative pH imaging in cells using confocal fluorescence lifetime imaging microscopy. Anal Biochem 227:302–308, 1995.

41. HC Gerritsen, R Sanders, A Draaijer. Confocal fluorescence lifetime imaging of ion concentrations. Proc SPIE Int Soc Opt Eng 2329:260–267, 1994.

42. TWJ Gadella. Fluorescence lifetime imaging microscopy (FLIM): instrumentation and applications. In: WT Mason, ed. Fluorescent and Luminescent Probes for Biological Activity. 2nd ed. San Diego: Academic Press, 1999, pp 570–596.

43. NS White, RJ Errington, MD Fricker, JL Wood. Aberration control in quantitative imaging of botanical specimens by multidimensional fluorescence microscopy. J Microsc (Oxf) 181:99–116, 1996.

44. DA Agard, Y Hiraoka, P Shaw, JW Sedat. Fluorescence microscopy in three dimensions. Methods Cell Biol 30:353–377, 1989.

45. BA Scalettar, JR Swedlow, JW Sedat, DA Agard. Dispersion, aberration and deconvolution in multi-wavelength fluorescence images. J Microsc 182:50–60, 1996.

46. WA Carrington, RM Lynch, EDW Moore, G Isenberg, KE Fogarthy, SF Fay. Superresolution three-dimensional images of fluorescence in cells with minimal light exposure. Science 268:1483–1487, 1995.

47. JA Thomas, RN Buchsbaum, A Zimniak, E Racker. Intracellular pH measurements in Ehrlich ascites tumor cells utilizing spectroscopic probes generated in situ. Biochemistry 18:2210–2216, 1979.

48. MR James-Kracke. Quick and accurate method to convert BCECF fluorescence to pHi: calibration in three different types of cell preparations. J Cell Physiol 151: 596–603, 1992.

49. G Boyarsky, C Hanssen, LA Clyne. Inadequacy of high K^+ nigericin for calibrating BCECF. I. Estimating steady-state intracellular pH. Am J Physiol 271:C1131–C1145, 1996.

50. MD Fricker, NS White, G Obermeyer. pH gradients are not associated with tip growth in pollen tubes of *Lilium longiflorum*. J Cell Sci 15:1729–1740, 1997.

51. G Boyarsky, C Hanssen, LA Clyne. Superiority of in vitro over in vivo calibrations of BCECF in vascular smooth muscle cells. FASEB J 10:1205–1212, 1996.

52. C Owen. Comparison of spectrum-shifting intracellular pH probes 5' (and 6')-carboxy-10-dimethylamino-3-hydroxyspiro [7*H*-benzo[c]xanthene-7,1' (3'*H*)-isobenzofuran]-3'-one and 2',7'-biscarboxyethyl-5 (and 6)-carboxyfluorescein. Anal Biochem 204:65–71, 1992.

53. O Seksek, N Henry-Toulme, F Sureau, J Bolard. SNARF-1 as an intracellular pH indicator in laser microspectrofluorometry: a critical assessment. Anal Biochem 193:49–54, 1991.

54. B Müller-Röber, T Ehrhardt, G Plesch. Molecular features of stomatal guard cells. J Exp Bot 49:293–304, 1998.

55. MR Blatt, G Thiel. Hormonal control of ion channel gating. Annu Rev Plant Physiol Plant Mol Biol 44:543–567, 1993.

56. HR Irving, CA Gehring, RW Parish. Changes in cytosolic pH and calcium of guard cells precede stomatal movement. Proc Natl Acad Sci USA 89:1790–1794, 1992.

57. CA Gehring, HR Irving, RW Parish. Effects of auxin and abscisic acid on cytosolic calcium and pH in plant cells. Proc Natl Acad Sci USA 87:9645–9649, 1990.

58. MR Blatt, F Armstrong. K^+ channels of stomatal guard cells: abscisic acid–evoked

control of the outward rectifier mediated by cytoplasmic pH. Planta 191:330–341, 1993.

59. H Miedema, SM Assmann. A membrane delimited effect of internal pH on the K$^+$ outward rectifier of *Vicia faba* guard cells. J Membr Biol 154:227–237, 1996.

60. G Thiel, MR Blatt, MD Fricker, IR White, P Millner. Modulation of K$^+$ channels in *Vicia* stomatal guard cells by peptide homologs to the auxin-binding protein C terminus. Proc Natl Acad Sci USA 90:11493–11497, 1993.

61. MR Blatt, G Thiel. K$^+$ channels of stomatal guard cells: bimodal control of the K$^+$ inward-rectifier evoked by auxin. Plant J 5:55–68, 1994.

62. F Armstrong, J Leung, A Grabov, J Brearley, J Giraudat, MR Blatt. Sensitivity to abscisic acid of guard-cell K$^+$ channels is suppressed by abi1-1, a mutant *Arabidopsis* gene encoding a putative protein phosphatase. Proc Natl Acad Sci USA 92: 9520–9524, 1995.

63. MR Blatt, A Grabov. Signalling gates in abscisic acid–mediated control of guard cell ion channels. Physiol Plant 100:481–490, 1997.

64. MP Leube, E Grill, N Amrhein. ABI1 of *Arabidopsis* is a protein serine/threonine phosphatase highly regulated by the proton and magnesium concentration. FEBS Lett 424:100–104, 1998.

65. E Grill, A Himmelbach. ABA signal transduction. Curr Opin Plant Biol 1:412–418, 1998.

66. S Hoth, I Dreyer, P Dietrich, D Becker, B Müller-Röber, R Hedrich. Molecular basis of plant-specific acid activation of K$^+$ uptake channels. Proc Natl Acad Sci USA 94:4806–4810, 1997.

67. DP Taylor, J Slattery, AC Leopold. Apoplastic pH in corn root gravitropism: a laser scanning confocal microscopy measurement. Physiol Plant 97:35–38, 1996.

68. A Grabov, MR Blatt. Co-ordination of signalling elements in guard cell ion channel control. J Exp Bot 49:351–360, 1998.

69. G Thiel, EAC MacRobbie, MR Blatt. Membrane transport in stomatal guard cells: the importance of voltage control. J Membr Biol 126:1–18, 1992.

70. B Linder, K Raschke. A slow anion channel in guard cells, activating at large hyperpolarization, may be principal for stomatal closing. FEBS Lett 313:27–30, 1992.

71. A Grabov, J Leung, J Giraudat, MR Blatt. Alteration of anion channel kinetics in wild type and abi-1 transgenic *Nicotiana benthamiana* guard cells by abscisic acid. Plant J 12:203–213, 1997.

72. JI Schroeder, S Hagiwara. Repetitive increases in cytosolic Ca^{2+} of guard cells by abscisic acid activation of non-selective Ca^{2+} permeable channels. Proc Natl Acad Sci USA 87:9305–9309, 1990.

73. A Grabov, MR Blatt. Membrane voltage initiates Ca^{2+} waves and potentiates Ca^{2+} increases with abscisic acid in stomatal guard cells. Proc Natl Acad Sci USA 95: 4778–4783, 1998.

74. I Staxén, C Pical, LT Montgomery, JE Gray, AM Hetherington, MR McAinsh. Abscisic acid induces oscillations in guard-cell cytosolic free calcium that involve phosphoinositide-specific phospholipase C. Proc Natl Acad Sci USA 96:1779–1784, 1999.

75. JM Ward, ZM Pei, JI Schroeder. Roles of ion channels in initiation of signal transduction in higher plants. Plant Cell 7:833–844, 1995.

76. B Schulz-Lessdorf, R Hedrich. Protons and calcium modulate SV-type channels in the vacuolar-lysosomal compartment—channel interaction with calmodulin inhibitors. Planta 197:655–671, 1995.
77. SR Muir, D Sanders. Pharmacology of Ca^{2+} release from red beet microsomes suggests the presence of ryanodine receptor homologs in higher plants. FEBS Lett 395: 39–42, 1996.
78. Y Wu, J Kuzma, E Marechal, R Graeff, HC Lee, R Foster, NH Chua. Abscisic acid signalling through cyclic ADP-ribose in plants. Science 278:2126–2130, 1997.
79. H Knight, AJ Trewavas, MR Knight. Calcium signalling in *Arabidopsis thaliana* responding to drought and salinity. Plant J 12:1067–1078, 1997.
80. CW Taylor, A Richardson. Structure and function of inositol trisphosphate receptors. Pharmacol Ther 51:97–137, 1991.
81. GJ Allen, D Sanders. Control of ionic currents in guard cell vacuoles by cytosolic and luminal calcium. Plant J 10:1055–1069, 1996.
82. R Malho, ND Read, AJ Trewavas, MS Pais. Calcium channel activity during pollen tube growth and reorientation. Plant Cell 7:1173–1184, 1995.
83. CA Gehring, HR Irving, RW Parish. Gibberellic acid induces cytoplasmic acidification in maize coleoptiles. Planta 194:532–540, 1994.
84. PC Bethke, RC Schuurink, RL Jones. Hormonal signalling in cereal aleurone. J Exp Bot 48:1337–1356, 1997.
85. SJ Swanson, RL Jones. Gibberellic acid induces vacuolar acidification in barley aleurone. Plant Cell 8:2211–2221, 1996.
86. S Heimovaara-Dijkstra, JC Heistek, M Wang. Counteractive effects of ABA and GA_3 on extracellular and intracellular pH and malate in barley aleurone. Plant Physiol 106:359–365, 1994.
87. S Heimovaara-Dijkstra, J Mundy, M Wang. The effect of intracellular pH on the regulation of the Rab 16 and the α-amylase 16-4 promoter by abscisic and gibberellic acid. Plant Mol Biol 27:815–820, 1995.
88. AC Scott, NS Allen. Changes in cytosolic pH within *Arabidopsis* root columella cells play a key role in the early signaling pathway for root gravitropism. Plant Physiol 121:1291–1298, 1999.
89. Y Mathieu, A Kurkdjian, H Xia, J Guern, A Koller, MD Spiro, M O'Neill, P Albersheim, A Darvill. Membrane responses induced by oligogalacturonides in suspension-cultured tobacco cells. Plant J 1:33–343, 1991.
90. C Lamb, RA Dixon. The oxidative burst in plant disease resistance. Annu Rev Plant Physiol Plant Mol Biol 48:251–276, 1997.
91. A Levine, R Tenhaken, R Dixon, C Lamb. H_2O_2 from the oxidative burst orchestrates the plant hypersensitive disease resistance response. Cell 79:583–593, 1994.
92. E Kombrink, IE Somssich. Defense responses of plants to pathogens. Adv Bot Res 21:1–34, 1995.
93. T Jabs, M Tschöpe, C Colling, K Hahlbrock, D Scheel. Elicitor-stimulated ion fluxes and O_2^- from the oxidative burst are essential components in triggering defense gene activation and phytoalexin synthesis in parsley. Proc Natl Acad Sci USA 94:4800–4805, 1997.
94. J Ebel, A Mithöfer. Early events in the elicitation of plant defense. Planta 206: 335–348, 1998.

95. RE Kneusel, U Matern, K Nicolay. Formation of *trans*-caffeoyl-CoA from *trans*-4-coumaroyl-CoA by Zn^{2+}-dependent enzymes in cultured plant cells and its activation by an elicitor-induced pH shift. Arch Biochem Biophys 269:455–462, 1989.

96. W Roos. Confocal pH topography in plant cells—acidic layers in the peripheral cytoplasm and the apoplast. Bot Acta 105:253–259, 1992.

97. K Kuchitsu, Y Yazaki, K Sakano, N Shibuya. Transient cytoplasmic pH change and ion fluxes through the plasma membrane in suspension-cultured rice cells triggered by *n*-acetylchitooligosaccharide elicitor. Plant Cell Physiol 38:1012–1018, 1997.

98. Y Mathieu, FJ Sanchez, MJ Droillard, D Lapous, C Lauriere, J Guern. Involvement of protein phosphorylation in the early steps of transduction of the oligogalacturonide signal in tobacco cells. Plant Physiol Biochem 34:399–408, 1996.

99. D Lapous, Y Mathieu, J Guern, C Lauriere. Increase of defense gene transcripts by cytoplasmic acidification in tobacco cell suspensions. Planta 205:452–458, 1998.

100. DY He, Y Yazaki, Y Nishizawa, R Takai, K Yamada, K Sakano, N Shibuya, E Minami. Gene activation by cytoplasmic acidification in suspension-cultured rice cells in response to the potent elicitor, *N*-acetylchitoheptaose. Mol Plant Microbe Interact 11:1167–1174, 1998.

101. H Frohnmeyer, A Grabov, MR Blatt. A role for the vacuole in auxin-mediated control of cytosolic pH by *Vicia* mesophyll and guard cells. Plant J 13:109–116, 1998.

102. C Moyen, E Johannes. Systemin transiently depolarizes the tomato mesophyll cell membrane and antagonizes fusicoccin-induced extracellular acidification of mesophyll tissue. Plant Cell Environ 19:464–470, 1996.

103. BC Gibbon, DL Kropf. Cytosolic pH gradients associated with tip growth. Science 263:1419–1421, 1994.

104. DL Kropf, CA Henry, BC Gibbon. Measurement and manipulation of cytosolic pH in polarizing zygotes. Eur J Cell Biol 68:297–305, 1995.

105. GD Robson, E Prebble, A Rickers, S Hosking, DW Denning, APJ Trincl, W Robertson. Polarized growth of fungal hyphae is defined by an alkaline pH gradient. Fungal Genet Biol 20:289–298, 1996.

106. T Roncal, UO Ugalde, A Irastorza. Calcium-induced conidiation in *Penicillium cyclopium*: calcium triggers cytosolic alkalization at the hyphal tip. J Bacteriol 175:879–886, 1993.

107. CL Bachewich, IB Heath. The cytoplasmic pH influences hyphal tip growth and cytoskeleton related organization. Fungal Genet Biol 21:76–91, 1997.

108. A Herrmann, HH Felle. Tip growth in root hair cells of *Sinapis alba* L.: significance of internal and external Ca^{2+} and pH. New Phytol 129:523–533, 1995.

109. B Kost, P Spielhofer, NH Chua. A GFP–mouse tubulin fusion protein labels plant actin filaments in vivo and visualizes the actin cytoskeleton in growing pollen tubes. Plant J 16:393–401, 1998.

110. J Poenitz, W Roos. A glucose-activated electron transfer system in the plasma membrane stimulates the H^+-ATPase in *Penicillum cyclopium*. J Bacteriol 176:5429–5438, 1994.

111. M Hofmann, R Ehness, TK Lee, T Roitsch. Intracellular protons are not involved in elicitor dependent regulation of mRNAs for defence related enzymes in *Chenopodium rubrum*. J Plant Physiol 155:527–532, 1999.

112. MA Messerli, KR Robinson. Cytoplasmic acidification and current influx follow growth pulses of *Lilium longiflorum* pollen tubes. Plant J 16:87–91, 1998.

113. SN Shabala, IA Newman, J Morris. Oscillations in H^+ and Ca^{2+} ion fluxes around the elongation region of corn roots and effects of external pH. Plant Physiol 113: 111–118, 1997.

114. AJ Trewavas, R Malho. Ca^{2+} signalling in plant cells: the big network! Curr Opin Plant Biol 1:428–433, 1998.

115. AJ Trewavas, R Malho. Signal perception and transduction: the origin of the phenotype. Plant Cell 9:1181–1195, 1997.

116. H Felle. Auxin causes oscillations of cytosolic free calcium and pH in *Zea mays* coleoptiles. Planta 174:495–499, 1988.

117. H Felle. Cytoplasmic free calcium in *Riccia fluitans* L. and *Zea mays* L.: interaction of Ca^{2+} and pH? Planta 176:248–255, 1988.

118. JM Ward, JI Schroeder. Calcium activated H^+ channels and calcium induced calcium release by slow vacuolar ion channels in guard cell vacuoles implicated in the control of stomatal closure. Plant Cell 6:669–683, 1994.

119. T Kinoshita, M Nishimura, KI Shimazaki. Cytosolic concentration of Ca^{2+} regulates the plasma membrane H^+-ATPase in guard cells of faba bean. Plant Cell 7:1333–1342, 1995.

120. B Lino, VM Baizabal-Aguirre, LEG de la Vara. The plasma membrane H^+-ATPase from beet root is inhibited by a calcium-dependent phosphorylation. Planta 204: 352–359, 1998.

121. JM Andersland, MV Parthasarathy. Conditions affecting depolymerization of actin in plant homogenates. J Cell Sci 104:1273–1279, 1993.

122. C Mazars, L Thion, P Thuleau, A Graziana, MR Knight, M Moreau, R Ranjeva. Organization of cytoskeleton controls the changes in cytosolic calcium of cold-shocked *Nicotiana plumbaginifolia* protoplasts. Cell Calcium 22:413–420, 1997.

123. S Vamos, LW Welling, B Wiegmann. Fluorescent analysis in polarized MDCK cell monolayers: intracellular pH and calcium interactions after apical and basolateral stimulation with arginine vasopressin. Cell Calcium 19:307–314, 1997.

124. B Schulz-Lessdorf, G Lohse, R Hedrich. GCAC1 recognizes the pH gradient across the plasma membrane: a pH sensitive and ATP dependent anion channel links guard cell membrane potential to acid and energy metabolism. Plant J 10:993–1004, 1996.

125. DP Blowers, AJ Trewavas. Second messengers: their existence and relationship to protein kinases. In: W Boss, DJ Moore, eds. Second Messengers in Plant Growth and Development. New York: Alan R Liss, 1989, pp 1–28.

126. G Tena, JP Renaudin. Cytosolic acidification but not auxin at physiological concentration is an activator of MAP kinases in tobacco cells. Plant J 16:173–182 1998.

127. AV Vener, PJM Vankan, A Gal, B Andersson, I Ohad. Activation deactivation cycle of redox-controlled thylakoid protein phosphorylation—role of plastoquinol bound to the reduced cytochrome *bf* complex. J Biol Chem 270:2525–2532, 1995.

128. H Felle. pH as a second messenger in plants. In: W Boss, DJ Moore, eds. Second Messengers in Plant Growth and Development. New York: Alan R Liss, 1989, pp 145–166.
129. J Guern, Y Mathieu, S Thomine, JP Jouanneau, JC Beloeil. Plant cells counteract cytoplasmic pH changes but likely use these pH changes as secondary messages in signal perception. Curr Top Plant Biochem Physiol 11:249–269, 1992.

5

pH as a Signal and Regulator of Membrane Transport

Hubert H. Felle
Justus Liebig University, Giessen, Germany

1 INTRODUCTION

In higher plants and fungi many essential transport processes are in some way linked to a transmembrane proton gradient that is built up and maintained by a single transporter, the H^+-ATPase (proton pump). Typically, the circulation of protons or, more generally, that of electrical charge represents the electrochemical driving belt by which a variety of matter is translocated against its concentration gradient. Because this primary active H^+ transport is the dominating force, it is not surprising that a good deal of the cell's energy is fed into such action [1,2]. Not all plants live in this proton world. With a few exceptions (*Acetabularia* has a Cl^- import pump), marine algae posses Na^+ export pumps, which enables them to deal directly with the high external Na^+ concentration. Higher plants or fungi that secondarily experience such salt stress remain with the proton pump and solve that problem through reducing their relative permeabilities (e.g., P_{Na}/P_K) and possibly through exchanging Na^+ for H^+ [3]. During the conquest of land, the salt stress ceased to exist, and plants switched to H^+ as the master ion. The reason for this far-reaching step we can only speculate on, but it is very likely based on the necessity to deal with large water potential gradients between the cell interior and the atmosphere. This forced the plants to create a space in which an aqueous layer adjacent to the plasma membrane could provide the conditions required for transport and a variety of extracellular enzymatic reactions to take place in an osmotically and structurally protected area. The exchange of Na^+ for H^+ was energetically favorable for two major reasons:

1. Whereas in seawater high external $[Na^+]$ provides a sufficient inwardly directed electrochemical Na^+ gradient to drive (co-)transport, such a gradient cannot be built up in plants because $[Na^+]$ is usually higher in cells than on the outside, and even a substantial membrane potential, built up by an Na^+ pump, would yield an electrochemical Na^+ gradient that is only a fraction of the electrochemical H^+ gradient built up by a proton pump.
2. The choice of H^+ instead of Na^+ also implies an osmotic argument. Because protons are osmotically inactive, any membrane transport that exchanges inorganic ions for H^+ will not just exchange charges but will cause water transport, an important issue with far-reaching consequences (e.g., stomatal movement).

In this chapter, an attempt is made to point out the numerous functions protons may have as a regulator of transport, as a transportee, as a part of electrochemical driving forces, and as a signal on either side of the plasma membrane. It elucidates, at least partly, the complexity of an apparently simple measured parameter, the pH.

1.1 pH as a General Mediator and Regulator

As life is based on aqueous chemistry and water spontaneously ionizes, protons cannot be excluded from their milieu; their activity must be regulated (and buffered). Besides being both the substrate and product of metabolic pathways, protons have the apparent potential to communicate information about the cellular energy balance to enzymes and structures that may not share other common effectors. Because small pH changes may have large effects on the activity of enzymes, membrane transporters included, the potential impacts of pH on transport processes are manifold and, in principle, need not be mediated by specialized receptors.

The term regulation can be defined rather narrowly, meaning a feedback loop in which one or more processes depend on preferably one (master) regulator. Usually, such a (quite unsatisfactory) definition of regulation is not adopted in the biological literature. We often use "regulation" as equivalent to "control" or use it in a even wider sense, meaning little more than "influence," while trying to explain an involvement of a certain effector in a regulation process that we do not understand. As most of us have adopted different understandings of a certain term, this custom of nonprecise description or characterization of (cellular) processes necessarily leads to misunderstandings or even misconceptions. In the case of protons, these problems might be exacerbated by the fact that the free proton concentration in any cellular compartment is the product of a complicated network of H^+-consuming and H^+-producing (metabolic) processes. Thus, the actual status of pH as a master regulator may not be undisputed. It appears that in many instances pH is a tool of regulation rather than the regulator itself, and often pH is little more than a "condition" in which processes take place. The latter point also involves the problem that pH changes at either side of a membrane always have a retroactive effect on the mechanism that created these changes, thus making it even more difficult to determine causalities.

2 BIOCHEMICAL IMPLICATIONS

2.1 Proton-Protein Interactions

Not counting the cell wall materials, cells are to a large extent made of proteins, which constitute more than half of the dry weight of the cell. Proteins determine the shape and structure of the cell and also serve as instruments of molecular recognition and catalysis. Many of the bonds in a long polypeptide chain allow free rotation of the atoms they join, giving the protein backbone great flexibility. In principle, then, any protein molecule can adopt an enormous number of different shapes or conformations. However, under biological conditions, most polypeptide chains fold into only one of these conformations because the side chains of the different amino acids associate with one another and with water to form various weak noncovalent bonds. The chemical properties of a protein depend almost entirely on its exposed surface residues, which are able to form different types of weak noncovalent bonds with other molecules; these bonds in many cases depend on the pH of the immediate surrounding, whether in a surface cavity or just in an aqueous layer. An effective interaction of a protein molecule with another one (ligand) requires that a number of weak bonds be formed simultaneously between them. Therefore, the only ligands that can bind tightly to a protein are those that fit precisely onto its surface.

The binding site usually takes the form of a cavity formed by a specific arrangement of amino acids on the protein surface. As water molecules tend to form hydrogen bonds, they will compete with ligands for selected side chains on the protein surface. This surface has a unique chemical reactivity that depends not only on which amino acid side chains are exposed but also on their exact orientation relative to each other. Because of this, even two slightly different conformations of the same protein molecule may differ drastically in their chemistry.

Enzymes are almost always allosteric proteins that can exist in two different conformations. One is the active conformation that binds substrate at its active site and catalyzes its conversion; the other is the inactive conformation that tightly binds the final product at a different place on the protein, the regulatory site. Such sites can be occupied, for instance, by Ca^{2+} and are very likely pH sensitive as well. The changes produced by one ligand can affect the binding of a second ligand, thereby providing a mechanism for regulating various cell processes. Such changes in protein shape are often driven in a unidirectional manner by the expenditure of chemical energy. For example, by coupling allosteric changes to ATP hydrolysis, proteins can do useful work, such as generating a mechanical force (muscle) or pumping ions across a membrane.

2.2 Ion-Induced pH Changes

An interesting question appears to be what determines the pH of a solution and in particular that of a cellular compartment following transport of H^+ or other ions. This almost trivial question summarizes some fundamental properties of water and acid-base relations of transmembrane ion transport [4]. The dissociation constant of water is 10^{-14}. Thus, if only H^+ were transported across a membrane (which of course is not the case), most of them would vanish right away by forming water. Because of electroneutrality requirements, the H^+ transport has to be balanced. This usually occurs with ions from salts that have higher dissociation constants than water (strong ions). Because of electroneutrality, when strong ions are translocated from one compartment into another, an increase in cation concentration, for instance, must be paralleled by a decrease in $[H^+]$ or an increase in $[OH^-]$, resulting in alkalization. Similar considerations hold for translocation of strong anions (e.g., Cl^-). According to Stewart [5,6], the consequences of transmembrane ion transport are

$$[H^+] = \sqrt{[H_2O] \times 10^{-14} \frac{([C^+] - [A^-])^2}{4}} - \frac{([C^+] - [A^-])}{2} \qquad (1)$$

where $[C^+] - [A^-]$ = strong ion concentration difference. When this difference becomes more positive (an increase of cation concentration or a decrease of anion concentration in the respective compartment), an alkalization occurs, and vice versa. Although the requirement of electroneutrality seems to be just stating the obvious, the fact that two ions being transported (e.g., K^+ or Cl^-) arose from compounds with rather different dissociation constants makes all the difference. Thus, H^+ transport itself does not imply pH changes, but transport of strong ions may. This is a rather important point because it means that pH changes within a given compartment need not be initiated by proton transport but could arise from membrane translocation of strong ions such as K^+, Na^+, Ca^{2+}, or Cl^-. The importance of this notion for cellular or apoplastic pH regulation is obvious. An indirect consequence of this may be that some of the ion/H^+ cotransport reported in the literature in fact may be mediated not by *one* molecule but by two different transporters just linked energetically.

3 pH AS A DRIVING FORCE

3.1 The Proton-Motive Force (pmf)

Much of substrate transport is coupled to the free proton concentration on either side of a membrane or to its movement across a membrane. Because the transport of protons across a membrane includes not only the translocation of the H^+ per se but also that of electrical charge, the driving force for H^+-coupled transport is the electrochemical proton gradient ($\Delta\mu_H^+/F$), consisting of the actual proton gradient (ΔpH) across the membrane plus the electrical potential difference ΔE_m:

$$\Delta\mu_H^+/F = \Delta E_m - 59\Delta pH (mV) \tag{2}$$

This force either can drive ATP synthesis in so-called energy-conserving membranes such as thylakoids, cristae, and inner bacterial membranes or, built up by ATP hydrolysis, can drive H^+-coupled membrane transport. The pmf as an entity is a thermodynamic force from which the maximal possible distribution of any given transportee across a membrane can be calculated. Clearly, in a membrane with numerous transporters working at the same time, any value calculated from Eq. (1) is of academic interest under most conditions, as all transport substrates driven by the pmf or by its individual forces will participate in the overall driving force according to their concentrations and permeabilities. The pmf itself does not give information about the velocity of substrate transport; moreover, both pH gradient and electrical membrane potential difference will influence the kinetics of substrate transport in a different manner.

3.2 Consequences for Transport across Different Membranes

Because the pmf is composed of the electrical potential difference and the pH gradient across a given membrane, the poise on the two forces will differ depending on the membrane location within the cell body or according to its physiological importance. For instance, the pmf across thylakoid membranes will consist mainly of a pH gradient, whereas across mitochondrial cristae membranes the electrical potential difference will be the preponderant force. Similarly, the pmf across the tonoplast will mainly consist of a pH gradient, whereas the membrane potential will represent the larger portion of pmf across the plasma membrane. For the plasma membrane, the pmf is variable and will depend on the pH of the immediate surrounding of the cells, be it the apoplast of a leaf, of a root or the aqueous milieu of fresh water or seawater. Thus, the membrane-related composition of the pmf implies the most likely transport direction. Proton-mediated transport of sugars and amino acids, for instance, would be directed from the cytoplasm into the vacuole only (H^+ antiport), whereas an H^+ symport would be the dominant transport type for these substances across the plasma membrane from outside into the cytoplasm (Fig. 1). Similar considerations hold for all intracellular compartments and their respective membranes. Because of this preference in transport direction, substrates translocated that way will remain in their respective compartments and travel in the opposite direction only in case of an inversion of the driving force or after a chemical conformation change that would alter the driving force conditions. Thus, the pmf acts not only as a driving force but also as a retrieval force for organic compounds, which otherwise would be lost through leakage.

Whereas the cytoplasmic pH is tightly regulated and usually does not permit physiological pH variations exceeding about ± 0.3 pH units (i.e., a change of the free [H^+] by a factor of 2), the pH of the vacuole and that of the apoplast (although regulated as well) may undergo much larger fluctuations, be it in the short term (seconds, minutes) or in the longer term (hours, days). This implies that pH-regulated transport at the cytoplasmic side of the plasma membrane or of the tonoplast should require more sensitive sites of the transport protein than at the far side.

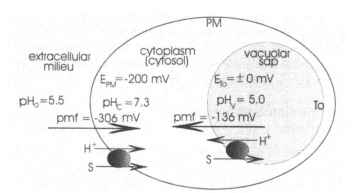

FIGURE 1 Schematic illustration of the proton-motive force (pmf) at the plasma membrane (PM) and at the tonoplast (To), respectively, calculated from transmembrane pH gradients and membrane potentials typically found in plant cells, using Eq. (2). Across the plasma membrane the pmf is always directed from the external milieu toward the cytoplasm, leading to a potential accumulation of the substrate S within the cytoplasm (symport). At the tonoplast, the pmf is almost completely dependent on a pH gradient, giving rise to transport of S from the cytoplasm into the vacuole, while H^+ moves in the opposite direction (antiport).

4 H⁺ AS A TRANSPORT SUBSTRATE

Whereas enzymes in a solution accelerate reaction rates without shifting equilibria, transporters, being embedded in a membrane, render a transport direction to the transported substrate and thus do shift equilibria. Protons not only biochemically influence the transporters but are themselves one of (sometimes) several transport substrates (see cotransport) and as such influence the activity of the transport enzyme in a Michaelis-Menten manner.

4.1 H⁺-ATPases (H⁺ Pumps)

Also called H^+ pumps, H^+-ATPases in plants and fungi are the primary active transporters that build up the electrochemical driving force for most other matter to be transported across membranes, be it inorganic or organic in nature. The activity of these proton pumps can be regulated directly by manipulating the pH at the hydrolyzing side but also by changing the pH from outside. For instance, decreasing the pH from the cytoplasmic side will activate the pump, proton extrusion will increase, and the membrane will be hyperpolarized. Quite conveniently, this can be (and has been) done by adding weak acids to the external medium. According to the existing pH gradient and to the pK_a of the weak acid, the protonated acid will cross the plasma membrane and release protons into the cytoplasm and thereby acidify it. Weak bases will have the opposite effect, alkalize the cytoplasm, deactivate the pump, and thus depolarize the cell.

Although this procedure is widely used, the regulation of the pump by thermodynamic forces is less familiar to researchers: the energy liberated through the hydrolysis of ATP is converted into transport energy, i.e., into an energy gradient across a membrane. This free energy (ΔG), expressed in electrical units ($\Delta G/F$), can be used by the ATPase to transport across the membrane n (1 or more) H^+ per ATP split. The more protons that are transported per ATP, the less potential difference can be built up. Without considering the transmembrane pH gradient, the reversal potential of an ATPase (E_p) can be expressed by

$$E_p = \Delta G/nF \qquad (3)$$

where p stands for pump and n for the amount of H^+ transported per ATP hydrolyzed. Here E_p is the voltage at which electrical and chemical forces are in balance, i.e., zero current. If the voltage were forced more negative than E_p, the ATPase would become an ATP synthase, a situation known from energy-conserving membranes. Actually, facultative bacteria use the ATPase as ATP synthase in the presence of a sufficient oxygen supply but hydrolyze ATP during oxygen shortage, thus building up an electrical potential gradient that can be used for the transport of metabolizable organic matter, part of which is converted to ATP. At this point the question can be asked whether the plasma membrane ATPase could also be used in that way. There are two reasons why not: first, the

ATPase is the primary pump that builds up the energy gradient by which all other transport is driven. Because there is no electron transport or any other process that would exceed the energy gradient built up with ATP, there is no driving force that would drive the ATPase in the other direction. Second, plasma membrane ATPases usually have stoichiometries of one H^+ per ATP hydrolyzed. According to Eq. (3), in the absence of a transmembrane proton gradient, a membrane potential of -400 to -500 mV would be needed to drive the ATPase as ATP synthase, an electrical field strength that may be hazardous to the membrane. Introduction of a transmembrane pH gradient will change Eq. (3) to

$$E_p = \Delta G / nF + 59\Delta pH \tag{3a}$$

Therefore, at the plasma membrane, the reversal potential of the H^+ pump will become less negative when the external conditions are more acidic than the cytosol, a frequent physiological situation that, however, will depolarize the membrane and thus has consequences for the subsequent transport involved. Equation (3) also implies that the stoichiometry n will influence the reversal potential for the pump and hence the resulting resting potential.

The reversal potential of a pump can be obtained experimentally by superimposing electrical pulses across the plasma membrane [7,8]. As Fig. 2 shows, this results in current-voltage characteristics (I-V curves) that cut through the voltage axis at the respective membrane potential (zero current). Because the total current across the membrane (i_m) consists of the pump current (outwardly directed H^+ current) and the leak current (i_d), represented by the sum of all electrical charge driven either into the cell (positive) or out of the cell (negative), blocking the pump will yield the I-V characteristics of the leak only. Graphical subtraction will then give the pump current (i_p), which, in the physiologically relevant voltage range, is rather independent of the membrane voltage; i.e., fluctuations in the membrane potential do not impair the pump's transport yield.

4.2 H$^+$ Cotransport

While H^+ is the primary transport substrate for the H^+ pump, it is the driver ion in carrier-mediated cotransport. Cations basically would not need protons to be transported against their concentration gradient, although there are some instances in which additional driving force is needed, as with the H^+/K^+ symport that, as a high-affinity transporter, is essential to overcome low K^+ situations. Other transportees, such as anions (e.g., Cl^-, NO_3^-) or electrically neutral substrates (e.g., sugars), need H^+ as a driver ion, either to be driven by the proton motive force into the cell or to be kept within to prevent loss.

Being intrinsically passive, such transport would prevail without pump activity for some time, for as long as an electrochemical proton gradient exists across the membrane. Although the transport effectivity is rather low under such

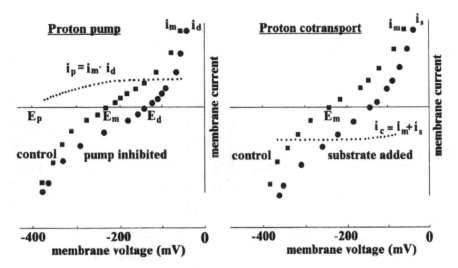

FIGURE 2   Half-schematic presentation of current-voltage relationships of the proton pump and proton cotransport (both dotted). The pump characteristic (left) was obtained by graphical subtraction of the test curve (inhibited pump) from the control curve; the cotransport characteristic (right) was obtained by graphical addition of the test curve to the control curve. The pump current is thus outwardly directed (H⁺ export; hyperpolarizing), cutting through the voltage axis at the reversal potential (*E*p); the cotransport is a positively charged inward current (H⁺ accompanying the substrate, depolarizing). In the physiological voltage range, both are relatively independent of *V* and are thus current sources.

conditions, cotransporters are functionally tightly interconnected with the primary H^+ pump; i.e., the activity of the H^+ pump determines the transport capacity of the cotransporter. In other words, a high turnover of H^+ results in a high cotransport effectivity. In fact, *I-V* curves of cotransporters are so-called current sources (Fig. 2) within the physiological voltage range (as pumps are), which is indicative of a high charge flow (capacity), largely independent of the membrane potential [9].

Because cotransport is a function of pH in different (mostly overlapping) ways, cotransport does not always show a clear-cut pH relationship and usually cannot be used for analysis and in most cases not for the identification of the existence of H^+ cotransport per se. An elegant way to identify the direct involvement of H^+ in substrate transport is the demonstration of transmembrane effects. Figure 3A shows that alanine efflux across the plasma membrane critically depends on the transmembrane pH gradient. Complicating for transport studies is the fact that, e.g., amino acids show a pH-dependent change in their level of

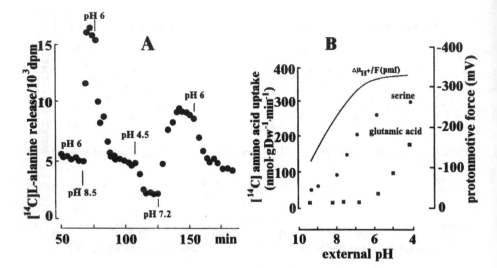

FIGURE 3   (A) Influence of external pH on the efflux of [^{14}C]alanine from green *Riccia fluitans* thalli. The data demonstrate a so-called transmembrane effect (trans-inhibition), which is a good indication of a carrier-mediated H$^+$ cotransport. (B) Uptake of [^{14}C]serine and [^{14}C]glutamic acid into thalli of the liverwort *Riccia fluitans* as a function of external pH and of the proton-motive force ($\Delta\mu_{H^+}/F$), as indicated.

protonation, which then alters the percentage of protonated/unprotonated species to be recognized by the transporter. Thus, it can be demonstrated that the glutamic acid anion apparently is not recognized by the respective carrier, whereas the protonated form, representing a so-called neutral amino acid, is. To demonstrate this, the dependence of amino acid uptake on pH and the proton-motive force is shown in Fig. 3B.

4.3   H$^+$ Channels?

The transmembrane F$_0$ unit of the mitochondrial ATP synthase can be thought of as a channel through which protons are driven according to their electrochemical gradient (pmf) to provide the energy for the formation of ATP at the F$_1$ unit. Whereas in this way the protons crossing the membrane transfer their energy into chemical energy (ATP), an open channel with no "cap" would just dissipate energy. Thus, a channel permitting protons to reenter a cell after being exported actively (ATPase) would be counterproductive to energy conservation. What happens when protons can cross more or less freely is demonstrated when protonophores such as carbonyl cyanide *m*-chlorophenylhydrazone (CCCP) are added.

The cytoplasmic pH drops quite rapidly and, after due time, approximates the external pH. It is conceivable, however, that proton channels exist in a well-regulated state but are kept closed most of the time. Their opening could be used for rapid proton fluxes, a process in which fast pH changes (cytosolic acidification) could be used as a signal. As energy requirements obviously make the existence of channels with high H^+ permeability in a plasma membrane unlikely, how is the flux equilibrium attained during constant H^+ extrusion by the H^+ pump? How do protons reenter the cell? It is suggested that H^+ cotransporters play an important role therein, in that they transport either organic substrates (sugars, amino acids) or inorganic ions, mainly anions that otherwise cannot cross the plasma membrane for energetic reasons (negative charge). That this is indeed the case can be demonstrated by exchanging Cl^- for gluconate, a largely membrane-impermeable anion. In a cell suspension in which a pH equilibrium has been attained, the exchange of Cl^- for gluconate immediately leads to further acidification of the medium because the proton entry along with the Cl^- ions is reduced.

An interesting feature is observed when the pH at the surface of roots is tested in the presence of different external pH values. It can be shown with pH-sensitive microelectrodes that the surface pH is clearly different from the bulk pH (see also Sec. 6.4). Although this difference can be more than one pH unit when the bulk pH is around 7, it becomes smaller with decreasing bulk pH and becomes zero around pH 5, the pH of the root cortex apoplast [10]. However, when the bulk pH is decreased below 5, the surface pH remains higher than the bulk pH. If proton channels were to permit passive H^+ influx, which should increase with dropping pH, such a result would not be possible. This indicates clearly that apoplastic pH and surface pH are well regulated to a value that permits transport to take place relatively independently of bulk conditions.

5 CONTROL OF MEMBRANE TRANSPORT (NOT H^+) BY EXTERNAL AND INTERNAL pH

5.1 Channels

Inward-rectifying K^+ channels are subject to control by external pH but in a manner that is fundamentally different from the influence cytosolic pH has, indicating that these channels possess two separate sites for sensing pH. Blatt [11] demonstrated that reducing the external pH from 7.4 to 5.5 enhanced the inwardly directed K^+ current five- to sevenfold and displaced its voltage dependence by roughly $+45$ mV. Although the tested pH range appears too wide for physiologically meaningful apoplastic conditions, titration of the current against external pH would suggest H^+ binding to a single site [11]. Increasing membrane voltage (inside negative) was accompanied by an increase in the apparent pK_a as if H^+ binding were favored by an electrical field driving H^+ to a site deep within the membrane. These characteristics pose a striking contrast to actions on K^+ and other channels in animal cells for which H^+ is often a potent blocker [12,13], but they make good physiological sense in an H^+-coupling membrane: the sensitivity to external pH may be seen as a measure to ensure that K^+ uptake cannot short-circuit other energetic demands on the H^+-ATPase and membrane voltage. It is interesting that parallel sensitivities to cytosolic pH and to external pH have been found for guard cell anion channels [14].

5.2 Weak Acids and Bases

Because uncharged molecules generally cross a membrane much more easily than ions, the pH gradient across any membrane has consequences for the passive distribution of weak acids and bases. Relevant in this context are, for instance, organic acids, CO_2, and NH_4^+. Organic acids with pK_a values around pH 5 will, of course, exist largely dissociated within the cytoplasm at pH 7.3 but will be protonated by roughly 50% in the acidic vacuole or the apoplast. Apart from the possibility that carriers might exist for organic acids such as malate at the tonoplast, by altering the pH of the acidic compartment (vacuole), the translocation of the uncharged form can be regulated to some extent by changing the concentration of the membrane-permeant molecule. Similar considerations hold for the apoplast and plasma membrane. Due to the transmembrane pH gradient, the loss of organic acids penetrating the membrane would be small but could become substantial through anion channels, as external alkalization following the activation of such channels indicates. This loss could, in principle, be reversed by uptake through H^+ cotransport, as shown for Cl^- [15].

Transport of NH_4^+/NH_3 (pK 9.25) is a special case of pH-dependent translocation with high physiological and nutritional relevance. As an uncharged weak base, NH_3 would penetrate any membrane rapidly, as demonstrated repeatedly [16,17]. As long as the external pH is alkaline, NH_3 penetrating the plasma membrane tends to alkalize the cytosol, regardless of whether NH_3 comes from an aqueous NH_4^+ solution [18] or through flushing it as gas. As to our knowledge the apoplast in root and shoot is acidic, flushing of NH_3 will increase the apoplastic pH but not the cytosolic pH. Actually, the predominating NH_4^+ at acidic pH will be imported electrophoretically via a specific uniporter but also by K^+ channels. NH_4^+, after entering the cytoplasm, will dissociate and liberate protons, thus acidifying the cytoplasm.

Uncouplers such as CCCP and 2,4-dinitrophenol (DNP) are a special case of weak acids. Unlike most weak acids, for which the membrane permeability of the protonated form is much higher than that of the anion, with uncouplers both the dissociated and the undissociated forms permeate the membrane relatively well. This has important consequences at the cellular level. With ordinary weak acids, the cytosolic acidification is a clear function of the acid pK_a and of the final transmembrane pH gradient. The process of cytosolic acidification is finished as soon as the transmembrane gradients and dissociation conditions match. With uncouplers, the anions are driven out of the cell by the membrane potential as H^+ ions are taken up and returned into the cell, a procedure that finally results in complete loss of the transmembrane pH gradient. Because ATP synthesis depends on an electrochemical H^+ gradient across the crista membranes, uncouplers prevent ATP synthesis, which is commonly known. Due to the negative membrane potential, anions are not taken up passively. Therefore, the concentration range in which an uncoupler can be used depends on the external pH and the pK_a of the respective uncoupler. It is commonly believed that CCCP is far more effective than DNP. It can be shown, however, that this observation is solely the consequence of the pK_a of DNP, which is about two pH units more acidic (about pH 4) than that of CCCP (about pH 6). Thus, at an external pH of 7, 10 μM CCCP strongly depolarizes the plasma membrane and acidifies the cytosol, whereas at the same concentration DNP has no measurable effect. As soon as the pH is lowered to 5, DNP works because the amount of protonated DNP has risen by a factor of about 100.

5.3 Membrane Permeability and H^+ Permeability

The selectivity of membrane permeability has often been ascribed to the presence of fixed charges that attract, and allow passage of, oppositely charged ions. Ionic permeability should vary with pH when the structural groups are weakly ionized. Thus, an increase of the H^+ concentration at the membrane's surface would lead to an increase in anionic and a decrease in cationic permeability.

The problem of defining H^+ permeability as a consequence of a channel-mediated H^+ flux has already been discussed. Like the term "regulation," the term "permeability" is not always understood identically everywhere. It can be used in a wider sense, with permeability meaning only how fast (the unit of permeability is that of a velocity) a particle (charged or uncharged) can cross a membrane. A narrower and more widely accepted definition would use the term permeability only for ions (usually inorganic ions). Applying the constant field condition, the Goldman equation would then yield permeability ratios (relative permeabilities). As pH has a profound influence on the diffusion potential of a membrane, including H^+ ions in the Goldman equation to calculate relative permeabilities toward other ions would be a logical consequence. Because of the usually low H^+ activities, one of the immediate outcomes of such a procedure would be the finding of enormously high ratios in favor of H^+. This, however, would make some sense only if it could be demonstrated that the pH effects are truly the result of H^+ moving through the membrane as other ions do.

6 TRANSMEMBRANE pH REGULATION

6.1 Transmembrane pH Changes

Although cytoplasmic pH regulation is dealt with in Chapter 3 of this book, it is necessary to mention some aspects thereof here because transport across a membrane involves the cytoplasmic as well as the apoplastic side (or internal sides in intracellular compartments). Because protons influence transporters biochemically, but at the same time are transportees, the knowledge of the pH on both sides of the membrane is of particular importance. There is considerable confusion about the origin or nature of pH changes on either side of a membrane during transport and regulatory processes. Consider the following situations:

1. The H^+ pump is stimulated by an acid load, e.g., by a weak acid that is added externally in an experimentally closed system. According to the existing external pH and the pK_a of the acid, the protonated acid will cross the membrane and acidify the cytosol [19–21]. Due to the increased free $[H^+]$, the pump will be stimulated, proton extrusion will increase, and the plasma membrane will be hyperpolarized. The external medium, however, will not acidify (in spite of the pump's stimulation) but will become more alkaline (depending on the buffer capacity) because more protons are imported through the acid translocation than are exported by the pump.

2. The H^+ pump is stimulated (not by an acid load but by fusicoccin): proton extrusion increases, external pH decreases; the cytosolic pH, however, does not necessarily alkalize [22–24] because the cytosolic pH is kept constant by a stable metabolic network. Basically, however, the development of the cytosolic pH depends critically on the external ion composition and on the ions countertransported to keep the charge balance [25]. Thus, cytosolic pH may alkalize, acidify, or not change, all of which has been demonstrated.

3. The H^+ pump is stimulated through cytosolic acidification caused by a metabolic process (not membrane transport): due to the pump stimulation, the membrane potential hyperpolarizes and the external medium is acidified.

4. The H^+ pump is inhibited (e.g., vanadate) or deactivated otherwise (e.g., by cyanide): the inhibition by vanadate will not change the cytosolic pH, and the external pH may alkalize. Cyanide acidifies the cytosolic pH by 0.5 to 0.6 units, not through stopping pump activity but for metabolic reasons (anoxic switch) [20,26–28]; the external pH alkalizes.

Although it can be predicted that with a stimulated H^+ pump the external pH will decrease (except in case 1 above), with a deactivated pump such a prediction is not so straightforward. It is clear, of course, that deactivation of the H^+ pump as a part of an equilibrium situation will temporarily destabilize this equilibrium. However, alkalization will occur only if the reentry of protons continues. Because pump deactivation is accompanied by a decrease in driving force, the reentry of protons, e.g., through cotransporters, will likewise be hampered. Therefore, if alkalization occurs, it will be short lived and may be physiologically irrelevant.

Another point is important in this context: even when active and passive fluxes across the plasma membrane are in equilibrium, there is no reason to assume that the same amount of protons that left the cells should reenter. Quite the opposite. The proton pump has two import functions: (1) to build up a transmembrane energy gradient and (2) to reduce the intracellular proton load. Thus, it can be safely assumed that the protons that reenter do this not on their own but bring into the cell metabolizable compounds (e.g., sugars), molecules or ions, that otherwise would not have been able to enter the cytoplasm (e.g., NO_3^-).

6.2 The Role of the Apoplastic pH

Apoplastic pH is more than just "external pH." As external pH, we usually consider the pH of the bulk solution, which, however, for regulation of membrane transport in most cases is of limited importance. Because of their cell walls, plant cells are able to create a specific microenvironment close to the plasma membrane, the composition of which determines the activity of the transporters and the amount of fixed charges. This environment is a physically protected space and an area where ion exchange takes place due to the negative charges of the polygalacturonic acid. These interactions between cations (K^+, Ca^{2+}) and protons in the apoplastic space play an important role in all transport [29]. On the one hand, this space makes the plasma membrane relatively insensitive to changes in the bulk solution, but on the other hand, transport of ions allows the possibility of changing local concentrations of one ion and, in turn, altering the concentration of the other. That means that regulation of a cation channel need not be done sterically but could be achieved through altering the pH, which will, by binding or releasing cations from their fixed charges, change the concentration of ions to be transported. In principle, however, the apoplastic space is *the* most relevant area for providing the transportees and the regulation of the forces to mediate their translocation. Figure 4 demonstrates this regulation; nitrate, an important N source, is taken up by the roots through H^+ symport that only temporarily changes the pH of the apoplast.

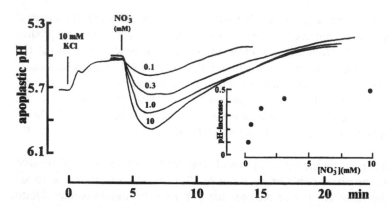

FIGURE 4 Apoplastic pH of a maize root cortex (fourth cell layer), measured with a pH-sensitive microelectrode, as a function of externally added NO_3^- at the indicated concentrations. Because of the ion-exchanger properties of the apoplast, the addition of cations causes acidification, so 10 mM KCl was added prior to the tests to adjust the apoplastic pH. Then NO_3^- was added in the presence of the KCl. The kinetics demonstrate concentration-dependent alkalizations, which are indicative of H^+ cotransported with the nitrate. The inset shows that the alkalization follows Michaelis-Menten kinetics.

6.3 The Problem of Local pH Domains and Unstirred Layers

Heterogeneity in the distribution of protons has to be considered at different levels: organs, organelles, and localized areas of membranes [30]. The pH domains spreading through a whole organ were demonstrated in *Elodea* leaves that (in light) display an alkaline upper surface and an acidic lower one [31]. Local pH domains are probably due to a special distribution of H^+ exchangers at the plasma membrane, as demonstrated in alkaline and acidic bands of characean cells [32]. At a lower level of integration, proton microdomains are intensively studied in thylakoid membranes with respect to their contribution to the pmf driving ATP synthesis.

Unstirred layers exist at every membrane as regions where the rate of laminar flow is restricted by interactions with a solid surface. In these layers, solutes move only by diffusion and represent physiological buffer regions between the membrane and the bulk solution. The thickness of the cell walls determines the dimension of the unstirred layer. Although there have been great advances in our technical abilities to determine the (pH) condition of such areas, there remains uncertainty about the actual pH at the binding site, which basically determines the transport activity. This holds true even more so for intracellular sites that are less accessible to investigations, although one intuitively would expect that, because of the short distances and cytosolic buffering, the actual pH differences should be less than the ones observed between the cell wall and the bulk solution. Nevertheless, it can be expected that pH domains could influence transport across membranes and could be involved in cell signaling, local enzyme activation, and in the local organization and movements of cells. To investigate relationships of bulk solution, unstirred layer, and apoplastic pH, roots have been a standard object in recent years. Figure 5 (see also Chapter 12) demonstrates that the pH on the root surface does not correspond to the bulk solution pH and is not constant along the root but yields a profile that apparently reflects the instantaneous physiological condition (e.g., growth or cell division) of the underlying root tissues. Thus, the surface pH may be considered as a buffer zone between the bulk solution (the rhizosphere) and the apoplast of the root cortex. The pH in the root apoplast may be measured with the techniques mentioned briefly in Sec. 6.4. However, the microelectrode approach is invasive here, as the electrodes have to be inserted through cells before a radial cell wall is hit. Nevertheless, continuous pH measurements within the root apoplast, carried out with simultaneous growth experiments, demonstrate the regulatory role of pH for growth. It has

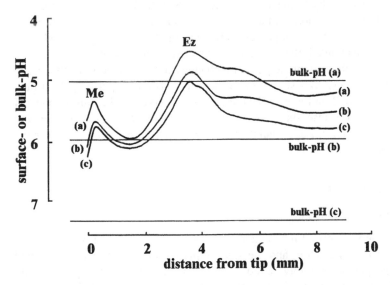

FIGURE 5 Comparison of the bulk pH and the surface pH at a corn root, measured with a pH-sensitive microelectrode. The curves demonstrate that the surface pH along a root differs for the different zones (Me, meristem; Ez, elongation zone) and may also be considerably different from the bulk solution pH, given as a solid line.

been shown that the pH in the elongation zone very much correlates with the elongation growth of that zone because the acidic peak coincides with the area of the highest growth intensity [33]. The question of the extent to which acid growth considerations hold could be raised (see Chapter 6).

6.4 The Leaf Apoplast

The pH of the leaf apoplast can be measured indirectly by extracting apoplastic fluid following infiltration or more directly by using fluorescent dyes [34,35]. A truly noninvasive way is to insert blunt pH-selective microelectrodes through open stomata and, after making electrical contact, monitor continuously the apoplastic pH or changes thereof due to the prevailing physiological conditions. In doing so, it has been demonstrated that the apoplastic pH of leaves is relatively low (4.8 in *Bromus erectus*, 5.0 in *Vicia faba*) and appears well regulated despite a buffer capacity that is about 10-fold lower than the cytosolic one, i.e., around 5 mM. As the apoplastic fluid of a leaf is confined to a rather thin film and thus small ion fluxes would have a massive influence on their concentration, the regulation of the ion activity and hence also that of pH must be rather strict. In fact, this can be demonstrated in two ways. First, fumigation with NH_3 showed that alkalization due to NH_4^+ formation had two components: a rapid one (seconds), mainly caused by primary dissolution of NH_3 in the apoplastic fluid and subsequent NH_4^+ formation, and a slower one (minutes), which involved NH_4^+ transport and very likely intracellular NH_4^+ consumption. Although the NH_3 fumigation was maintained, the pH increase leveled off after a few minutes [36]. Second, based on the membrane potential changes and the ion currents that cause them, light-dark transitions should have major effects on apoplastic pH. In fact, only relatively small and slow pH changes are observed, indicating strong pH regulation [36]. Because the plasma membrane transporters experience any changes in ion activity, such regulation of the apoplastic pH is essential for the leaf physiology.

7 pH SIGNALING

7.1 pH as a Messenger

Because the term "second messenger" is used for Ca^{2+}, inositol trisphosphate, diacylglycerol, cyclic AMP, and the like, implying a certain function in signal transduction, pH is preferably called a "cellular messenger." The pH not only carries information across membranes but also mediates information between molecules and transfers it between compartments and from cell to cell. In the green alga *Eremosphera viridis*, the acidification of the cytoplasm induces an "action-potential-like response" [37,38], corresponding to an increase in the conductivity of K^+ channels. This suggests that in plant cells, as in excitable animal cells, cytosolic pH modifies the properties of ion channels [12] and plays an important role in the signal transduction from photosystems in the chloroplasts to K^+ channels in the plasma membrane.

In plant hormone action, the pH may play a crucial role. Ever since the formulation of the acid growth theory [39,40], the role of pH as a messenger ion within the apoplast has been discussed [41]. Auxin acidifies the cytoplasm of maize coleoptiles in an oscillatory manner, accompanied by membrane potential and free $[Ca^{2+}]$ oscillating with similar frequency, from which a role of pH as a cellular messenger was suggested [26,42–44]. Because elongation growth of the grass coleoptile also proceeds in waves and depends on the activity of the proton pump, the pH changes (or their origin) may be the basis for such behavior. The idea that pH plays an important role in auxin signaling has been corroborated in experiments where the H^+ signal was disrupted by buffering of cytoplasmic pH [45]. No effect of auxin on the inwardly directed K^+ flux was observed under these conditions. Inversely, cytosolic acidification achieved by weak acid loading promoted elongation growth [23,42] and stimulated influx of K^+ and inactivated K^+ efflux [11,46,47].

In stomatal movements, a role of cytoplasmic pH in signaling was recognized after discovering abscisic acid (ABA) action was strictly associated with cytoplasmic alkalization of 0.2 to 0.4 units [44–46,48]. ABA-induced stomatal closure was blocked when the cytoplasmic pH was clamped at a constant low value. On the molecular level, this closure is initiated by the export of anions and of K^+, the latter of which is strongly activated by increased cytoplasmic pH, suggesting it as a major target of pH signaling [11,49]. ABA also changes the extracellular pH. Using pH-sensitive microelectrodes, which were inserted through open stomata, Felle et al. [50] demonstrated recently that ABA, fed through the transpiration stream, *transiently* alkalized the apoplast around the guard cells by 0.3 to 0.4 units. Interestingly, darkness alkalized the apoplast *constantly* by roughly 0.2 pH units. If channels are to be affected by pH changes, two points are evident: the first is that the ABA- and darkness-induced alkalizations differ in their kinetics. Because with ABA the pH change runs in reverse, it must be the initial pH changes that act as a trigger. Second, the pH changes in the apoplast, which accompany stomatal closure, are small, meaning that potentially affected channel proteins must recognize these changes.

7.2 pH as a Messenger in Defense and Symbiosis

When plant cells are attacked by fungi, bacteria, or viruses, they produce a variety of defense reactions, including the expression of defense-related genes and production of antimicrobial substances such as phytoalexins. For instance, oligosaccharides derived from the cell surface of pathogenic microorganisms act as signal molecules that induce various defense responses [51,52]. Early responses are changes in external as well as cytoplasmic pH. Within seconds of the encounter with the pathogen, the cells react with external alkalization and internal acidification [53,54], which, however, are seemingly not correlated. The current understanding is that cytosolic acidification activates defense genes that encode proteins that provide adequate responses to microbial attack. Here, the cytoplasmic pH change is transient and hence fulfills the requirements of a real messenger, with the H^+ concentration being elevated briefly to turn a molecular switch but then dropping again to permit regular metabolic and transport processes to proceed [53–55].

In symbiotic systems such as the legume-rhizobial partnership, the cytoplasmic pH response to Nod factors (lipochitooligosaccharides) is not an acidification but a persistent alkalization [56]. Accepting the notion that acidification activates defense genes, a persistent alkalization would block such action, exactly what is needed to prevent defense reactions and to initiate symbiotic interrelation.

So far, it remains open whether the extracellular pH changes (alkalization) in response to elicitors are a signal. Being part of the early response, which is initiated by Ca^{2+} influx and subsequent anion efflux, it could be that this rapid change in extracellular condition at the cell surface provides conditions that are unfavorable to potential attackers. An alkalization of 0.3 to 0.6 units or more [57] would clearly decrease the proton-motive force needed for optimal transport and possibly shift the pH away from the optima of cell wall–digesting enzymes. This would win some time while producing a second wave of defense measures.

7.3   pH and Ca^{2+}

In animal cells, changes in cytosolic pH and Ca^{2+} are often associated [58]. Correlative evidence shows that in some systems the cytosolic Ca^{2+} rises when the cytoplasm is acidified. As there are only few data available for plants [26,59,60], the origin of these correlations and the physiological importance for the organism involved are not yet clear. Apart from direct pH effects on Ca^{2+}-binding proteins, it could be that pH changes within the cytoplasm, or even more so locally in membrane pockets, could be responsible for pH-induced changes in $[Ca^{2+}]$ or any other ion. Non–channel-mediated Ca^{2+} transport across membranes appears regularly linked to protons [61]. Accumulation of Ca^{2+} in the vacuole seems a result of an nH^+/Ca^{2+} antiport, although the stoichiometry remains a matter of debate. From vesicle work, it has been suggested that the stoichiometry should be $1:1$. However, thermodynamic considerations based on measurements of vacuolar free $[Ca^{2+}]$ and pH would point to two or even three H^+ ions transported per one Ca^{2+} ion. Due to the high inwardly directed electrochemical $[Ca^{2+}]$ gradient across the plasma membrane, a Ca^{2+}-ATPase on its own in most instances would not be able to export Ca^{2+} sufficiently, which was the reason why it was suggested that protons should be countertransported with the Ca^{2+} ions [61].

8 CONCLUSIONS

As the effects of protons on the organism are manifold and may proceed simultaneously in a complicated network wherein many processes depend on pH, it is difficult to attribute a general function to the pH. The question is not whether pH is a regulator and a signal but when and under which conditions. When is a pH change a consequence of preceding transport or metabolic processes, and when is a pH change the precondition, or even the trigger, of following processes? As cellular reactions are usually fast and take place almost simultaneously, the answers to these questions may not always be easily accessible and are hence prone to misinterpretations. To prevent this, causalities have to be checked out first, as well as the origin of a pH change: is the observed effect due to H^+ transport across a membrane and, if so, which membrane? Is the effect due to a metabolic shift or simply the result of other ions or molecules entering the compartment in question? In considering all of these possibilities before the background of the actual physiological situation, causalities can be found without actually having to resolve pH-related processes on a defined time scale.

REFERENCES

1. CL Slayman, WS Long, CY-H Lu. The relationship between ATP and an electrogenic pump in the plasma membrane of *Neurospora crassa*. J Membr Biol 14:305–338, 1973.
2. HH Felle. A study of the current-voltage relationships of electrogenic active and passive membrane elements in *Riccia fluitans*. Biochim Biophys Acta 646:151–160, 1981.
3. JA Fernandez, MG Sanchez, HH Felle. Physiological evidence for a proton pump and sodium exclusion mechanisms at the plasma membrane of the marine angiosperm *Zostera marina* L. J Exp Bot 50:1763–1768, 1999.
4. CI Ullrich, AJ Novacky. Recent aspects of ion-induced pH changes. Curr Top Plant Biochem Physiol 11:231–248, 1992.
5. PA Stewart. How to Understand Acid-Base. A Quantitative Acid-Base Primer for Biology and Medicine. New York: Elsevier North Holland, 1981.
6. PA Stewart. Modern quantitative acid-base chemistry. Can J Physiol Pharmacol 61: 1444–1461, 1983.
7. D Gradmann, U-P Hansen, CL Slayman. Reaction-kinetic analysis of current-voltage relationships for electrogenic pumps in *Neurospora* and *Acetabularia*. In: A Kleinzeller, F Bronner, eds. Current Topics in Membranes and Transport. New York: Academic Press, 1982, pp 258–276.

8. D Gradmann, U-P Hansen, WS Long, CL Slayman, J Warncke. Current-voltage relationships for the plasma membrane and its principal electrogenic pump in *Neurospora crassa*: I. Steady state conditions. J Membr Biol 39:333–367, 1978.

9. HH Felle. Steady state current-voltage characteristics of amino acid transport in rhizoid cells of *Riccia fluitans*. Is the carrier negatively charged? Biochim Biophys Acta 772:307–312, 1984.

10. HH Felle. The apoplastic pH of the *Zea mays* root cortex as measured with pH-sensitive microelectrodes: aspects of regulation. J Exp Bot 49:987–995, 1998.

11. MR Blatt. K^+ channels of stomatal guard cells. Characteristics of the inward rectifier and its control by pH. J Gen Physiol 99:615–644, 1992.

12. W Moody. Effects of intracellular H^+ on the electrical properties of excitable cells. Annu Rev Neurosci 7:257–278, 1984.

13. J Tytgat, B Nilius, E Carmeliet. Modulation of the T-type cardiac Ca^{2+} channel by changes in proton concentration. J Physiol (Lond) 96:973–990, 1990.

14. B Schultz-Lessdorf, G Lohse, R Hedrich. GCAC1 recognizes the pH gradient across the plasma membrane: a pH-sensitive and ATP-dependent anion channel links guard cell membrane potential to acid and energy metabolism. Plant J 10:993–1004, 1996.

15. HH Felle. The H^+/Cl^- symporter in root-hair cells of *Sinapis alba*. An electrophysiological study using ion-selective microelectrodes. Plant Physiol 106:1131–1136, 1994.

16. NA Walker, FA Smith, MJ Beilby. Amine uniport at the plasmalemma of charophyte cells. II. Ratio of matter to charge transport and permeability of free base. J Membr Biol 49:283–296, 1979.

17. D Kleiner. The transport of NH_3 and NH_4^+ across biological membranes. Biochim Biophys Acta 639:41–52, 1981.

18. A Bertl, HH Felle, F-W Bentrup. Amine transport in *Riccia fluitans*. Cytoplasmic and vacuolar pH recorded by a pH-sensitive microelectrode. Plant Physiol 76:75–78, 1984.

19. A Roos, WF Boron. Intracellular pH. Physiol Rev 61:296–434, 1981.

20. D Sanders, CL Slayman. Control of intracellular pH. Predominant role of oxidative metabolism, not proton transport, in the eucaryotic micro-organism *Neurospora*. J Gen Physiol 80:377–402, 1982.

21. E Johannes, HH Felle. Implications for cytoplasmic pH, proton-motive force, and amino-acid transport across the plasmalemma of *Riccia fluitans*. Planta 172:53–59, 1987.

22. A Bertl, HH Felle. Cytoplasmic pH of root hair cells of *Sinapis alba* recorded by a pH-sensitive micro-electrode. Does fusicoccin stimulate the proton pump by cytoplasmic acidification? J Exp Bot 36:1142–1149, 1985.

23. B Brummer, A Bertl, I Potrykus, HH Felle, RW Parish. Evidence that fusicoccin and indole-3-acetic acid induce cytosolic acidification of *Zea mays* cells. Fed Eur Biochem Soc 189:109–114, 1985.

24. LD Talbott, PM Ray, JKM Roberts. Effect of indoleacetic acid and fusicoccin-stimulated proton extrusion on internal pH of pea internode cells. Plant Physiol 87:311–316, 1988.

25. CI Ullrich, J Guern. Ion fluxes and pH changes induced by trans-plasmalemma electron transfer and fusicoccin in *Lemna gibba* L. (strain G1). Planta 180:390–399, 1990.

26. HH Felle. Cytoplasmic free calcium in *Riccia fluitans* L. and *Zea mays* L.: interaction of Ca^{2+} and pH? Planta 176:248–255, 1988.

27. GG Fox, NR McCallan, RG Ratcliffe. Manipulating cytoplasmic pH under anoxia: a critical test of the role of pH in the switch from aerobic to anaerobic metabolism. Planta 195:323–330, 1995.

28. JKM Roberts, J Callis, D Wemmer, V Walbot, O Jardetzky. Mechanism of cytoplasmic pH regulation in hypoxic maize root tips and its role in survival under hypoxia. Proc Natl Acad Sci USA 81:3379–3383, 1984.

29. C Grignon, H Sentenac. pH and ionic conditions in the apoplast. Annu Rev Plant Physiol Plant Mol Biol 42:103–128, 1991.

30. J Guern, HH Felle, Y Mathieu, A Kurkdjian. Regulation of intracellular pH in plant cells. Int Rev Cytol 127:111–173, 1991.

31. HBA Prins, JFH Snel, PE Zanstra, RJ Helder. The mechanism of bicarbonate assimilation by the polar leaves of *Potamogeton* and *Elodea*. CO_2 concentrations at the leaf surface. Plant Cell Environ 5:207–214, 1982.

32. WJ Lucas. Photosynthetic assimilation of exogenous HCO_3^- by aquatic plants. Annu Rev Plant Physiol 34:71–104, 1983.

33. WS Peters, HH Felle. The correlation of surface pH and elongation growth in maize roots. Plant Physiol 121:905–912, 1999.

34. B Hoffmann, R Plänker, K Mengel. Measurements of the pH in the apoplast of sunflower leaves by means of fluorescence. Physiol Plant 84:146–153, 1992.

35. KH Mühling, C Plieth, U-P Hansen, B Sattelmacher. Apoplastic pH of intact leaves of *Vicia faba* as influenced by light. J Exp Bot 46:377–382, 1995.

36. S Hanstein, HH Felle. The influence of atmospheric NH_3 on the apoplastic pH of green leaves: a non-invasive approach with pH-sensitive microelectrodes. New Phytol 143:333–338, 1999.

37. K Köhler, W Steigner, J Kolbowski, U-P Hansen, W Simonis, W Urbach. Potassium channels in *Eremosphera viridis*. II. Current- and voltage-clamp experiments. Planta 167:66–75, 1986.

38. W Steigner, K Köhler, W Simonis, W Urbach. Transient cytoplasmic pH changes in correlation with opening of potassium channels in *Eremosphera*. J Exp Bot 39:23–26, 1988.

39. A Hager, H Menzel, A Krauss. Versuche und Hypothese zur Primärwirkung des Auxins beim Streckungswachstum. Planta 100:47–75, 1971.

40. RE Cleland. Auxin-induced hydrogen ion excretion from *Avena* coleoptiles. Proc Natl Acad Sci USA 70:3092–3093, 1973.

41. RE Cleland. Kinetics of hormone-induced H^+ excretion. Plant Physiol 58:210–213, 1976.

42. HH Felle, B Brummer, A Bertl, RW Parish. Indole-3-acetic acid and fusicoccin cause cytosolic acidification of corn coleoptile cells. Proc Natl Acad Sci USA 83:8992–8995, 1986.

43. HH Felle. pH as a second messenger in plants. In: WF Boss, DJ Morré, eds. Second Messengers in Plant Growth and Development. New York: Alan R Liss, 1989, pp 145–166.

44. CA Gehring, DA Williams, SH Cody, RW Parish. Phototropism and geotropism in

maize coleoptiles are spatially correlated with increases in cytosolic free calcium. Nature 345:528–530, 1990.

45. MR Blatt, G Thiel. K$^+$ channels of stomatal guard cells: bimodal control of the K$^+$ inward-rectifier evoked by auxin. Plant J 5:55–68, 1994.

46. MR Blatt, F Armstrong. K$^+$ channels of stomatal guard cells: abscisic acid–evoked control of the outward rectifier mediated by cytoplasmic pH. Planta 191:330–341, 1993.

47. A Grabov, MR Blatt. Parallel control of the inward-rectifier K$^+$ channel by cytosolic free Ca^{2+} and pH in *Vicia* guard cells. Planta 201:84–95, 1997.

48. HR Irving, CA Gehring, RW Parish. Changes in cytosolic pH and calcium of guard cells precede stomatal movements. Proc Natl Acad Sci USA 89:1790–1794, 1992.

49. H Miedema, SM Assmann. A membrane-delimited effect of internal pH on the K$^+$ outward rectifier of *Vicia faba* guard cells. J Membr Biol 154:227–237, 1996.

50. HH Felle, S Hanstein, R Steinmeyer, R Hedrich. Dynamics of ionic activities in the apoplast of the sub-stomatal cavity of intact *Vicia faba* leaves during stomatal closure evoked by ABA and darkness. Plant J 24:297–304, 2000.

51. CA Ryan, EE Farmer. Oligosaccharide signals in plants: a current assessment. Annu Rev Plant Physiol Plant Mol Biol 42:651–674, 1991.

52. T Boller. Chemoreception of microbial signals in plant cells. Annu Rev Plant Physiol Plant Mol Biol 46:189–214, 1995.

53. K Kuchitsu, Y Yazaki, K Sakano, N Shibuya. Transient cytoplasmic pH change and ion fluxes through the plasma membrane in suspension-cultured rice cells triggered by N-acetylchitooligosaccharide elicitor. Plant Cell Physiol 38:1012–1018, 1997.

54. Y Mathieu, A Kurkdjian, H Xia, J Guern, A Koller, MD Spiro, M O'Neill, P Albersheim, A Darvill. Membrane responses induced by oligogalacturonides in suspension-cultured tobacco cells. Plant J 1:333–343, 1991.

55. E Minnami, K Kuchitsu, D-Y He, H Kouchi, N Midoh, Y Ohtsuki, N Shibuya. Two novel genes rapidly and transiently activated in suspension-cultured rice cells by treatment with N-acetylchitoheptaose, a biotic elicitor for phytoalexin production. Plant Cell Physiol 37:563–567, 1996.

56. HH Felle, E Kondorosi, A Kondorosi, M Schultze. Rapid alkalinization in alfalfa root hairs in response to rhizobial lipochitooligosaccharide signals. Plant J 10:295–301, 1996.

57. G Felix, M Regenass, T Boller. Specific perception of subnanomolar concentrations of chitin fragments by tomato cells: induction of extracellular alkalinization, changes in protein phosphorylation, and establishment of a refractory state. Plant J 4:307–316, 1993.

58. WB Busa, R Nucitelli. Metabolic regulation via intracellular pH. Am J Physiol 246:409–438, 1984.

59. CA Gehring, HR Irving, RW Parish. Effects of auxin and abscisic acid on cytosolic calcium and pH in plant cells. Proc Natl Acad Sci USA 87:9645–9649, 1990.

60. A Grabov, MR Blatt. Parallel control of the inward-rectifier K$^+$ channel by cytosolic free Ca^{2+} and pH in *Vicia faba* guard cells. Planta 201:84–95, 1997.

61. AJ Miller, D Sanders. The energetics of cytosolic calcium homeostasis in fungal cells. Plant Physiol Biochem 27:551–556, 1987.

6

The Role of the Apoplastic pH in Cell Wall Extension and Cell Enlargement

Robert E. Cleland
University of Washington, Seattle, Washington

1 THE ROLE OF WALL EXTENSIBILITY IN PLANT CELL ENLARGEMENT

One of the features that distinguishes plants from animals is that almost all plant cells undergo considerable enlargement after their production by mitosis; in many cases, cells can enlarge to over 100-fold greater volume than the original cell. The final size and shape of a plant are determined by the extent of this cell enlargement process. For example, if the cells in the stem of a redwood tree were restricted to the length of meristematic cells, the height of a tree with the same number of cells as a 300-foot redwood would be less than 2.5 feet!

Plant cells are like inflated balloons, with the pressurized cell contents counteracted by a resistant cell wall. The cell can take up water and enlarge only if the wall is stretched [1]. An extension of a cell can be brought about, in theory, by either an increase in the internal pressure (the turgor pressure) or an increase in the extensibility of the wall [1]. In practice, the many studies on cell enlargement have shown that when there is an increase or decrease in the rate at which cells enlarge, it is due primarily to a change in wall extensibility and only occasionally to a change in turgor pressure [1,2]. So what controls the amount of wall extensibility?

The primary wall of a plant is a complex mixture of polysaccharides [3,4]. In general, it consists of cellulose microfibrils, oriented perpendicular to the direction of cell expansion, held together by xyloglucan and other hemicellulose cross-links. The extensibility of the wall increases when load-bearing bonds in the wall are cleaved (wall loosening) [1,2]. For example, the xyloglucan and hemicellulose cross-links between the microfibrils can be broken, allowing the cellulose microfibrils to move apart from each other and thus the wall to extend. In the living cell, there is also continual synthesis of new cell wall, primarily next to the plasma membrane, which produces new wall cross-links [4]. Cell wall extension consists of a series of extension events. Each starts with the cleavage of a load-bearing cross-link. Extension of a small bit of wall then occurs until stopped by another cross-link taking the stress and restricting further wall extension [1]. Wall extensibility, then, is a measure of the rate at which these load-bearing cross-links are cleaved. This, in turn, depends on the activity of the wall-loosening proteins (WLPs) in the wall. Wall-loosening proteins, like all proteins, have an optimal pH for their activity and usually are inactive or barely active at pH values far removed from the optimal pH. As a result, the apoplastic pH (pHa) is an important factor in determining the activity of the wall-loosening proteins and thus the rate of cell extension.

2 APOPLASTIC pH AND WALL EXTENSIBILITY; THE COMPETING THEORIES

There are at least three ways in which the WLPs and pHa could interact to cause a change in the rate of cell expansion. The amount of the WLPs might remain constant, while a change in pHa changes the activity of the WLPs and thus the rate of wall loosening. In 1937, Ruge [5] first suggested that the extensibility of the cell wall is controlled by the apoplastic pH and that cell wall extension is induced by an acidic pHa. This idea was largely ignored until 1970 and 1971, when Rayle and Cleland [6] and Hager et al. [7] independently proposed that the activity of the wall-loosening proteins was primarily a response to acidification of the apoplast. They proposed that cells induce cell wall loosening by excreting protons into the wall, making the wall pH acidic. This "acid-growth theory" states [8] that the pH optimum for the wall-loosening proteins is in the acidic range (<5). In nonexpanding cells, the pHa is above the pH range needed for active wall-loosening activity. Cell expansion is initiated by the release of protons into the wall. This results in the pHa being lowered into the range where the wall-loosening proteins are active. For this mechanism to work, the pH optimum of the WLPs must be acidic, and the pHa must be able to decrease into this acidic range in order to induce cell elongation. This theory has been controversial, with evidence presented both for it [8] and against it [9]. It is also possible for a change in pHa to be the factor that regulates the rate of cell expansion, even when the pH optimum of the WLP is in a more neutral range. In this case, growth could be induced by either an increase or a decrease in pHa as long as the pHa started below or above the optimal pH value of the WLP [10].

The second possibility is that changes in the amounts of WLPs, rather than changes in pHa, are responsible for changes in the rate of cell elongation [2,11]. As described in Sec. 5, there is an increasing amount of data showing that such changes in WLPs occur during plant development. This theory assumes that the pHa will be in the range where the WLPs are active and that the pHa does not change in response to factors that cause an increase or decrease in the rate of cell expansion.

A third possibility is that wall loosening is due to intercalation of new cell wall material into the preexisting cell wall [12]. The new wall material and the enzymes needed to intercalate that material into the wall would be delivered by Golgi vesicles. The WLPs would then cleave existing bonds in order to intercalate the new material into the walls. The rate of wall loosening, in this case, would be primarily determined by the rate of supply of these new cell wall components.

In order to distinguish between these competing possibilities, certain information is needed. First, the identity of the wall-loosening proteins must be determined. For each WLP, its pH profile for activity must be known, and the amount of the WLP and the timing of changes that occur when the rate of cell expansion changes have to be determined. Second, the apoplastic pH must be determined, and changes in pHa must be correlated with changes in the rate of cell expansion.

3 WALL-LOOSENING PROTEINS

3.1 How to Detect WLPs

The identity of the proteins responsible for wall loosening has been a matter of considerable speculation over the years. Only in the past few years has the picture become clearer [11], primarily due to the development of an in vitro wall extension assay. When intact plant cell walls, freed from their cell contents, are subjected to an applied stress at neutral pH values, they undergo a viscoelastic extension [13]. The extension is proportional to log time, with the result that for all practical purposes, extension has effectively ceased after 15 to 30 min. In 1970, Rayle et al. [14] showed that as long as the wall proteins are not denatured, a change in the solution pH from 7 to 3 causes the walls to resume extension (Fig. 1). As long as the stress on the walls is maintained, the walls continue to extend at a constant rate for at least several hours. Cosgrove [15] then adapted

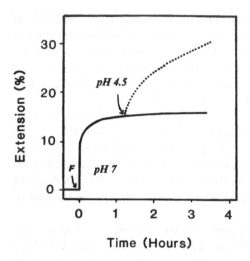

FIGURE 1 The in vitro extension of cell walls in response to tension and to acid. Frozen-thawed *Avena* coleoptile walls were subjected to 20 g force in a pH 7 buffer (F). The walls underwent viscoelastic extension, which was nearly completed after 1 h. When the solution was replaced by a new pH 3.0 solution, further wall extension occurred, which continued for up to 24 h.

this technique to produce an in vitro assay for WLFs. In this assay, tissues are frozen and thawed, and the native wall proteins are either extracted or denatured. The isolated walls are subjected to an applied stress at an appropriate pH until the viscoelastic extension is essentially complete. Under these conditions, no change in pH will cause further extension. But if a putative WLP is added at a pH at which it is active, extension of the walls will resume. The rate of this additional extension is a measure of the activity of the WLP.

It is important to recognize the limitations of this in vitro assay for WLPs. If wall loosening requires the input of energy, such as ATP, or some cell wall precursor, such as xyloglucan components, it cannot occur in this assay. Likewise, if a WLP is used up in the wall-loosening process and must be continuously supplied, this in vitro wall-loosening assay may fail to detect it.

A second way to identify a protein as being a WLP is to use transgenic plants in which the protein is either overexpressed or has its expression suppressed [e.g., 16]. If the protein is a WLF, overexpression should lead to enhanced endogenous growth, whereas suppression should result in dwarfism.

A final method is to make use of specific antibodies to the putative WLP. If the antibodies effectively inhibit cell elongation, it is reasonable to conclude that the protein in question might be a WLP [17].

3.2 The Putative Wall-Loosening Proteins

Many proteins have been suggested as possible wall-loosening proteins, but significant evidence exists for only three families of proteins: the expansins, xyloglucan endotransglycosylase (XET), and the polysaccharide endoglycosidases, including cellulase and β-(1,3)-D-glucanases [11]. Each of these might cleave bonds that are potentially load-bearing bonds in the cell wall, and each has been shown to be present in elongating tissues. In order to assess the role of each of these WLPs, two pieces of information are needed: their ability to cause wall extension in an in vitro extension assay and the pH optimum and pH values at which the proteins are effectively inactive.

3.2.1 Expansins

The expansins are a large family of closely related small ($\sim$ 30 kDa) proteins that all have the ability to induce acid-mediated in vitro extension of isolated stem and coleoptile cell walls [11]. They are separated into two groups, the α-expansins and the β-expansins. Most plants contain multiple α-expansins, which have a high degree of sequence relatedness; for example, 24 α-expansin genes are known from *Arabidopsis*. The β-expansins are more common in the grasses and include the grass pollen allergens.

Expansins have no known enzymatic activity [11]. They are, however, the only proteins known to cause long-term acid-induced wall extension in the in vitro assay [18]. They appear to be breaking hydrogen bonds, probably between the xyloglucans and cellulose [19,20]. Although the pH profiles for the different expansins can be slightly different, in general the pH optimum is 3.5–4.5. At pH 5.5 there is almost no activity, and at pH 3.0 the activity is slightly reduced [21]. Expansins are concentrated in the regions of cell elongation [22], and addition of expansins to excised *Arabidopsis* hypocotyls [11], tomato shoot meristems [23], or tobacco suspension culture cells [24] caused significant cell expansion. The data strongly support the role of expansins as WLPs in plants.

3.2.2 Xyloglucan Endotransglycosylase

XETs are a family of relatively small ($\sim$33 kDa) proteins that catalyze the cleavage of xyloglucan chains, with the addition of one of the resulting half-chains to the nonreducing end of a second xyloglucan chain or to water [25,26]. This can lead to a lengthening or shortening of the xyloglucan chain, depending on the acceptor. XET could be cleaving load-bearing xyloglucan chains between cellulose microfibrils and reforming them in a nonstressed configuration [27]. Support for the idea that XET might be a WLP comes from the concentration of XET in regions of cell elongation [e.g., 28] and colocalization of XET with its substrate [29]. However, when XET was tested in the in vitro extension assay, it failed to cause any wall loosening [30], although it did render the walls more responsive to elongation promoted by expansins. It is uncertain, therefore, whether XET is truly a WLP.

Unlike expansins, which are active under acidic conditions, XET has a pH optimum of 5.5–5.8 and has less than 25% activity at pH 4.0 or 7.5 [25,26].

3.2.3 Polysaccharide Endoglycosylases

Cell walls are composed of a variety of complex polysaccharides containing a number of different sugars and often highly branched [3,4]. A wide variety of endoglycosylases are known [31] that have the capacity to cleave these polysaccharide chains. Although many of these enzymes have been isolated from fungi and bacteria, higher plants also possess quite an array. Two, in particular, have attracted attention as possible WLPs; the β-(1,4)-D-glucanase (cellulase) that attacks cellulose and xyloglucan and a β-D-glucanase that cleaves the β-(1-3,1-4)-D-glucan polymers found primarily in grass cell walls [32].

Cellulases are capable of hydrolyzing both cellulose and the glucan backbone of xyloglucans [33]. In most cases, the pH optimum is in the 5.5–6.0 range. It would seem logical that this enzyme would have the ability to loosen cell walls, as it attacks two of the main components of the cross-linked walls. However, addition of cellulases to peeled *Avena* coleoptiles failed to cause any elongation in either the presence or absence of auxin [34]. When tested on isolated cell walls in the in vitro extension assay [18], cellulases caused no extension for hours, after which the tissues broke. The cellulase treatment did enhance the ability of added expansin to cause wall loosening. Although these results are not conclusive, they cast doubt on the ability of cellulases to act as WLPs, although they may contribute to the wall-loosening process mediated by a different protein. On the other hand, the inhibition of cell elongation caused by a mutation in the *KORRIGAN* (*KOR*) gene [35], which codes for a β-1,4-D-glucanase, points to some important role for this enzyme in cell elongation.

Several endoglucanases that have the ability to degrade the β-(1-3,1-4)-D-glucans of grass cell walls have been studied in some detail [36]. The pH optimum for one endoglucanase from corn seedlings is 5.0, with over 50% of optimal activity at pH 4.0 and 6.0 [37]. The ability of these enzymes to cause wall loosening in the in vitro assay has not yet been tested, but the fact that antibodies specific to at least the mixed link endoglucanase can block auxin-induced growth [17,38] suggests that these enzymes can act as WLPs.

4 APOPLASTIC pH AND HOW TO MEASURE IT ACCURATELY

The activity of the putative wall-loosening proteins, described in the previous section, depends on the apoplastic pH. If the pHa is not at the optimum for the WLP, a change in pHa can promote or inhibit cell elongation, depending on whether the pH change brings the pHa closer to or farther from the optimal pHa for that protein. The optimal pH values for the putative WLPs vary considerably; pH 4.5 and below for expansin, 5.5 for XET, and 5.0 to 6.5 for the various endoglycosylases. It is essential, therefore, to know what the pHa actually is.

A number of different methods have been employed to determine the pHa (see Chapter 11 for a more extensive analyses of these methods). Most, at best, give only an approximation of this parameter. One method is to centrifuge growing tissues to remove the "apoplastic solution" [39]. Centrifugation will remove liquid from the vascular tissue and intercellular spaces but cannot remove it from the walls, especially from the outer cell layers that control elongation growth [40]. In addition, this liquid will be altered by water moving out of the cells in response to the centrifugal force. There is no reason to believe that the pH of the centrifugate will accurately reflect the actual pHa of growing tissues.

A second, commonly employed method (the solution pH method) is to let a tissue come to equilibrium with the surrounding solution and assume that the solution pH will be that of the apoplast [41]. For aerial portions of a plant (stem, coleoptile, or leaf), this requires removal of the cuticle, as that protective layer is an effective barrier to the movement of protons [42]. This is usually accomplished by abrasion of the epidermis or by simply stripping off the epidermal cell layers. Unfortunately, both methods cause damage to the tissue and at least partly compromise the growth process [43]. In addition, this would measure only the pHa of the outermost cell layer; there is no reason to believe that there is a uniform pHa within a tissue. Finally, both theoretical analysis [44] and actual measurements [45] have shown that the pH of the bulk solution can be greatly different from that of the apoplast.

A variant on the solution pH method is the "null method" [46]. Here a growing tissue, with the cuticle barrier removed, is incubated in buffers of varying pH and the growth rate is determined. In theory, the solution that causes no increase or decrease in growth rate should have the same pH as the pHa. But not only do all the problems with the solution pH method apply here as well, but to what control should the growth rate be compared? It cannot be the growth of sections with an intact cuticle because removal of the cuticle itself changes the growth rate [43]. Moreover, it cannot be the growth rate of sections incubated in distilled water because of ionic and osmotic differences. In practice, the lack of a satisfactory baseline growth rate with which to make comparisons makes it impossible to use this technique to get an actual value for the pHa. Both the solution method and the null method are best used to determine the direction in which pHa changes when the growth rate is changed by external agents such as hormones.

A better technique is to use pH microelectrodes, placed as close to the walls as possible. Felle [45] used pH microelectrodes to measure the pH at varying positions in the solution, at the surface of a corn root, and in the apoplast of

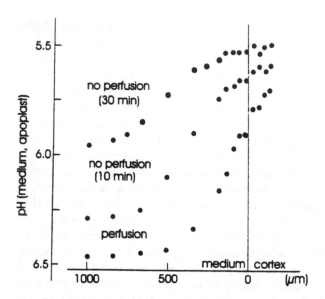

FIGURE 2 Variation in pH in the medium surrounding corn roots and in the apoplastic pH. The pH-sensitive microelectrodes were used to measure the pH in the pH 6.5 medium at varying distances from the root surface and in the apoplast of the cortex. Measurements were made with perfusion of the medium and without perfusion for 10 and 30 min. A pH value taken more than 500 μm from the root surface provides no measure of the pHa of the root tissue. (From Ref. 45.)

several cortical cell layers (Fig. 2). In a perfused pH 6.5 solution, the pH was around 6.4 until within 500 μm of the root, then dropped to about 5.9 at the root surface and was even lower in the cortical apoplast. The actual pH that was recorded, however, still depended upon the pH of the bathing solution [47]; for example, the measured pH at the same place on the root surface was 4.25 in a pH 4.2 solution, 5.45 in a pH 6.5 solution, and 6.05 in a pH 7.8 solution.

The most direct method for measuring pHa of a growing tissue is to infiltrate into the walls a pH-sensitive fluorescent dye and then measure the fluorescence at two wavelengths by confocal laser scanning microscopy. Ideally, the probe should be ratioable, so that the pH will be independent of the concentration of the fluorochrome, and attached to a complex that prevents its entry into cells [48]. To date, this procedure has received only limited use. Hoffmann et al. [49] used a fluorescein isothiocyanate (FITC)-dextran to show that the pHa of mature sunflower leaves is about 6.2. Gilroy et al. [50], using Oregon Green attached to the cellulose binding domain, recorded a pHa of 4.7 for the cortical cells in the elongation zone of *Arabidopsis* roots.

Even this technique fails to address one significant problem; this is the fact that the pH within the apoplast of any cell is nonuniform. In the region close to the fixed negative charges of the pectic substances, called the Donnan free space (DFS), the pH is lower than that of the rest of the apoplast (the water free space, WFS) due to trapping of protons in the DFS in order to maintain electroneutrality. Sentenac and Grignon [51] pointed out that the pH of the DFS can be as much as 3 pH units lower than that of the adjacent WFS! It is not clear whether the pH-sensitive fluorescent dyes measure the pH of the WFS, the DFS, or a combination of the two. However, it is not known whether the wall-loosening proteins lie in the DFS or the WFS and thus whether they are reacting to the pH of the DFS or the WFS. In conclusion, it appears that we do not have the technical capacity, at present, to provide an accurate value of the apoplastic pH, although approximations and trends can be determined.

5 THE RELATIONSHIP BETWEEN pHa AND GROWTH

Cell walls contain a variety of putative wall-loosening proteins, each of which responds to the pHa to which it is exposed. When the growth rate of any tissue changes, it could be due to a change in pHa, a change in amount of WLPs, or both. A variety of conditions cause changes in the growth rates of stems, coleoptiles, roots, and leaves. So, what is the role of pHa in controlling the growth rate?

Because the identity of the WLPs is still uncertain and we cannot yet determine pHa accurately and convincingly, it might seem that it would be impossible to answer this question. But, in fact, a great deal of information is available. This information comes from the correlations that exist between the growth rate, the apparent pHa, and the amount of each putative WLP. The correlations have been obtained for different areas along a growing organ or at different times after the growth rate is altered by hormones or light. This information will be summarized for coleoptiles, stems, roots, and dicot leaves.

5.1 Coleoptiles

Addition of an auxin to slowly growing coleoptile sections results in the initiation of rapid cell elongation after a lag of only about 10 min [52]. This initiation of rapid growth is paralleled by a decrease in the apparent pHa [53,54]. For example, in *Avena* coleoptiles auxin caused the apparent pHa to decrease from 5.6 to 4.5 with a lag of about 10 min [55]. Peters et al. [55] measured the change in growth rate and surface pH for corn coleoptile sections and demonstrated a remarkable correlation over time (Fig. 3). The fungal toxin fusicoccin (FC) induces even more rapid growth than auxin in *Avena* coleoptiles and also more vigorous proton efflux [56]. Addition of acidic buffers to coleoptile sections causes rapid elongation for several hours [6,7]. There have been considerable disagreements about the exact pHa in both the absence and presence of auxin [8,9], but there can be

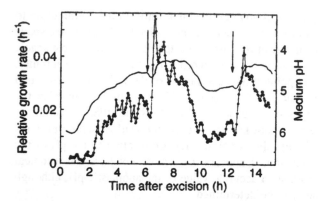

FIGURE 3 The correlation between the medium pH surrounding abraded corn coleoptile segments and the growth rate measured on the same samples. The pH (solid line) was measured in a 1 mM KCl, 0.1 mM NaCl, 0.1 mM CaCl$_2$ solution. Indoleacetic acid (IAA) (1 μM) was added at times indicated by the arrows. Note that the maize coleoptile undergoes a spontaneous increase in growth rate, in the absence of auxin, which begins after about 1 h and is mirrored by a comparable increase in acidity in the medium. (From Ref. 55.)

little doubt that the initiation of rapid cell elongation occurs only when the pHa is < 5.5 and that auxins induce a rapid acidification of the apoplast.

These data support the idea that the initiation of rapid cell elongation in coleoptiles is controlled by the WLP expansin. Expansins are present in coleoptiles [57] and require a pHa < 5.5 for activity. The effect of auxin on expansin synthesis is not known for coleoptiles, but it is unlikely that the initiation of rapid growth within 10 min of the addition of auxin can be due to any real increase in amount of expansins in the walls. On the other hand, the increased growth rate of rice coleoptiles in response to submergence or ethylene is accompanied by a comparable increase in expression of two expansin genes [58].

Cells in coleoptiles continue to elongate in vivo for several days in response to auxins. It is possible that other WLPs may play a more major role in this continued expansion. Cleland [10] showed, for example, that the apparent pH optimum for auxin-induced growth of coleoptiles shifted from 4.5 to 5.5 after several hours. Because coleoptile walls are rich in the β-(1-3,1-4)-glucans, the endoglucanase that hydrolyzes this polymer might be important. Whereas the isolated endoglucanase from corn seedlings has a sharp pH optimum of 5.0 [37], the autolysis of isolated maize coleoptile walls, presumably mediated by this enzyme, has a much broader pH profile, with an optimum at 5.5–6.0 [59]. The enzyme activity of the endoglucanase increases during the period of in vivo coleoptile growth but peaks after growth ceases [36]. However, the fact that specific antibodies to this enzyme block auxin-induced growth of maize coleoptile cells indicates that this endoglucanase participates in cell elongation [17,38].

In conclusion, the initiation of rapid elongation in coleoptiles by auxin seems to be due to a decrease in pHa into the acidic range caused by auxin [8]. The prolonged growth of coleoptiles may require expansins and other wall-loosening proteins, such as the endoglucanase, and may occur at a less acidic pHa. More accurate measurements of the pHa, using infiltrated apoplastic pH-sensitive fluorescent dyes, are badly needed.

5.2 Stems

Considering the large number of available mutants that interfere with stem elongation [60], it is surprising that we know considerably less about the control of cell elongation in stems as compared with coleoptiles. A variety of external agents affect the growth rate; growth is promoted by auxins, gibberellins, and brassino-steroids and inhibited by cytokinins, ethylene, abscisic acid, and both red and blue light [61]. However, the effect of most of these agents on pHa is unknown. For example, elongation of cucumber hypocotyls is strongly inhibited within 1 min after exposure to blue light [62]. Wall extensibility and the activity of the plasma membrane H^+-ATPase show parallel declines, strongly suggesting that the growth inhibition is due to an increase of pHa to a value at which expansins are no longer active [63]. But this must remain a hypothesis until actual pHa measurements are made.

Auxin causes a rapid and extensive decrease in the apparent pHa of a variety of different hypocotyl and epicotyl tissues [64]. In general, the recorded values are higher than those for coleoptiles (a pH of 6.2 in the absence of auxin and 5.0 in its presence). Both growth and proton excretion are induced by FC, and acidic buffers promote the growth of dicot stem sections as well [65]. In gravitropic curvature of sunflower stems, the faster growing lower side is more acidic than the slower growing upper side [66]. Taken as a whole, the data indicate that agents that cause the apparent pHa to decrease to < 5.5 promote cell elongation, whereas agents that cause an increase in pHa inhibit growth. This is consistent with expansins being the main wall-loosening proteins in stems. However, changes in pHa are not always responsible for changes in stem growth rate. The growth of lettuce hypocotyls is strongly promoted by gibberellin (GA), but GA apparently does not cause any change in pHa [67].

All growing stem tissues tested so far contain expansins in the walls and extend in response to acidic solutions in the in vitro extension test [11,68]. Expansins are concentrated in the growing tissues of cucumber stems [11,68] and rice internodes [22]. In *Arabidopsis* hypocotyls, expansins appear to be present in suboptimal quantities, as addition of exogenous expansin to hypocotyl sections promoted their growth [69]. Auxin causes a slow increase in expansin activity in cucumber stems [70]. In the internodes of deepwater rice, there is a close correlation between the growth rate and the content of expansins, whether measured along the internode, in response to submergence [71], or in response to GA [22]. These data suggest that an increase in expansin activity contributes to the growth of stems. However, the cessation of elongation must be due to some other cause, as expansin activity remained high in cucumber hypocotyls even after the cells lost their ability to elongate [72].

XET activity is also concentrated in regions of cell elongation [26]. In rice, two XET genes are up-regulated in the elongation zone by GA and brassinosteroids in parallel with the promotion of growth [73].

In conclusion, it appears that rapid changes in the rate of cell elongation in stems may be primarily controlled by changes in pHa and long-term changes can be regulated by changes in activity of the WLPs.

5.3 Roots

Measurement of the pH along the surface of roots has shown that significant differences exist, which can be correlated with cellular activities [45,47,74]. In general, the pH is more acidic in the central elongation zone (CEZ), where the bulk of cell elongation occurs. In the meristem and distal elongation zone (DEZ), where cell elongation commences, the pH is higher. When cell elongation was inhibited by auxin or KCN, the pH of the elongation zone increased significantly [47]. These measurements have been confirmed [75] with the use of pH-sensitive fluorescent dyes infiltrated into the walls of *Arabidopsis* roots. In addition, it was shown that, when gravistimulated [50], there was a rapid acidification of the pHa on the upper side and an alkalinization on the lower side of the *Arabidopsis* roots.

There are also changes in WLP activities that correlate with growth. Isolated corn root walls can undergo acid-mediated in vitro extension, and the content of extractable expansin is greater in the rapidly growing apical regions than in the slower growing basal regions [76]. Similar differences are found for the XET activity of these roots [77].

In conclusion, changes in pHa, acting through expansins, appear to control the rapid growth responses of the CEZ of roots, while changes in expansin and/or XET activities modulate these responses. In the DEZ, in contrast, there is no direct evidence to indicate that expansins play a major role, and the higher pHa that exists there [50] and the insensitivity of the growth there to added auxin [78] suggest that XET might be the WLP there and that the control of growth in this region might be due to changes in amount of this enzyme.

5.4 Leaves

Cell enlargement in dicot leaves is promoted by light, GA, and cytokinins (CKs), whereas in grass leaves the promotion is primarily by GAs [79]. The growth rate of maize leaves shows a strong correlation with the pH at the leaf surface [80].

Injection into the leaf of strong buffers at pH 5.5 or above inhibited leaf growth, whereas buffers at pH 4.5 were without effect on normal growth [81]. With both bean leaves [82] and pea leaves [83], a concurrent acidification by the epidermal cells and the onset of cell enlargement occur upon illumination. Using the fluorescent dye FITC-dextran infiltrated into *Vicia faba* leaves, Mühling and Läuchli [84] showed that the pHa drops rapidly from 5.9 to 5.2 upon illumination. Light appears to cause the pHa of dicot leaf cells to decrease into the range where expansins are active. On the other hand, neither GA nor CK caused any acidification by bean leaf cells, even though both promoted leaf expansion [85]. These agents do not act via an acidification of the apoplast.

Putative expansin activity has been shown to exist in bean leaf walls [82]. In the aquatic angiosperm *Rumex palustris*, flooding promotes the growth of the leaf lamina, probably via ethylene and GA [86]. This growth promotion is accompanied by a large increase in expansins. In barley leaves, the promotion of growth by GA is accompanied by an increase in transcripts for three XET genes [87]. It would be interesting to know whether the promotion of bean leaf expansion by GA and CK involved an increase in either expansin or XET. In conclusion, the limited data that now exist suggest that light stimulates leaf growth by a change in pHa, whereas other agents such as GA may increase the amount of the WLPs instead.

6 CONCLUSIONS

1. Cell enlargement is primarily controlled by the rate at which the cross-links in the wall are cleaved by wall-loosening proteins (WLPs).
2. The rate of wall loosening is controlled by the amount of the WLPs and the apoplastic pH (pHa) in which they reside.
3. Strong evidence supports the expansins as WLPs. The pHa optimum for expansins is in the 3.5–4.5 range. Weaker evidence suggests that both xyloglucan endoglycosylase (XET), with a pH optimum of 5.5–5.8, and an endo-β-D-glucanase, with a pH optimum of 5.0–6.0, can act as WLPs.
4. A number of methods exist that are suitable for measuring the direction of change in the pHa, but most provide unreliable measurements of the pHa itself. The use of pH-sensitive fluorescent dyes and confocal microscopy provides the most reliable pHa values.
5. Auxin initiates rapid elongation in coleoptiles by acidifying the apoplast. Expansins are certainly WLPs in coleoptiles. and there is good evidence that the endoglucanases also play an important role.
6. Rapid changes in elongation rate in stems are primarily due to changes in pHa. Changes in WLPs over time, especially expansins and XET, may regulate the long-term growth rates.
7. In roots, the pHa is most acidic in the region of maximal cell elonga-

tion. Although rapid changes in growth rate, such as occur in gravitropism, may be due to changes in pHa, the longer term change in elongation rate that occur during normal root development may be due more to changes in WLPs.

8. In leaves, light-induced leaf expansion occurs in response to an acidification of the apoplast, but expansion induced by other agents such as GA may be due primarily to changes in WLPs.

9. There is a great need for more widespread and more accurate measurements of pHa in order to assess the role of pH in the control of cell enlargement.

ABBREVIATIONS

CEZ   central elongation zone of a root
DEZ   distal elongation zone of a root
DFS   Donnan free space
pHa   apoplastic pH
WFS   water free space
WLP   wall-loosening protein
XET   xyloglucan endotransglycosylase

REFERENCES

1. RE Cleland. Cell wall extension. Annu Rev Plant Physiol 22:197–222, 1971.
2. DJ Cosgrove. Wall extensibility: its nature, measurement, and relationship to plant cell growth. New Phytol 124:1–23, 1993.
3. NC Carpita, DM Gibeaut. Structural models of primary cell walls in flowering plants: consistency of molecular structures with the physical properties of the walls during growth. Plant J 3:1–30, 1993.
4. CT Brett, KW Waldron. Physiology and Biochemistry of Plant Cell Walls. 2nd ed. London: Chapman & Hall, 1996.
5. U Ruge. Untersuchen über den Einflux des Heteroauxins auf das Streckungwachstum des Hypokotyls von *Helianthus annuus*. Z Bot 31:1–56, 1937.
6. DL Rayle, RE Cleland. Enhancement of wall loosening and elongation by acid solutions. Plant Physiol 46:250–253, 1970.
7. A Hager, H Menzel, A Krauss. Versuche und Hypothese zur Primärwirkung des Auxins beim Streckungswachstum. Planta 100:47–75, 1971.
8. DL Rayle, RE Cleland. The acid growth theory of auxin-induced cell elongation is alive and well. Plant Physiol 99:1271–1274, 1992.
9. U Kutschera. The current status of the acid-growth hypothesis. New Phytol 126:549–569, 1994.
10. RE Cleland. Auxin-induced growth of *Avena* coleoptiles involves two mechanisms with different pH optima. Plant Physiol 99:1556–1561, 1992.

11. DJ Cosgrove. Enzymes and other agents that enhance cell wall extensibility. Annu Rev Plant Physiol Plant Mol Biol 50:391–417, 1999.
12. H Söding. Über die Wachstumsmechanik der Haferkoleoptile. Jahrb Wiss Bot 79: 231–255, 1934.
13. RE Cleland. The mechanical behaviour of isolated *Avena* coleoptile walls subjected to constant stress. Plant Physiol 47:805–811, 1971.
14. DL Rayle, PM Haughton, RE Cleland. An in vitro system that simulates plant cell extension growth. Proc Natl Acad Sci USA 67:1814–1817, 1970.
15. DJ Cosgrove. Characterization of long-term extension of isolated cell walls from cucumber hypocotyls. Planta 177:121–130, 1989.
16. K Herbers, HJ Flint, U Sonnewald. Apoplastic expression of the xylanase and β-(1-3,1-4) glucanase domains of the *xynD* gene from *Ruminococcus flavefaciens* leads to functional polypeptides in transgenic tobacco plants. Mol Breeding 2:81–87, 1996.
17. S Mutaftschiev, R Prat, M Peirron, G Devilliers, R Goldberg. Relationship between cell-wall β-1,3-endoglucanase activity and auxin-induced elongation in mung bean hypocotyl segments. Protoplasma 199:49–56, 1997.
18. DJ Cosgrove, DM Durachko. Autolysis and extension of isolated walls from growing cucumber hypocotyls. J Exp Bot 45:1711–1719, 1994.
19. S McQueen-Mason, DJ Cosgrove. Disruption of hydrogen bonding between wall polymers by proteins that induce plant wall extension. Proc Natl Acad Sci USA 91: 6574–6578, 1994.
20. SEC Whitney, MJ Gidley, SJ McQueen Mason. Probing expansin action using cellulose/hemicellulose conjugates. Plant J 22:327–332, 2000.
21. S McQueen-Mason, DM Durachko, DJ Cosgrove. Two endogenous proteins that induce cell wall extension in plants. Plant Cell 4:1425–1433, 1992.
22. H-T Cho, H Kende. Expression of expansin genes is correlated with growth in deep-water rice. Plant Cell 9:1661–1671, 1997.
23. AJ Fleming, S McQueen-Mason, T Mandel, C Kuhlemeier. Induction of leaf primordia by the cell wall protein expansin. Science 276:1415–1418, 1997.
24. BM Link, DJ Cosgrove. Acid growth response and α-expansins in suspension cultures of *Nicotiana tabacum* L., cv. BY2. Plant Physiol 118:907–916, 1998.
25. SC Fry, RC Smith, KF Renwick, DJ Martin, SK Hodge, KJ Matthews. Xyloglucan endotransglycosylase, a new wall-loosening enzyme activity from plants. Biochem J 282: 821–828, 1992.
26. K Nishitani. The role of endoxyloglucan transferase in the organization of plant cell walls. Int Rev Cytol 173:157–206, 1997.
27. JB Passioura, SC Fry. Turgor and cell expansion: beyond the Lockhart equation. Aust J Plant Physiol 19:565–576, 1992.
28. C Cataló, JKC Rose, A. Bennett. Auxin regulated and spatial localization of an endo-1-4-β-D-glucanase and a xyloglucan endoglycosylase in expanding tomato hypocotyls. Plant J 12:417–426, 1997.
29. K Vissenberg, IM Martinez-Vilchez, J-P Verbelen, JG Miller, SC Fry. In vivo colocalization of xyloglucan endotransglycosylase activity and its donor substrate in the elongation zone of *Arabidopsis* roots. Plant Cell 12:1229–1237, 2000.
30. S McQueen-Mason, SC Fry, DM Durachko, DJ Cosgrove. The relationship between

xyloglucan endotransglycosylase and in vitro cell wall extension. Planta 190:327–331, 1993.

31. DJ Huber, DJ Nevins. Partial purification of endo- and exo-β-D-glucanase enzymes from Zea mays L. seedlings and their involvement in cell wall autohydrolysis. Planta 151:206–214, 1981.

32. DJ Huber, DJ Nevins. Preparation and properties of a β-D-glucanase for the specific hydrolysis of β-D-glucans. Plant Physiol 60:300–304, 1977.

33. DA Brummell, CC Lashbrook, AB Bennett. Plant endo-1,4-β-D-glucanases. Structure, properties, and physiological functions. ACS Symp Ser 566:100–129, 1994.

34. AW Ruesink. Polysaccharidases and the control of cell wall elongation. Planta 89:95–107, 1969.

35. F Nichol, I His, A Jauneau, S Vernhettes, H Canut, H Höfte. A plasma membrane–bound putative endo-1,4-β-glucanase is required for normal wall assembly and cell elongation in Arabidopsis. EMBO J 17:5563–5576, 1998.

36. M Inouhe, DJ Nevins. Changes in the activities and polypeptide levels of exo- and endoglucanases in cell walls during developmental growth of Zea mays coleoptiles. Plant Cell Physiol 39:762–768, 1991.

37. R Hatfield, DJ Nevins. Purification and properties of an endoglucanase isolated from the cell walls of Zea mays seedlings. Carbohydr Res 148:265–278, 1986.

38. M Inouhe, DJ Nevins. Inhibition of auxin-induced cell elongation of maize coleoptiles by antibodies specific for cell wall glucanases. Plant Physiol 96:426–431, 1991.

39. ME Terry, RL Jones. Effect of salt on auxin-induced acidification and growth by pea internode sections. Plant Physiol 68:59–64, 1981.

40. U Kutschera, R Bergfeld, P Schopfer. Cooperation of epidermis and inner tissues in auxin-mediated growth of maize coleoptiles. Planta 170:168–180, 1987.

41. RE Cleland. Auxin-induced hydrogen ion excretion from Avena coleoptiles. Proc Natl Acad Sci USA 70:3092–3093, 1973.

42. SA Dreyer, V Seymour, RE Cleland. Low proton conductance of plant cuticles and its relevance to the acid-growth theory. J Exp Bot 34:676–680, 1983.

43. C Branca, D Ricci, M Bassi. Epidermis integrity and epicotyl growth in Azuki bean. J Plant Growth Regul 7:95–109, 1988.

44. C Grignon, H Sentenac. pH and ionic conditions in the apoplast. Annu Rev Plant Physiol Plant Mol Biol 42:103–128, 1991.

45. H Felle. The apoplastic pH of the Zea mays root cortex as measured with pH-sensitive microelectrodes: aspects of regulation. J Exp Bot 49:987–995, 1998.

46. MJ Vesper. Use of a pH-response curve for growth to predict apparent wall pH in elongating segments of maize coleoptiles and sunflower epicotyls. Planta 166:96–104, 1985.

47. WS Peters, H Felle. The correlation of profiles of surface pH and elongation growth in maize roots. Plant Physiol 121:905–912, 1999.

48. S Gilroy. Fluorescence microscopy of living plant cells. Annu Rev Plant Physiol Plant Mol Biol 48:165–190, 1997.

49. B Hoffman, R Plänke, K Mengel. Measurement of pH in the apoplast of sunflower leaves by means of fluorescence. Physiol Plant 84:146–153, 1992.

50. S Gilroy, JM Fasano, R Hirsch, P Minnich, SJ Swanson. pH signaling in the gravitropic response of Arabidopsis roots. Gravit Space Biol Bull 14:47, 2000.

51. H Sentenac, C Grignon. A model for predicting ionic equilibrium concentrations in cell walls. Plant Physiol 68:415–419, 1981.

52. PM Ray, AW Ruesink. Kinetic experiments on the nature of the growth mechanism in oat coleoptile cells. Dev Biol 4:377–397, 1962.

53. RE Cleland. Kinetics of hormone-induced H⁺ excretion. Plant Physiol 58:210–213, 1976.

54. M Jacobs, PM Ray. Rapid auxin-induced decrease in free space pH and its relationship to auxin-induced growth in maize and pea. Plant Physiol 58:203–209, 1976.

55. WS Peters, H Lüthen, M Böttger, H Felle. The temporal correlation of changes in apoplast pH and growth rate in maize coleoptile segments. Aust J Plant Physiol 25: 21–25, 1998.

56. RE Cleland. Fusicoccin-induced growth and hydrogen ion excretion in *Avena* coleoptiles: relation to auxin responses. Planta 128:201–206, 1976.

57. Z-C Li, DM Durachko, DJ Cosgrove. An oat coleoptile wall protein that induces wall extension in vitro and that is antigenically related to a similar protein from cucumber hypocotyls. Planta 191:349–356, 1993.

58. J Huang, T Takano, S Akita. Expression of α-expanson genes in young seedlings of rice (*Oryza sativa* L). Planta 211:4657–473, 2000.

59. M Inouhe, DJ Nevins. Auxin-enhanced glucan autohydrolysis in maize coleoptile cell walls. Plant Physiol 96:285–290, 1991.

60. JB Reid, SH Howell. Hormone mutants and plant development. In: PJ Davies, ed. Plant Hormones. Physiology, Biochemistry and Molecular Biology. 2nd ed. Dordrecht: Kluwer Academic Press, 1995, pp 448–485.

61. A von Arnim, X-W Deng. Light control of seedling development. Annu Rev Plant Physiol Plant Mol Biol 47:215–243, 1996.

62. DJ Cosgrove, PB Green. Rapid suppression of growth by blue light. Biophysical mechanism of action. Plant Physiol 68:1447–1453, 1981.

63. EP Spalding, DJ Cosgrove. Mechanism of blue-light-induced plasma-membrane depolarization in etiolated cucumber hypocotyls. Planta 188:199–205, 1992.

64. J Mentze, B Raymond, JD Cohen, DL Rayle. Auxin-induced H⁺ secretion in *Helianthus* and its implications. Plant Physiol 60:509–512, 1977.

65. DL Rayle, RE Cleland. Evidence that auxin-induced growth of soybean hypocotyls involves proton excretion. Plant Physiol 66:433–437, 1980.

66. TJ Mulkey, KM Kuzmanoff, ML Evans. Correlations between proton-efflux patterns and growth patterns during geotropism and phototropism in maize and sunflower. Planta 152:239–241, 1981.

67. DA Stuart, RL Jones. The role of acidification in GA- and fusicoccin-induced elongation growth of lettuce hypocotyl sections. Planta 142:135–145, 1978.

68. SJ McQueen-Mason, F Rochange. Expansins in plant growth and development: an update on an emerging topic. Plant Biol 1:19–25, 1999.

69. RC Moore, D Flecker, DJ Cosgrove. Expansin action on cells with tip growth and diffuse growth. J Cell Biochem Suppl 21A:457, 1995.

70. MW Shieh, J Shi, DJ Cosgrove. Developmental, hormonal and light regulation of the transcript for the cell wall loosening protein expansin. Plant Physiol 114:S85, 1997.

71. HT Cho, H Kende. Expansions and internodal growth of deepwater rice. Plant Physiol 113:1145–1151, 1997.
72. SJ McQueen-Mason. Expansions and cell wall extension. J Exp Bot 46:1639–1650, 1995.
73. S Uozu, M Tanaka-Ueguchi, H Kitano, K Hattori, M Matsuoka. Characterization of XET-related genes in rice. Plant Physiol 122:853–860, 2000.
74. M Kollmeier, H Felle, WJ Horst. Genotypical difference in aluminum resistance of maize are expressed in the distal part of the transition zone. Is reduced basipetal auxin flow involved in inhibition of root elongation by aluminum? Plant Physiol 122:945–956, 2000.
75. EB Blancaflor, SJ Swanson, JM Fasano, P Dowd, T-H Kao, S Gilroy. Altered apoplastic and cytoplasmic pH during gravitropsim in the *Arabidopsis* root cap and elongation zone. Am Soc Plant Physiol Meeting Abstracts, No 510, p 112, 2000.
76. Y Wu, RE Sharp, DM Durachko, DJ Cosgrove. Growth maintenance of the maize primary root at low water potentials involves increases in cell-wall extension properties, expansin activity, and wall susceptibility to expansins. Plant Physiol 111:765–772, 1996.
77. Y Wu, WG Spollen, HE Sharp, PR Hetherington, SC Fry. Root growth maintenance at low water potentials. Increase in activity of xyloglucan endotransglycosylase and its possible regulation by abscisic acid. Plant Physiol 106:607–615, 1994.
78. H Ishikawa, ML Evans. The role of the distal elongation zone in the response of maize roots to auxin and gravity. Plant Physiol 102:1203–1210, 1993.
79. E Van Volkenburgh. Leaf expansion—an integrating plant behaviour. Plant Cell Environ 22:1463–1473, 1999.
80. E Van Volkenburgh, JS Boyer. Inhibitory effects of water deficit on maize leaf elongation. Plant Physiol 77:190–194, 1985.
81. L Bogoslavsky, PM Neumann. Rapid regulation by acid pH of cell wall adjustment and leaf growth in maize plants responding to reversal of water stress. Plant Physiol 118:701–709, 1998.
82. E Van Volkenburgh, RE Cleland. Proton excretion and cell expansion in bean leaves. Planta 148:273–278, 1980.
83. R Stahlberg, E Van Volkenburgh. Light effect on membrane potential, apoplastic pH and cell expansion in leaves of *Pisum sativum* L, var. Argenteum. Role of plasma-membrane H^+-ATPase and photosynthesis. Planta 208:188–195, 1999.
84. KH Mühling, A Läuchli. Light-induced pH and K^+ changes in the apoplast of intact leaves. Planta 212:9–15, 2000.
85. TG Brock, RE Cleland. Role of acid efflux during growth promotion of primary leaves of *Phaseolus vulgaris* L. by hormones and light. Planta 177:476–482, 1989.
86. WH Vriezen, B De Graaf, C Mariani, LACJ Voesenk. Submergence induces expansin gene expression in flooding tolerant *Rumex palustris* and not in flooding-intolerant *Rumex acetosa*. Planta 210:956–963, 1999.
87. PHD Schønmann, RC Lang, V Lóng, PR Matthews, PM Chandler. Expression of XET-related genes and its relation to elongation in leaves of barley (*Hordeum vulgare* L). Plant Cell Environ 20:1439–1450, 1997.

7

Mechanisms and Physiological Roles of Proton Movements in Plant Thylakoid Membranes

W. S. Chow
Australian National University, Canberra, Australia

Alexander B. Hope
Flinders University, Adelaide, South Australia, Australia

1   INTRODUCTION

1.1   The Arrangement of Electron Transfer Components in the Thylakoid Membrane

Photosynthetic electron transport is the light-driven transfer of electrons from water molecules, through two photosystems (PSs) and other redox components, to nicotinamide adenine dinucleotide phosphate ($NADP^+$), forming NADPH. The protons released during the oxidation of water and plastoquinol drive the synthesis of adenosine triphosphate or ATP (photophosphorylation). The free energy in NADPH and ATP is finally stored in carbohydrates.

   Electron flow to $NADP^+$ and proton-driven ATP synthesis occur in or

---

Dedicated to the memory of Sir Rutherford (Bob) Robertson.

across the inner chloroplast membranes, called thylakoids; these are flattened sacs with an internal and an external aqueous phase—the lumen and stroma, respectively. Figure 1 shows the general arrangement of protein complexes in thylakoid membranes of the chloroplasts of higher plants and algae. There are four membrane-spanning protein complexes, three of which are involved in electron transport, namely PS II, cytochrome bf, and PS I complexes. The cytochrome bf complex functions between the two photosystems, accepting electrons from PS II via plastoquinol (PQH_2) and passing electrons on to PS I via plastocyanin (PC). The ATP synthase catalyzes ATP formation.

In PS II, as excitons are trapped in P680, thought to be a special pair of chlorophyll molecules in the reaction center, electron transfer begins with an electron going to a bound quinone acceptor, Q_A, leaving P680 oxidized. That

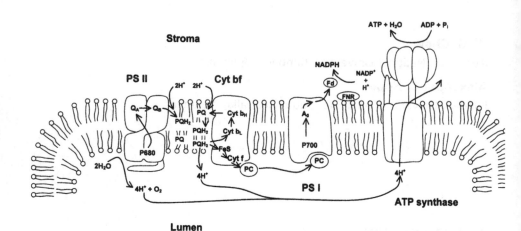

FIGURE 1 Schematic diagram of the electron transport chain in thylakoid membranes and the associated deposition of protons in the internal aqueous space (lumen) of thylakoids. The electrons, originating from split water molecules, are transported through PS II, the cyt bf complex, PS I, and other carriers before being finally accepted by $NADP^+$ to form NADPH. The protons are deposited in the lumen partly as the result of water oxidation and partly due to oxidation of plastoquinol (PQH_2). At the same time, protons are taken up/consumed in the aqueous phase (stroma) external to the thylakoids. The protons that are accumulated in the lumen drive ATP synthesis and serve certain protective functions (see text). For simplicity, the full subunit structures of the complexes, e.g., the light-harvesting subunits of the photosystems, are not shown. Also for simplicity, the (Q_o or p) site on the cyt bf complex where PQH_2 is oxidized (located closer to the lumen) and the (Q_r or n) site where PQ is reduced (located near the stroma) are not labeled. (Adapted from Ref. 75.)

hole is filled by an electron from the Mn cluster in the oxygen-evolving complex (OEC) via a tyrosine Tyr z (see Fig. 2). It takes four such sequences and the energy of four photons to produce one oxygen molecule from two water molecules:

$$2H_2O \rightarrow 4e^- + O_2 + 4H^+$$

We note in passing that starting from a dark-adapted state, it takes only three excitations to generate oxygen, because the OEC and Mn cluster do not relax to the completely reduced state. These states are termed S states [1]. In the dark,

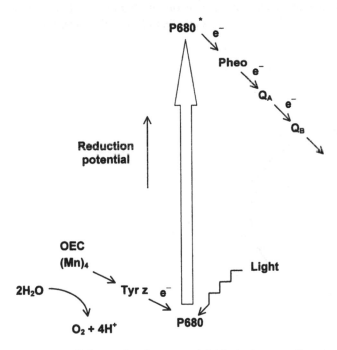

FIGURE 2 Schematic diagram showing electron flow in PS II along a chain of components arranged qualitatively on a reduction potential scale. P680* denotes the first singlet excited state of P680, induced upon absorption of a photon (hollow arrow). The electron stripped from P680* travels "downhill" through pheophytin (Pheo), the first bound quinone acceptor (Q_A), the second bound quinone acceptor (Q_B), and then to the general plastoquinone pool. On the "donor side," the hole (P680+) is filled by an electron from a tyrosine (Try z), which in turn regains its electron from the oxygen-evolving complex (OEC) containing four Mn ions. The Mn cluster extracts four electrons in a stepwise manner from two water molecules, releasing one O_2 molecule.

the OEC and Mn cluster are in the S_1 state instead of the completely reduced S_0 state. Each successful excitation advances the "clock" by one S state and thereby releases one electron to P680$^+$ (Fig. 3). When the clock reaches S_4 (after three flashes given to a dark-adapted sample), an O_2 molecule is released, while S_4 advances spontaneously to S_0. From S_0 it takes four excitations to again yield the maximal amount of oxygen, this period-of-four oscillation continuing with decreasing amplitude due to a gradual loss of synchronization of the population of S states.

The reactions beginning with oxidation of P680 and resulting in the oxidation of water are said to occur on the donor side of PS II. It should be pointed out that all of the oxygen, released as a by-product, originates from water, not from CO_2. That water oxidation takes place so readily in illuminated chloroplasts and at physiological temperatures is a remarkable feat of evolution over the past 3.5 billion years.

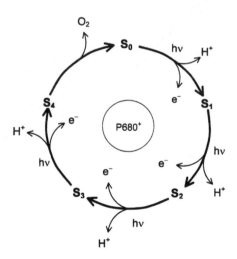

Figure 3 Cycling of the S states in the OEC and Mn cluster on consecutive excitations of P680. In the dark, the OEC and Mn cluster are in the S_1 state. Each successful excitation advances the "clock" by one S state, releasing one electron to P680$^+$ (via a tyrosine, not shown) and one proton (as observed in isolated cores of PS II or in thylakoids isolated from plants grown in intermittent light [3]). The transition from to S_3 to S_4 to S_0 requires one excitation of P680 only. The pattern of proton release is more complex in suspensions of normal thylakoids, being dependent on external pH and not necessarily integral in nature.

2 PROTON MOVEMENTS ASSOCIATED WITH THE LIGHT REACTIONS

2.1 Protons Are Deposited in the Lumen as a Result of Water Splitting

Leaving aside the details of the precise reactions in the OEC, it is seen that protons are seemingly a waste product, left in the lumenal spaces of the thylakoids as the S states advance, lowering the pH there. Over the years there has been considerable variation in opinion about the way proton deposition (first directly observed by Ausländer and Junge [2] using neutral red as a pH indicator) is associated with the removal of electrons from water molecules. Figure 3 shows a pattern of proton deposition associated with S-state transitions, typical of isolated cores of PS II and of thylakoids isolated from plants grown in intermittent light. In suspension of normal thylakoids, however, the pattern is considerably more complex; the extent of proton release during the cycle depends on the pH of the suspension and may result in a nonintegral pattern [3]. This complexity may be due to proton transfers to the pH-sensitive, measuring dye molecules as an electrostatic response of amino acids in the vicinity of the reaction center [4].

2.2 Protons Are Taken up from the Stroma as Quinone Is Reduced

On the acceptor side of PS II, slower (less than a millisecond) electron transfers occur from the first quinone acceptor Q_A to the second quinone acceptor Q_B, the latter being bound to a site in the D1 subunit in the PS II reaction center (Fig. 1). When Q_B has received two electrons, it takes up two protons from the adjacent stroma. Overall, the reaction is

$$PQ + 2e^- + 2H^+ \rightarrow PQH_2$$

It may be seen that the light reactions in PS II lead to both a separation of charge (an electric potential difference) and a pH difference across thylakoid membranes, which together go toward forming the proton motive force (pmf), a central parameter of Mitchell's chemiosmotic hypothesis [5,6]. As is well known, the pmf drives ATP synthesis or photophosphorylation. Proton movements in other reactions also contribute to the pmf as described subsequently.

2.3 Proton Movements Associated with Cytochrome *bf* Activity

Protons deposited in the lumen when water molecules are split and proton uptake from the stroma when Q_B is reduced represent one mechanism of proton transfer across the thylakoid membrane. Plastoquinol oxidation and plastoquinone reduction in the cytochrome bf complex also lead to proton transfer into the lumen. The latter mechanism will now be described.

2.3.1 Protons Are Deposited in the Lumen as Plastoquinol Is Oxidized at Q_o Sites at Cytochrome bf Complexes

The plastoquinol is released from the Q_B site, whose binding is specific for the quinone form, and is mobile in the membrane phase. Random diffusion in a space crowded with membrane-spanning complexes (or ''supercomplexes'' [7]) soon (in approximately 1–2 ms) brings the PQH_2 near a cytochrome bf complex, which has sites that bind PQH_2 specifically (Q_o or p sites). These sites are located near the lumenal side of the complex. If conditions are right, the PQH_2 is oxidized— electrons are donated. As the protons leave PQH_2 (probably two at a time, i.e., after the second of the two electrons), being water seeking and highly membrane insoluble, they are deposited in the lumenal space, again tending to lower the pH there (Fig. 1).

2.3.2 Protons Are Taken up from the Stroma as Plastoquinone Is Reduced at Q_r Sites at Cytochrome bf Complexes

The oxidized PQ generated at the Q_o sites becomes unbound and the quinone ring of the PQ may move to be bound at a Q_r (n) site near the stromal side of the complex; such sites are PQ specific, like Q_B sites (certain inhibitors bind to both). An electron that has previously moved within the bf complex to cyto- chrome b_H (for high potential) is available for transfer to PQ to form a radical anion species $PQ^{\cdot-}$. However, electron paramagnetic resonance (epr) experiments do not detect such radicals, so either such a transfer does not occur (because of an unfavorable difference in reduction potential) or such a form of PQ is short lived. Possibly a proton is taken up from the stroma to form PQH, which would be epr silent. Alternatively, because bf complexes are dimeric, a second electron may be available to PQ at Q_r sites, upon which two protons are absorbed from the stroma to regenerate PQH_2. In any case, when a second PQH_2 from PS II is oxidized, a second electron becomes available in the b-cytochrome pair in a cyt b subunit to reduce PQ fully at Q_r. The complete process is a Q cycle (Fig. 4), and the outcome is that for two PQH_2 oxidized, four protons are deposited in the lumen and two taken up from the stroma, while one PQH_2 is regenerated by the cycle. Furthermore, also counting the protons from water oxidation, the overall ratio of protons deposited to electrons passing through from PS II to PS I is three, which has implications for the balance between the reducing power (NADPH) and ATP available to fix CO_2.

    In the oxidation of plastoquinol, the loss of the two electrons and accompa- nying protons is a concerted reaction [8]. If the same applies for the reduction

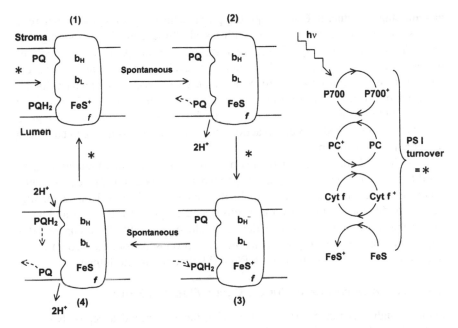

FIGURE 4 The Q cycle of electron flow in the cyt *bf* complex. In (1) and (2), PQH$_2$ is oxidized at the Q$_o$ site near the lumen-facing region of the cyt *bf* complex: two H$^+$ ions are deposited in the lumen, while one electron reduces the iron-sulfur to FeS and the second electron reduces the high-potential cyt *b* to b_H^- via the low-potential cytochrome $b(b_L)$. In (3) and (4), a second PQH$_2$ is oxidized, depositing two more protons in the lumen, while one electron reduces FeS$^+$ and the second electron together with that on b_H^- reduces PQ at the Q$_r$ site, near the stroma side of the cyt *bf* complex. FeS$^+$ is regenerated from FeS by PS I turnover, as shown on the right and by the * sign on the left. PS II turnover (one PQH$_2$ becomes available after each second turnover of a PS II) may be needed to prime the cycle if all *bf* cytochromes are oxidized. In the steady state, each electron reducing P700$^+$ (via cyt *f* and plastocyanin PC) leads to the deposition of two H$^+$ in the lumen.

of PQ at Q$_r$ sites, then two electrons must be transferred simultaneously, upon which two protons are taken up to complete the reaction.

2.3.3 A Possible Proton Pump in Cytochrome *bf* Complexes, in Addition to Q-Cycle Proton Translocation

The proton movements so far described are "redox linked" in the sense that they occur as a result of an electron transfer event; the electron and proton are at some stage together in PQH_2. A proton pump is the term for a process whereby a vectorial ion movement is coupled with a scalar biochemical reaction (e.g., the H^+-ATPase), and the coupled processes are separate in space. When electrons traverse the cytochrome *bf* complexes during Q cycling, charge separation occurs additional to that caused in the reaction centers (referred to particularly for PS II earlier but occurring in PS I as well). This charge movement is detectable in experiments measuring the "electrochromic shift" [9], an absorbance change at 515–520 nm unrelated to redox reactions that also cause absorbance changes in the green part of the spectrum.

Joliot and Joliot [10] have shown with *Chlamydomonas* mutants lacking PS II that the electrochromic signal associated with cytochrome *bf* electron transfers is too large to be accounted for in the conventional Q-cycle scheme, suggesting additional charge movement, envisaged as proton release in a channel facing the lumen. More direct evidence would be desirable to strengthen this hypothesis. In this model, proton channels are proposed from the stroma to the interior of *bf* complexes and from a PQ oxidation site to the lumen.

2.4 Proton Movements during Cyclic Electron Flow

During conditions that restrict the use of reduced terminal acceptor NADPH, such as low CO_2 or insufficient ATP, electrons from NADPH or from reduced Fd may cycle back in the paths indicated in Fig. 5. Because the cyclic flow includes plastoquinone and the cytochrome *bf* complex, protons are translocated, forming a pmf. The resulting pmf supports the synthesis of ATP. The importance of cyclic electron flow in ATP synthesis is particularly obvious in C_4 plants [11,12] in which at least five ATP and two NADPH molecules are required per CO_2 fixed. In C_3 plants, even if the contribution of cyclic electron flow to ATP synthesis is small, it may be significant in maintaining a sufficiently, but not excessively, acidic pH in the lumen for the prevention of light-induced inactivation of PS II [13]. Further discussion of mechanisms of photoinactivation of PS II and its prevention follows in Sec. 3.4.4.

3 THE pH AND pH-DEPENDENT PROCESSES IN CHLOROPLAST COMPARTMENTS

3.1 Stromal pH in the Light and Dark

In isolated intact chloroplasts in the dark, the pH in the stroma is below the pH of the suspension medium (7.6) by about 0.1–0.6 units, the difference decreasing with an increase in external $[K^+]$ from 20 to 100 mM [14]. This pH difference is attributed to a Donnan distribution of ions across the chloroplast envelope [15]. In the light, the stromal pH rises to about 8, depending again on $[K^+]$ [14].

Light-induced proton translocation into the thylakoid lumen is an obvious

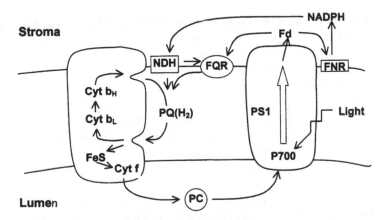

FIGURE 5 Cyclic electron flow around PS I and the cyt *bf* complex. Excitation of PS I oxidizes P700, initiating an electron transfer as indicated by the hollow arrow. Reduced ferredoxin (Fd) and NADPH feed electrons back to the PQ pool via NAD(P)H dehydrogenase (NDH) and ferredoxin-quinone reductase (FQR). PQH$_2$ is oxidized at the Q$_o$ site of the cyt *bf* complex as described for Fig. 4. Plastocyanin (PC) shuttles the electron to P700$^+$ to complete the cycle. (Adapted from Refs. 76–78.)

mechanism of alkalization of the stroma, as already described. In addition, other processes at the chloroplast envelope may help to maintain an alkaline pH at about 8 for optimal photosynthesis. Two such processes have been proposed. The first is a passive efflux of H$^+$ from the stroma in exchange for an influx of K$^+$ in the presence of a high [K$^+$] outside the envelope; this passive exchange is reversibly mediated by Mg^{2+}, Ca^{2+}, or Mn^{2+} but not Ba^{2+} [16]. The second is an oligomycin-sensitive, active efflux of H$^+$ in exchange for an influx of K$^+$, mediated by an ATPase in the chloroplast envelope [16]. All three mechanisms may contribute to alkalization of the stroma in the light, but it appears that an active efflux of protons from the stroma to the cytosol may be more important for sustained alkalization of the stroma to compensate for inward proton leakage across the envelope [17].

3.2 Lumenal pH in Light and Dark

3.2.1 The Meaning of pH in the Lumen

In a space dominated by protruding charged proteins, where electric double layers overlap and the proton buffering capacity is large, estimating the lumenal pH is not easy. Use of weak base fluorescing dyes was once popular, but such techniques overestimated the pH difference across the thylakoid membranes [18]. It is likely that in intact plants in moderate light the lumenal pH does not drop much below 6 (ΔpH = 2.0–2.5). If lumenal pH is unlikely to be lower than 5.5 (see later), how abundant are hydrogen *ions* in the lumen? Assuming 10^5 chlorophyll molecules per thylakoid [19] and a lumenal volume [20] of 0.007 m^3 (mol Chl)$^{-1}$, it can be estimated that at a lumenal pH of 5.5, there is on average less than one hydrogen ion per thylakoid, all the other protons in the lumen being covalently bound. A way to circumvent this conceptual difficulty may be to consider a large number of thylakoids, which together contain an average concentration of proton ions, in a state of flux in the light.

3.2.2 Estimates of Lumenal pH

The magnitude of lumenal pH, and hence ΔpH, is often estimated in isolated thylakoids using probe molecules (e.g., radiolabeled or fluorescent amines) that distribute between the external medium and the lumenal spaces according to the pH in both compartments (for a review see Ref. 21). These in vitro studies yield lumenal pH values that tend to be low (3 or 4 pH units below the external pH). Such pH gradients are likely to be overestimates due, for example, to interactions between probe molecules and lumenal components or surfaces [18,22]. It is also likely that feedback processes in vivo will limit excessive acidification of the lumen.

Kramer et al. [22] advanced other arguments, based on examples from the literature, for lumenal pH being only moderately acidic in the light: (1) the capacity for oxygen evolution decreases when the lumenal pH is below 5.5, so a low lumenal pH is to be avoided; (2) plastocyanin is unstable at pH < 5.5 in some plants; (3) cyt f reduction is slowed down when thylakoids are suspended in a medium at pH ranging from 5.5 to 6.5, so it is desirable to operate at higher lumenal pH values; and (4) deepoxidation of violaxanthin in the dark is maximal in a medium at pH 5.5 to 6.0, at which dissipation of excitation energy as heat will prevent excessive acidification of the lumen. Experiments supporting points (3) and (4) are based on the assumption that the lumenal pH and external pH are identical in the presence of an uncoupler or in the dark [23,24]. However, deducing lumenal pH by assuming it to be identical to that of the medium is problematic because the lumen is a Donnan phase capable of sequestering cations, including protons. For example, thylakoids suspended in the dark at pH 6.6 may have a lumenal pH of 6.1 [20]. Nevertheless, the permissible lower limit of lumenal pH in the light is likely to be about 5.5.

3.2.3 Control of Electron Transfer Rates by Lumenal pH Directly

Interactions between various electron transfer partners, such as plastocyanin and cytochrome f, are electrostatic in nature [25]. The charge on the proteins, and hence the probability of electrostatic attraction leading to electron transfer, is dependent on the local pH (in the lumen in the case of PC/cyt f). With effective pK values around 4.8 (PC carboxyl residues), it is obvious that a lumenal pH around 5 will result in protonation of about half the PC acidic groups and in consequent lowering of the collision rates and therefore electron transfer rate. However, this particular control mechanism is unlikely to come into operation in intact chloroplasts in the first place. The reason is that, before the pH could drop so low, the oxidation of plastoquinol at Q_o sites would have declined in rate, being controlled by a pK of around 6 [22,26,27]. The acidification of the lumen would thus be limited.

3.3 "Photosynthetic Control" by ΔpH (or, More Generally, by the PMF)

If protons accumulate in the lumen without being effluxed, eventually electron transport will slow down because of the property of quinones that electron and proton losses or gains are strictly coupled. The main efflux route for protons from the lumen is via the ATP synthase complexes, in which one ATP molecule is synthesized after the passage of four protons through the rotor [28] of the ATP synthase motor. Thus, when the ATP/ADP ratio rises (lack of CO_2, whose fixation uses ATP), proton efflux declines because the proton motive force is unable to exceed the "phosphorylation potential" (free energy of the reaction ADP + P_i → ATP + H_2O), and electron transfers slow down. This concept is elegantly supported by the action of ionophores that induce membrane conductance to protons (gramicidins) or enable proton/cation antiport (nigericin). Such "uncouplers" stimulate the rate of electron transfer because they abolish the strict relationship between electron transfer and proton deposition.

3.4 Other Processes Dependent on the Lumenal pH, Stroma pH, or the PMF

3.4.1 Control of Carbon Assimilation by the Stromal pH

The pH in the stroma during illumination has a strong influence on the rate of carbon assimilation. If the stromal pH is decreased by lowering the pH of the suspension medium containing intact chloroplasts, rates of both oxygen evolution [16] and CO_2 fixation [14,29] are decreased. Thus, maximal rates of CO_2 fixation occur at a stromal pH of 8.0 [29], although the pH optimum can be broadened and extended to lower external pH values with increased [K^+] in the suspension medium [14].

From studies of levels of metabolites in intact chloroplasts, it was shown that fructose bisphosphatase and sedoheptulose bisphosphatase are two of the pH-sensitive steps in the carbon assimilation process [30–32]. In addition, a lowered stromal pH may adversely affect other enzymes involved in photosynthesis, including the thylakoid membrane–spanning ATP synthase that has a pH optimum as high as 8.5. Thus, maintaining an appropriate stromal pH is essential for optimal rates of photosynthesis.

3.4.2 Control of Chloroplast-Encoded Protein Synthesis

In addition to optimizing photosynthetic rate, a slightly alkaline stromal pH has an important role in enabling photoinactivated PS II complexes to be repaired. During normal photosynthesis, even in low light, photoinactivation of PS II occurs because of "wear and tear." Photoinactivation of PS II is thought to be due to the strongest oxidant in photosynthesis, P680$^+$, which is needed to oxidize water molecules but which can also oxidize other neighboring groups with a small but not insignificant probability [33–35]. Further, after a PS II complex is photoinactivated, it may generate singlet excited oxygen molecules, which are highly oxidizing and may cause further damage [36]. Recovery of PS II from photoinactivation necessitates de novo protein synthesis, particularly that of D1 protein in the PS II reaction center [37–41].

During normal photosynthesis or when leaves are exposed to high irradiances far in excess of photosynthetic requirements, photoinactivation and repair of PS II occur simultaneously at a rate that depends on irradiance. It has been estimated that leaf discs from capsicum plants grown at 500 µmol photons $m^{-2}s^{-1}$, when exposed to 900 µmol photons $m^{-2}s^{-1}$ while floating on water at 25°C, have a turnover rate of 0.34 µmol PS II $m^{-2}h^{-1}$ (equivalent to 10^7 PS II $mm^{-2}s^{-1}$) or about one third of the concentration of PS II in an hour [42]. If repair by de novo synthesis of D1 protein is inhibited, for example, by lincomycin [43], net photoinactivation is greatly exacerbated. This demonstrates the importance of D1 protein synthesis in sustaining a functional population of PS II complexes. Not only does the synthesis of D1 protein depend on the supply of ATP, which in turn depends on the transthylakoid pH gradient, but also it is optimal only over a narrow range of stromal pH from 7.6 to 7.9 [44]. It is clear, therefore, that an appropriate alkaline stromal pH is essential for routine repair of photoinactivated PS II complexes.

3.4.3 Passive Redistribution of Ions Other than H$^+$ in the Lumen

The deposition of protons in the lumen is active in the sense that it is driven by light. Other ions such as Mg^{2+}, Cl^-, and K^+/Na^+ redistribute passively between the stroma and the lumen as simultaneously required by acid-base equilibrium, Donnan equilibrium, osmotic equilibrium, and electroneutrality. The extents of redistribution of these ions were first predicted [20] and subsequently confirmed experimentally [45].

3.4.4 The Dependence of Dissipation of Excess Excitation Energy on Lumenal pH

A leaf acclimated to high light conditions typically has a maximal photosynthetic rate of about 40 μmol O_2 m^{-2}s^{-1}. Because the evolution of 1 O_2 molecule requires about 10 photons [46], an absorbed irradiance of about 400 μmol photons m^{-2}s^{-1} would be theoretically sufficient to sustain such a photosynthetic rate. The irradiance in full sunlight, about 2000 μmol photons m^{-2}s^{-1}, is far in excess of this. The great excess of absorbed photons, if not dissipated harmlessly, can cause damage to the photosynthetic apparatus, as described earlier. Plants have, therefore, evolved a number of "photoprotective" mechanisms for mitigating light-induced damage [39]. In the present context, the photoprotective mechanism(s) triggered by acidification of the thylakoid lumen (i.e., by the "energized state") will be considered.

Excitation energy associated with singlet excited states of chlorophyll is dissipated by (1) photochemical conversion that yields useful reducing equivalents, (2) wasteful heat dissipation, and (3) to a minor extent, emission of chlorophyll *a* fluorescence or generation of long-lived triplet excited states (Fig. 6). Given that these pathways compete with one another, increased dissipation of excitation by either direct heat loss or photochemical conversion occurs at the expense of chlorophyll fluorescence emission, resulting in quenching of chlorophyll fluorescence. Murata and Sugahara [47] were the first to show quenching of chlorophyll fluorescence associated with the energized state of thylakoids: uncouplers such as methylamine and carbonylcyanide *m*-chlorophenylhydrazone, which abolish the energized state, enhance chlorophyll fluorescence emission. Thus, a state of low chlorophyll fluorescence yield is generally indicative of increased thermal deexcitation of antenna pigments. Such a state, with enhanced thermal dissipation, should confer protection against light-induced damage of chloroplasts. Indeed, Fork et al. [48] proposed that formation of the energized state could serve as a photoprotective mechanism in plants. Similarly, Krause and Behrend [49] showed that light-induced damage to isolated chloroplasts was exacerbated by uncouplers. Subsequently, the uncoupler nigericin was shown to increase light-induced damage in leaves [50,51].

A number of mechanisms have been proposed for quenching of excitation energy associated with the energized state of chloroplasts. A detailed discussion of the mechanisms is outside the scope of this chapter but can be found in other reviews [52–55]. Following are outlines of four ways in which an acidic lumen pH may bring about photoprotection of the photosynthetic apparatus.

Conformational Changes in Antenna Pigment-Protein Complexes Dependent on Lumen pH. By far, the majority of the protons accumulated in the lumen during illumination are involved in protonation of protein residues such as car-

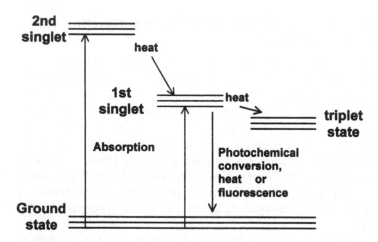

FIGURE 6 Excited states of a chlorophyll molecule and pathways of dissipation. Absorption of blue light and red light gives rise to the second and first excited states, respectively. Transition from the second to the first excited state occurs by heat dissipation. The first excited state can be dissipated by useful photochemical conversion, wasteful heat dissipation, and, to a small extent, chlorophyll fluorescence emission or conversion to a long-lived triplet state. These pathways compete with one another, so that, for example, increased heat dissipation leads to quenching of fluorescence emission.

boxylate groups. The membrane-spanning, light-harvesting chlorophyll a/b–protein complex (LHCIIb) that serves PS II is a major component of the thylakoid membrane and an obvious target for protonation. In light-harvesting chlorophyll-protein complexes, chlorophyll molecules are noncovalently ligated to protein structures, usually at a distance optimal for photosynthetic efficiency. When the lumen pH is sufficiently acidic, however, protonation of protein carboxylate groups will lead to aggregation of two or more neighboring chlorophyll molecules [53]. As postulated by Horton et al. [53], this interaction then leads to the formation of multimers and enhanced thermal dissipation of excitation energy (concentration quenching [56]). Dissipation of excitation energy in the peripheral antennae will then limit the amount of excitation energy reaching the PS II reaction center, thereby minimizing damage to it.

A variation of the preceding mechanism has been proposed by Crofts and Yerkes [57]. The authors reason that the minor chlorophyll-protein complexes, rather than the much more abundant LHCIIb, are the more critical sites for protonation. They suggest that the ligation of chlorophyll is changed upon protonation of these glutamate residues, leading to formation of an exciton-coupled dimer with a neighboring pigment; such "statistical dimers" possess additional energy levels, allowing thermal pathways for deexcitation, as in normal concentration quenching [56]. Given that these minor chlorophyll-protein complexes provide the connection for excitation energy to be funneled from LHCIIb toward the reaction center, their state of protonation may have a profound effect on the flow of excitation energy.

Both variations of this mechanism may occur, however. Mutants [58] and intermittent-light-grown plants [59] that lack LHCIIb, but contain the minor chlorophyll-protein complexes, have decreased "nonphotochemical quenching" of chlorophyll fluorescence and, by inference, of excitation energy. Hence, efficient thermal dissipation of excitation energy probably requires a complete light-harvesting system and lumen acidification as well as deepoxidation of xanthophyll pigments, as described next.

Deepoxidation of Xanthophyll Pigments Induced by Acidic Lumenal pH. The enzyme violaxanthin deepoxidase converts violaxanthin first to antheraxanthin and then further to zeaxanthin in antenna pigment-protein complexes. The enzyme, the activity of which is measured by an absorbance change at 505 nm associated with violaxanthin deepoxidation in isolated thylakoids, begins to be activated at an external pH less than 6.3 and shows maximal activity at pH 5.8 [24]. Zeaxanthin has its lowest excited, singlet state below that of chlorophyll [60], and therefore excitation energy could be transferred from chlorophyll to zeaxanthin, and to some extent to antheraxanthin, to be directly dissipated as heat. That carotenoids are potential sinks for excitation energy is shown by the short lifetime (approximately 10 ps) of their singlet excited states [61,62]. Indeed, quenching of chlorophyll fluorescence by β-carotene (which, like zeaxanthin, has 11 conjugated double bonds) has been observed in benzene [63]. However, for zeaxanthin to act as a quencher in thylakoids, it apparently has to bind to a structural niche that is exposed upon protonation of a chlorophyll-protein complex; only then will there be molecular contact between chlorophyll and zeaxanthin [54]. Hence, an acidic lumenal pH is required for energy dissipation via zeaxanthin. Energy dissipation via zeaxanthin in the antenna of PS II will short-circuit the normal flow of excitation energy to the reaction center, helping to minimize light-induced damage.

Thermal Dissipation by PS II Reaction Centers That Are Rendered Inactive by an Acidic Lumen. Weis and Berry [64] proposed that acidification of the lumen leads to conversion of PS II from a state with high photochemical and fluorescence yields to a quenched state of low photochemical and fluorescence yields but with high rates of thermal dissipation. It appears that an acidic lumen induces the release of one Ca^{2+} from each PS II [65], slowing down electron donation to, and prolonging the lifetime of, the quencher P680$^+$. Charge recombination between P680$^+$ and a prereduced quinone in the *reaction center* is then thought to lead to thermal dissipation of excitation energy [65], as distinct from thermal dissipation in light-harvesting *antennae* described earlier.

    Photoinactivated PS II Reaction Centers as Dissipators of Excitation Energy. Despite the photoprotective strategies available to plants, there is a low but not insignificant probability that a PS II complex is damaged (photoinactivated) during normal photosynthesis, as mentioned before in relation to the need for continual repair. Once photoinactivated, a PS II complex is unable to evolve oxygen.

    It has been proposed that a photoinactivated PS II, from which the chlorophyll fluorescence is strongly quenched, may efficiently dissipate excitation energy as heat, thereby preventing further damage to itself [66]. Öquist et al. [67] suggested as well that a photoinactivated PS II complex may even protect its functional neighbor if exciton transfer is permitted. Recently, it was observed that after prolonged high light stress, which resulted in the photoinactivation of more than 70% of the PS II complexes, the remaining functional PS II complexes did indeed seem to be protected, even when repair was inhibited by the presence of lincomycin [42]. Thus, at least in severely photoinhibited leaves, a proportion of functional PS II complexes is sustained, presumably via dissipation of excitation energy by connected, nonfunctional neighbors. Such a residual population of functional PS II, critical for recovery when favorable conditions return, may be crucial for the survival of the leaf and, indeed, the plant.

    This photoprotective mechanism, which operates only when a substantial fraction of functional PS II has been lost, appears also to require lumen acidification (H.-Y. Lee, Y.-N. Hong, and W. S. Chow, unpublished data). Consistent with this idea, the uncoupler nigericin impaired the operation of the mechanism, and even high concentrations of CO_2 seemed to limit lumen acidification to such an extent as to interfere with this "last-ditch" protective mechanism.

3.5 Localized Proton Domains

So far, the bulk lumenal aqueous phase has been considered to be the sole repository for the protons deposited during photosynthetic electron flow. Other experimental findings, however, point to an additional proton reservoir—metastable proton-buffering domains located in thylakoid membrane (see Ref. 68 and references therein). Such proton reservoirs are revealed by the release of protons upon adding uncouplers to dark-adapted thylakoids.

    These localized protons in the buffering domains, like the delocalized protons in the bulk lumen, are able to support ATP formation, and perhaps both localized and delocalized coupling can occur in isolated thylakoids. Quantifying their concentration in the domains, however, is problematic. Dilley et al. [68] pointed out that an advantage of plants utilizing protons in localized membrane domains is that overacidification of the bulk lumenal phase can be avoided. At the same time, protons could still be delivered rapidly along the thylakoid surface/domains to the ATP synthase, much in the same way that proton migration along the surface of bacteriorhodopsin micelles is faster than transfer from the surface to the bulk solution [69].

4 ANTIPORTS AND ION CHANNELS INVOLVING PROTONS

4.1 K^+/H^+ Exchange and Regulation of Stromal pH

A reversible H^+/K^+ exchange, dependent on Mg^{2+}, has been shown to affect the pH in the stroma [16]. Other studies [70] using the K^+ channel blocker tetraethylammonium indicated that K^+ is likely to move, in a uniport fashion, into or out of the stroma through a monovalent cation channel in the envelope. This channel also allows Na^+ to pass, such that when external Na^+ is substituted by the less permeant $Tris^+$, H^+ efflux cannot readily occur, resulting in inhibition of photosynthesis because of insufficient stromal alkalization [70].

4.2 Ca^{2+}/H^+ Antiport

When thylakoid membranes are fused with azolectin liposomes and enlarged to form giant thylakoid/liposome vesicles, ion permeation through the membrane can be investigated by the patch-clamp technique. In this way, Enz et al. [71] observed three different types of voltage-dependent ion channels—putatively, a potassium channel, a chloride channel, and an Mg^{2+}/Ca^{2+} channel—which could be distinguished on the basis of their unit conductance and ion specificity. To circumvent a general problem of reconstitution studies with spinach thylakoid membrane fragments, namely that artificial channels may be generated after reconstitution, Pottosin and Schönknecht [72] have characterized a cation-selective channel in isolated patches from osmotically swollen, but otherwise native, spinach thylakoids. This channel was permeable to K^+ as well as to Mg^{2+} and Ca^{2+} but not to Cl^-. Its total open probability increased at more positive membrane potentials. This channel probably allows redistribution of cations in response to proton deposition in the lumen during illumination [72].

Although the channel allows Mg^{2+} and Ca^{2+} to permeate the thylakoid membrane, these ions are required in different compartments of the chloroplast. Mg^{2+} is a main counterion that compensates for protons deposited in the lumen [20,45], and light-induced increase of stromal Mg^{2+} apparently occurs in the concentration range where it may be needed to activate stromal enzymes during photosynthesis [73]. On the other hand, Ca^{2+} is required in the lumen during illumination for proper functioning of the OEC. In addition, its concentration in the stroma has to be kept low, otherwise it inhibits CO_2 fixation. How then does the chloroplast maintain an influx of Ca^{2+} into the lumen in the light?

Ettinger et al. [74] have observed that thylakoids are capable of accumulating up to 30 mmol Ca^{2+} (mol Chl)$^{-1}$, equivalent to about 4 mM in the lumen, from an external concentration of 15 µM. The uptake was not as rapid as the efflux of Mg^{2+}, but it occurred steadily for about 10 min. Therefore, they proposed that uptake of Ca^{2+} was subsequent to the lumen acidification and the efflux of Mg^{2+} and K^+: the steady-state concentration of protons established in the lumen drives a Ca^{2+}/H^+ antiport system to sequester Ca^{2+} in the lumen. The resultant high lumenal $[Ca^{2+}]$ and low stromal $[Ca^{2+}]$ optimize functioning of the OEC and stromal enzymes.

5 CONCLUDING REMARKS

Light reactions and ion exchanges set up conditions of optimal pH in both stroma and thylakoid lumen. The stromal pH is optimal for activities of the carbon fixation enzymes/ATP synthase and for chloroplast-encoded protein synthesis. The lumenal pH is optimal for ready reactions between certain electron transfer components (such as between plastoquinol and the cytochrome *bf* complex) and for photoprotective processes that help plants to deal with excess light, while being sufficient to drive ATP synthesis. Without such optimization of stromal and lumenal pH values, as may happen in environmental conditions that cause uncoupling, the photosynthetic apparatus cannot function efficiently and may even suffer severe damage. In this sense, pH regulation in the chloroplast compartments is of paramount importance.

ACKNOWLEDGMENTS

ABH is grateful for the award of a Visiting Fellowship by the Research School of Biological Sciences, Australian National University. We thank Zhi-Ren Chow for drawing the diagrams.

ABBREVIATIONS

ATP   adenosine 5'-triphosphate
cyt   cytochrome
ΔpH   pH difference across the thylakoid membrane
D1 protein   the PS II reaction center protein subunit that is turned over most rapidly
FNR   ferredoxin-NADP$^+$ reductase
FQR   ferredoxin-quinone reductase
LHCIIb   the major light-harvesting chlorophyll *a/b*–protein complex in PS II
NAD   nicotinamide adenine dinucleotide

NADP$^+$, NADPH   oxidized and reduced nicotinamide adenine dinucleotide phosphate, respectively
NDH   NAD(P)H dehydrogenase
OEC   oxygen-evolving complex
P680 and P700   a special pair of chlorophyll molecules that acts as the
primary electron donor in the PS II and PS I reaction centers, respectively
PC   plastocyanin
Pheo   pheophytin
pmf   proton motive force
PQ   plastoquinone
PS   photosystem
Q_A and Q_B first and second quinone electron acceptor in PS II, respec-
tively
Q_o site on the cyt bf complex where PQH_2 is oxidized
Q_r site on the cyt bf complex where PQ is reduced
Tyr z   tyrosine that donates an electron to P680$^+$
S states   the states in which positive charge is accumulated in PS II

REFERENCES

1. B Kok, B Forbush, M McGloin. Cooperation of charges in photosynthetic oxygen
    evolution. Photochem Photobiol 11:457–475, 1970.
2. W Ausländer, W Junge. Neutral red, a rapid indicator for pH changes in the inner
    phase of chloroplasts. FEBS Lett 59:310–315, 1975.
3. J Lavergne, W Junge. Proton release during the redox cycle of the water oxidase.
    Photosynth Res 38:279–296, 1993.
4. M Haumann, W. Junge. Protons and charge indicators in oxygen evolution. In: DR
    Ort, CF Yocum, eds. Oxygenic Photosynthesis: The Light Reactions. Dordrecht:
    Kluwer, 1996, pp 165–192.
5. P Mitchell. Chemiosmotic coupling in oxidative and photosynthetic phosphoryla-
    tion. Biol Rev 41:445–502, 1966.
6. P Mitchell. Keilin's respiratory chain concept and its chemiosmotic consequences.
    Science 206:1148–1159, 1979.
7. J Lavergne, P Joliot. Dissipation in bioenergetic electron transfer chains. Photosynth
    Res 48:127–138, 1996.
8. DM Kramer, AR Crofts. The concerted reduction of the high- and low-potential
    chains of the bf complex by plastoquinol. Biochim Biophys Acta 1183:72–84, 1993.
9. HT Witt. Energy conservation in the functional membrane of photosynthesis. Analy-
    sis by light pulse and electric pulse methods. The central role of the electric field.
    Biochim Biophys Acta 505:355–427, 1979.
10. P Joliot, A Joliot. In vivo analysis of the effect of dicyclohexylcarbodiimide on
    electron and proton transfers in cytochrome bf complex of Chlorella sorokiniana.
    Biochemistry 37:10404–10410, 1998.
11. RT Furbank, CLD Jenkins, MD Hatch. C_4 photosynthesis: quantum requirements,
    C_4 acid overcycling and Q-cycle involvement. Aust J Plant Physiol 17:1–7, 1990.

12. DS Bendall, RS Monasse. Cyclic photophosphorylation and electron transport. Biochim Biophys Acta 1229:23–38, 1995.

13. U Heber, DA Walker. Concerning a dual role of coupled cyclic electron transport in leaves. Plant Physiol 100:1621–1626, 1992.

14. B Demmig, H Gimmler. Properties of the isolated intact chloroplast at cytoplasmic K^+ concentrations. I. Light-induced cation uptake into intact chloroplasts is driven by an electrical potential difference. Plant Physiol 73:169–174, 1983.

15. H Gimmler, G Schäfer, U Heber. Low permeability of the chloroplast envelope towards cations. In: Proceedings of the 3rd International Congress on Photosynthesis Research. Vol 3. Amsterdam: Elsevier, 1974, pp 1381–1392.

16. WJ Maury, SC Huber, DE Moreland. Effects of magnesium on intact chloroplasts. II. Cation specificity and involvement of the envelope ATPase in (sodium) potassium/proton exchange across the envelope. Plant Physiol 68:1257–1263, 1981.

17. U Heber, HW Heldt. The chloroplast envelope: structure, function and role in leaf metabolism. Annu Rev Plant Physiol 32:139–168, 1981.

18. AB Hope, DB Matthews. Adsorption of amines to thylakoid surfaces and estimations of ΔpH. Aust J Plant Physiol 12:9–19, 1985.

19. W Junge, HT Witt. On the ion transport system of photosynthesis—investigations on a molecular level. Z Naturforsch 23B:244–254, 1968.

20. WS Chow, AB Hope. Light-induced pH gradients in isolated spinach chloroplasts. Aust J Plant Physiol 3:141–152, 1976.

21. U Pick, RE McCarty. Measurement of membrane pH. Methods Enzymol 69:538–546, 1980.

22. DM Kramer, CA Sacksteder, JA Cruz. How acidic is the lumen? Photosynth Res 60:151–163, 1999.

23. JN Nishio, J Whitmarsh. Dissipation of the proton electrochemical potential in intact chloroplasts. II. The pH gradient monitored by cytochrome *f* reduction kinetics. Plant Physiol 101:89–96, 1993.

24. EE Pfündel, RA Dilley. The pH dependence of violaxanthin deepoxidation in isolated pea chloroplasts. Plant Physiol 101:65–71, 1993.

25. AB Hope. Electron transfers amongst cytochrome *f*, plastocyanin and photosystem I: kinetics and mechanisms. Biochim Biophys Acta 1456:5–26, 2000.

26. AN Tikhonov, GB Khomutov, EK Ruuge, LA Blumenfeld. Electron transport control in chloroplasts. Effects of photosynthetic control monitored by the intrathylakoid pH. Biochim Biophys Acta 637:321–333, 1981.

27. AB Hope, DB Matthews, P Valente. Effects of pH on the kinetics of reactions in and around the cytochrome *bf* complex in an isolated system. Photosynth Res 42:111–120, 1994.

28. H Noji, R Yasuda, M Yoshida, K Kinosita Jr. Direct observation of the rotation of F1-ATPase. Nature 386:299–302, 1997.

29. K Werdan, HW Heldt, M Milovancev. The role of pH in the regulation of carbon fixation in the chloroplast stroma. Studies on CO_2 fixation in the light and the dark. Biochim Biophys Acta 396:276–292, 1975.

30. P Purczeld, CJ Chon, AR Portis, HW Heldt, U Heber. The mechanism of the control of carbon fixation by the pH in the chloroplast stroma. Studies with nitrite-mediated proton transfer across the envelope. Biochim Biophys Acta 501:488–498, 1978.

31. U Enser, U Heber. Metabolic regulation by pH gradients. Inhibition of photosynthesis by indirect proton transfer across the chloroplast envelope. Biochim Biophys Acta 592:577–591, 1980.

32. UI Flügge, M Freisl, HW Heldt. The mechanism of the control of carbon fixation by the pH in the chloroplast stroma: studies with acid mediated proton transfer across the envelope. Planta 149:48–51, 1980.

33. LK Thompson, GW Brudvig. Cytochrome b-559 may function to protect photosystem II from photoinhibition. Biochemistry 27:6653–6658, 1988.

34. J Barber. Molecular basis of the vulnerability of photosystem II to damage by light. Aust J Plant Physiol 22:201–208, 1995.

35. JM Anderson, Y-I Park, WS Chow. Unifying model for the photoinactivation of photosystem II in vivo under steady-state photosynthesis. Photosynth Res 56:1–13, 1998.

36. E Hideg, T Kálai, K Hideg, I Vass. Photoinhibition of photosynthesis in vivo results in singlet oxygen production. Detection via nitroxide-induced fluorescence quenching in broad bean leaves. Biochemistry 37:11405–11411, 1998.

37. O Prášil, N Adir, I Ohad. Dynamics of photosystem II: mechanism of photoinhibition and recovery processes. In: J Barber, ed. The Photosystems: Structure, Function and Molecular Biology. Amsterdam: Elsevier, 1992, pp 295–348.

38. E-M Aro, I Virgin, B Andersson. Photoinhibition of photosystem II. Inactivation, protein damage and turnover. Biochim Biophys Acta 1143:113–134, 1993.

39. WS Chow. Photoprotection and photoinhibitory damage. Adv Mol Cell Biol 10: 151–196, 1994.

40. CB Osmond. What is photoinhibition? Some insights from comparison of shade and sun plants. In: NR Baker, JR Bowyer, eds. Photoinhibition of Photosynthesis: From Molecular Mechanisms to the Field. Oxford: Bios Scientific, 1994, pp 1–24.

41. A Melis. Photosystem II damage and repair cycle in chloroplasts: what modulates the rate of photodamage in vivo? Trends Plant Sci 4:130–135, 1999.

42. H-Y Lee, Y-N Hong, WS Chow. Photoinactivation of photosystem II complexes and photoprotection by non-functional neighbours in *Capsicum annuum* L. leaves. Planta 212:332–342, 2001.

43. E-M Aro, S McCaffery, JM Anderson. Photoinhibition and D1 protein degradation in peas acclimated to different growth irradiances. Plant Physiol 103:835–843, 1993.

44. D Bhaya, AT Jagendorf. Optimal conditions for translation by thylakoid-bound polysomes from pea chloroplasts. Plant Physiol 75:832–838, 1984.

45. WS Chow, G Wagner, AB Hope. Light-dependent redistribution of ions in isolated spinach chloroplasts. Aust J Plant Physiol 3:853–861, 1976.

46. O Björkman, B Demmig. Photon yield of O_2 evolution and chlorophyll fluorescence characteristics at 77K among vascular plants of diverse origins. Planta 170:489–504, 1987.

47. N Murata, K Sugahara. Control of excitation transfer in photosynthesis. III. Light-induced decrease of chlorophyll *a* fluorescence related to photophosphorylation system in spinach chloroplasts. Biochim Biophys Acta 189:182–192, 1969.

48. DC Fork, S Bose, SK Herbert. Radiationless transitions as a protective mechanism against photoinhibition in higher plants and algae. Photosynth Res 10:327–333, 1986.

49. GH Krause, U Behrend. ΔpH-dependent chlorophyll fluorescence quenching indicating a mechanism of protection against photoinhibition of chloroplasts. FEBS Lett 200:298–302, 1986.

50. E Ögren. Prediction of photoinhibition of photosynthesis from measurements of fluorescence quenching components. Planta 184:538–544, 1991.

51. Y-I Park. Antenna size dependency of photoinactivation of photosystem II in light-acclimated pea leaves. Plant Physiol 115:151–157, 1997.

52. B Demmig-Adams, WW Adams III. Photoprotection and other responses of plants to high light stress. Annu Rev Plant Physiol 43:599–626, 1992.

53. P Horton, AV Ruban, RG Walters. Regulation of light harvesting in green plants. Annu Rev Plant Physiol 47:655–684, 1996.

54. AM Gilmore. Mechanistic aspects of xanthophyll cycle–dependent photoprotection in higher plants chloroplasts and leaves. Physiol Plant 99:197–209, 1997.

55. KK Niyogi. Photoprotection revisited: genetic and molecular approaches. Annu Rev Plant Physiol Mol Biol 50:333–359, 1999.

56. GS Beddard, G Porter. Concentration quenching in chlorophyll. Nature 260:366–367, 1976.

57. AR Crofts, CT Yerkes. A molecular mechanism of q_E-quenching. FEBS Lett 352:265–270, 1994.

58. H Lokstein, H Härtel, P Hoffmann, P Woitke, G Renger. The role of light-harvesting complex II in excess excitation energy dissipation: an in vivo fluorescence study on the origin of high-energy quenching. J Photochem Photobiol B 26:174–185, 1994.

59. P Jahns, GH Krause. Xanthophyll cycle and energy-dependent fluorescence quenching in leaves from pea plants grown under intermittent light. Planta 192:176–182, 1994.

60. HA Frank, A Cua, V Chynwat, AJ Young, D Goztola, MR Wasielewski. Photophysics of the carotenoids associated with the xanthophyll cycle in photosynthesis. Photosynth Res 41:389–395, 1994.

61. H Hashimoto, Y Koyama. The C=C stretching Raman lines of β-carotene isomers in the S_1 state as detected by pump-probe resonance Raman spectroscopy. Chem Phys Lett 154:321–325, 1989.

62. H Hashimoto, Y Koyama. Raman spectra of all-$trans$-β-apo-8′-carotenal in the S_1 and T_1 states: a picosecond pump-and-probe technique using ML-QS pulse trains. Chem Phys Lett 162:523–527, 1989.

63. HA Frank, A Cua, V Chynwat, AJ Young, Y Zhu, RE Blankenship. Quenching of chlorophyll excited states by carotenoids. In: P Mathis, ed. Photosynthesis: From Light to Biosphere. Vol IV. Dordrecht: Kluwer, 1995, pp 3–7.

64. E Weis, JA Berry. Quantum efficiency of photosystem II in relation to 'energy'-dependent quenching of chlorophyll fluorescence. Biochim Biophys Acta 894:198–208, 1987.

65. A Krieger, E Weis. The role of calcium in the pH-dependent control of photosystem II. Photosynth Res 37:117–130, 1993.

66. GH Krause. Photoinhibition of photosynthesis. An evaluation of damaging and protective mechanisms. Physiol Plant 74:566–574, 1988.

67. G Öquist, WS Chow, JM Anderson. Photoinhibition of photosynthesis represents a mechanism for long-term regulation of photosystem II. Planta 186:450–460, 1992.

68. RA Dilley, SM Theg, WA Beard. Membrane-proton interactions in chloroplast bioenergetics: localized proton domains. Annu Rev Plant Physiol 38:347–389, 1987.

69. P Scherrer, U Alexiev, T Marti, HG Khorana, MP Heyn. Covalently bound pH-indicator dyes at selected extracellular or cytoplasmic sites in bacteriorhodopsin micelles and its delayed transfer from surface to bulk. Biochemistry 33:13684–13692, 1994.

70. W Wu, GA Berkowitz. Stromal pH and photosynthesis are affected by electroneutral K^+ and H^+ exchange through chloroplast envelope ion channels. Plant Physiol 98: 666–672, 1992.

71. C Enz, T Steinkamp, R Wagner. Ion channels in the thylakoid membrane (a patch-clamp study). Biochim Biophys Acta 1143:67–76, 1993.

72. II Pottosin, G Schönknecht. Ion channel permeable for divalent and monovalent cations in native spinach thylakoid membranes. J Membr Biol 152:223–233, 1996.

73. AR Portis, HW Heldt. Light dependent changes of the Mg^{2+} concentration in the stroma in relation to the Mg^{2+} dependency of CO_2 fixation in intact chloroplasts. Biochim Biophys Acta 449:434–446, 1976.

74. WF Ettinger, AM Clear, KJ Fanning, ML Peck. Identification of a Ca^{2+}/H^+ antiport in the plant chloroplast thylakoid membrane. Plant Physiol 119:1379–1385, 1999.

75. DR Ort, CF Yocum. Electron transfer and energy transduction in photosynthesis: an overview. In: DR Ort, CF Yocum, eds. Oxygenic Photosynthesis: The Light Reactions. Dordrecht: Kluwer, 1996, pp 1–9.

76. J Ravenel, G Peltier, M Havaux. The cyclic electron pathways around photosystem I in *Chlamydomonas reinhardtii* as determined in vivo by photoacoustic measurements of energy storage. Planta 193:251–259, 1994.

77. HV Scheller. In vitro cyclic electron transport in barley thylakoids follows two independent pathways. Plant Physiol 110:187–194, 1996.

78. T Endo, H Mi, T Shikanai, K Asada. Donation of electrons to plastoquinone by NAD(P)H dehydrogenase and by ferredoxin-quinone reductase in spinach chloroplasts. Plant Cell Physiol 38:1272–1277, 1997.

8

Dynamics of H⁺ Fluxes in Mitochondrial Membrane

Francis E. Sluse
University of Liege, Liege, Belgium

Wiesława Jarmuszkiewicz
Adam Mickiewicz University, Poznan, Poland

1 INTRODUCTION

1.1 Mitochondria as H⁺ Gradient Generator [1,2]

According to the chemiosmotic energy transduction concept, respiring mitochondria generate an H⁺ electrochemical gradient ($\Delta\mu H^+$) across the inner membrane. This occurs when electrons pass along the electron transport chain that consists of respiratory enzyme complexes made of electron carriers. These complexes have a specific orientation in the membrane in order to carry out the vectorial net H⁺ movement outward from the mitochondria during the oxidoreduction reactions. The ATP synthase uses the H⁺ gradient to make ATP inside the mitochondria with stoichiometric uptake of external H⁺. Thus, the energy released by the redox reaction is first transferred into a transmembrane H⁺ gradient and then trapped by the synthesis of ATP. The overall energy conservation requires that redox reactions and ATP hydrolysis are coupled to H⁺ pumping; thus, the ATP synthase is a reversible H^+-translocating ATPase. If one chemical reaction occurs without H^+ pumping, the energy conservation is abolished and the redox Gibbs free energy is dissipated. As the H^+ gradient links respiration and phosphorylation, this implies that protons pumped out by the respiratory complexes and used by ATP synthase are free to diffuse in the aqueous phases. Thus, a so-called delocalized proton circuit exists, requiring the inner mitochondrial membrane to have a low H^+ conductance to avoid short-circuiting, i.e., breaking the energetic coupling between the ATP synthase and the respiratory H^+ pumps. As this coupling is indirect, the H^+ gradient may be diverted to drive other processes such as ion transport systems and heat production. Numerous carriers exist in the inner membrane to allow reducing metabolites to enter the matrix space and to integrate metabolism of mitochondria and cytoplasm (Fig. 1).

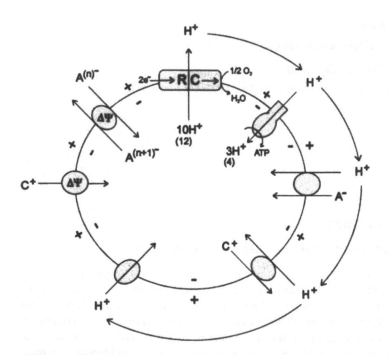

FIGURE 1 The delocalized proton circuits in mitochondria. Respiratory chain (RC) generates an H^+ electrochamical gradient: $\Delta\mu H^+ = -F\Delta\Psi + 2.3\ RT\Delta pH$, where $\Delta\Psi = \Psi_{in} - \Psi_{out}$ and $\Delta pH = pH_{out} - pH_{in}$. This proton gradient is used (clockwise from RC) by the ATP synthase ($\Delta\mu H^+$), by proton-anion (A^-) electroneutral symport carriers (ΔpH), by cation (C^+)/H^+ exchangers (electroneutral, ΔpH or electrogenic, $\Delta\mu H^+$), by uncoupling protein ($\Delta\mu H^+$), by cation uniport channels ($\Delta\Psi$), and by electrogenic anion antiport carriers ($\Delta\Psi$).

1.2 Outline of the Chapter

The aim of this chapter is to gather updated understanding of H^+ flux–linked processes in mitochondria. Enormous progress has been made in our knowledge during the past 5–10 years through direct molecular approaches. New understanding at the level of molecular mechanisms has been reached, especially for redox H^+ pumps and ATP synthase. Conversely, the lack of three-dimensional structures of the ion carriers of the inner mitochondrial membrane hampers the description of their molecular mechanism. The family of membrane carriers has attracted significant attention because of (1) the description of the whole protein family in yeast following the full genome sequencing and (2) the impact of the discovery of uncoupling protein in plants.

2 PROTON ELECTROCHEMICAL GRADIENT BUILDUP

2.1 Plant Respiratory Chain: Shared and Peculiar Properties

The mitochondrial energetics is based on the organized function of transmembrane high-order structured respiratory complexes with a specific orientation in the inner membrane (Fig. 2). The classical mammalian respiratory chain is composed of three H^+-pumping complexes: NADH:ubiquinone oxidoreductase (complex I), ubiquinol:cytochrome c oxidoreductase (complex III), and cytochrome c oxidase (complex IV). Succinate:ubiquinone oxidoreductase (complex II) transfers electrons from succinate to ubiquinone but does not translocate protons, thereby feeding only electrons into the respiratory chain.

Plant mitochondria deviate from this basic setup by the presence of four additional NAD(P)H dehydrogenases and an alternative ubiquinol oxidase, of which none is employed in proton gradient buildup. Two of the additional dehydrogenases, one specific for NADH and one for NADPH, are directed toward the matrix space and compete with complex I for NADH produced inside the mitochondria. The two other dehydrogenases are directed toward the cytosol and can oxidize directly cytosolic NADH and NADPH without operation of substrate shuttles as in animal mitochondria. The electron input at this level of the respiratory chain creates special conditions for complex I, which is in competition for NADH oxidation to contribute to H^+ gradient buildup. On the other hand, the complexity of electron feeding at the level of ubiquinone potentially increases the electron input downstream. The alternative ubiquinol oxidase can be regarded as an uncoupled pathway spoiling the redox Gibbs free energy of the ubiquinol-oxygen (QH_2-O_2) span responsible for a decrease in ATP synthesis yield from ubiquinol oxidation. The protons required for O_2 reduction are taken from the matrix and, as there is no H^+ translocation, the oxidation of ubiquinol by the alternative oxidase must release protons in the matrix.

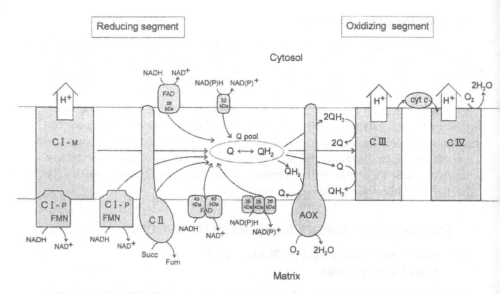

FIGURE 2 The respiratory chain of plant mitochondria. The respiratory chain is divided into a reducing segment (left part), which reduces ubiquinone (Q), and an oxidizing segment (right part), which oxidizes ubiquinol (QH₂) and reduces oxygen. The reducing segment consists of complex I (C I–M, membrane arm and C I–P, peripheral arm); complex II (C II); two NADH dehydrogenases, one on each side of the membrane (FAD 58 kDa outside and FAD 43 kDa dimer inside); and two NAD(P)H dehydrogenases, one internal (26 kDa trimer) and one external (32 kDa). The Q pool is considered to be a common source of Q or QH₂ for all enzymes implicated in Q reduction or QH₂ oxidation. The oxidizing segment consists of complex III (C III), peripheral cytochrome *c* (cyt *c*), complex IV (C IV), and alternative ubiquinol oxidase (AOX). Fum, fumarate; succ, succinate; FMN, flavin mononucleotide; FAD, flavine adenine dinucleotide; NAD⁺, NADH, nicotinamide adenine dinucleotide and its reduced form; NADP⁺, NADPH, nicotinamide adenine dinucleotide phosphate and its reduced form.

Thus, the plant respiratory chain presents a complex electron input system reducing ubiquinone (Q) and a double-branched ubiquinol (QH₂) oxidizing pathway, consisting of the cytochrome pathway and the alternative oxidase. Both systems, upstream and downstream of UQ, can modulate H⁺ fluxes outside the mitochondria and so can affect the ATP synthesis yield. Indeed, transfer of electrons from NADH to O₂ can be totally uncoupled from H⁺ pumping by exclusion of complex I and the cytochrome pathway.

2.2 Complex I

2.2.1 Molecular Organization [3,4]

In term of complexity and size, the plant complex I resembles the mammalian and *Neurospora* enzymes. The 32 resolved polypeptides from potato complex I have a total molecular mass of 891 kDa. It contains flavin mononucleotide (FMN) as prosthetic group and several (up to seven) iron-sulfur (Fe/S) centers as electron carriers, and it is inhibited by rotenone. Part of the subunits is encoded by mitochondrial DNA. Seven mitochondrial genes found in mammals and *Neurospora* are present in plants, which possess two additional mitochondrial genes. Only a few nuclear genes (five) encoding complex I have been characterized, and they are implicated in NADH binding and Fe/S cluster liganding. The nine mitochondrial-encoded subunits, together with the five nuclear-encoded subunits, make up a bacterial-like minimal complex I (14 subunits, 525 kDa) that represents the basic evolutionarily conserved subunit organization of complex I. Thus, it can be assumed that oxidoreduction function and H⁺ pumping follow the same mechanism within prokaryotic and eukaryotic organisms.

Complex I is an L-shaped structure with two domains, a membrane arm and a peripheral arm. The peripheral arm contains the NADH binding site, Fe/S clusters, and FMN. Its polypeptides have no transmembrane α-helix. The peripheral arm has NADH-ferricyanide oxidoreductase activity and contains the catalytic site that transfers electrons from NADH to FMN. The polypeptides in the membrane arm contain many transmembrane helices and internal Q molecules but no redox group. It has been proposed that the peripheral arm can exist separately with a sufficient life span in plants, catalyzing an FMN-dependent rotenone-insensitive NADH dehydrogenation that donates electrons to the Q pool without H⁺ pumping.

2.2.2 Function

Proton translocation is coupled to electron transfer in complex I. This coupling has been modeled from kinetic, spectroscopic, and inhibition data, but verification of these models awaits relevant precise structural information. Anyway, in the matter of stoichiometry, current data suggest that complex I translocates four H⁺ per electron pair. The basic mechanism of H⁺ motion can be either a "loop" or a purely conformational pump. The loop mechanism, first proposed by P. Mitchell, has a strict stoichiometry of one H⁺ per electron per loop. In a loop, the electrons cross the membrane twice, first in one direction via an [H] carrier and then via a pure electron carrier in the other direction, with the release of one H⁺ on the positive side of the membrane (outside). Conversely, when the electron is transferred from an electron carrier to an [H] carrier, an H⁺ is taken up from the negative side (inside). The conformational H⁺ pump is proposed to coordinate protonation-deprotonation (change in pK_A) induced by oxidoreduction and associated with conformational changes, ensuring the directionality of H^+ transport and thus making the protonated site alternatively exposed to either side of the membrane. Protonated sites do not need to be redox centers and the stoichiometry can be 1, 2, 3, n H^+ per electron. To account for a 4 $H^+/2$ e^- stoichiometry at the level of complex I and as all the redox groups are situated in the peripheral arm and should not be implicated in transmembrane H^+ movement, it seems that only the direct pumping mechanism should work.

2.2.3   Mechanistic Models [5,6]

The common feature of mechanistic models proposed by Brandt [5] and by Dutton and colleagues [6] is only one iron-sulfur cluster (N-2) and ubiquinone molecules contributing to H^+ translocation. The N-2 cluster is situated close to the membrane domain at the interface between the two arms of the enzyme and may be reduced by NADH through FMN and several Fe/S clusters. Three different ubiquinone binding sites take part in the overall energy conversion and are carried by the membrane arm of the complex. One ubiquinone is close to the N-2 cluster and can receive an electron from it.

The main features of the Brandt model are shown in Fig. 3A. The pK_A of the bound QH_2 at center PI is believed to be lowered, allowing one spontaneous deprotonation with release of one H^+ in the cytosol. When cluster N-2 (inner side) is reduced by an electron, its pK_A increases, allowing extraction by the cluster of the second H^+ from QH^- bound at center PI (outer side). When it reduces Q bound at center NA (inner side), the H^+ is forced out toward the external positive side of the membrane through a proton channel. This leads to Q^{2-} at center PI, which can reduce the semiquinone ($Q^{\cdot-}$, formed at center NA) to QH_2 with the uptake of two H^+ from the matrix. Then the semiquinone formed at center PI reduces Q at center NB to form semiquinone. The whole sequence is repeated when N-2 is again reduced by an electron, and now the semiquinone bound at center NB is fully reduced to QH_2 with uptake of two H^+ from the matrix. Thus, the $4H^+/2e^-$ are realized by redox-linked protonations and deprotonations of ubiquinone at the three binding sites PI, NA, and NB that can exchange with the Q pool.

In the Dutton model, there are two quinone binding sites that can exchange Q/QH_2 with the membrane pool, Q_{nz} that can receive H^+ from matrix and Q_{nx} that can release H^+ into the cytosol (Fig. 3B). A third quinone (Q_{ny}), non–pool exchangeable, occupies a third site (between the two other quinone binding sites) that can assume either of two different geometries. One conformation provides access to the protons on the matrix side of the membrane through a channel, and the second conformation provides access to H^+ on the cytosolic side. The catalysis proceeds as follows:

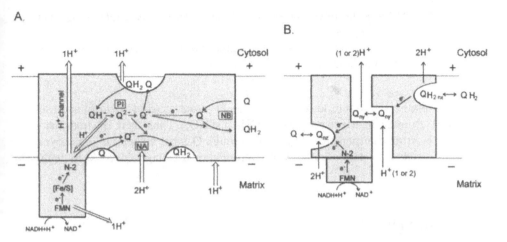

Figure 3 Mechanistic models of complex I. (A) Model simplified with permission from Brandt [5] with 1 e⁻ stoichiometry. Complex I is made of two domains, a peripheral protruding arm that contains FMN, [Fe/S] clusters, N-2 Fe/S cluster, and a membrane domain that carries three ubiquinone binding sites or centers: the inner side center A (NA), the outer side center PI, and the internal membrane center NB. Open arrows indicate proton movements. (B) Model simplified with permission from Dutton et al. [6] with 1 e⁻ stoichiometry. The membrane domain contains two ubiquinone binding sites that can exchange Q or QH_2 with the membrane pool, the matricial site Q_{nz} that can receive protons from the matrix, and the cytosolic site Q_{nx} that can release H⁺ into the cytosol. The third site Q_{ny}, in the middle of the membrane, cannot exchange its quinone with the Q pool but can receive from matrix or release in cytosol 1 or 2H⁺ through two half H⁺ channels according to its conformation.

1. The N-2 cluster is reduced by NADH.
2. N-2 reduces Q in the Q_{nz} site to semiquinone.
3. Semiquinone in the Q_{nz} is reduced (one e⁻) by QH_2 fixed in the Q_{ny} site, and it binds 2H⁺ drawn from the matrix; at the same time, semiquinone fixed in Q_{ny} releases one or two H⁺ through the cytosolic channel.
4. Semiquinone in the Q_{ny} site changes conformation with access to protons from the matrix.
5. In this geometry, QH_2 in site Q_{nx} can reduce (one e⁻) semiquinone in the Q_{ny} site with capture of one or two H⁺ from the matrix channel to form QH_2.

6. At the same time, one or two H^+ are released from QH_2 in Q_{nx} site into the cytosol.

7. QH_2 at site Q_{ny} moves in its original position, a second e^- from N-2 can enter, and steps 1 to 5 are repeated.

The overall reaction translocates four or six H^+ from matrix to cytosol and uses two H^+ from the matrix to reduce Q to QH_2 at site Q_{nz}. Thus, there is a transmembrane electron transport from Q_{nx} to Q_{nz} through Q_{ny} that makes a quinone cycle and Q_{ny} acts as a proton pump with conformational change linked to its redox state.

2.3 Complex III

2.3.1 Composition and Structure [2,7]

The composition of complex III is well characterized in prokaryotes, yeast, mammals, and plants. In eukaryotes, it consists of 10 to 11 different subunits, but only 3 are implicated in electron transfer and energy conservation. In *Paracoccus denitrificans*, the bc_1 complex is constituted of only these three redox center–holding subunits that represent the evolutionarily conserved basic organization: the diheme cytochrome b, the cytochrome c_1, and the Rieske iron-sulfur protein. Cytochrome b has eight transmembrane helices with the two hemes b sandwiched between helices 2 and 4. It is the only subunit of complex III that is encoded by the mitochondrial DNA. Cytochrome c_1 has a globular structure incorporating heme c_1 and extending into the aqueous phase on the cytosolic face of the membrane and is anchored to the membrane via its hydrophobic C-terminus. The Fe/S protein (ISP) carries a Rieske-type iron-sulfur center (2Fe/2S) into its globular structure located on the cytoplasmic side and is membrane anchored via its hydrophobic N-terminus. The other seven to eight subunits are "core proteins" that surround the metalloprotein nucleus, but two of them that face the matrix space are homologous to mitochondrial processing peptidases that function in protein import. In potato, 10 subunits are well resolved by blue native electrophoresis of mitochondrial proteins and show the primary sequence similarity with corresponding subunits of the yeast and bovine bc_1 complexes. However, in potato, complex bc_1 not only conserves redox energy but also has itself the mitochondrial processing peptidase activity involved in removing the mitochondrial presequences. These proteolytic properties are very similar to those of processing peptidase complex localized in the matrix of yeast and mammal mitochondria. Thus, complex III in plants may be a bifunctional enzyme at least through part of its core subunits. Moreover, cytochrome bc_1 complex is a stable dimer, with each unit having a molecular mass of around 240 kDa.

Cytosol

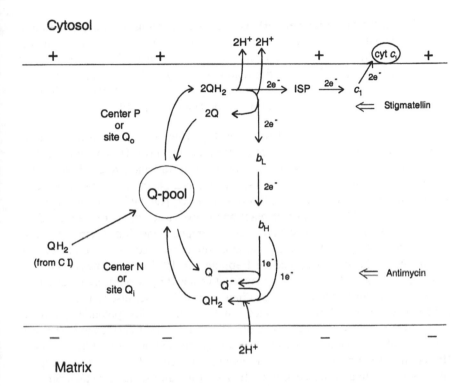

Matrix

FIGURE 4 The Mitchell ubiquinone cycle. The Q cycle is drawn within the thickness of the membrane. ISP, the Rieske iron-sulfur protein; b_L and b_H, low- and high-potential hemes b; c_1, heme c_1; cyt c, cytochrome c; C I, complex I; stigmatellin, an inhibitor of the QH_2 oxidation site (center P or site Q_o); antimycin, an inhibitor of the Q reduction site (center N or site Q_i); Q^-, semiquinone.

2.3.2 Function: The Q Cycle [7,8]

The Q cycle can be depicted as in Fig. 4 in a way that evidences its two key features:

1. The two distinct ubiquinone reaction centers on opposite sides of the membrane, the QH_2 oxidation center called center P (positive side of the membrane), or site Q_o (outer side) or QO (oxidation site) and the Q-reduction center called center N (negative side of the membrane), or site Q_i (inner side) or QR (reduction site). QH_2 oxidation is linked to H⁺ release on the positive side (cytosol) and Q reduction to H⁺ uptake on the negative side of the membrane (matrix).

2. At the Q_o site, when QH_2 is oxidized, electrons are transferred onto
    two different acceptors, i.e., the low-potential ($E_{m,7} \approx -20$ mV) heme
    b_L and the Rieske Fe/S center ($E_{m,7} \approx +290$ mV). This bifurcation of
    electron flow is the central feature of the Q cycle resulting in vectorial
    H^+ translocation as it allows the recycling of every "second" electron
    from site Q_o to site Q_i through the heme b_L-b_H path.

For a complete cycle as depicted in Fig. 4, the electron transfer reaction
sequence has to be realized twice: at center P (site Q_o) two QH_2 are oxidized in
a sequence delivering, one by one, two electrons to ISP and two electrons to b_L-
b_H, and releasing four H^+ in the outer space; at center N (site Q_i) one Q is reduced
sequentially by the two e^- coming through the b_L-b_H path taking up two H^+ from
the matrix space. The Q-reduction reaction involves a semiquinone species ($Q^{\cdot-}$)
that is formed during the first half-cycle by trapping the first electron coming
through the b_L-b_H path. The semiquinone is further reduced by a second electron
originating in the same way and by trapping two H^+ from the matrix. On the
contrary, the chemistry of QH_2 oxidation at center P by two different acceptors
is a unique reaction and its exact mechanism is still under discussion. It is this
unique reaction that forces one electron to enter the low-potential pathway
(heme b) that is the electrogenic step driving vectorial H^+ translocation. The fact
that the center N inhibitor, antimycin, is a very efficient inhibitor of the complex
III activity in the presence of substrates and an excess of an electron acceptor
such as ferricyanide (i.e., conditions that lead to fully reduced cytochrome b)
demonstrates the obligatory character of the bifurcation of electron flow. How
this tight control of the reaction in center P is ensured remains a crucial point
in energy conservation at the level of complex III. It may involve the formation
and movement of semiquinone within the center P after the release of the first
electron.

2.3.3 Bifurcation of Electrons and Protein Domain Mobility [9]

It seems that the bifurcation of electron flow at center P (site Q_o) is controlled
during steady state by the first deprotonation of QH_2 and that the hydrophilic
domain of the Rieske protein is found in two positions in structural studies that
could allow bifurcation of electron flow. Indeed, the crystallographic data indicate
that the position of the Rieske 2Fe/2S center relative to the other metals varies
with the presence of inhibitors of site Q_o (e.g., stigmatellin). The globular domain
containing the Fe/S center can rotate by about 60°. This mobility of the Rieske
extrinsic domain that holds the 2Fe/2S cluster could allow the Fe/S center to
move (after its reduction by QH_2 with the first electron) about 2 nm away from
site Q_o and to come near to cytochrome c_1, delivering this electron to it. This
movement may not only allow the shuttle of one electron from the site of ubiqui-

nol oxidation to the site of cytochrome c_1 reduction but also force the second electron to cycle through the b hemes to center N. Thus, the unique bifurcation of electron flow occurring in complex bc_1 seems to be associated with a protein domain movement that would be unique among redox protein complexes.

Another major conclusion arising from crystallographic studies is that the dimeric structure is essential to the chemical reaction mechanism of the bc_1 complex. Indeed, the extrinsic domain of ISP, implicated in the bifurcation of electron flow, has no contact with the other extrinsic domains of a monomer but it is close to the heme of cytochrome c_1 of the other monomer in the dimer. In fact, the iron-sulfur cluster of one monomer is located in the dimer in such a way that it can transfer electrons from center P to the cytochrome c_1 in the other monomer.

2.4 Complex IV

2.4.1 Composition and Structural Information [2,4,10]

Cytochrome c oxidase has a number of subunits ranging from 7 to 13 in eukaryotes and from 2 to 4 in prokaryotes. Potato complex IV is resolved into 10 subunits and seems to be similar to cytochrome c oxidase of fungi (11 subunits) and bovine oxidase (13 subunits). The three largest subunits (COX I, II, III) are mitochondrial encoded and homologous to the main subunits of *Paracoccus denitrificans* complex IV. The apparent molecular mass of potato complex IV is 160 kDa. The bovine cytochrome oxidase occurs as a homodimer, each monomer consisting of the 13 subunits with a total molecular mass of 204 kDa.

The general structure of cytochrome oxidase may be described as a catalytically active core formed by the 3 major subunits surrounded by a multiprotein matrix made of the 10 nuclear-encoded small subunits (for the bovine enzyme). The redox-active part contains several redox cofactors: two Cu atoms forming a dinuclear mixed-valence copper center (Cu_A) that is situated at the part of subunit II protruding into the cytoplasm and that is the first site to accept electrons from cytochrome c and one low-spin heme a (cytochrome a) and one heme a_3–Cu_B binuclear active center, both located in subunit I (Fig. 5). Electrons are transferred from Cu_A via heme a to the binuclear heme a_3–Cu_B center where oxygen reduction takes place.

The whole crystal structure of bovine heart cytochrome oxidase has revealed the 13 protein subunits, the presence of phospholipids, and the presence of two possible pathways for H⁺ pumping as well as possible channels for H⁺ needed for O_2 reduction, for removal of water, and for oxygen.

Two hydrophilic channels connect the active center of the enzyme to the aqueous phase of the matrix (Fig. 5). Channel D (with conserved aspartate at the matrix entry) and channel K (with conserved lysine in the middle) allow the H⁺ used in O_2 reduction to be taken up from the matrix. The same channels are used to pump H⁺ across the membrane. A conserved glutamate in the end of the D

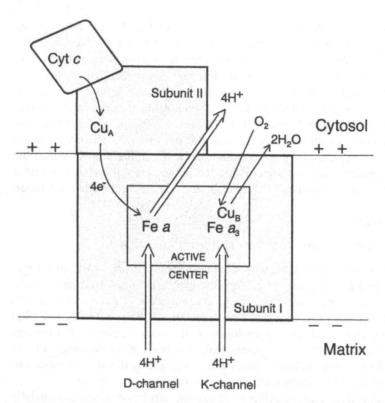

FIGURE 5 Catalytic heart of complex IV. Shown are cytochrome *c* (cyt *c*), peripheral part of subunit II that contains copper center Cu_A, intrinsic membrane part of subunit I, the other metal centers in the "active center," i.e., Fe *a*(heme *a*), Fe a_3 (heme a_3), and copper Cu_B. Open arrows indicate proton channels. For details see text.

channel is required for H^+-pumping activity. Thus, protons that are consumed in O_2 reduction and protons that are translocated across the membrane enter through D and K channels that are essential for the full catalysis. Translocated protons have to pass through a hydrophobic barrier within the enzyme to reach the external aqueous phase by a mechanism that is not yet known. Structural information has initially suggested that ligand coordination around Cu_B (three histidines) could be dynamic, as one of the histidines has a weak electron density in the crystal of the bacterial enzyme. This observation supported the proposal that two H^+ are simultaneously carried through the hydrophobic barrier by moving histidine, which is simply the only amino acid that can carry two H^+. However, more recent crystallographic studies of the *P. denitrificans* enzyme and the bovine enzyme have shown that all histidine ligands of Cu_B are visible and do not change upon reduction of the enzyme, refuting this mechanism. A new H^+-pumping mechanism coupled to O_2 reduction has been proposed based on recent structural data. It suggests that H^+ translocation can be achieved by an interplay between a conserved glutamate and the propionates of hemes a and a_3.

2.4.2 Catalytic Cycle [2,11,12]

Cytochrome c binds stoichiometrically to cytochrome c oxidase and donates its electrons on the cytoplasmic side of complex IV. These electrons are transferred to the active site where O_2 is reduced into two H_2O by four electrons and four H^+ that are taken from the matrix side. As this reaction is accompanied by the vectorial transport of four H^+ across the membrane from the matrix, and as cytochrome c donates its electrons from the outside, eight positive charges are translocated across the membrane per O_2 reduced.

The conventional catalytic cycle leading to O_2 reduction includes several intermediates of the enzyme (Fig. 6) starting from the fully oxidized enzyme (O) in which both metals of the active site are oxidized. Arrival of one e^- leads to the formation of the one-electron reduced enzyme (E). The next electron transfer produces a species (R) in which both metals are reduced. Species R can bind oxygen and the peroxy state P is reached via an intermediate A. The peroxy state exists in two forms, P_M and P_R (after the third reduction). The oxoferryl state F is generated after O—O bond splitting, uptake of two extra protons from matrix, pumping out of two H^+, and formation of one H_2O. Reduction by a fourth electron leads to a sequential uptake of two H^+ and release of the second water molecule and shift of the reactive center back to the O state via the hydroxy intermediate H together with the pumping out of two H^+. Thus, only transitions P to F and F to O are coupled to proton pumping and to water formation, and both steps pump two H^+ to fulfill the overall stoichiometry of four H^+ pumped per four e^-, suggesting that water formation is linked somehow to H^+ pumping. This has led to the formulation of a principle of electroneutrality, postulating that reduction of the binuclear center is accompanied by uptake of two H^+ to match this electroneutrality and that these two H^+ are expelled outside by electrostatic repulsion by the protons required afterward for water formation.

Unfortunately, this widely accepted model has been weakened by the demonstration that in P states the O—O bond was already broken, suggesting that the P states are a hydrogen-bonded oxoferryl form. Thus, water could be already formed during the O → P state transition, and the H^+ pump mechanism just described would be doubtful. A new mechanistic model has been proposed by Michel [11] and is based on a wealth of structural, spectroscopic, and mutagenesis data. In this model each of the four reductions of heme a during the catalytic cycle is coupled to uptake of one H^+ via the D pathway. These protons are temporarily stored in the regions of the heme a and a_3 propionates and are driven to

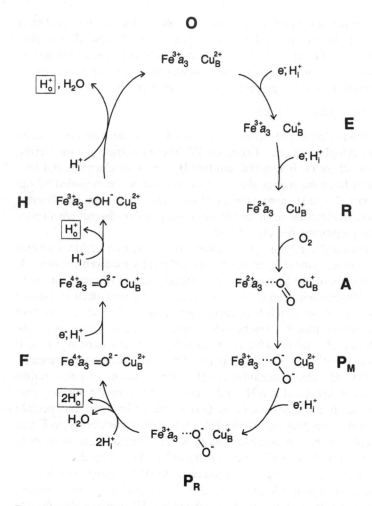

FIGURE 6 Catalytic cycle of cytochrome oxidase leading to oxygen reduction and proton pumping. For details of the mechanism see text. Fe^{n+} a_3; heme a_3; H_i^+, proton in matrix; H_o^o, proton in cytosol. (Adapted from Ref. 11.)

the outside by electrostatic repulsion from H^+ entering the active site during turnover. The first proton is pumped out by uptake of one proton via the K pathway during reduction (O → R transition), the second and third protons during the P → F transition, and the fourth one upon conversion of F to O. Details of this mechanism are extensively described by Michel [11], and atomic structures are assigned to each intermediate with potential H^+ accepting sites.

Despite the enormous progress made in the understanding of complex IV activity, the knowledge of the precise pathways and mechanism of H^+ translocation will require a huge amount of additional, highly sophisticated work.

2.5 Proton Pumping Pyrophosphatase [13]

Beside soluble pyrophosphatases, membrane-bound proton pumping pyrophosphatases (H^+-PP_iase) have been identified in several endomembranes, including the inner mitochondrial membrane. In higher plant mitochondria, H^+-PP_iase is able to synthesize PP_i in a respiration-dependent manner as well as to couple PP_i hydrolysis to the generation of $\Delta\mu H^+$ with an H^+/PP_i stoichiometry of two. The PP_i-dependent H^+ transport requires Mg^{2+}, is stimulated by K^+ and Rb^+, and is inhibited by Ca^{2+}. This activity is associated with a 35-kDa catalytic protein subunit that hydrolyzes only PP_i. Using submitochondrial particles, it has been shown that the PP_iase activity is localized on the inner surface of the inner membrane.

The mitochondrial PP_iase differs from the vacuolar PP_iase (V-PP_iase) by its sensitivity to inhibitors, by the molecular mass of the catalytic subunit (35 kDa versus 70 kDa for V-PP_iase), and by its H^+/PP_i stoichiometry (two instead of one for V-PP_iase). Moreover, polyclonal antibodies raised against each of them do not cross-react. Plant mitochondrial H^+-PP_iase appears to be arranged as a multisubunit catalytic head connected to a hypothetical proton channel and is similar to those from mammalian and yeast mitochondria.

The ability of mitochondrial PP_iase to pump H^+ outside poses the question of its functional relationship with the F_1F_0 ATPase and of its role in H^+ gradient generation or consumption. A functional link is proposed to exist between F_1F_0 ATPase, H^+-PP_iase, succinate dehydrogenase, and ADP/ATP carrier through the equilibrium between matrix free Mg^{2+} and Mg-PP_i, Mg-ATP, and Mg-ADP. Indeed, the Mg complexes are substrates of the phosphorylating enzymes, free ATP and ADP are substrates of the translocator, and free Mg^{2+} is a strong inhibitor of succinate dehydrogenase. H^+-PP_iase may act as a synthase and produce PP_i that may be exported to the cytosol by the ADP/ATP carrier. Therefore, H^+-PP_iase acting as a synthase or hydrolase, depending on the energetic state of the plant cell, could function as a buffer for $\Delta\mu H^+$.

3 REDOX FREE-ENERGY DISSIPATING PATHS IN PLANT RESPIRATORY CHAIN [14]

As already mentioned (Sec. 2.1), the plant respiratory chain has six ways to reduce ubiquinone from NADH and two ways to oxidize ubiquinol (Fig. 2). As extra electron pathways do not contribute to H^+ pumping, the redox free energy release during electron transfer is lost as heat. The role of these multiple ''nonphosphorylating enzymes'' in the energetic economy of plant cells is not yet clear. They could appear as wasteful systems except in specialized tissues where this release of heat induces thermogenesis that is used for reproductive purposes.

Function of these alternative electron pathways could provide metabolite flexibility. Indeed, plant mitochondria exist in cells containing chloroplasts that produce ATP and a large portion of their respiratory substrates. As the rate of phosphorylating respiration is regulated by the supply of ADP and as during photosynthesis the ATP/ADP ratio is high, the phosphate potential exerts a back-pressure on the H^+-pumping electron transfer respiratory chain causing an increase in the NADH/NAD$^+$ ratio. The scarcity of NAD$^+$ can slow down the Krebs cycle and glycolysis, so limiting the amount of carbon skeletons produced that are required in anabolic processes. Thus, alternative pathways appear to be able to shortcut high phosphate potential in order to provide the Krebs cycle and glycolysis with NAD$^+$ and increase carbon flow when the plants' needs for carbon skeletons increase. These needs increase mainly during growing and the stress response to wounding, cold exposure, nutrient limitations, and parasite attacks. Thus, alternative pathways could play a role in adaptation, resistance, and growth.

3.1 Alternative Non-H$^+$ Pumping NAD(P)H Dehydrogenases [15]

From available knowledge a clearer picture appears concerning the multiple enzymes implicated in the alternative NAD(P)H oxidations associated with plant mitochondria. Five routes have been proposed: three routes on the matrix side of the inner membrane and two routes on the cytosolic side, with various NADH and NADPH specificities (Fig. 2).

Direct evidence has been obtained for the presence of two external dehydrogenases peripherally bound to the inner membrane, one specific for NADH and the other relatively specific for NADPH. Activities of both NADH and NAD(P)H dehydrogenases are Ca^{2+} dependent and can be associated with 58- and 32-kDa enzymes, respectively. A 43-kDa polypeptide with NADH dehydrogenase activity has been isolated from soluble proteins of beetroot mitochondria. Its native molecular mass is 86 kDa and its location is on the matrix side of the inner membrane. It was concluded that this protein is a dimer responsible for matrix rotenone-insensitive NADH oxidation. An NAD(P)H dehydrogenase activity has also been identified showing equal affinity for NADH and NADPH that is associated with a trimer of a 26-kDa polypeptide.

Besides these four routes, several observations suggest the presence in plant mitochondria of a smaller form of complex I (400 kDa) that could catalyze NADH:Q reductase activity in a rotenone-insensitive way. This form could be an assembly intermediate (peripheral arm) of complex I (see Sec. 2.2).

How the electrons are partitioned between the numerous internal NADH oxidation pathways is not well understood. It seems that the rotenone-sensitive pathway (complex I fully assembled) has a higher affinity for NADH, so the other paths could act more efficiently only when a reducing power overload exists in the matrix. The existence of the two external pathways makes most plant mitochondria able to oxidize cytosolic NADH and NADPH, decreasing the reducing power in the cytosol without the need for substrate shuttle as in animal mitochondria. All these alternative NAD(P)H dehydrogenases seem to be inducible, species specific, or even tissue specific.

3.2 Alternative Ubiquinol Oxidase [16,17]

Cyanide-resistant respiration has been known for a long time. It was shown to occur during flowering development in aroids and to be thermogenic. This respiration was attributed to a cyanide-resistant quinol oxidase called the alternative oxidase (AOX). The production of monoclonal antibodies raised against *Sauromatum guttatum* AOX has allowed identification of AOX proteins in a wide variety of plants, in fungi, in trypanosomes, and in some protists such as amoebae, indicating that AOX is quite widespread among organisms except animals.

Isolation of complementary DNA (cDNA) and nuclear genes encoding AOX has led to the primary structure, topological model, and structural modeling. Sequence analysis allows the proposal of two membrane-spanning helices separated by 40 amino acids, including an amphipathic helix exposed to the cytosolic side and two hydrophilic regions of 100 amino acids on both COOH and NH_2 termini extending into the matrix. Cross-linking studies have shown that AOX exists as a homodimer of 65 kDa with two district states: an oxidized state covalently linked by an intermolecular disulfide bridge at the level of two cysteines situated in each N-terminal domain and a reduced state (more active enzyme) linked through noncovalent interactions. As AOX is believed to reduce oxygen to water involving a four-electron reaction, an active site containing transition metals has been predicted. The reducing-substrate binding site is proposed to be situated on the matrix side in a hydrophobic pocket between the two transmembrane helices containing three fully conserved residues that are potential ligands for ubiquinone. A cysteine residue that was claimed to be different from the cysteine implicated in the intermolecular disulfide bridge interacts with α-keto acid (pyruvate) with resulting activation of the enzyme through thiohemiacetal formation. This activation was dependent on the oxidation-reduction state of the intersubunit disulfide bond, more efficient in the reduced state. Site-directed mutagenesis of the two conserved cysteines (Cys-128 and Cys-78), heterologous expression in *Escherichia coli* cells of *Arabidopsis italiana* AOX, and cross-linking and activity assays have been used to show finally that the Cys-78 is involved in both regulatory disulfide bond formation and pyruvate stimulation [17]. It is remarkable that AOX from some microorganisms is regulated neither by α-keto acids nor by the redox state of the enzyme, which could indicate the absence of this cysteine.

Regulation of electron flow through AOX is very complex: the level of AOX activity is a function of the amount of enzyme in the membrane, its redox state, and its activation by α-keto acids that are not independent and of the amount of ubiquinone in the membrane and its redox state.

3.3 Electron Partitioning between Ubiquinol Oxidizing Pathways [16]

Interplay of the various levels of regulation influences not only AOX activity but also the way the electrons are distributed between the alternative and the cytochrome pathways, i.e., electron partitioning, and consequently the part of oxygen consumption building the H^+ electrochemical gradient. The only way to quantify electron partitioning between the branching terminal oxidases (i.e., the cytochrome and alternative oxidases) is to measure their actual contribution in the overall respiration. Measurements of respiration specifically sustained by the cytochrome pathway (in the presence of the AOX inhibitor, e.g., benzohydroxamate, BHAM) and by the alternative pathway (in the presence of cyanide or antimycin) do not reflect the true contribution of each pathway to the total respiration, as inhibition of one pathway inevitably forces the electrons into the other pathway.

Three different methods are available to study electron partitioning:

1. The kinetic approach based on the interplay between ubiquinol-oxidizing and ubiquinone-reducing electron fluxes in the overall steady-state respiratory rate and the redox state of ubiquinone.
2. The oxygen isotope differential discrimination based on the fractionation of oxygen isotopes (^{18}O, ^{16}O) that occurs during plant respiration because oxygen containing the lighter isotope reacts more readily than molecules containing the heavier isotope. A small difference exists in the oxygen isotope discrimination of AOX and cytochrome oxidase.
3. The ADP/O method based on the nonphosphorylating property of AOX involves pair measurements of the ADP/O ratios in the presence and in the absence of the AOX inhibitor and measurements of the overall phosphorylating respiration (state 3).

Each method has its own domain of application because of its own limitation or restraining basic hypothesis. The three methods have shown that AOX can be engaged in state 3 respiration when the cytochrome pathway is not saturated and thus can contribute efficiently to a decrease in the H^+/O ratio and in the rate of H^+ gradient buildup. Thus, in plant mitochondria energy released by redox reaction can or cannot be totally transferred into a transmembrane H^+ electrochemical gradient depending on both pathways used to reduce ubiquinone and the partitioning of ubiquinol electrons between the two terminal oxidizing pathways.

4 PROTON ELECTROCHEMICAL GRADIENT CONSUMPTION

After its transfer into a transmembrane electrochemical gradient, the free energy released by redox reactions can be trapped by the ATP synthesis to fulfill the overall energy conservation. The H$^+$ gradient can be also used in H$^+$- or $\Delta\Psi$-dependent transport processes and by the energy-linked pyridine nucleotide transhydrogenase that transfers hydride equivalents from NADH to NADP$^+$ with concomitant translocation of H$^+$ from cytosol to matrix. There is no indication of such an H$^+$ transhydrogenase in plant mitochondria. As a minor portion of mitochondrial proteins is encoded in the mitochondria, the other proteins (there may be over 1000 in higher organisms) must be imported from the cytosol as precursors. A membrane potential is necessary to stabilize the presequence on the inside of the inner membrane before translocation, processing, and assembly. This complicated process is outside the scope of this chapter.

4.1 Complex V: The ATP Synthase

4.1.1 Organization [2,4]

Complex V is a reversible enzyme that can synthesize ATP using $\Delta\mu H^+$ and can hydrolyze ATP to pump H$^+$ against the electrochemical gradient, as required by the chemiosmotic theory. ATP synthase exhibits a complex tripartite structure similar in all organisms that synthesize or cleave ATP coupled to H$^+$ translocation. It consists of an extrinsic catalytic domain (F$_1$) protruding in the matrix and an intrinsic membrane domain (F$_0$) connected by a narrow stalk. However, when exposed to a low-ionic-strength medium, the enzyme resolves into only two parts, the membrane-bound F$_0$ portion, which contains the H$^+$ channel, and a soluble part that is the catalytic component (F$_1$). The connecting stalk consists of components of both F$_1$ and F$_0$.

The bovine enzyme contains 16 different subunits for a total molecular mass larger than 500 kDa. The F$_1$ domain contains five different subunits with the stoichiometry $3\alpha:3\beta:1\gamma:1\delta:1\epsilon$. The α and β subunits are homologous, holding nucleotide binding sites, but only β subunits have catalytic activity. The ATP synthase inhibitor protein is loosely attached to F$_1$. The F$_0$ sector contains 10 subunits: a, b, c, d, e, f, g, OSCP (oligomycin sensitivity–conferring protein), A6L, and F6. Part of the new subunits discovered in a highly purified and functional synthase has been shown by immunological methods. The stalk is formed from part of the γ subunit and other subunits including OSCP, F6, β, and δ. In plants, the mitochondrial ATP synthase (from potato) has been resolved by blue native page electrophoresis. The intact complex (580 kDa) contains 13 subunits. The F_1 domain is made of six subunits analogous to those of the bovine F_1 and the F_0 part contains at least seven subunits. In yeast, ATP synthase consists of 13 different subunits, 5 in the F_1 sector and 8 in the F_0 sector. All the subunits are nuclear encoded except (1) in mammals, subunits a and A6L (genes *atp*6 and *atp*8) of F_0; (2) in plants, subunits of F_0 encoded by *atp*6 and *atp*9 and of F_1 encoded by *atp*1; and (3) in yeast, subunits of F_0 encoded by *atp*6, *atp*8, and *atp*9. All these subunits are encoded by the mitochondrial genome. In *E. coli*, the F_1F_0 ATP synthase contains eight subunits, five in the F_1 sector analogous to $\alpha\beta\gamma\delta\varepsilon$ and three in the F_0 sector (a, b, and c); again, the prokaryotic enzyme appears to be the minimal evolutionarily conserved system.

4.1.2 Kinetic Mechanism [18]

The mechanism by which ATP synthase couples the H^+ electrochemical gradient to synthesis of ATP was one of the most central questions in bioenergetics. First steps in elucidation of the mechanism have consisted of kinetic and binding studies that led to three important pieces of information: (1) ATP synthase catalytic subunits strongly interact, and the enzyme activity depends on these cooperative interactions; (2) catalytic activity follows a binding change mechanism in which energy from the H^+ gradient promotes the release of tightly bound ATP from the enzyme, and (3) the three β-subunit catalysis proceeds sequentially in an identical manner through conformational changes that facilitate the binding, interconversion, and ATP release steps (alternating site mechanism). This information, together with the observation that the γ subunit influences the catalytic events, was the basis of the concept of rotational catalysis (Fig. 7A) proposed in 1981. According to Boyer, each of the three active sites would pass through a cycle of three different states: "loose" in which ATP and ADP + P_i bind equally to the active site and can exchange rapidly with solution, "tight" in which ATP and ADP + P_i are in equilibrium at the active site and the ATP synthesized is tightly bound, and "open" in which ATP is expelled due to a conformational change induced by H^+ uptake. However, at any given moment the three catalytic sites would be in a different state.

FIGURE 7   Rotational catalysis of ATP synthase. (A) Complete cycle of alternating site mechanism. The $\alpha_3\beta_3$ hexamer is divided into three sectors each in a different state: O, "open"; L, "loose"; T, "tight". γ subunit (triangle) is rotating 120° clockwise, involving energy input in each step. (B) Structural aspects. Rotor is made of 12 subunits c, 1 γ, and 1 ε. Stator is composed of 1 subunit a, 2 b, 1 δ, and the $\alpha_3\beta_3$ hexamer. Subunit a contains the putative H^+ half-channels. (Redrawn from Ref. 22.)

A.

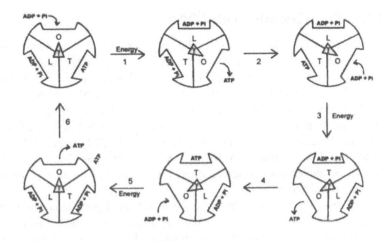

B.

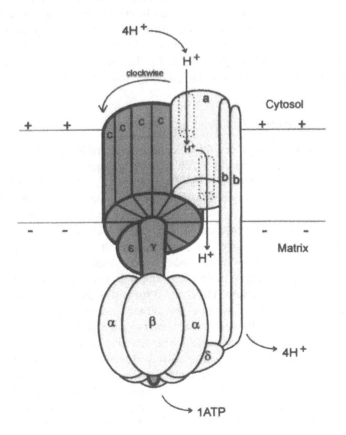

4.1.3   Structural Information [19,20]

The structural asymmetry of the β subunits revealed in the structure of the bovine F_1 ATPase correlates nicely with the sequential binding changes and nucleotide affinities of Boyer's mechanism. The structural asymmetry arises from interactions of the α-helical domain of the γ subunit, which fills the central cavity of the $\alpha_3\beta_3$ subassembly, with the COOH-terminal domain of the β subunits. The nucleotide-binding active sites are at the interfaces between α and β subunit C-terminal domains, predominantly in β subunits. The NH_2-terminal domains form a barrel of β-sheets that keeps the $\alpha_3\beta_3$ hexamer together. Rotation of the γ subunit inside the hexamer would facilitate the cooperative binding of the substrates and release of the product.

Subsequent to the bovine structure report, an elegant demonstration of rotational movement of subunits in the binding change mechanism was provided with *E. coli* enzyme. Based on the mitochondrial F_1 structure that shows a specific interaction between an α-helix containing a cysteine of the γ subunit and a loop of a β subunit, several residues of this loop of *E. coli* β subunit were substituted by cysteine. Addition of an oxidant induced formation of a specific disulfide bond between the β and γ subunits, with loss of activity that was restored by reduction of the disulfide. Although evidence for the rotational catalysis seemed convincing, such novel behavior in enzymology required a physical demonstration of rotation.

4.1.4   Rotational Catalysis [21–23]

Strong evidence that a central structure rotates inside F_1 during ATPase catalysis arises from experiments in which a probe is attached to the COOH-terminus of the γ subunit that protrudes from the center of F_1. This procedure has been used with immobilized chloroplastic F_1 labeled by fluorescent eosine derivatives at the penultimate cysteine residue of the γ subunit and with application of polarized absorption anisotropy relaxation after photobleaching. The results showed that the ATP-driven rotation of the γ subunit over three equidistantly spaced angular positions in the hexagon $\alpha_3\beta_3$ takes place. Another example of the procedure was applied to a single molecule of the $\alpha_3\beta_3\gamma$ subcomplex of F_1 ATPase from a thermophilic bacterium. The α and β subunits were engineered to contain NH_2-terminal polyhistidine tags to allow immobilization on nickel-coated beads, and a fluorescent actin filament was attached to the γ subunit with streptavidin. Addition of ATP caused rotation of the actin filament, as viewed by video microscopy, in discrete 120° steps in an anticlockwise direction when viewed from the "membrane" side. From load experiments (variable length of actin filament), the work associated with a 120° step can be estimated and suggests nearly 100% energy conversion from ATP to motion. These experiments confirmed the rotary mechanism at least in the direction of ATP hydrolysis and make ATPase the world's smallest rotary engine. Its mechanochemical behavior has been modeled, and simulation supports the high efficiency of the motor. The model proposes that ATPase converts energy of nucleotide binding to an elastic strain at the hinge between the moving lobes of the β subunit.

However, the preceding studies have not examined whether rotation of the F_1 subunit occurs in coupled F_0F_1 during ATP synthesis. The first clear indication that subunit rotation in F_0F_1 is an integral part of the coupling during ATP synthesis has been provided by a hybrid F_1/cross-linking approach. The F_1 hybrid made of a β/γ specific sulfide bridge and two epitope-tagged β (β flag) subunits in the two non–cross-linked positions was bound to F_0 in F_1-depleted membranes. After reduction, a brief exposure to conditions of ATP synthesis (ADP, P_i, NADH) followed by reoxygenation resulted in the appearance of the β flag in the β-γ cross-linked product, indicating rotation of γ relative to the β subunits. This rotation was inhibited when ADP, P_i, and NADH were omitted; when H^+ transport through F_0 was blocked with N,N'-dicyclohexylcarbodiimide (DCCD); and in the absence of ADP and P_i and presence of NADH. These results indicate that an H^+ electrochemical gradient alone (presence of NADH alone) is not sufficient to promote subunit rotation in F_0F_1 and that ADP and P_i must bind at a catalytic site on F_1 before H^+ can be transported. If these two events were not sequentially coupled, H^+ crossing F_0 could induce subunit rotation when the F_1 catalytic sites were empty, so the intrinsic coupling of F_0F_1 ATP synthase would be lost and energy wasted.

Rotational catalysis implies that the F_0 domain must contain a rotating structure responding to $\Delta\mu H^+$ and that its rotation must be coupled to the γ subunit. The structure of F_0 remains to be determined; nevertheless, the oligomeric ring of 12 subunits c is proposed to rotate with respect to a static a_1b_2 subcomplex and to drive the rotation of γ in F_1. Proton binding and release at a conserved carboxylic acid residue in the center of the membrane may drive the rotation of the c ring. The number of c subunits logically determined the H^+/ATP stoichiometry, which is 12 H^+ for 3 ATP synthesized. A second requirement for rotational catalysis is that some structures must prevent the 3α-3β ring from rotating when the c ring coupled to γ rotation occurs. These components would form a stator arm structure and are proposed to be subunit a, dimer of subunit b, and subunit δ that link the arm to $\alpha_3\beta_3$ by an interaction with one α subunit. Thus, the entire structure of F_0F_1 ATP synthase can be divided into two counterrotating parts: the stator consisting of a, 2b, δ, and $\alpha_3\beta_3$ and the rotor consisting of 12c, γ, and ε (Fig. 7B). Proton flow through the interface between a and c oligomer generates torque that drives the rotor and the stator in opposite directions.

The heart of the rotary mechanism may be the protonation-deprotonation of the conserved carboxylic groups of c subunits. The precise mechanism by which H^+ translocation through F_0 drives subunit rotation remains the main enigma in the system, and better understanding will await the determination of the three-dimensional structure of the entire synthase. Nevertheless, it has been proposed that the a subunit contains two half H^+ channels, each in contact with the opposite side of the membrane. An H^+ could traverse the membrane if it moves through one half-channel to the center of the membrane, binds to one carboxylic group of the c subunits, and then is carried to the other half-channel by rotation of the c subunit complex. As c subunits are anchored to γ and as a subunit is anchored to $\alpha_3\beta_3$ through b subunit, the rotation of c subunits relative to the a subunit in F_0 would drive the rotation of the γ subunit relative to $\alpha_3\beta_3$ (in F_1).

4.2 Mitochondrial Transport of Metabolites [24–26]

One of the basic postulates of the chemiosmotic theory is the presence of exchange-diffusion carrier systems that allow transport of ions through the inner mitochondrial membrane without collapsing the $\Delta\mu H^+$ needed for ATP synthesis.

The metabolic compartmentation within the eukaryotic cell into two or more compartments requires a continual traffic of metabolites and end products between organelles and cytosol and between the different organelles. Indeed, in animal mitochondria ATP synthesis occurs inside, whereas ATP is used mostly outside. So, ATP must be transported out and ADP, P_i, and oxidizable substrates must be transported in. In plant cells, metabolic interactions between chloroplasts, mitochondria, peroxisomes, and glyoxysomes require rapid exchanges of metabolites in order to link respiration and processes such as photosynthesis, photorespiration, lipid-sugar transformation, and nitrate assimilation, among others. As such metabolites are polar molecules and often need to move against a concentration gradient, energy is required. In mitochondria, the primary energy source for transport is $\Delta\Psi$ and/or a ΔpH gradient. Our purpose is not to describe extensively the family of specific carriers that catalyze the transport of metabolites across the inner mitochondrial membrane but to link their activity with the H^+ electrochemical gradient built up by the respiration.

4.2.1 Energetic Aspects

Several strategies exist to drive specific carrier activities. Some carrier activities are electrophoretic (i.e., they lead to net translocation of electric charge), such as adenine nucleotide carrier (AAC), aspartate/glutamate carrier (AGC), and uncoupling protein (UCP), with a transport mode that is either antiport (AAC and AGC) or uniport (UCP) and with a driving force that is $\Delta\Psi$ (AAC) or $\Delta\Psi + \Delta pH$ (AGC and UCP). Indeed, ADP^{3-}/ATP^{4-} exchange causes the exit of one negative charge, $Glu^{2-} + H^+/Asp^{2-}$ exchange is responsible for an H^+ reentry and an exit of one negative charge, and the UCP uniport catalyzes either H^+ uptake or the exit of the anionic form of a free fatty (see later). Some other carriers catalyze electroneutral transports because they are proton compensated, such as phosphate carrier (P_iC), pyruvate carrier (PYC), and glutamate carrier with a transport mode that can be symport (with one H^+) or antiport (with one hydroxyl) and with a driving force ΔpH.

Other carriers catalyze electroneutral exchanges, such as oxoglutarate carrier (OGC), dicarboxylate carrier (DIC), and citrate carrier (CIC), with an obvious antiport transport mode and with a driving force that is the electrochemical gradient of one substrate used to build the electrochemical gradient of the second substrate. However, these electroneutral exchange carriers are indirectly linked to the H^+ electrochemical gradient because they operate in sequence through the phosphate carrier: P_iC imports P_i requiring an H^+ gradient, DIC accumulates dicarboxylate in exchange with this P_i, and finally OGC and CIC exchange this dicarboxylate for external oxoglutarate and citrate, respectively. This is a cascade of osmo-osmotic energy couplings linked to the chemiosmotic coupling of the respiratory chain. Several other carriers translocate neutral molecules in either antiport or uniport mode and use the concentration gradient as the driving force (e.g., carnitine carrier, glutamine carrier). Carriers involved in oxidative phosphorylation are energy consuming. For example, the import of inorganic phosphate into the mitochondrial matrix plus the exchange of $ATP_{in}^{4-}/ADP_{out}^{3-}$ costs the equivalent of one H^+ uptake. Thus, this leads to an H^+/ATP ratio of five: four H^+ for ATP synthase + one H^+ (i.e., 20%) for translocation.

4.2.2 Structural Aspects

All these carriers belong to the mitochondrial carrier family (MCF), which has been completely described, thanks to genomic programs, in yeast, *Caenorhabditis elegans*, and partially in *Arabidopsis thaliana*. In yeast no fewer than 34 different carriers exist, most of them of unknown function. The MCF carriers have a similar primary structure (tripartite), a molecular mass of around 30 kDa, and a consensus sequence called the "mitochondrial energy signature." The three-dimensional structure of these carriers is not known because their high hydrophobicity has hindered crystallization. However, prediction of their transmembrane topology from sequence analysis has been confirmed by investigating the accessibility of polar parts on each side of the membrane. A structural model of the carrier protein of MCF consists of six transmembrane α-helices, three large loops on the matricial side, and two short loops on the cytosolic side with the NH_2 and COOH termini. Large matricial loops might be inserted into the channel made by α-helices as hairpin structures that could participate in the transport process. Several studies support a functional dimeric state into the membrane for carriers such as PiC, UCP, OGC, AAC and even tetrameric AAC, and multiple oligomeric form of OGC.

4.2.3 Kinetic Mechanism

The first attempts to determine the kinetic mechanism of an MCF member were undertaken in 1972 with intact mammalian mitochondria with OGC. Two-substrate initial-rate studies with a fast kinetic method have led to the conclusion that

OGC functions according to a double-binding-site mechanism with independent binding of internal and external substrates. Thus, both exchanged substrates form a ternary complex with the carrier, meaning that both internal and external sites are simultaneously accessible to substrates before the transport (limiting step) occurs. Two-substrate kinetic studies carried out since 1988 with carriers reconstituted in liposomes have confirmed the occurrence of the double-binding-site mechanism. Most of these studies have been performed with mammalian and yeast mitochondria. Plant carriers have been identified on the basis of their substrate and inhibitor specificities and, as in mammals and yeast, some are involved in oxidative phosphorylation, others in oxidizable substrate transport or amino acid and cofactor transport.

4.2.4 Plant Mitochondrial Carrier Family

The plant P_iC has the same characteristic as the other P_iC and seems to be highly expressed in developing plant organs that require energy. The plant mitochondria AAC exhibits substrate and inhibitor specificity similar to that of the other AAC, except that plant AAC is able to translocate GDP and GTP whereas mammalian AAC is only able to bind these nucleotides. In plants, ADP/ATP exchange also occurs in plastids, but this carrier has 12 potential transmembrane helices and has some similarities to bacterial adenylate carrier. Besides a major function in oxidative phosphorylation, the plant mitochondrial AAC seems to be involved in various physiological events such as cytoplasmic male sterility, uncoupling of mitochondria, and apoptosis. Most of the oxidizable substrate carriers isolated from other organisms have been detected in plant mitochondria (i.e., PYC, DIC, CIC, OGC) and are implicated in the transport of Krebs cycle intermediates. Their roles in plant cells are reviewed by Laloi [25], who also discussed their functional analogy with the carriers on the other organisms.

Animal mitochondria do not transport oxaloacetate. In contrast, in plant mitochondria the transport of oxaloacetate occurs via a carrier different from DIC and OGC. However, if its transport mode seems to be antiport, the nature of the countersubstrates is not yet clear. A mechanism by which amino acids are transported through the plant mitochondrial membrane is not yet well understood, even if their transport seems to be metabolically important as in the case of glycine, serine, and proline, for example. Plant mitochondria seem to have an AGC that could catalyze a glutamate-aspartate exchange that is not electrogenic as in mammal mitochondria, where it is implicated in the malate-aspartate shuttle transferring reducing equivalents through the inner membrane. In plant mitochondria, this transfer may be carried out by the oxaloacetate-malate shuttle and by the external NADH dehydrogenase (see earlier). Few data suggest that plant mitochondria transport cofactors such as NAD^+, coenzyme A, thiamine pyrophosphate, and tetrahydrofolate. The genome sequencing of A. thaliana will make it possible to identify all the sequences having the features of mitochondrial carrier family members, but it is obvious that the study of plant mitochondrial carriers is still at the beginning compared with that of mammalian carriers.

4.3 The Uncoupling Protein

4.3.1 Mechanism [27,28]

The major discovery in the MCF field occurred in 1995 when an uncoupling-like protein was observed in potato mitochondria. Uncoupling protein (UCP) is located in the inner mitochondrial membrane and its activity dissipates the proton electrochemical gradient. The action of UCP is to mediate free fatty acid (FFA)–activated H$^+$ reuptake driven by transmembrane electrical potential and pH difference. UCP is activated by FFA and is allosterically inhibited by purine nucleotides. For instance, addition of FFA to isolated mitochondria results in mitochondrial uncoupling revealed by an increase in resting respiration (state 4) and a decrease in $\Delta\Psi$. Further additions of GTP, which inhibits UCP, or of bovine serum albumin (BSA), which removes FFA, result in initial respiration and $\Delta\Psi$.

Two mechanisms describing H$^+$ reuptake by UCP and FFA activation are still under discussion (Fig. 8). One model proposes that UCP is a true H$^+$ carrier and that FFAs bind to the carrier in such a way that their carboxyl groups enter the carrier hydrophilic pore, where they serve as H$^+$ donors and acceptors to resident carboxyl groups of aspartate or glutamate residues. Histidine residues are proposed to be the final H$^+$ relay that liberates the H$^+$ in the mitochondrial matrix. The other model proposes that UCP is an FFA carrier that transports the deprotonated fatty acid from the matrix to the cytosol, where it is protonated. The protonated FFA crosses the membrane by diffusion, resulting in H$^+$ reuptake by the FFA-cycling process.

4.3.2 Distribution [25]

For more than 20 years, the mammalian UCP1 present in brown adipose tissue was believed to be unique and a late evolutionary acquisition required especially for thermogenesis in newborn, cold-acclimated, and hibernating mammals. Discovery of a UCP-like protein in potato tubers in 1995 and study of its activity in proteoliposomes demonstrating FFA-mediated H$^+$ transport similar to that described for UCP1 have strongly activated the UCP field. Moreover, the discovery of several novel uncoupling proteins (UCP2, UCP3, and UCP4) in various mammalian tissues has shown that UCP is widespread in tissues of higher animals and could have various physiological roles. The plant UCP is also rather ubiquitous in plants. It has been isolated from potato and tomato and immunologically detected in several climacteric fruits. Characterization and expression of UCP genes in plants have also been addressed in various tissues of potato and in *Arabidipsis thaliana*, where two different isoforms of UCP exist.

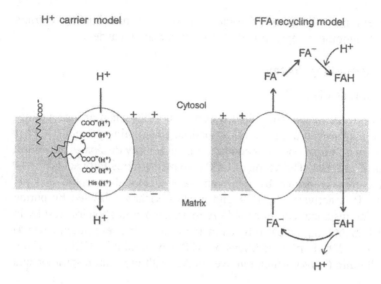

Figure 8 Two possible mechanisms of H⁺-reuptake by uncoupling protein. (Left) The true H⁺ carrier model with free fatty acid carboxylate as H⁺ relay serving as H⁺ donor/acceptor to resident carboxyl groups of the protein. Histidine residue (His) is shown as a final H⁺ relay that liberates H⁺ in the matrix (as proposed in Ref. 27). (Right) The free fatty acid recycling model (as proposed in Ref. 28).

Thus, the presence of UCP in plants seems to be widespread, suggesting various physiological roles, for instance, in fruit ripening and in protection against oxidative damage caused by free radicals. The recent immunodetection and functional characterization of a UCP-like protein in mitochondria from *Acanthamoeba castellanii*, which in molecular phylogenesis appears on a branch basal to the divergence points of plants, animals, and fungi and in mitochondria from the nonfermentative yeast *Candida parapsilosis*, prove that UCP-like proteins occur in the four eukaryotic kingdoms: animals, plants, fungi, and protists. These results demonstrate that UCPs emerged as specialized proteins for H⁺ recycling very early during phylogenesis before the major radiation of phenotypic diversity in eukaryotes ($>10^9$ years ago).

4.3.3 Proton Reuptake Partitioning between Uncoupling Protein and ATP Synthase [29]

Both uncoupling protein and ATP synthase are able to consume $\Delta\mu H^+$ built up by the protonophoric oxidoreductases of the respiratory chain and may be considered as two branching pathways: UCP is the $\Delta\mu H^+$ energy-dissipating path, and

ATP synthase is the $\Delta\mu H^+$ energy-conserving path. The ADP/O method (see Sec. 3.3) has been applied to calculate the contributions of UCP activity and ATP synthesis in state 3 respiration of green tomato mitochondria fully depleted of FFA, using pair measurements of ADP/O ratios in the absence or presence of various linoleic acid (LA, an abundant FFA in plants) concentrations. These measurements were made with succinate (plus rotenone) as an oxidizable substrate in the presence of benzohydroxamate (BHAM) to exclude the oxygen consumption by AOX and with increasing concentrations of n-butyl malonate (an inhibitor of succinate uptake) in order to decrease the rate of the quinone-reducing pathway. This approach has allowed description of how contributions of UCP and ATP synthase change with variations in the BHAM-resistant phosphorylating respiration: UCP activity, at a fixed LA concentration, remains constant, and ATP synthesis decreases linearly. These results show how efficiently UCP activity can divert energy from oxidative phosphorylation, especially when state 3 respiration is decreased.

Therefore, besides the redox free energy dissipating systems [NAD(P)H rotenone–insensitive dehydrogenases and AOX] able to decrease the buildup of an H⁺ electrochemical gradient, plant mitochondria possess UCP, which is able to dissipate this H⁺ gradient and is potentially able to decrease the ATP synthesis yield (per O_2 consumed), converting the osmotic energy into heat.

4.4 Mitochondrial Transport of Cations

There is an enormous literature on cation transport in mammal mitochondria, but it mainly concerns the functional properties of carriers, whereas little information is available about their structure.

Monovalent and divalent cations can be translocated in and out of mitochondria via exchangers (antiport process) or channels (uniport process) in the inner mitochondrial membranes, and the driving force of this transport can be $\Delta\Psi$ alone, ΔpH alone, or both together ($\Delta\mu H^+$). The cation transport is important for vital functions such as volume control of mitochondria and regulation of energy metabolism. Carriers and channels for K⁺ and Na⁺ are important in the volume homeostasis because the concentrations of these ions in the cytosol are very high (approximately 150 mM for [K⁺] and 5 mM for [Na⁺]). Indeed, if they were space distributed only according to electrochemical equilibrium for a membrane potential of 180 mV, they would reach internal concentration of 150 M for K⁺ and 5 M for Na⁺, leading to mitochondrial swelling and disruption.

4.4.1 Monovalent Cation Transport [30,31]

Monovalent cations are transported via antiporter (exchanger) systems (Fig. 9A). Two separate antiporters exist, the Na⁺-selective/H⁺ exchanger (NHE) that does not transport K⁺ and the K⁺-unselective/H⁺ exchanger (KHE) that transports K⁺

A.   B.

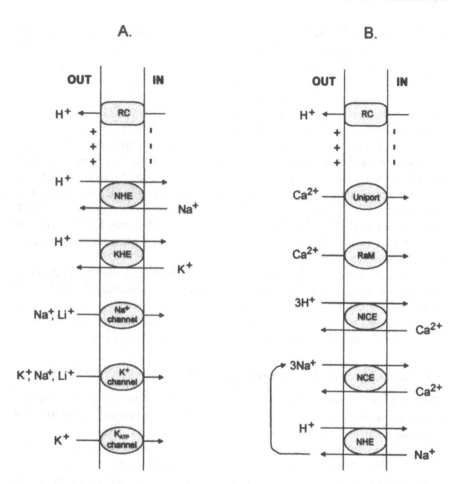

FIGURE 9 Mitochondrial transport of cations. (A) Transport of cations by NA+-selective/H+ exchanger (NHE), K+-unselective/H+ exchanger (KHE), Na+ channel, K+ channel, K+-selective channel sensitive to ATP (K_{ATP} channel). (B) Transport of Ca^{2+} by uniporter, rapid uptake mode mechanism (RaM), Na+-independent pathway for Ca^{2+} efflux (NICE), and Na+-dependent pathway for Ca^{2+} efflux (NCE) that is coupled with NHE activity. RC, respiratory chain.

and Na+. The Na+/H+ antiporter (NHE) is not inhibited by Mg^{2+} and its activity decreases at alkaline pH. The activity of this antiporter is probably implicated in Ca^{2+} cycling (see later). The K+/H+ antiporter (KHE) has a low selectivity for the monovalent cations exchanged for H+. It transports K+, Na+, Li+, and some organic cations, and it is inhibited by Mg^{2+} and activated by high pH.

Several channels (uniporters) transporting monovalent cations inside mitochondria exist in the inner mitochondrial membrane (Fig. 9A). Their activity occurs in energized mitochondria via an electrophoretic (i.e., net charge transport) mechanism. The Na^+ channel selective for Na^+ and Li^+ is competitively inhibited by Mg^{2+} and is activated by ATP. It has an optimal activity at 7.5–8.0 pH. Its activity increases exponentially with $\Delta\Psi$. The nonselective K^+ channel supports $\Delta\Psi$-dependent K^+, Na^+, and Li^+ electrophoretic flux inactivated by matrix Mg^{2+}. The Na^+ fluxes through the Na^+ channel and the nonselective K^+ channel are additive, suggesting the existence of two distinct paths. A K^+-selective channel sensitive to ATP (K_{ATP} channel) has been described and partially purified and reconstituted. It is inhibited by ATP and ADP in the presence of Mg^{2+} and also by palmitoyl coenzyme A. These inhibitors can be relieved by GTP and GDP. It is difficult to decide how many mitochondrial K^+ channels exist at present. What is sure is the existence of a native K^+ conductance and of a K^+ cycle that could allow tuning of mitochondrial volume at the expense of $\Delta\mu H^+$.

In plant mitochondria, the existence of a very active K^+/H^+ antiporter that can partially collapse ΔpH has been shown. Recently, the electrophoretic K^+ uptake by isolated plant mitochondria has been demonstrated, indicating the existence of a K_{ATP} channel that is inhibited by ATP and NADH but not by Mg^{2+} and is stimulated by superoxide anion and palmitoyl coenzyme A. The plant mitochondrial K_{ATP} channel is thus quite differently regulated than the mammalian K_{ATP} channel. Moreover, the K^+ cycle in plant mitochondria seems to be able to collapse $\Delta\mu H^+$ completely (full uncoupling), in contrast to the situation in mammals, where the maximal rate of the cycle is less than 20% of the maximal rate of H^+ ejection by the respiratory chain.

4.4.2 Divalent Cation Transport [30,32]

The Ca^{2+} distribution between cytosol and mitochondria is the result of a kinetic steady state of electrophoretic Ca^{2+} uptake and of Ca^{2+} efflux via two different exchangers. Their interplay allows maintaining a Ca^{2+} concentration gradient in respiring mitochondria that oscillates between 0 and 10 rather than 10^6 as calculated from thermodynamic considerations (for an electrophoretic Ca^{2+} uniporter with $\Delta\Psi = 180$ mV).

It seems that there are two pathways for Ca^{2+} uptake: the well-known Ca^{2+} uniporter and the rapid uptake mode mechanism (RaM) (Fig. 9B). The uniporter catalyzes a $\Delta\Psi$-driven electrophoretic uptake of Ca^{2+} that can be inhibited by ruthenium compounds. The Ca^{2+} uniporter is regulated by divalent cations that are themselves transported by this uniporter (Sr^{2+}, Mn^{2+}, Ba^{2+}, and lanthanides). They competitively inhibit Ca^{2+} uptake through interaction with the transport sites of the uniporter. Mg^{2+}, which is not transported by the uniporter, is a modulator that affects the kinetics of Ca^{2+} uptake by binding to regulatory sites (thereby changing the relationship between the rate and Ca^{2+} external concentration from hyperbolic to sigmoidal). Manganese can counteract the effect of Mg^{2+}, suggesting that Mn^{2+} can displace Mg^{2+} from its binding sites (regulatory sites) and therefore that Mn^{2+} can interact with both regulatory and transport sites. It has been shown that the Ca^{2+} uniporter can be inhibited by adenine nucleotides in the order ATP > ADP > AMP independently of ATP hydrolysis and of Ca^{2+} and Mg^{2+} concentrations. Finally, this uniporter does not appear to be readily reversible by depolarization and by reversal of the Ca^{2+} electrochemical gradient. The RaM mechanism was revealed by Ca^{2+} uptake studies when mitochondria were exposed to a series of Ca^{2+} pulses. This mechanism is also inhibited by ruthenium red but is stimulated by ATP and is not affected by Mg^{2+}, unlike the Ca^{2+} uniporter.

Ca^{2+} efflux could occur through two pathways: an Na^+-independent pathway for Ca^{2+} efflux (NICE) and an Na^+-dependent pathway for Ca^{2+} efflux (NCE) (Fig. 9B). The NICE pathway could be a $3H^+/Ca^{2+}$ exchanger that requires a transmembrane potential as a component of its driving force. This pathway is saturated at low Ca^{2+} loads and is slow. The NCE pathway is also not electroneutral and takes places through the Na^+/Ca^{2+} antiporter (with a stoichiometry of $3Na^+/Ca^{2+}$), requiring $\Delta\Psi$ for its activity.

Mitochondria take up and extrude Mg^{2+} via poorly defined pathways. These Mg^{2+} movements are respiration-dependent reactions (i.e., energy dependent). It seems that Mg^{2+} uptake occurs by nonspecific diffusion processes in response to a high transmembrane potential. Magnesium efflux may take place in exchange for external protons, so it requires $\Delta\mu H^+$. An alternative system for both uptake and release of Mg^{2+} is the reversible and electroneutral $ATP^{4-} Mg^{2+}/Pi^{2-}$ antiport catalyzed by one carrier of the MCF.

The electrophoretic nature of Ca^{2+} uptake and efflux processes and external H^+ consumption by NICE indicate that energy of $\Delta\mu H^+$ is required for Ca^{2+} uptake into mitochondria and for Ca^{2+} release from mitochondria. Moreover, coupled activity of the Na^+/H^+ exchanger (NHE) and of the Na^+-dependent pathway for Ca^{2+} efflux (NCE) also consumes external H^+ when Ca^{2+} is released. This concerted interplay between NHE and NCE can mediate physiological Ca^{2+} cycling and link it to the H^+ electrochemical gradient built up by respiration.

The close connection between respiration and Ca^{2+} movement suggests a role of Ca^{2+} in energy metabolism. Indeed, Ca^{2+} modulates the activities of three important matricial enzymes, pyruvate dehydrogenase, NAD^+-linked isocitrate dehydrogenase, and 2-oxoglutarate dehydrogenase, that are up-regulated by Ca^{2+} uptake. At present, it is quite sure that an important function of Ca^{2+} uptake by mitochondria is to transmit an increase in cellular energy demand. On the other hand, numerous observations indicate that mitochondria could play a role in cellular Ca^{2+} homeostasis.

Although calcium uptake has been demonstrated in plant mitochondria for a long time, its physiological significance is still questioned. There are several reasons for a lack of interest in Ca^{2+} transport in plant mitochondria, such as an apparent lower affinity of plant mitochondria for calcium and a supposed weak electrochemical gradient of Ca^{2+} across the mitochondrial membrane (compared with the plasma membrane, vacuole, and endoplasmic reticulum). Moreover, mitochondrial Ca^{2+} uptake varies between different plant species and tissues and with age in the same species. The low affinity for Ca^{2+} has led to the opinion that it is unlikely that mitochondria could participate in regulation of the cytoplasmic Ca^{2+} level. Studies using mitochondria from corn coleoptiles, pea stems, and Jerusalem artichoke tubers have led to different results. In corn mitochondria (1) respiration-coupled Ca^{2+} uptake cannot be dissociated from the simultaneous influx of phosphate and is sensitive to mersalyl and ruthenium red and (2) a ruthenium red–induced efflux of Ca^{2+} is not affected by mersalyl and can occur in the absence of phosphate movement. Thus, it is concluded that corn mitochondria possess an electrogenic influx pathway (i.e., mersalyl-sensitive Ca^{2+}/Pi cotransport) that is totally different from the electrogenic import systems of mammalian mitochondria and a phosphate-independent efflux pathway similar to the Na^{+}-independent Ca^{2+} efflux (NICE) of mammalian mitochondria. Results obtained with pea mitochondria indicate that phosphate is not required in Ca^{2+} uptake that occurs via an electrophoretic Ca^{2+} uniporter with a high affinity for this ion, whereas in artichoke mitochondria Ca^{2+} uptake is phosphate dependent.

It can be concluded that Ca^{2+} movements in plant mitochondria are much more variable than in mammalian mitochondria and deserve further studies for their full description.

5 FUNCTIONAL CONNECTION AND ROLE OF UNCOUPLING PROTEIN AND ALTERNATIVE OXIDASE [33]

Plant mitochondria have the puzzling property of possessing two free energy–dissipating systems that lead to the same final effect, i.e., a decrease in ATP synthesis yield. A crucial observation has been made in vitro with green tomato mitochondria fully depleted of FFA. Addition of linoleic acid (LA) in the presence of BHAM (an inhibitor of AOX) induces an increase in respiration that is the result of the UCP activity. This LA-induced respiration increases with LA concentration. On the contrary, the CN-resistant respiration that represents the AOX activity (in the absence of UCP activity) decreases with increasing LA concentration. These results show how an increase in FFA level in vivo could affect both of the energy-dissipating systems but in opposite directions: AOX could be progressively switched off by an increase in the FFA content in cells when in contrast UCP is switched on. They also indicate that AOX and UCP would never work together at their maximal activity.

As AOX and UCP activities seem to be mutually excluded in vitro, they could work sequentially during the life of plant cells according to their particular physiological state. Evolution of AOX and UCP activities and immunodetectable protein amounts has been studied during postharvest tomato ripening. Indeed, ripening of fruits provides an interesting model to study a relationship between AOX and UCP as thermogenesis occurs during ripening and as the FFA concentration increases in the postgrowing stage. Results clearly show that the level of AOX protein decreases from the green stage forward and parallels the decrease in AOX-sustained respiration, whereas the level of UCP protein decreases only at the orange stage as is the case for UCP-sustained respiration. Therefore, it was concluded that AOX is active mainly during the growing period and UCP start working in the postgrowth state.

The only obvious physiological function of AOX and UCP can be recognized in specialized plant and animal thermogenic tissues as heat generation related to increasing temperature as in spadices of Araceae during reproductive processes (AOX activity) and in mammalian brown adipose tissue (UCP activity). In nonthermogenic plant tissues and in unicellulars, the role of AOX and/or UCP is not fully understood. In nonthermogenic tissues, AOX and UCP could have a subtle role in energy metabolism control, working as safety valves when overloads in redox potential and/or in phosphate potential occur. These overloads are consequences of imbalance between reducing substrate supply and energy and carbon demand for biosynthesis, both being coupled by the respiratory chain activity. As operation of AOX can directly diminish the reducing power rise, providing NAD^+ to the Krebs cycle and glycolysis, and as operation of UCP can directly induce a drop in phosphate potential, they could theoretically correct the imbalance.

6 CONCLUDING REMARKS

Proton fluxes in mitochondria sustain an impressive variety of processes that are implicated not only in the organellar physiology but also in the general homeostasis of the cell. Even if the new acquired knowledge is important, numerous areas remain obscure and will need a tremendous amount of work. Some fields in mitochondrial research have progressed almost to the molecular mechanism level of understanding (e.g., oxidative phosphorylation), whereas other domains (e.g., metabolite and ion traffic across the membrane) remain far behind. Better understanding of the latter is essential to integrate mitochondrial functions into the whole cell life.

ACKNOWLEDGMENTS

We thank Dr. C. Sluse-Goffart for a critical reading of the manuscript. Because of space limitations, it was not possible to include a comprehensive list of references for all of the work discussed.

ABBREVIATIONS

AAC   adenine nucleotide carrier
AGC   aspartate/glutamate carrier
AOX   alternative oxidase
BHAM   benzohydroxamate
CIC   citrate carrier
complex I   NADH:ubiquinone oxidoreductase
complex II   succinate:ubiquinone oxidoreductase
complex III   ubiquinol:cytochrome c oxidoreductase
complex IV   cytochrome c oxidase
DCCD N,N'-dicyclohexylcarbodiimide
DIC   dicarboxylate carrier
$\Delta\mu H^+$ H⁺ electrochemical gradient
$E_{m,7}$ midpoint potential at pH 7
FAD   flavine adenine dinucleotide
FFA   free fatty acid
FMN   flavin mononucleotide
ISP   the Rieske iron-sulfur protein
KHE   K⁺-unselective/H⁺ exchanger
LA   linoleic acid
NAD⁺, NADH   nicotinamide adenine dinucleotide and its reduced form
NADP⁺, NADPH   nicotinamide adenine dinucleotide phosphate and its
   reduced form
NCE   Na⁺-dependent pathway for Ca^{2+} efflux that is coupled with NHE
   activity
NHE   NA⁺-selective/H⁺ exchanger
NICE   Na⁺-independent pathway for Ca^{2+} efflux
OGC   oxoglutarate carrier
PiC   phosphate carrier
pK_A pH of half dissociation of acidic groups
PYC   pyruvate carrier
Q   ubiquinone
QH_2 ubiquinol
QH_2-O_2 ubiquinol-oxygen
RaM   rapid uptake mode mechanism
RC   respiratory chain
UCP   uncoupling protein

REFERENCES

1.   P Mitchell. Coupling of phosphorylation to electron and proton transfer by a chemi-osmotic type of mechanism. Nature 191:144–148, 1961.

2. M Saraste. Oxidative phosphorylation at the fin de siecle. Science 283:1488–1493, 1999.
3. AG Rasmusson, V Heiser, E Zabaleta, A Brennicke, L Grohmann. Physiological, biochemical and molecular aspects of mitochondrial complex I in plants. Biochim Biophys Acta 1364:101–111, 1998.
4. F Vedel, É Lalanne, M Sabar, P Chétrit, R De Paepe. The mitochondrial respiratory chain and ATP synthase complexes: composition, structure and mutational studies. Plant Physiol Biochem 37:629–643, 1999.
5. U Brandt. Proton-translocation by membrane-bound NADH:ubiquinone-oxidoreductase (complex I) through redox-gated ligand conduction. Biochim Biophys Acta 1318:79–91, 1997.
6. PL Dutton, CC Moser, VD Sled, F Daldal, T Ohnishi. A reductant-induced oxidation mechanism for complex I. Biochim Biophys Acta 1364:245–257, 1998.
7. P Mitchell. Possible molecular mechanisms of the protonmotive function of cytochrome systems. J Theor Biol 62:327–367, 1976.
8. U Brandt. Energy conservation by bifurcated electron-transfer in the cytochrome-bc_1 complex. Biochim Biophys Acta 1275:41–46, 1996.
9. Z Zhang, L Huang, VM Shulmeister, YI Chi, KK Kim, LW Hung, AR Crofts, EA Berrt, SH Kim. Electron transfer by domain movement in cytochrome bc_1. Nature 392:677–684, 1998.
10. T Tsukihara, H Aoyama, E Yamashita, T Tomizaki, H Yamaguchi, K Shinzawa-Itoh, R Nakashima, R Yaono, S Yoshikawa. The whole structure of the 13-subunit oxidized cytochrome c oxidase at 2.8 Å. Science 272:1136–1144, 1996.
11. H Michel. The mechanism of proton pumping by cytochrome c oxidase. Proc Natl Acad Sci USA 95:12819–12824, 1998.
12. GT Babcock, M Wikstrom. Oxygen activation and the conservation of energy in cell respiration. Nature 356:301–309, 1992.
13. A Vianello, F Macrì. Proton pumping pyrophosphatase from higher plant mitochondria. Physiol Plant 105:763–768, 1999.
14. S Mackenzie, L McIntosh. Higher plant mitochondria. Plant Cell 11:571–585, 1999.
15. KL Soole, RI Menz. Functional molecular aspects of the NADH dehydrogenases of plant mitochondria. J Bioenerg Biomembr 27:397–406, 1995.
16. FE Sluse, W Jarmuszkiewicz. Alternative oxidase in the branched mitochondrial respiratory network: an overview on structure, function, regulation, and role. Braz J Med Biol Res 31:733–747, 1998.
17. DM Rhoads, AL Umbach, CR Sweet, AM Lennon, GS Rauch, JN Siedow. Regulation of the cyanide-resistant alternative oxidase of plant mitochondria. Identification of the cysteine residue involved in alpha-keto acid stimulation and intersubunit disulfide bond formation. J Biol Chem 273:30750–30756, 1998.
18. PD Boyer. The ATP synthase—a splendid molecular machine. Annu Rev Biochem 66:717–749, 1997.
19. JP Abrahams, AG Leslie, R Lutter, JE Walker. Structure at 2.8 Å resolution of F_1-ATPase from bovine heart mitochondria. Nature 370:621–628, 1994.
20. TM Duncan, VV Bulygin, Y Zhou, ML Hutcheon, RL Cross. Rotation of subunits during catalysis by *Escherichia coli* F_1-ATPase. Proc Natl Acad Sci USA 92:10964–10968, 1995.

21. R Yasuda, H Noji, K Kinosita Jr, M Yoshida. F1-ATPase is a highly efficient molecular motor that rotates with discrete 120 degree steps. Cell 93:1117–1124, 1998.
22. Y Zhou, TM Duncan, RL Cross. Subunit rotation in *Escherichia coli* F_0F_1-ATP synthase during oxidative phosphorylation. Proc Natl Acad Sci USA 94:10583–10587, 1997.
23. T Elston, H Wang, G Oster. Energy transduction in ATP synthase. Nature 391:510–513, 1998.
24. FE Sluse. Mitochondrial metabolite carrier family, topology, structure and functional properties: an overview. Acta Biochem Pol 43:349–360, 1996.
25. M Laloi. Plant mitochondrial carriers: an overview. Cell Mol Life Sci 56:918–944, 1999.
26. B el Moualij, C Duyckaerts, J Lamotte-Brasseur, FE Sluse. Phylogenetic classification of the mitochondrial carrier family of *Saccharomyces cerevisiae*. Yeast 13:573–581, 1997.
27. E Winkler, M Klingenberg. Effect of fatty acids on H⁺ transport activity of the reconstituted uncoupling protein. J Biol Chem 269:2508–2515, 1994.
28. KD Garlid, DE Orosz, M Modriansky, S Vassanelli, P Jezek. On the mechanism of fatty acid–induced proton transport by mitochondrial uncoupling protein. J Biol Chem 271:2615–2620, 1996.
29. W Jarmuszkiewicz, AM Almeida, CM Sluse-Goffart, FE Sluse, AE Vercesi. Linoleic acid–induced activity of plant uncoupling mitochondrial protein in purified tomato fruit mitochondria during resting, phosphorylating, and progressively uncoupled respiration. J Biol Chem 273:34882–34886, 1998.
30. P Bernardi. Mitochondrial transport of cations: channels, exchangers, and permeability transition. Physiol Rev 79:1127–1155, 1999.
31. D Pastore, MC Stoppelli, N Di Fonzo, S Passarella. The existence of the K⁺ channel in plant mitochondria. J Biol Chem 274:26683–26690, 1999.
32. M Zottini, D Zannoni. The use of fura-2 fluorescence to monitor the movement of free calcium ions into the matrix of plant mitochondria (*Pisum sativum* and *Helianthus tuberosus*). Plant Physiol 102:573–578, 1993.
33. FE Sluse, W Jarmuszkiewicz. Activity and functional interaction of alternative oxidase and uncoupling protein in mitochondria from tomato fruit. Braz J Med Biol Res 33:259–268, 2000.

9

H+ Fluxes in Nitrogen Assimilation by Plants

Fernando Gallardo and
Francisco M. Cánovas

Andalusian Institute of Biotechnology, University of Málaga,
Málaga, Spain

1 COMPARTMENTALIZATION OF NITROGEN ASSIMILATION IN PLANTS

Nitrogen assimilation is a multistep process that involves a set of enzymatic reactions located in different subcellular compartments and plant organs (Fig. 1). Although most plants are able to absorb and incorporate both inorganic and organic nitrogen compounds, the inorganic forms present in the soil represent the main nitrogen source in most habitats. A general scheme of the inorganic nitrogen assimilation requires the distinction of the sources implicated in the supply of inorganic nitrogen. In addition to the primary sources, in which nitrogen is incorporated into plants as nitrate or ammonium from water or soil or as dinitrogen in symbiotic plants with nitrogen-fixing bacteria, a number of metabolic processes such as photorespiration, the biosynthesis of phenylpropanoids, or the catabolism of amino acids supply ammonium ion as a secondary source. These secondary nitrogen sources are of great relevance in some tissues and phases of the plant life cycle. Thus, after germination, most nitrogen is mobilized from storage proteins in the embryo and cotyledons to support the early plant development. This nitrogen mobilization involves protein degradation and amino acid deamination, releasing ammonium ion that will be used in plants as a unique nitrogen source

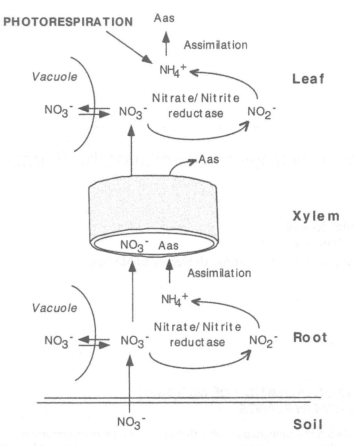

FIGURE 1 General scheme of nitrogen assimilation in higher plants. Only nitrate assimilation is considered for simplicity. Surplus of nitrate in root cells can be stored in the vacuole or translocated to the shoot via xylem. Nitrate reduction to ammonium ion through nitrate and nitrite reductase can operate in both root and photosynthetic cells. Photorespiration is a main source of ammonium ion in photosynthetic cells of C_3 plants. Aas, amino acids.

during the first stages of plant development. Ammonium ion is also produced in the mitochondria of photosynthetic cells by the reaction catalyzed by glycine decarboxylase, a step of the photorespiratory pathway that is due to the oxygenase activity of ribulose bisphosphate carboxylase/oxygenase (Rubisco). The ammonia produced during photorespiration can be quantitatively more important than the ammonia produced through the reduction of nitrate in the leaf of herbaceous C_3 plants [1,2]. Another secondary source that can be quantitatively important in a plant cell comes from the reaction catalyzed by phenylalanine-ammonia lyase, a key step in the biosynthesis of phenylpropanoids that may also be relevant in specialized cells, such as those involved in lignin biosynthesis.

The complexity of the sources of inorganic nitrogen has required the development by the plants of a robust system able to incorporate and assimilate the inorganic nitrogen in response to variations of environmental conditions and to the physiological and metabolic changes associated with the development of the different plant organs.

Regardless of its origin, inorganic nitrogen has to be reduced to ammonium ion to be incorporated in the pool of organic nitrogen–containing molecules in plants. Because of the biological relevance of these reactions, a large research effort was made in the molecular characterization of the enzymes involved. These enzymes include (1) those involved in nitrate and dinitrogen reduction into ammonium ion and (2) the enzymes implicated in the incorporation of reduced nitrogen into the amino acids glutamine and glutamate.

Nitrate, the major nitrogen form in agricultural soil, is reduced to ammonium in partial reduction steps catalyzed by nitrate and nitrite reductases. Both enzymes are metalloenzymes that can be considered as authentic electron transport chains. Nitrate reductase, a cytosolic enzyme, transfers electrons from NADH to nitrate to produce nitrite; nitrite is toxic to the cell, and an efficient reduction to ammonium ion is required to prevent its accumulation. This second step is catalyzed by the plastidic enzyme nitrite reductase, which uses reduced ferredoxin as an electron donor and exhibits a high affinity for nitrite. These two reactions take place in root as well as in mesophyll cells and involve the participation of two different subcellular compartments, the cytosol and the plastid. Once the nitrate is in the cell, its reduction to ammonium is only one possible fate. Depending on the environmental conditions and the type of plant, nitrate can be stored in the vacuole of epidermal and cortex root cells and, in the case of a surplus in the cell, can also be transported to the shoot via xylem for its reduction in photosynthetic mesophyll cells. In herbaceous plants, most inorganic nitrogen assimilation from the primary source takes place in the leaf, whereas in woody plants both root and leaf organs contribute significantly to the assimilation of inorganic nitrogen.

Once inorganic nitrogen has been reduced to ammonium, its incorporation into amino acids is catalyzed by the glutamine synthetase (GS)/glutamate synthase (GOGAT) cycle (Fig. 2). In this pathway, GS catalyzes the incorporation of ammonium into glutamine, and GOGAT facilitates the transfer of the amido group from glutamine to glutamate. These two reactions play a relevant biological role because glutamine and glutamate are the organic nitrogen donors for the synthesis of major nitrogen compounds in plants including other amino acids, nucleic acid bases, polyamines, and chlorophyll.

GS exists as two main isoenzyme forms (isoforms) located in the cytosol

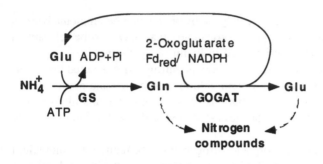

Figure 2 The glutamine synthetase (GS)/glutamate synthase (GOGAT) cycle in higher plants. The amino acids glutamine and glutamate are the main nitrogen donors for the biosynthesis of nitrogen compounds. The stoichiometry of the cycle is shown in the lower part of the figure. Net biosynthesis of nitrogen compounds requires a net supply of energy, reducing power, and 2-oxoglutarate as carbon skeleton.

(GS1) and chloroplast (GS2), which have nonoverlapping roles in mature plant organs [3]. The chloroplastic isoenzyme is responsible for the assimilation of ammonium from primary sources and the photorespiration process, and GS1 is mainly expressed in vascular tissue, where it may be involved in the biosynthesis of glutamine for transport.

GOGAT exists as two enzymes that differ in their source of reductants for catalysis: NADH or reduced ferredoxin (Fd). In photosynthetic tissues, the plastidic Fd-GOGAT is the main GOGAT enzyme, and NADH-GOGAT is mainly implicated in glutamate biosynthesis in nonphotosynthetic tissues and nodules of leguminous plants. The global stoichiometry of GS/GOGAT reactions shows the relevance of the generation of reduction power (NADH and reduced Fd), energy (ATP), and carbon skeletons (2-oxoglutarate) in the assimilation of nitrate. A similar scheme can be described for the assimilation of ammonium ions, with the difference that neither nitrate reductase nor nitrite reductase activities are required.

For the assimilation of molecular nitrogen (dinitrogen), the process implies the generation of a new organ, the nodule, where endosymbiotic bacteroids produce the dinitrogenase complex involved in the reduction of dinitrogen to ammonium.

This chapter deals with the regulation of nitrate assimilation and amino acid biosynthesis. Although the knowledge of most aspects is still fragmentary, the assimilation of inorganic nitrogen depends on the existence of H^+ gradients, mainly in the steps involving transport of molecules through a biological membrane and those requiring energy as ATP and reductant power. The relevance of proton flux in nitrogen fixation in the context of the role of pH in the legume-rhizobia symbiosis is treated in Chapter 15 and, therefore, is not discussed here.

2 REGULATION OF INORGANIC NITROGEN UPTAKE AND TRANSPORT

The availability of inorganic nitrogen in the soil is a limiting factor for plant growth and development. In most agricultural soils, nitrate is the major inorganic nitrogen form [4], but its availability to the plant varies because of agricultural practices, seasonal and spatial variations due to several factors (change in temperature, rain, flooding, etc.), and its depletion by microorganisms living in soil. Because nitrate cannot diffuse through biological membranes, plants have developed an efficient system for nitrate uptake to allow their survival and adaptation to changes in nitrate availability.

Uptake of nitrate by the root cells is the balance between the active nitrate influx and the passive efflux into the soil. It has been calculated that nitrate efflux can represent about 40% of the influx rate, which may represent a regulatory control of the net uptake. Although biochemical studies indicate that nitrate efflux is saturable and inducible by high concentrations of nitrate [5,6], very little is known about the molecular characteristics of the protein(s) involved. More attention has been devoted to the transport system implicated in the nitrate influx into the root cells. The nitrate transport into the root is saturable, even when nitrate concentration is high in the soil or in the apoplast. The kinetics of [^{13}N]nitrate influx in barley plants have shown the existence of a transport process that exhibits hyperbolic kinetics with a low value of K_m; this transport has been termed the high-affinity transport system (HATS) [7,8]. Interestingly, the rate of uptake by barley plants increases with the concentration of nitrate and duration of treatment, indicating the existence of another transport system of high affinity and inducible by nitrate. According to their kinetic characteristics, a total of three transport systems have been identified in barley so far, the two already mentioned with high affinity for nitrate and termed CHATS, for constitutively expressed HATS (K_m ranging from 6 to 20 μM), and IHATS, for inducible HATS (K_m ranging from 20 to 100 μM). The third transport system exhibited a low affinity for nitrate (LATS) and was detected in barley plants that were not treated with nitrate previously, suggesting that it is expressed in a constitutive manner [8].

In all cases, the transport of nitrate is associated with the alkalization of the external medium (Fig. 3), suggesting that other ions are cotransported with

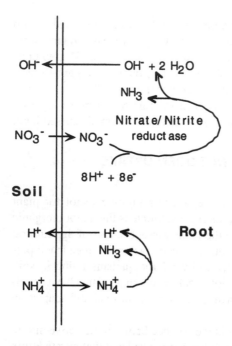

FIGURE 3 Nitrate uptake and assimilation and ammonium ion uptake by the root are associated with changes in the pH of the external medium.

nitrate across the plasma membrane. These empirical observations can be explained if OH⁻ or bicarbonate ions are transferred to the medium in an antiport mechanism with nitrate or if protons are cotransported into the cell with nitrate [9]. Empirical measurements of membrane potential have also indicated that the transport of nitrate into the root cells is associated with the depolarization of the plasma membrane due to the entry of positive charges into the cell. According to these data, the cotransport of one nitrate ion and two protons has been suggested in both inducible and constitutive transport systems [7]. This assumption has been confirmed by functional characterization of heterologously expressed nitrate transporters in *Xenopus* oocytes [10]. The measurement of membrane potential with microelectrodes and the compartmental analysis of N isotopic exchanges have shown that nitrate concentration in the cytosol ranges from 2 to 40 mM (reviewed in Ref. 9). Taking empirical and thermodynamic considerations together, it has been calculated that nitrate transport requires energy. A cost of 1 to 3 mol of ATP per mol of nitrate transported has been estimated in barley roots, which represents 10–20% of the total energy required for the assimilation in the roots [11]. The energy is supplied by the proton motive force that is maintained by the plasma membrane ATPase. Because the efflux of nitrate into the soil increases with an increase in the concentration of nitrate, the dissipation of energy associated with the efflux of nitrate could be considered as a mechanism for sensing the changes in nitrate concentration as suggested for other futile cycles in several metabolic pathways.

Uptake of inorganic nitrogen in the form of ammonium ion is believed to be quantitatively relevant in plants adapted to acidic soils or in anaerobic conditions such as result from flooding. In these environmental conditions, nitrification is reduced or absent and, hence, ammonium ions are not oxidized to nitrate. Studies with soybean plants revealed that the absorption of ammonium ion is faster than the absorption of nitrate in nitrogen-depleted plants, which suggests a preference for ammonium over nitrate as the inorganic nitrogen source [12]; however, uptake of ammonium ion by roots is associated with the acidification of the external medium, which in turn provokes the inhibition of inorganic nitrogen uptake and a decrease of the biomass accumulation in comparison with plants treated with nitrate [13]. Therefore, the use of ammonium as a nitrogen source is limited to special situations, and nitrate is the major form of inorganic nitrogen in the soil and used by plants.

Another difference with respect to nitrate concerns the concentration and distribution of ammonium ion in the root. The concentration of the ammonium ion in root cells appears to be in the μM range, and because it is not transported to the shoot via xylem, it has to be incorporated in the root cell by the cytosolic GS and eventually stored in the vacuole. It has been calculated that in rice, cultivated in the presence of ammonium as the sole nitrogen source, 41% of the ammonium transported in the cells remains in the cytosol, 19% is metabolized, 20% is stored in the vacuole, and the remaining ammonium is transferred to the medium by an efflux process [14,15]. Thus, similarly to nitrate uptake, net ammonium uptake by roots is a balance of the influx and efflux processes. In this case, the efflux of ammonium appears to be independent of the ammonium concentration in the soil or in the cytosol, which is usually low; in contrast, it depends on the N status of plants, determined by the growing conditions [14].

Kinetic studies have showed the existence of two different transport systems implicated in the ammonium influx in rice. One of the transport systems operates at a low concentration of ammonium ion (transport system of high affinity), and the second one operates at a high concentration (transport of low affinity) [14]. The transport of the ammonium ion also provokes the depolarization of the plasma membrane, an event that is overcome by inhibitors of ATP synthesis, H⁺-ATPase inhibitors, or protonophores. Two possible transport mechanism have been proposed for the high-affinity influx of ammonium: (1) a proton/ammonium ion symport and (2) the existence of a specific NH_4^+-ATPase [16]. Available results indicate a direct or indirect dependence of the high-affinity ammonium transport on the proton motive force. At high external concentration of ammonium ion (low-affinity transport), the entry into the cell appears to correspond to a passive electrogenic uniport [17], which may be associated with a K^+ channel [18]; nevertheless, the existence of a specific channel responsible for the low-affinity transport of ammonium ion cannot be ruled out [16].

The purification and characterization of the proteins involved in nitrate and ammonium transport are difficult tasks because of the existence of the different transport systems and the fact that these proteins are associated with the membrane, which complicates their isolation and maintenance in an active form in vitro. Therefore, genetic approaches have proved to be more useful for the isolation of the genes encoding the transport systems and the molecular and functional characterization of the proteins.

The identification of genes encoding nitrate transporters was based on the characterization of mutant plants defective in nitrate assimilation by selecting plants with low growth on nitrate and exhibiting resistance to chlorate (a nonselective herbicide that serves as the nitrate analogue) due to its reduction to chlorite by nitrate reductase. Chlorite is a strong oxidant that is highly toxic to plant cells. These strategies resulted in the identification of mutants affected in nitrate uptake. The characterization of the mutants and the target genes has revealed the existence of a family of transporters conserved among higher plants. The genes isolated can be grouped in two families termed NRT1 and NRT2, which are present in a variety of organisms, including higher plants, photosynthetic bacteria, and fungi [7]. The two families exhibit no sequence similarity and encode hydrophobic proteins with 12 putative membrane-spanning regions. Similarly, gene expression in both families is induced by nitrate. Functional analysis of the putative nitrate transporters has been achieved in some cases by expression of the cloned gene in a heterologous system. Thus, one member of the NRT1 family from *Arabidopsis*, termed CHL1, was expressed in *Xenopus* oocytes, conferring the capacity for nitrate uptake. The depolarization of the oocyte membrane confirmed that CHL1 protein is an electrogenic nitrate transporter. The functional analysis of CHL1 in *Xenopus* oocytes suggested that CHL1 is a low-affinity nitrate transporter (LATS) [19]; however, its nitrate-inducible expression in roots [20] does not fit with the constitutive characteristic of LATS, indicating that another, as yet unknown, gene(s) encodes components of the constitutive LATS [19]. Other members of the NTR1 family also include peptide and amino acid transporters, with at least one gene in *Brassica* being characterized as a dual histidine/nitrate transporter [21], suggesting that several members of the NTR1 family might be involved in nitrate uptake in root cells.

The genes of the NTR2 family were first discovered in *Aspergillus* [22]; however, a series of related genes has been identified in higher plants, cyanobacteria, and yeast [7,23]. The expression analysis of these genes indicated correspondence with the nitrate-inducible and high-affinity nitrate transporters (HATS), although a constitutive gene belonging to this family was also isolated from *Arabidopsis thaliana* [24]. Studies in *Chlamydomonas* have shown the existence of three different nitrate/nitrite transport systems belonging to the NTR2 family [25] that are regulated by nitrate. One of them consists of bispecific nitrate/nitrite transporters, and the others are specific for either nitrate or nitrite [26].

The molecular characterization of transporters involved in ammonium ion uptake was approached by complementation of yeast mutants affected in uptake of ammonium ions [27]. Three functional ammonium transporters corresponding to the high-affinity transport system were identified in *Arabidopsis*. The three transporters, encoded by *AtaMT1;1*, *AtaMT1;2*, and *AtaMT1;3*, are expressed in root; their affinity for ammonium ($\sim$ 0.5 to 40 μM) covers the range of the ammonium ion concentration typically found in soils, suggesting that these transporters are involved in the ammonium ion uptake by roots. Interestingly, the transcript level of the *AtaMT1;1* gene is also abundant in other plant organs such as stem and leaf, indicating a possible role for this transporter in the reimport of ammonium ions that leaked out from the cell during the photorespiration process. In addition, Southern blot analysis of the *Arabidopsis* genome using a 600-bp probe from the conserved region of *AtaMT1;1* suggests the existence of at least three additional homologous genes. The existence of several ammonium transporter genes could be required in the root to sense small changes in external ammonium ion concentration [28].

In conclusion, both kinetic and expression studies indicate that inorganic nitrogen uptake is regulated by the availability of nitrate and ammonium ion in the soil. In both cases, transport of inorganic nitrogen into the roots seems to be associated with membrane depolarization and with a proton-symport mechanism. Nevertheless, besides the progress made on the molecular characterization of nitrogen transporters, further work would be necessary to fit the biochemical and kinetic data with the knowledge acquired from the characterization of the genes encoding the nitrogen transporters and with their expression pattern. In addition, little is known about the proteins involved in the translocation of nitrate into the vacuole or in the loading into xylem and the proteins involved in transport processes in photosynthetic tissues that might represent important steps in the regulation of nitrogen assimilation in plants.

3 pH CONTROL IN NITROGEN ASSIMILATION AND AVAILABILITY OF CARBON SKELETONS FOR AMMONIUM ASSIMILATION

The assimilation of nitrogen into amino acid requires the provision of carbon skeletons in the form of keto acids. The stoichiometry of the GS/GOGAT cycle clearly shows that the incorporation of ammonium for the net synthesis of glutamate requires the supply of 2-oxoglutarate (Fig. 2). This keto acid is provided by the reaction catalyzed by isocitrate dehydrogenase (IDH). Two different IDHs that exist in plant cells differ in the pyridine nucleotide they use as cosubstrate, $NADP^+$ or NAD^+, and also in their subcellular location in the cell. The mitochondrial NAD^+-IDH is the enzyme involved in the Krebs cycle, while $NADP^+$-dependent IDH exists in different subcellular compartments such as cytosol, chloroplast, peroxysome, and mitochondria. The cytosolic $NADP^+$-IDH is the most active IDH enzyme in both angiosperms and gymnosperms, and it has been considered the main enzyme involved in the assimilation of ammonium and synthesis of glutamate when large quantities of the 2-oxoglutarate are required [29]. Such a hypothesis is based, among others facts, on the existence of cytosolic aconitase in plant cells [29]. Therefore, the supply of 2-oxoglutarate through a cytosolic pathway involving aconitase and $NADP^+$-IDH is an alternative to the Krebs cycle enzymes for the provision of carbon skeletons for the assimilation of ammonium ions (Fig. 4).

Regardless of its origin, the supply of 2-oxoglutarate (in the form of citrate or 2-oxoglutarate) for amino acid biosynthesis requires an increase in the flow of carbon metabolites through respiratory pathways; therefore, to avoid the depletion of intermediates in the Krebs cycle, it is necessary to allow the regeneration of oxaloacetate. This requirement must be met by increasing the flux via glycolysis and increasing the production of phosphoenolpyruvate and its conversion to oxaloacetate by the carboxylation reaction catalyzed by cytosolic phosphoenolpyruvate carboxylase (PEPC). Oxaloacetate can easily be converted into malate by the action of cytosolic malate dehydrogenase. Malate and oxaloacetate generated in the cytosol can be transported into the mitochondria to warrant the functioning of the Krebs cycle during a net demand for carbon skeletons for ammonia assimilation and amino acid biosynthesis. Therefore, on balance, the incorporation of ammonium ion into glutamate requires the feeding of the Krebs cycle via the carboxylation of phosphoenolpyruvate. The relevance of the reaction catalyzed by PEPC is supported by reports describing the increase in PEPC activity during nitrogen assimilation in algae and higher plants [30,31]. The stoichiometry already described is balanced during the cell life because surplus of malate and citrate can be stored appropriately in the vacuole and used through the Krebs cycle on demand for ammonia assimilation. The storage of malate and also citrate in the vacuole depends on the transport of protons by the tonoplast H^+-ATPase (see Chapter 1); therefore it is difficult to assess whether feeding malate into the Krebs cycle can influence cytosolic pH.

Regulation of PEPC by pH has been well established in C_4 plants, in which PEPC represents a key step in primary carbon assimilation [32]. In C_4 plants, the enzyme is a tetrameric protein that is allosterically regulated by photosynthesis-related metabolite effectors that activate (e.g., glucose 6-phosphate) or inhibit (L-malate) the enzymatic activity. In addition, PEPC is subjected to a posttranslational modification affecting its catalytic properties (decrease in its sensitivity to L-malate). This modification corresponds to the phosphorylation of a serine resi-

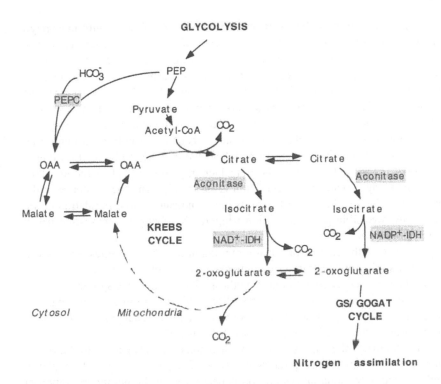

FIGURE 4 Generation of carbon skeletons for nitrogen assimilation requires the increase in the flux through respiratory pathways. Phosphoenolpyruvate carboxylase (PEPC), aconitase, and isocitrate dehydrogenases (IDH) are supposed to be main enzymes controlling the supply of 2-oxoglutarate for ammonia assimilation and amino acid biosynthesis. GS, glutamine synthetase; GOGAT, glutamate synthase; OAA, oxaloacetate.

due in the amino-terminal moiety of enzyme subunits. Interestingly, the serine residue is conserved among higher plants including C_3 and CAM plants. Therefore, the regulation of PEPcase by reversible phosphorylation of the protein appears to be a universal mechanism in plants.

Some research effort has been focused on characterization of the factors involved in the phosphorylation of PEPC. In both intact leaves and mesophyll cell protoplasts, it has been suggested that PEPC phosphorylation is stimulated through a light-dependent transduction cascade [33–35]. Two protein kinases differing in their requirement for Ca^{2+} have been shown to phosphorylate PEPC. The calcium-independent kinase has been considered the best candidate for the regulation of PEPC in leaves because its activity exhibits light dependence in vivo and sensitivity to inhibitors of photosynthesis [36]. Studies with mesophyll protoplasts from the C_4 plant *Digitaria sanguinalis* have shown an increase in the phosphorylation state of the protein and an increase in the activity of the calcium-independent protein kinase when protoplasts are treated with methylamine or ammonium chloride in the presence of light [37]. The early events in the activation of PEPC kinase by light are alkalization of cytosolic pH and calcium mobilization from vacuoles. The characterization of the phosphorylation state of PEPC in the presence of inhibitors of photosynthetic electron transport and ATP synthesis also suggests that production of ATP and/or NADPH by photosynthesis is required for the activation of the transduction pathway [37]. Although most of the knowledge about PEPC regulation is for the C_4 form, the existence of the conserved serine residue in the C_3 forms could indicate similar regulation of the enzyme by reversible phosphorylation triggered by changes in cytosolic pH in nonphotosynthetic cells [38].

Studies of the role of pH in the regulation of other enzymes involved in nitrogen assimilation are scarce and in most cases restricted to the proton concentration changes associated with photosynthesis. Thus, glutamine synthetase activity in tomato leaves is regulated during light-dark transitions, which might be due to the availability of substrates (ATP-Mg) and pH changes in the chloroplastic stroma associated with the photosynthesis [39].

In summary, the supply of carbon skeletons for nitrogen assimilation requires the contribution of respiratory pathways. The participation of both photosynthesis and mitochondrial respiration is necessary for the net incorporation of inorganic nitrogen into the organic nitrogen pool of molecules in plants. The control of cytosolic and stroma pH could be a relevant point in the regulation of ammonium assimilation because alkalization of the cytosol and stroma may affect the flow of carbon metabolites through respiratory pathways by activating cytosolic PEPC and plastidic glutamine synthetase, respectively.

4   PROTON FLUX AND AMINO ACID TRANSPORT SYSTEMS

Once the inorganic nitrogen has been assimilated into amino acids, they have to be transported to the different tissues and cells for the biosynthesis of proteins and other nitrogen-containing molecules. Two kinds of amino acid transport can be considered: (1) transport from cell to cell, which takes place through the plasmodesmata (symplastic transport) or through the intercellular spaces (apoplastic transport), and (2) long-distance transport to other plant organs, mediated by the xylem and the phloem. The xylem translocates water and nutrients from the roots, and the phloem transports the photoassimilates generated in the mesophyll cells to sink tissues. Because the xylem consists of nonliving lignified tubes, the most interesting transport process is related to the distribution via the phloem sieve elements of amino acids synthesized in the leaf. The sieve elements consist of elongated cells, without nuclei, connected by plates, which enables the diffusion of the photoassimilates loaded into the companion cells in the sink tissues. Two methods of phloem loading have been considered [40]: a symplastic method, via plasmodesmata, which does not involve transporters, and an apoplastic loading found in cereals, Solanaceae, and other plants, in which photoassimilates are first transported from the parenchyma cells of source tissues to the apoplast and then transported into the companion phloem cells involving membrane transporters. The final loading into the sieve element occurs as a diffusive process via plasmodesmata.

The knowledge of these transport processes is limited because of the difficulties related to the isolation, characterization, and localization of the proteins involved and also the complex experimental procedures involved in collection of phloem sap samples. Studies regarding the transport of the main photoassimilates (sucrose and amino acids) from the apoplasts to the companion phloem cells indicated the existence of a proton symport mechanism. The characterization of amino acid transport systems has been approached in a few plant models. Early studies of castor bean cotyledons indicated that amino acid transport was associated with depolarization of the plasma membrane [41]. This early evidence has been confirmed using plasma membrane vesicles from sugar beet leaves demonstrating that amino acid transport is electrogenic and corresponds to a secondary active process that uses the proton gradient created by the plasma membrane ATPase [42,43].

The biochemical analysis of the amino acid transport in vesicles has allowed the identification of four symport systems. Two of them are involved in the translocation of neutral amino acids and exhibit different amino acid specificities. The other two correspond to an acidic symport and a basic symport system [43]. The identification and purification of these transporters is also a very complicated task, but some progress has been made at the molecular level by complementation of yeast mutants deficient in the transport of amino acids. Several cDNAs corresponding to amino acid permease have been isolated from *Arabidopsis* [44,45]. The analysis of some of these sequences has revealed that amino acid transporters are membrane proteins with 9–12 membrane-spanning domains with no relevant homology with other known transporters [46]. Two of the cDNAs isolated appear to correspond to general amino acid transporters; the functional characterization of other isolated genes will allow the future identification of new transporters.

REFERENCES

1. AJ Keys, IF Bird, ML Cornelius, PJ Lea, RM Wallsgrove, BJ Miflin. The photorespiratory nitrogen cycle. Nature 275:741–743, 1978.

2. BJ Miflin, PJ Lea. Amino acids and derivatives. In: BT Miflin, ed. The Biochemistry of Plants. New York: Academic Press, 1980, pp 169–202.

3. P Lea. Primary nitrogen metabolism. In: PM Dey, JB Harbone, eds. Plant Biochemistry. San Diego: Academic Press, 1997, pp 273–313.

4. JD Wolt. Soil Solution Chemistry. New York: John Wiley & Sons, 1994.

5. M Aslam, RL Travis, DW Rains. Evidence for substrate induction of a nitrate efflux system in barley roots. Plant Physiol 112:1167–1175, 1996.

6. JP Grouzis, P Pouliquin, J Rigaud, C Grignon, R Gibrat. In vitro study of passive nitrate transport by native and reconstituted plasma membrane vesicles from corn root cells. Biochim Biophys Acta 1325:329–342, 1997.

7. NM Crawford, ADM Glass. Molecular and physiological aspects of nitrate uptake in plants. Trends Plant Sci 3:389–395, 1998.

8. MY Siddiqi, ADM Glass, TJ Ruth, TW Rufty. Studies on the uptake of nitrate in barley. Plant Physiol 93:1426–1432, 1990.

9. C Grignon, J-B Thibaud, T Lamaze. Transport du nitrate par la racine. In: J-F Morot-Gaudry, ed. Assimilation de l'azote chez les plantes. Aspects physiologique, biochimique et moléculaire. Paris: INRA Editions, 1997, pp 27–43.

10. SJ Smith, J-J Zhou, A Miller. Characterisation of nitrate transport using nitrate-selective microelectrodes and heterologous expression in *Xenopus* oocytes. In: FM Cánovas, FJ Florencio, eds. Avances en el metabolismo del nitrógeno: bioquímica, fisiología y biología molecular. Málaga: Servicio de Publicaciones de la Universidad de Málaga, 2000, pp 345–350.

11. AJ Bloom, SS Sukrapanna, R Warner. Root respiration associated with ammonium and nitrate absorption and assimilation by barley. Plant Physiol 99:1294–1301, 1992.

12. CH Saravitz, S Cahillou, J Musset, CD Raper, J-F Morot-Gaudry. Influence of nitrate on uptake of ammonium by nitrogen-depleted soybean: is the effect located in roots or shoot? J Exp Bot 45:1575–1584, 1994.

13. S Chaillou, JK Vessey, J-F Morot-Gaudry, CD Raper, LT Henry, JP Boutin. Expression of characteristics of ammonium nutrition as affected by pH of the root medium. J Exp Bot 42:189–196, 1991.

14. MY Wang, MY Siddiqi, ADM Glass. Ammonium uptake by rice roots. I. Fluxes and subcellular distribution of $^{13}NH_4^+$. Plant Physiol 103:1249–1258, 1993.

15. MY Wang, MY Siddiqi, JJ Ruth, ADM Glass. Ammonium uptake by rice roots. II. Kinetics of $^{13}NH_4^+$ influx across the plasmalemma. Plant Physiol 103:1259–1267, 1993.

16. MY Wang, ADM Glass, JE Shaff, LV Kochian. Ammonium uptake by rice roots III. Electrophysiology. Plant Physiol 104:899–906, 1994.

17. WR Ullrich, M Larsson, C-M Larsson, S Lesch, A Novacky. Ammonium uptake in *Lemna gibba* G1, related membrane potential changes, and inhibition of anion uptake. Physiol Plant 61:369–376, 1984.

18. DP Schachtman, SJ I., WJ Lucas, JA Anderson, RF Gaber. Expression of an inwardly-rectifying potassium channel by the *Arabidopsis* KAT1 cDNA. Science 258:1654–1658, 1992.

19. N-C Huang, C-S Chiang, NM Crawford, Y-F Tsay. *CHL1* encodes a component of the low-affinity nitrate uptake system in *Arabidopsis* and shows cell type–specific expression in roots. Plant Cell 8:2183–2191, 1996.

20. Y-F Tsay, JI Schroeder, KA Feldmann, NM Crawford. A herbicide sensitivity gene *CHL1* of *Arabidopsis* encodes a nitrate-inducible nitrate transporter. Cell 72:705–713, 1993.

21. JJ Zhou, FL Theodoulou, I Muldin, B Ingemarsson, AJ Miller. Cloning and functional characterization of a *Brassica napus* transporter that is able to transport nitrate and histidine. J Biol Chem 273:12017–12023, 1998.

22. AG Brownlee, HN Arst. Nitrate uptake in *Aspergillus nidulans* and involvement of the third gene of nitrate assimilation gene cluster. J Bacteriol 155:1138–1146, 1983.

23. H Zhang, G Leggewie, M Hansen, AJ Jennings, BG Forde. The NRT2 family of high-affinity nitrate transporters. In: FM Cánovas, FJ Florencio, eds. Avances en el metabolismo del nitrógeno: bioquímica, fisiología y biología molecular. Málaga: Servicio de Publicaciones de la Universidad de Málaga, 2000, pp 97–102.

24. R Wang, NM Crawford. Genetic identification of a gene involved in constitutive, high affinity, nitrate transport in *Arabidopsis thaliana*. Proc Nat Acad Sci USA 93:9297–9301, 1996.

25. A Galván, A Quesada, E Fernández. Nitrate and nitrite are transported by different specific transport systems and by a bispecific transporter in *Chlamydomonas reinhardtii*. J Biol Chem 271:2088–2092, 1996.

26. A Quesada, J Hidalgo, E Fernández. Three Nrt2 genes are differentially regulated in *Chlamydomonas reinhardtii*. Mol Gen Genet 258:373–377, 1998.

27. O Ninnemann, JC Jauniaux, WB Frommer. Identification of a high affinity NH₄⁺ transporter from plants. EMBO J 13:3464–3471, 1994.

28. S Gazzarrini, L Lejay, A Gojon, O Ninnemann, WB Frommer, N von Wirén. Three functional transporters for constitutive, diurnally regulated, and starvation-induced uptake of ammonium into *Arabidopsis* roots. Plant J 11:937–947, 1999.

29. RD Chen, P Gadal. Do the mitochondria provide the 2-oxoglutarate needed for glutamate synthesis in higher plants? Plant Physiol Biochem 28:141–146, 1990.

30. GC Valerberghe, KA Schuller, RG Smith, R Feil, WC Plaxton, DH Turpin. Relationship between NH₄⁺ assimilation rate and in vivo phosphoenolpyruvate carboxylase. Plant Physiol 94:284–290, 1990.

31. G Le Van Quay, C Foyer, M-L Champigny. Effect of light and NO₃⁻ on wheat leaf phosphoenolpyruvate carboxylase activity. Plant Physiol 97:1476–1482, 1991.

32. R Chollet, J Vidal, MH O'Leary. Phosphoenolpyruvate carboxylase: a ubiquitous, highly regulated enzyme in plants. Annu Rev Plant Physiol Plant Mol Biol 47:273–298, 1996.

33. J-A Jiao, R Chollet. Light activation of maize phosphoenolpyruvate carboxylase protein-serine kinase activity is inhibited by mesophyll and bundle sheath–directed photosynthesis inhibitors. Plant Physiol 98:152–158, 1992.

34. N Bakrim, C Echevarria, C Crétin, M Arrio-Dupont, J-N Pierre, J Vidal, R Chollet, P Gadal. Regulatory phosphorylation of *Sorghum* leaf phosphoenolpyruvate carboxylase, identification of the protein-serine kinase and some elements of the signal-transduction cascade. Eur J Biochem 204:821–830, 1992.

35. J-N Pierre, V Pacquit, J Vidal, P Gadal. Regulatory phosphorylation of phosphoenolpyruvate carboxylase in protoplasts from *Sorghum* mesophyll cells and the role of pH and Ca²⁺ as possible components of the light-transduction pathway. Eur J Biochem 210:531–537, 1992.

36. B Li, R Chollet. Resolution and identification of phosphoenolpyruvate carboxylase

protein-kinase polypeptides and their reversible light activation in maize leaves. Arch Biochem Biophys 307:416–419, 1993.

37. N Giglioli-Guivarch, J-N Pierre, S Brown, R Chollet, J Vidal, P Gadal. The light-dependent transduction pathway controlling the regulatory phosphorylation of C_4 phosphoenolpyruvate carboxylase in protoplast of *Digitaria sanguinalis*. Plant Cell 8:573–586, 1996.

38. L Lepiniec, E Keryer, H Phillippe, P Gadal, C Crétin. The phosphoenolpyruvate carboxylase gene family of *Sorghum*: structure, function and molecular evolution. Plant Mol Biol 21:487–502, 1993.

39. FM Cánovas, C Ávila, JR Botella, V Valpuesta, I Núñez de Castro. Effect of light-dark transition on glutamine synthetase activity in tomato leaves. Physiol Plant 66: 648–652, 1986.

40. W Eschrich, J Fromm. Evidence for two pathways of phloem loading. Physiol Plant 90:699–707, 1994.

41. B Etherton, B Rubinstein. Evidence for amino acid–H^+ cotransport in oat coleoptiles. Plant Physiol 61:933–937, 1978.

42. DR Bush, Z-C Li. Proton-coupled amino acid symports in the plant plasma membrane. In: BK Singh, HE Flores, JC Shannon, eds. Biosynthesis and Molecular Regulation of Amino Acids in Plants. Rockville, Maryland. American Society of Plant Physiologists, 1992, pp 121–127.

43. DR Bush. Proton-coupled sugar and amino acid transporters in plants. Annu Rev Plant Physiol Plant Mol Biol 44:513–542, 1993.

44. WB Frommer, S Hummel, JW Riesmeier. Expression cloning in yeast of a cDNA encoding a broad specificity amino acid permease from *Arabidopsis thaliana*. Proc Nat Acad Sci USA 90:5944–5948, 1993.

45. LC Hsu, TJ Chiou, L Chen, DR Bush. Cloning a plant amino acid transporter by functional complementation of a yeast amino acid transport mutant. Proc Nat Acad Sci USA 90:7441–7445, 1993.

46. WB Frommer, M Kawart, B Hirner, WN Fisher, S Hummel, O Ninnemann. Transporters for nitrogenous compounds in plants. Plant Mol Biol 26:1651–1670, 1994.

10

Crassulacean Acid Metabolism: A Special Case of pH Regulation and H⁺ Fluxes

Karl-Josef Dietz and Dortje Golldack
University of Bielefeld, Bielefeld, Germany

1 INTRODUCTION

Crassulacean acid metabolism (CAM) is apparently the most conspicuous performance of plants involving alternating production and consumption of hydrogen ions. The sweet taste of leaves at the end of the light period and their acid taste with immediate pheophytin production during homogenization at the end of the dark period provide direct evidence for "acid metabolism." Thus, CAM is defined as massive diurnal fluctuation of titratable acidity [1]. However, it should be made clear at this point that there are many metabolic processes that also involve considerable or even higher rates of H^+ production or H^+ translocation than CAM. In such cases, H^+ metabolism may be exceedingly high when related to shorter time periods, the volume of specialized cells, or the membrane area of specific organelles. Examples are apoplastic phloem loading, nitrate assimilation, and stomatal movement. This may be illustrated with one example: the typical rates of nocturnal CO_2 fixation of CAM plants are in the range of 1 to 5 μmol CO_2 m^{-2} s^{-1}. Assuming a 10-h night period and using the high rate of 2 μmol CO_2 m^{-2} s^{-1}, it is calculated that about 72 mmol malic acid accumulates in leaves per 1 m^2 of surface area. Simultaneously, 144 mmol of H^+ are produced if starch is used for phosphoenolpyruvate (PEP) synthesis, protons being partly buffered but mainly pumped into the vacuole.

Although a number of productive CAM species exist, they essentially do not accumulate more acid than species considered to be nonproductive CAM species. Their increased carbon assimilation is mainly due to parallel use of CAM and the C_3 pathway of photosynthesis. Rates of CAM-dependent accumulation of H^+ can be compared with those occurring during assimilate export in C_3 and C_4 photosynthesis, which may proceed at rates as high as 50 μmol CO_2 m^{-2} s^{-1}. Assuming a 12-h photoperiod, 80% export of produced assimilates from the leaves, and an export mechanism using an H^+/sucrose symporter, 144 mmol H^+ m^{-2}s^{-1} have to be pumped back from the phloem to the apoplast to maintain the required pH gradient. This is likely to be achieved via a smaller membrane surface area than nocturnal storage of malic acid in the mesophyll vacuoles of CAM plants. The model calculation illustrates that CAM is not unique in terms of biochemical rates. Conversely, CAM is unique and important as an ecophysiological mechanism to increase productivity and survival of plants under adverse environmental conditions. In this chapter we first summarize the distribution of CAM among plants, address the basic biochemical features of CAM, and then describe the molecular genetics of CAM expression, with emphasis on inducible CAM plants.

2 THE ECOPHYSIOLOGY OF CAM PLANTS, THEIR MORPHOLOGY, AND THEIR PHYLOGENETIC DISTRIBUTION

2.1 Ecophysiological Significance of CAM

Crassulacean acid metabolism as a CO_2-concentrating mechanism in photosynthetic tissue allows plant adaptation to dry climates as well as environments with a limited carbon supply. CAM-expressing plants are primarily succulents and epiphytes inhabiting arid tropical and subtropical regions. Besides, CAM is also found in a number of aquatic vascular plants. CAM has a wide taxonomic distribution in plants, being expressed in 33 families with at least 328 genera. CAM plants occur in the classes Lycopodiopsida, Filicopsida, and Gnetopsida (ferns) and are spread throughout the Liliopsida (monocots, e.g., Agavaceae, Orchidaceae, Bromeliaceae) and Magnoliopsida (dicots, e.g., Aizoaceae, Cactaceae, Clusiaceae, Crassulaceae) [2]. Apparently, CAM evolved frequently and independently in the plant kingdom and is a remarkable example of evolutionary convergence. From that, it can be concluded that expression of CAM does not depend on the employment of new metabolic pathways but mainly involves regulatory aspects to increase activities of ubiquitous housekeeping enzymes, for example, those involved in PEP synthesis from starch or, as shown in some cases, gene duplication in the genome during phylogenesis and development of specific promoters such as the promoter for the CAM-specific isoform of PEP carboxylase [3]. The use of general and ubiquitous mechanisms in CAM includes the reactions of H^+ homeostasis.

Initially, CAM was exclusively considered as an adaptation to water shortage. Nocturnal uptake of CO_2 at low temperature and high atmospheric water vapor saturation decreases water loss per unit assimilatory carbon gain, thus increases the water use efficiency of CAM plants compared with C_3 and C_4 plants. Typical values were given by Winter [4]: the C_3-plant *Gossypium hirsutum* incorporated 3.4 mmol CO_2 per mol H_2O lost by transpiration. The equivalent values were 6.7 for the C_4 plant *Zea mays* and 10.0 for *Opuntia inermis* in the CAM state but only 5.0 for *Opuntia* performing C_3 photosynthesis in the light. Identification of aquatic plants such as *Isoëtes howellii* that efficiently perform CAM although water is available in almost unlimited quantities questioned the exclusive function of CAM in improving the water use efficiency under water-limited growth conditions [5]. In these cases, CAM is suggested to play an additional role by increasing carbon efficiency at sites of low carbon availability. Aquatic plants cease to perform CAM when the carbon supply is increased [6]. It is concluded that CAM is an important adaptive mechanism not only during water-limited growth but also under conditions of carbon limitation. Aquatic CAM plants are discussed in more detail in the following.

2.2 The Morphology of Typical CAM Plants

The morphological characteristic of most CAM plants is distinct leaf or stem succulence [7]. This feature provides the capacity to store water and large amounts of the nocturnally synthesized organic acids, particularly malic acid. The low root-to-shoot ratio of many CAM plants maintains limited water loss from the succulent photosynthetic tissue through the roots to the soil [8]. The distinct morphology and low day-time transpiration due to the CAM-specific stomatal opening at night enable CAM plants to have high water use efficiency in dry environments, resulting in higher productivity compared with C_3 and C_4 plants under these growth conditions.

2.3 Facultative CAM Plants

In addition to obligate CAM plants, there is a wide range of species with facultative and transient CAM expression. Interestingly, performance of CAM is not uniform in plant families, but even members of one genus can carry out either C_3 or CAM photosynthesis constitutively or be facultative CAM plants. The occurrence of CAM can be induced ontogenetically, by day length, by seasons, or by environmental stress such as reduced water availability. Obligate CAM plants, such as *Bryophyllum (Kalanchoë) fedtschenkoi*, exhibit CAM constitutively without developmental or environmental triggering. *B. fedtschenkoi* has been shown to continue with the CAM-specific diurnal CO_2 assimilation even when grown without day-night periods in continuous light or darkness [9,10].

Developmental shifts from C_3 photosynthesis to CAM metabolism are known, for instance, from *Peperomia camtotricha* (Piperaceae [11]), *Kalanchoë daigremontiana* (Crassulaceae [12]), and *Mesembryanthemum crystallinum* (Aizoaceae [13]). In these plants CAM expression occurs with maturing. Interestingly, CAM develops gradually in these plants, with low activities of CAM enzymes and diurnal organic acid fluctuations in the youngest leaves and highest activities in the oldest leaves [14]. Although the process of CAM expression is ontogenetically determined in these plants, CAM induction can also occur in response to environmental factors. *Kalanchoë blossfeldiana* (Crassulaceae) exhibits CAM with increasing age when grown under a long-day regime, but the shift from C_3 to CAM metabolism is stimulated by short-day treatment [15] independent of age. In *M. crystallinum*, CAM is induced in response to stress factors, such as increased salinity and drought [16]. *Peperomia scandens* exhibits induction of CAM on exposure to water stress [14]. Whereas the stress-induced CAM expression in *M. crystallinum* and *P. scandens* is reversible up to a certain age and the plants return to C_3 photosynthesis when grown under nonstress conditions again [17], the short-day stimulated shift to CAM in *K. blossfeldiana* is not reversible by returning to long-day treatment [15].

As a representative for another type of facultative CAM plants, *Sedum telephium* (Crassulaceae) exhibits C_3 photosynthesis when grown under well-watered conditions without any developmental shift to CAM, but CAM is induced by water stress and day-length changes [18,19]. In the following sections, induction and features of crassulacean acid metabolism in the facultative CAM plants *Clusia* and *M. cystallinum* and in aquatic macrophytes are reviewed.

2.3.1 Clusia

Clusia is a genus of about 150 species [20] that is found in the tropics in very diverse habitats, ranging from montane forests and deciduous forests to rocky shorelines sand dunes and savannas [20,21]. The species of *Clusia* are trees and shrubs that grow mainly as epiphytes or hemiepiphytes strangling their host plants and growing roots to the soil. Alternatively, *Clusia* may also germinate in soil and develop directly into nonepiphyte trees [22]. The occurrence of constitutive CAM has been reported, e.g., for *Clusia rosea* [21], and constitutive C_3 photosynthesis has been observed in *Clusia multiflora* [22]. Most *Clusia* species are, however, facultative CAM plants and respond rapidly and reversibly to changes of environmental conditions with shifts between C_3 photosynthesis and CAM but without convincing evidence for ontogenetic CAM induction. In their natural environment, transition between C_3 photosynthesis and CAM is primarily seasonally induced and responds mainly to changes in water availability, temperature, and light intensity. When studied in its natural habitat in a rain forest in Panama, an increase in temperature or daily photosynthetic photon flux density resulted in increasing CAM activity in *Clusia uvitana*. Lowest nocturnal acidification has been observed in *C. uvitana* and *C. minor* during the wet season, whereas during the dry season the nocturnal carbon assimilation and titratable tissue acidity were highest [19]. Interestingly, well-watered *C. minor* did not shift from C_3 photosynthesis to CAM when exposed to high light intensities, indicating that the effects of high irradiance and temperature observed in *Clusia* might be as well due to reduced water availability [22]. In contrast to many other CAM species, *Clusia* is characterized by a number of specific features of the metabolic reactions involved in CAM.

2.3.2 Mesembryanthemum crystallinum

In the facultative halophyte *M. crystallinum* (common ice plant, Aizoaceae) responses to salt and water stress have probably been most intensively investigated on the molecular, physiological, and biochemical levels. Thus, the plant serves not only as a model halophyte but also as a model CAM plant. *M. crystallinum* is native to, e.g., southern Africa, the Mediterranean, the Pacific coastline of the United States, and Australia [16]. In their natural habitats, the annual plants germinate during a short rainy season and experience increasing temperature, increasing salinity, and increasing drought during their development. The plant's life cycle is characterized by distinct morphological changes. Juvenile plants up to the age of 5 weeks develop large succulent leaves with the epidermal bladder cells—a main morphological characteristic of *M. crystallinum*—still appressed to the surface. Mature plants show a changed morphology with developing shoots from the axes of the juvenile leaves and small succulent secondary leaves as well as fully developed epidermal bladder cells. When flowering, the vegetative tissue of shoots and leaves gradually dies while the flowers and later the seed capsules are covered with large epidermal bladder cells and stay photosynthetically active.

There are three main mechanisms that enable *M. crystallinum* to cope with the increasing salinity and drought stress during its developmental cycle:

Metabolic transition from C_3 photosynthesis exhibited in the juvenile stage to the more water-conserving CAM,

Accumulation of sodium that is excessively taken up into the vacuolar compartment,

Osmotic balancing due to synthesis and cytoplasmic accumulation of osmoprotective metabolites.

Synthesis of the compatible solutes D-ononitol and D-pinitol and of proline is induced in response to salinity stress in *M. crystallinum* independent of age [16]. Increased concentrations of the osmoprotective solutes balance the increasing osmotic potential caused by vacuolar sodium sequestration in salinity-stressed

M. crystallinum. Further on, the osmoprotective compounds stabilize proteins structurally, thus maintaining metabolic reactions in the salt-stressed cells. The synthesis of compatible solutes is transcriptionally activated as shown, e.g. for *myo*-inositol *O*-methyltransferase, the key enzyme of the biosynthesis of D-ononitol and D-pinitol [23,24].

Knowledge of uptake and transport of sodium chloride in *M. crystallinum* is mainly restricted to the physiological and biochemical level. Independent of the developmental stage, sodium taken up is not stored in the root but is transported very efficiently to the leaves, reaching concentrations of up to 1.0 M in the epidermal bladder cells [16]. Sodium/proton antiport at the tonoplast has been induced by salinity stress in *M. crystallinum*, indicating that vacuolar sodium accumulation is driven by a secondary active Na^+/H^+ antiporter [25]. The activity of the vacuolar-type H^+-ATPase increases under salt stress in *M. crystallinum* [26,27], whereas the activity of the second vacuolar proton pump, the vacuolar pyrophosphatase, decreases under salt stress [28]. According to these data, the V-ATPase is supposed to be the primary proton pump energizing the sodium transport into the vacuole.

2.3.3 CAM in Aquatic Plants

Expression of CAM has been reported for 69 species of aquatic macrophytes out of 180 aquatic species analyzed [29]. CAM is distributed in freshwater species of diverse families ranging from monocotyledonous families including *Vallisneria* (Hydrocharitaceae [30]) and *Sagittaria* (Alismataceae [31] and Dicotyledonae such as *Littorella* (Plantaginaceae [31]) to aquatic fern relatives of the Isoetaceae [5]. Aquatic CAM plants mainly inhabit mesotrophic seasonal pools and oligotrophic lakes. They are distributed in North America and Europe as well as in the tropics and subtropics. The main characteristics of shallow and mesotrophic seasonal pools are substantial diurnal fluctuations of pH and CO_2 availability [32] due to intense photosynthetic CO_2 consumption during the day, resulting in increasing pH that is followed by the nighttime repiratory carbon release. At rising pH, bicarbonate is the predominant species of dissolved inorganic carbon, whereas CO_2 levels are reduced and CO_2 is completely absent at alkaline pH. In contrast to a number of aquatic plants that also use bicarbonate as a carbon supply for photosynthesis (e.g., *Elodea canadensis* [29]), aquatic CAM plants are not able to take up bicarbonate as shown, e.g., for *Isoëtes macrospora* [33] and *Crassula helmsii* [34]. Oligotrophic lakes with their low biomass do not show pronounced diurnal changes in CO_2 availability and pH but are characterized by considerably lower CO_2 concentrations than mesotrophic systems, resulting in CO_2 availability that may be limiting for photosynthesis as well [35,36]. Thus, CAM is an adaptation to low daytime availability of CO_2 in certain aquatic habitats, enabling aquatic CAM plants to maintain photosynthesis under conditions of limiting CO_2.

The crassulacean acid metabolism expressed in aquatic plants is characterized by the same biochemical reactions found in terrestrial CAM plants. Nighttime degradation of starch has been reported for *Isoëtes howelii* and *Isoëtes bolanderi* [37], indicating that starch serves as a substrate for glycolytic phosphoenolpyruvate synthesis. Nocturnal accumulation of substantial amounts of malic acid has been observed in *Vallisneria spiralis* [38]. Acid accumulation of up to 300 mmol protons per kg fresh weight has been reported for photosynthetic tissue of aquatic CAM plants in nighttime [31]. In *I. howellii*, *Crassula aquatica*, and *Littorella uniflora* substantial activities of the NADP malic enzyme have been found, indicating that aquatic CAM plants are plants of the NADP malic enzyme type [31]. In contrast to terrestrial CAM plants, the aquatic CAM plants continue CO_2 uptake in daytime and assimilate the carbon via C_3 photosynthesis. This behavior is mainly due to a morphological characteristic of these plants—submerged leaves either lack functional stomata or are astomatous. Interestingly, when aquatic CAM plants become amphibious, the aerial parts of the leaves shift gradually from CAM to C_3 photosynthesis and often develop functional stomata, whereas submerged parts of the leaves continue with CAM [29].

3 pH CHANGES DURING CAM

3.1 Diurnal pH Changes in Leaf Tissue of CAM Plants

The pH of whole tissue homogenates of CAM plants differs between predusk and predawn. The diurnal differences in pH are in the order of 2–3.5. Due to significant buffering capacity, the more relevant physiological parameter is the titratable acidity of the tissue. CAM plants accumulate 100 to 300 μmol H^+ per g fresh weight during the night [31,39,40]. In many cases accumulation of malic acid fully accounts for the increase in tissue acidity during the night, assuming a stoichiometry of 2 protons per 1 malic acid being synthesized and stored in the vacuole [41]. In other species, additional nocturnal accumulation of citric acid serves as intermediate carbon storage. As will be discussed subsequently, breakdown of citric acid to pyruvate releases three molecules of CO_2, whereas only one mole of CO_2 is released from one mole of malate. However, due to the synthetic pathway, citrate synthesis does not allow net carbon gain during the night. In this context it is interesting to note that the organic acid pool present at the end of the night frequently is not decarboxylated completely during the day. Arata et al. [42] have suggested that this is necessary to balance cations present in the vacuole and to avoid alkalization of the vacuolar sap. Species of *Clusia* can accumulate higher amounts of organic acids than CAM plants of other genera, with nocturnal titratable acidity reaching 1410 mM protons in *Clusia minor* [43], the value corresponding to a vacuolar pH of about 3.

3.2   Four Time Phases of CAM

Initially based on gas exchange measurements, CAM is usually divided into four phases that reflect specific metabolic states of the plants [1,44]: phase I characterizes the nocturnal uptake of CO_2 by PEP carboxylase and organic acid accumulation. Phase II is a short intervening phase of both PEP carboxylation and Rubisco-mediated carbon assimilation early after onset of illumination. Closed stomates, decarboxylation of organic acids, and highly active Rubisco define phase III. In phase IV at the end of the light period, stomates are open and carbon assimilation proceeds via normal C_3 photosynthesis. The expression of these phases is modulated by growth conditions. For example, phase IV will not occur under conditions of severe water deficit but may be dominant under optimal growth conditions, particularly in plants with intermediate CAM and C_3 metabolism. Naturally, the phases of gas exchange describe states of carbon metabolism and concomitantly characterize states of distinct metabolic regulation.

3.3   Subcellular pH Changes

The typical CAM plant is succulent with fleshy leaves or stems. On the subcellular level, the vacuole makes up by far the largest volume of the cells, with up to 90% or more of the total. Under these cellular conditions it is difficult to measure cytoplasmic pH. Marigo et al. [45] employed the diffusion rate of the weak acid 5,5-dimethyl-2-oxazolidine-2,4-dione (DMO) to study relative changes of cytoplasmic pH during CAM of *Kalanchoë*. Although DMO diffusion rates showed distinct differences between the various phases of CAM, the approach did not allow calculation of cytoplasmic pH values. However, it appears to be generally accepted that the cytoplasmic pH is maintained slightly alkaline, even in the presence of highly acidified vacuolar sap [42]. The bulk of the malic acid present in leaf tissue of CAM plants is contained in the mesophyll vacuole as shown, for example, for *Sedum telephium* by comparing malic acid contents of whole tissue and isolated vacuoles during the time course of a day [46].

4   THE BIOCHEMISTRY OF CAM

Figure 1 summarizes the basic biochemical steps of CAM. The key step is the fixation of HCO_3^- by PEP carboxylase at night. In order to provide sufficient amounts of substrate for carboxylation, glycolysis must be activated to form PEP from carbohydrates. The product of PEP carboxylation, oxaloacetate, is reduced to malate by NADH malate dehydrogenase. The steady-state pool of oxaloacetate (OAA) changes during the dark and light phases of CAM [40] and is higher than in C_3 plants, where concentrations of 10 to 100 μM OAA have been reported [47]. The equilibrium of the oxaloacetate reduction reaction at pH 7 is 3×10^{-5} and, therefore, malate is present at a concentration orders of magnitude higher

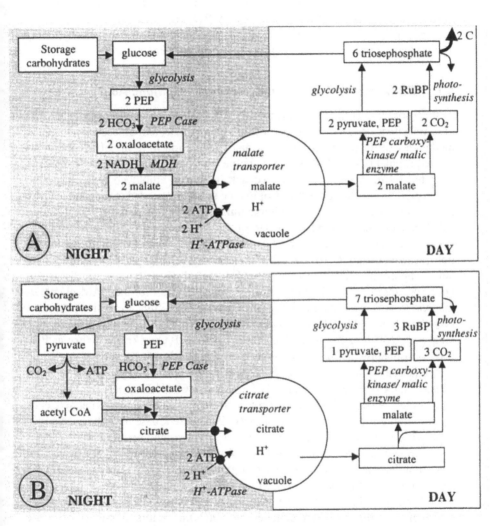

FIGURE 1   Schematic depiction of the basic reactions of CAM. (A) CAM with malic acid storage, (B) CAM with citric acid storage. The scheme is not intended to depict complete stoichiometric carbon and metabolite flow.

than OAA. Malate is stored in the vacuole that acidifies during the night. During the light phase, malate released from the vacuole is decarboxylated by either malic enzyme or PEP carboxykinase. CO_2 is fixed by Rubisco and assimilated in the photosynthetic carbon reduction cycle. PEP and pyruvate are converted to triosephosphate, which is consumed in the synthesis of storage carbohydrates. Some of the major steps will be discussed in more detail.

4.1 PEP Carboxylation and the Synthesis of Malic Acid

PEP carboxylase is the key enzyme of CAM. PEP Case is a homodimer of 220 kDa. Its activity depends on Mg^{2+} or Mn^{2+}. pH, malate, and glucose-6-phosphate modulate PEP Case activity, but the major regulation in terms of adaptation to the requirements of CAM is its reversible phosphorylation [48]. PEP Case is inhibited by malate in a typical feedback inhibition employed to decrease the activity of the anaplerotic pathway of malate synthesis needed to replete the citric acid cycle. Such feedback inhibition is unlikely to be useful in cells performing CAM because malate synthesis is part of the nocturnal acid metabolism and has to proceed even at elevated cytosolic malate concentrations. Protein phosphorylation during the night desensitizes PEP Case toward malate. In fact, the phosphorylation-dephosphorylation reaction is under control of a circadian oscillator and not a response to the light-dark transition. For PEP Case of the CAM plant *Kalanchoë fedtschenkoi*, the K_i(malate) increased from 0.3 mM to about 3 mM upon phosphorylation [49]. The phosphorylation state of PEP Case is adjusted by the relative activities of a protein phosphatase and a PEP Case kinase. Whereas PEP Case phosphatase activity appears to be unchanged during the diurnal cycle, the kinase activity and its messenger RNA (mRNA) are present only during the night phase of the circadian oscillator [50,51]. The PEP Case kinase is a Ca^{2+}-independent protein kinase of 30–31 kDa roughly composed of the minimal sequence motifs required for protein kinase activity. PEP Case kinase phosphorylates PEP Case with high specificity and apparently is not regulated by modulators. Therefore, expressional control and high rates of turnover represent the decisive parameters that determine PEP Case kinase activity, the phosphorylation state of PEP Case, and hence the malate sensivity of PEP Case [48].

Some data indicate that the coupling of PEP Case kinase expression to the circadian oscillator may be modulated by metabolic factors. Possible factors are the cytosolic concentration of malate, the activity of tonoplast efflux, and the cytosolic pH and are presently under investigation. PEP Case kinase shows a pronounced pH optimum at pH 7, which may be a mechanism to reduce PEP Case activity when the cytosolic pH turns acidic [52]. In plants with inducible CAM, such as *Mesembryanthemum crystallinum*, CAM expression, the daily pH changes, and the increase in PEP Case activity are strongly correlated as shown under various growth and developmental conditions. Enzyme activity of PEP

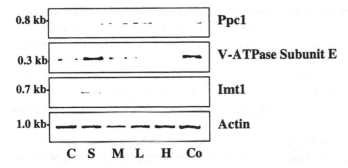

FIGURE 2 Analysis of mRNA accumulation of PEP carboxylase (Ppc1), V-AT-Pase subunit E, and *myo*-inositol *o*-methyltransferase (Imt1) in leaf tissue of 5-week-old *Mesembryanthemum crystallinum* by reverse transcription–polymerase chain reaction (RT-PCR). C, Control; S, 6 h salt stress of 400 mM NaCl; M 6 h osmotic stress with 700 mM mannitol; L, 6 h light stress with doubling of the light irradiance; H, 6 h heat stress at 42°C; Co, 6 h cold treatment at 4°C.

carboxylase as well as transcription and translation of the CAM-specific PEP carboxylase isoform Ppc1 increase in well-watered mature plants of *M. crystallinum* [53,54]. Whereas in seedlings of *M. crystallinum* Ppc1 is only slightly expressed in response to salt stress, the transcriptional inducibility of the CAM-specific gene increases with plant age [53]. Next to exposure to increased salinity, expression of Ppc1 is also inducible by osmotic stress, high irradiance, and cold stress, but it is not synthesized in response to heat stress (Fig. 2). The salinity-induced transition from C_3 photosynthesis to CAM is reversible. In plants that are transferred to well-watered growth conditions again, PEP carboxylase transcript amounts, protein, and activity decrease [17,53]. The correlation between CAM expression and PEP Case protein and activity is also observed in other facultative CAM plants, for example, *Clusia uvitana* [55].

4.2 The Pathway of Citric Acid Synthesis

Malic acid is the dominant organic acid synthesized and stored at night in all CAM plants. However, levels of some other organic acids have been shown to undergo small day-night changes. Conversely, in the genus *Clusia* substantial accumulation of citrate of up to 100 mM and more occurs during the night [19,56,57]. Citrate synthesis depends on transport of malate or oxaloacetate into the mitochondria (Fig. 1B). Malate is oxidized to oxaloacetate by mitochondrial MDH. Citrate is formed by condensation of acetyl-CoA with oxaloacetate by citrate synthase. Acetyl-CoA must be synthesized by glycolysis and pyruvate dehydrogenase complex. Thus, from one glucose, one molecule of PEP for carboxylation, one acetyl-CoA for citrate synthesis, and one CO_2, two $NADH+H^+$ and one ATP are formed. Additional ATP is synthesized from the liberated $NADH+H^+$ in the respiratory electron transport. Nocturnal accumulation of citrate does not lead to net carbon gain because for each CO_2 fixed by PEP carboxylation, one CO_2 is generated during decarboxylation of pyruvate. The benefit of citrate synthesis may be the supply of extra ATP for malate compartmentation [15].

The ratio of malic acid to citric acid accumulation depends on environmental conditions, with increasing citric acid amounts in response to drought stress [58], high night temperatures, and high daytime light intensities [22]. It has been discussed that citric acid accumulation protects against photoinhibition due to the energy conversion connected with the citrate decarboxylation in the day. The hypothesis needs to be proved experimentally [20].

4.3 Vacuolar Storage of Organic Acids

4.3.1 The Organic Acid Transporters
at the Tonoplast Membrane

Malic acid is the major organic acid formed and stored in CAM plants. The pK values of the two carboxylic groups of malic acid are 3.3 and 5.4. Consequently, malic acid is mainly present in the dianionic state in the cytosol at physiological pH values of about 7.2 [59]. Accumulation of malate in the vacuolar lumen depends on at least two transport systems, a malate transporter and an electrogenic proton pump, and a trapping mechanism. In isolated vacuoles of barley mesophyll cells, Mg-ATP activates malate transport across the tonoplast membrane [60]. The transported malate species is probably malate^{2-}. Protonation at acidic vacuolar pH will produce Hmalate and H_2malate, which are trapped inside the vacuole. A current model suggests that malate^{2-} equilibrates between the cytoplasm and the vacuolar lumen in response to the electrical gradient [41]. The electrical potential difference between the vacuole and the cytoplasm appears to be small, i.e., in the range of 10 to 40 mV, negative at the cytoplasmic face, mediating a 2- to 20-fold accumulation of malate^{2-} in the vacuole.

Figure 3 summarizes the transport processes involved in malate accumulation. Assuming a cytoplasmic pH of 7.25 and a vacuolar pH of 4 at the end of the night, the model calculation shows a gradient of 310 mM total vacuolar malate to 1 mM cytoplasmic malate. The energetics of malate accumulation in the vacuole of *Kalanchoë* has been calculated by Lüttge et al. [61] and was consistant with a stochiometry of two H^+ per one malate and one ATP. The affinity of the malate transport system is similar in plants with C_3 and CAM photosynthesis. Malate transport was half-saturated at a malate concentration of 2.5 mM in barley and at 2.7–3.0 mM in *Kalanchoë*. In most tested species the K_M was in the range

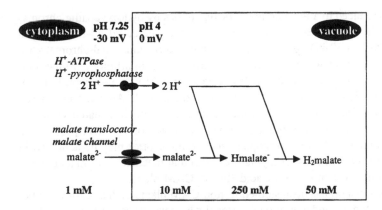

FIGURE 3 Simplified model of malate transport at the tonoplast membrane (see Ref. 40). Malate is transported as malate^{2-} via a translocator or channel protein. The electrical membrane potential and pH gradient determine the distribution of malate^{2-} across the tonoplast and the representation of the various malate species in the vacuolar lumen. The concentrations of the vacuolar malate species were calculated exemplarily under the given conditions. The cytoplasmic Hmalate concentration is in the lower micromolar range and the cytoplasmic H$_2$malate concentration in the nanomolar range.

1 to 4 mM [27,41,60]. The maximum storage capacity of vacuoles for malate may be determined by the tonoplast permeability for H$_2$malate. H$_2$malate was suggested to be the predominant malate species released from the vacuole during phase 3 of CAM, which is characterized by rapid decarboxylation of malate [39]. Thus, maximal acidification of CAM vacuoles may reflect the equilibrium between uptake of malate^{2-} and passive efflux of H$_2$malate. Moreover, the acidification and osmotic pressure associated with the nocturnal accumulation of organic acids in the vacuolar lumen may play a role as an oscillatory generator [62]. Such an endogenous generator is needed to understand the free-running endogenous rhythm in CAM, including acid metabolism and regulatory gene expression, for example, for the PEP Case kinase.

Employing a model of lipid membrane order, Neff et al. [62] were able to simulate a hysteresis switch in tonoplast structure at the end of the night period, which in turn could trigger the metabolic transition from CAM phase 1 to phases 2 and 3. The tonoplast malate transporter has not yet been characterized at the gene level. Two approaches have been undertaken to identify the transporter protein, namely affinity labeling and chromatographic purification using a reconstitution system for tracking malate transport activity. By starting from purified tonoplast membranes of *Kalanchoë daigremontiana*, the complexity of tonoplast proteins was reduced to a few candidate polypeptides by hydroxyapatite chromatography, while simultaneously the specific activity increased 44-fold [63]. In successive work, a 32-kDa polypeptide was purified by additional chromatography on an anion exchanger and benzene tricarboxylic acid affinity chromatography [64]. A photolyzable malate analogue labeled a 37-kDa polypeptide in tonoplast preparations of the C_3 plant *Catharanthus* [65]. An antibody raised against this 37-kDa polypeptide inhibited malate transport in isolated vacuoles. The protein appears to constitute a tetrameric complex of 160 kDa [66]. Thus, molecular cloning of the plant tonoplast malate transporter may be achieved in the near future and will allow studying the molecular mechanism of malate translocation as a major component of acid flux in CAM plants.

Citrate transport has been studied in non-CAM plants such as barley, rubber tree, lime, and tomato and exhibited $\Delta\mu$-dependent characteristics similar to those of malate transport [67,68]. Photoaffinity labeling has allowed identification of a tonoplast-associated citrate binding protein, which is a novel class of protein but not the transporter [69].

4.3.2 The Vacuolar H^+-ATPase and H^+-Pyrophosphatase

Transport and accumulation of organic acids in the vacuole are not energized directly but depend on the transtonoplast electrical potential [63,68] and the acidification of the vacuolar lumen as trapping mechanisms. Two electrogenic proton pumps reside at the tonoplast of plant cells, the V-type H^+-ATPase and the H^+-translocating pyrophosphatase. The V-type H^+-ATPase appears to be the predominant proton pump involved in energizing transtonoplast transport of organic acids. Bremberger and Lüttge [70] described a decline in H^+-PP$_i$ase during induction of CAM in the facultative CAM plant *Mesembryanthemum crystallinum*. Conversely, V-type H^+-ATPase is up-regulated at the mRNA and protein levels and enzyme activity in CAM plants [27].

The V-ATPase is an H^+ pump ubiquitously distributed among all eukaryotes. The complex comprises about 12 different polypeptide subunits with the suggested stochiometry of $A_3B_3CDEFG_3Hac_6de$, all of which have been cloned from plants by now. The bold letters stand for subunits suggested to form the stator, the letters in italics those constituting the putative rotor during ATP-dependent proton pumping [71]. From CAM plants, cDNA sequence information is available for all subunits [72,73, Kluge, Golldack, Dietz, unpublished].

Up-regulation of gene expression for subunits of H^+-ATPase is an important part of establishing CAM by increasing pump protein density per tonoplast membrane cross section [74] and is discussed subsequently. However, little is known about the biochemical regulation of V-ATPase activity in general and in CAM in particular. The V-ATPase activity is stimulated with increasing ATP concentration and inhibited by ADP and P_i, a modulation that may be sufficient

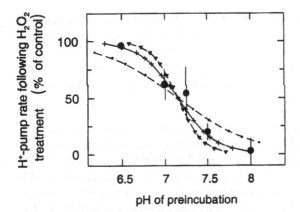

FIGURE 4 pH dependence of V-ATPase inactivation by H_2O_2. Barley tonoplast vesicles were tested for bafilomycin-sensitive ATP hydrolysis in the presence of 2 mM H_2O_2 (●). Other conditions were as in Dietz et al. [75]. Calculations were performed to fit the experimental data by a single SH group (◆) and two (+) or three cooperative SH groups (∇). The best fit was obtained assuming two sensitive target residues. Error bars indicate ± standard deviation of means.

to understand light-dependent acidification of vacuoles of mesophyll cells of the C_3 plant barley [75]. Subcellular metabolite levels during the diurnal CAM cycle are not available. Therefore, the cytoplasmic adenylate status during the night and its effect on V-ATPase activity are unknown. Redox regulation is another mechanism that should be discussed, although its physiological significance is not clear. Oxidative inactivation of V-ATPase, for instance, by the nonphysiological oxidant H_2O_2, shows pronounced pH dependence, which is likely to reflect the thiolate formation at alkaline pH. At acidic pH values, H_2O_2 had no effect on bafilomycin-sensitive ATP hydrolysis (Fig. 4) [107]. Such a type of pH profile could be necessary to prevent redox inactivation of the V-ATPase when the cytoplasm has acidified and may also be important in CAM.

4.4 The Decarboxylation Reaction and H⁺ Fluxes

During the light phase, stored organic acids are released from the vacuole. Lüttge and Smith [39] suggested passive efflux of the rather lipophilic form H_2malate as the mechanism of acid release from the vacuole. Other export mechanisms must be involved at less acidic vacuolar pH. Two groups of CAM plants are distinguished on the basis of the decarboxylating enzyme:

1. PEP carboxykinase (PEPCK) releases CO_2 and PEP from oxaloacetate and ATP. PEPCK-CAM occurs in the families Asclepiadaceae, Bromeliaceae, Euphorbiaceae, and Portulaceae.
2. Malic enzyme (ME) synthesizes pyruvate and CO_2 from malate. ME-CAM is found in the families Aizoaceae, Cactaceae, Crassulaceae, and Orchidaceae [76,77].

In plant families of higher taxonomic levels such as Asparagales and Caryophyllales, both ME-CAM and PEPCK-CAM plants are observed simultaneously. The energetic costs of ME-CAM, with 5.2 ATP and 3.5 NADPH per assimilated CO_2, are slightly lower than those of PEPCK-CAM, where 5.9 ATP and 3.9 NADPH are needed per CO_2 [15]. Substantial increases in the activities of NAD malic enzyme and NADP malic enzyme have been observed in CAM-performing *Mesembryanthemum crystallinum* [78]. A study by Christopher and Holtum [77] showed that there is no general relationship between storage carbohydrates and the type of decarboxylation reaction. PEPCK-CAM plants do not always store extrachloroplastic sugars and carbohydrate polymers, and ME-CAM plants do not exclusively store chloroplastic starch and glucans. For example, ME-CAM plants such as *Sansevieria hahnii* accumulate sucrose and the PEPCK-CAM plant *Hoya carnosa* starch.

In *Clusia rosea*, phosphoenolpyruvate carboxykinase activity has been observed, whereas the activities of the NADP malic enzyme and the NAD malic enzyme were not detectable and negligible, respectively [79]. Besides, Borland et al. [21] detected the phosphoenolpyruvate carboxykinase protein in CAM-performing *Clusia aripoensis*, *Clusia minor*, and *C. rosea* by Western blot analysis, but only trace amounts of the NADP malic enzyme and NAD malic enzyme, indicating that *Clusia* is a PEP carboxykinase type of CAM plant. In *C. aripoensis*, *C. minor*, and *C. rosea* amounts of phosphoenolpyruvate carboxylase and of the phosphoenolpyruvate carboxykinase proteins increased in response to drought stress as studied by immunoblotting experiments [21]. These data indicate a direct regulation of CAM capacity by the amounts of PEP carboxylase and phosphoenolpyruvate carboxykinase in *Clusia* [21].

5 GENE EXPRESSION DURING CAM

The transition from C_3 photosynthesis to CAM is accompanied by transcriptional induction of a number of CAM-specific genes as well as transcriptional activation of genes that are also involved in C_3 metabolic reactions, for example, in carbon metabolism. Regulation of CAM activity occurs on different molecular levels and includes CAM-specific gene expression as well as posttranscriptional and posttranslational processes. In *Mesembryanthemum crystallinum*, genes transcriptionally activated in CAM-performing plants are, e.g., NADP malic enzyme [80],

NADP malate dehydrogenase [81], glyceraldehyde-3-phosphate dehydrogenase [82], and the vacuolar ATPase subunits A, B, E, F, c [72,83–85]. CAM-specific gene expression has been studied most intensely for PEP carboxylase, the key enzyme of CAM catalyzing the nocturnal carboxylation of PEP to oxaloacetate.

In CAM-performing plants, the transcript level, the protein amount, and the activity of PEP carboxylase increased, indicating that the activity of the enzyme is regulated by de novo synthesis. In *Peperomia camtotricha*, with increasing leaf age parallel increases have been reported for PEP carboxylase mRNA and protein amounts, PEP carboxylase activity, as well as the CAM activity measured as leaf acidification [11,14].

In *Kalanchoë blossfeldiana*, PEP carboxylase is encoded by a small gene family with two isogenes that are specific to C_3 photosynthesis and two isogenes that have been suggested to be CAM specific [86]. In *Vanilla planifolia* (Orchidaceae) one PEP carboxylase isoform has been found by Northern blot analysis in CAM-performing tissue and another isoform that is mainly expressed in nonphotosynthetic organs [87].

In *M. crystallinum*, two isogenes of PEP carboxylase are expressed that share 83% identity at the amino acid level [3]. One of the isoforms, Ppc2, is synthesized in plants performing either C_3 photosynthesis or CAM and is supposed to be a housekeeping enzyme. In contrast, synthesis of a CAM-specific isoform, Ppc1, is strongly induced by environmental stress (e.g., increased salinity [3]). Interestingly, there are no substantial sequence homologies of the Ppc2 promoter region and the stress-inducible Ppc1 promoter. Elements such as TATA, CAAT, and GT motifs as well as consensus abscisic acid (ABA) response elements have been identified in the Ppc1 5′-flanking region but are absent in the Ppc2 promoter sequences [88]. Regulation of Ppc1 transcription by interaction of specific *cis*-acting elements with *trans*-acting transcription factors is likely but needs to be confirmed on the molecular level [16,89]. Corresponding to ontogenetically controlled induction of CAM in *M. crystallinum*, the expression of Ppc1 is developmentally regulated. The CAM-specific gene is not expressed in seedlings and juvenile plants but is detectable in the leaf tissue of mature *M. crystallinum* [85]. In our experiments Ppc1 synthesis could be induced by salinity stress in *M. crystallinum* starting at the age of 3 weeks but the highest amounts of transcripts could be found in salt-stressed mature plants. Next to exposure to increased salinity, other environmental factors induce Ppc1 synthesis in *M. crystallinum* (e.g., osmotic stress, increased light intensity, and cold stress; Fig. 2).

Posttranscriptional regulation of the activity of CAM-specific enzymes by increased mRNA stability has been reported, e.g., PEP carboxylase in *M. crystallinum* [90]. Posttranslational processes influencing the activity of PEP carboxylase are diurnal modifications of the phosphorylation state of the enzyme. In the constitutive CAM plant *Bryophyllum (Kalanchoë) fedtschenkoi*, PEP carboxylase is dephosphorylated in the day and the enzyme activity is sensitively inhibited by malate [49]. Another enzyme that is phosphorylated in the nighttime and dephosphorylated in the daytime in CAM plants is the PEP carboxykinase that catalyzes the decarboxylation of oxaloacetate to PEP in the day [91].

6 SIGNALS AND CAM

A number of studies focused on the elucidation of signal transduction events involved in induction of CAM-specific gene expression and the metabolic reactions of CAM. The data provided indicate that CAM is controlled by complex signaling events that are not well understood yet. Control of enzyme activities by substrates has been demonstrated, e.g., PEP carboxylase and PEP carboxylase kinase, which are regulated by glucose-6-phosphate and malate [50,92].

Next to allosteric effects, plant growth regulators have been shown to play an important role in CAM induction. In the facultative CAM plant *Portulacaria afra* (Portulacaceae), treatment of leaves with abscisic acid (ABA) induced daytime stomatal closure and increase of PEP carboxylase activity [1,14,93]. For *Mesembryanthemum crystallinum*, it has been reported that exogenous application of ABA increased activities of NADP malic enzyme and PEP carboxylase [94]. Thomas et al. [95] have demonstrated that treatment of *M. crystallinum* with ABA does not induce PEP carboxylase transcription when the plants are grown under low light, whereas ABA application combined with exposure to high light induces transcription and translation of PEP carboxylase [96]. Interestingly, the 5'-flanking genomic sequence of the CAM-specific Ppc1 contains putative ABA-responsive elements [97]. However, when synthesis of ABA was inhibited, similar amounts of PEP carboxylase protein accumulated under salt stress compared with plants without treatment with the inhibitor [95]. These data indicate that ABA is not primarily triggering CAM induction, although it might be one of the signaling elements involved in CAM stimulation. Application of cytokinin to roots of *M. crystallinum* induced CAM activity and accumulation of PEP carboxylase transcripts, whereas treatment of leaves with cytokinin inhibited transcription of PEP carboxylase [98,99]. Accumulation of Ppc1 transcripts in *M. crystallinum* decreased when detached leaves were treated prior to salinity stress with okadaic acid to inhibit protein phosphatase 1 and 2A, with the calcium chelator EGTA, and with inhibitors of Ca^{2+}- and calmodulin-dependent protein kinases such as lavendustin [100]. Ppc1 transcription was stimulated by application of the Ca^{2+} ionophore ionomycin, the Ca^{2+}-ATPase inhibitor thapsigargin, and the inhibitors of protein phosphatase 2B ascomycin and cyclosporin A [100]. In our experiments to dissect signaling pathways of the different salt response mechanisms in *M. crystallinum* we found that transcription of *myo*-inositol *o*-methyltransferase (Imt1), Ppc1, and V-ATPase subunit E could be blocked in detached leaves by treatment with the activator of trimeric G proteins mastoparan.

Application of the calcium chelator EGTA and membrane-permeable EGTA/AM inhibited induction of Ppc1 in salinity-stressed detached leaves but did not affect transcription of IMT1 and V-ATPase subunit E [85].

7 CAM AND pH-STAT

Regulation and balancing of intracellular pH are of vital importance for maintenance of metabolic reactions and transport processes. The main components of cytoplasmic pH control are the so-called biophysical pH-stat and the biochemical pH-stat (see Ref. 101 and also Chapter 3). The biophysical pH-stat consists of ATP-driven efflux of protons mediated by, e.g., plasma membrane H^+-ATPase (see Chapter 1) and vacuolar H^+-ATPase (see Chapter 2). The biochemical pH-stat mechanisms are reversible metabolic reactions consuming and delivering protons and are supposed to be important for short-term balancing of intracellular pH and for its fine control [102,103].

The most accepted model for the biochemical pH-stat is based on the mechanism of pH-dependent carboxylation-decarboxylation of malate. Its significance is confirmed by the fluctuations of intracellular malate observed in response to intracellular pH changes [104]. According to this model, an increase of cytoplasmic pH stimulates the activity of PEP carboxylase, malic acid is synthesized, and the pH is decreased. The activity of PEP carboxylase is feedback inhibited by malate and is reduced by decreasing pH. In addition, the activity of malic enzyme increases and the cytoplasmic pH increases with decarboxylation of malate (Fig. 5) [101,103]. Sakano [104] did model calculations on the fluxes of the pH-stat mechanism and concluded that the H^+ amounts produced and consumed

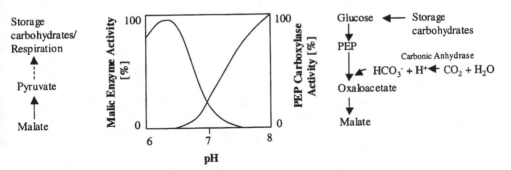

FIGURE 5 Model of biochemical pH-stat based on the reactions catalyzed by PEP carboxylase and malic enzyme in plants performing C_3 photosynthesis. (Modified from Refs. 103 and 104.)

in the reactions catalyzed by PEP carboxylase and malic enzyme were not sufficient to explain the pH changes observed in parallel to the measured intracellular changes of malate concentration. However, the model became more conclusive with the proton-producing reactions of glycolysis and carbonic anhydrase. The proton-consuming reactions of respiration following malate decarboxylation were taken into account in the acidification and the alkalization step, respectively. In CAM-performing plants, transport of protons into the vacuole mediated by the vacuolar H^+-ATPase regulates the cytoplasmic pH and generates the proton motive force for the accumulation of organic acids in the vacuole [105].

In contrast to these biophysical pH-stat mechanisms, biochemical pH-stat reactions are less likely to be involved in CAM regulation. As already described in Sec. 4 of this chapter, CAM-specific isoforms of PEP carboxylase show low sensitivity to feedback inhibition by malate in their active, phophorylated state in the nighttime. PEP carboxylases of a number of CAM-performing plants have a low pH optimum of about 6.0 to 6.5 compared with the pH optimum of about 8.0 in plants performing C_3 photosynthesis [101,103]. Consistent with this observation, in *Kalanchoë tubiflora* increases in cytoplasmic pH cause inhibition of malate synthesis, whereas an increase in malate concentration with increasing pH would be expected for C_3 metabolism type of PEP carboxylase as described earlier [45,106].

8 PERSPECTIVES

CAM is a photosynthetic mechanism of carbon concentration that has attracted the attention of scientists for decades. However, in contrast to our extensive knowledge of the biochemistry, physiology, and ecology of CAM, the role of pH is still not satisfactorily understood in regulating CAM, triggering partial reactions, and possibly altering gene expression. This has been due partly to the experimental challenges of studying pH fluctuations as well as metabolite pools on the cellular and subcellular level in living tissue. Experimental approaches to analyze effects of pH on CAM by extracellular modifications have demonstrated that the cytoplasmic pH is well balanced in plants and is not significantly changed in response to extracellular fluctuations [45,106]. Thus, the proper site of sensing pH changes could be the tonoplast (see Ref. 62). A powerful tool to extend studies to the subcellular level will be the use of fluorescent pH indicators combined with confocal laser scanning microscopy as reviewed by Roos (Chapter 4). Generation of transgenic plants with silenced key enzymes of CAM (e.g., PEP carboxylase, malic enzyme, PEP carboxykinase, and vacuolar ATPase) will help to dissect the involvement of biochemical and biophysical pH-stat mechanisms in CAM. Future work on pH and CAM will focus on the question of whether pH acts as a second messenger in CAM and on identifying signaling events triggered by pH changes on the molecular level.

REFERENCES

1. IP Ting. Crassulacean acid metabolism. Annu Rev Plant Physiol 36:595–622, 1985.
2. JAC Smith, K Winter. Taxonomic distribution of crassulacean acid metabolism. In: K Winter, JAC Smith, eds. Crassulacean Acid Metabolism: Biochemistry, Ecophysiology and Evolution. Berlin: Springer-Verlag, 1996, pp 427–436.
3. JC Cushman, G Meyer, CB Michalowski, JM Schmitt, HJ Bohnert. Salt stress leads to differential expression of two isogenes of phosphoenolpyruvate carboxylase during Crassulacean acid metabolism induction in the common ice plant. Plant Cell 1:715–25, 1989.
4. K Winter. Crassulacean acid metabolism. In: Barber J, Baker NR, eds. Photosynthetic Mechanisms and the Environment. Amsterdam: Elsevier Science Publishers, 1985, pp 329–387.
5. JE Keeley. *Isoëtes howellii*: a submerged aquatic CAM plant? Am J Bot 68:420–424, 1981.
6. O Hostrup, G Wiegleb. The influence of different CO_2 concentrations in the light and the dark on diurnal malate rhythm and phosphoenolpyruvate carboxylase activities in leaves of *Littorella uniflora* Aschers. Aquat Bot 40:91–100, 1991.
7. JA Teeri, SJ Tonsor, M Turner. Leaf thickness and carbonisotope composition in the Crassulaceae. Oecologia 50:367–369, 1981.
8. PS Nobel, GB North. Features of roots of CAM plants. In: K Winter, JAC Smith, eds. Crassulacean Acid Metabolism: Biochemistry, Ecophysiology and Evolution. Berlin: Springer-Verlag, 1996, pp 266–280.
9. W Cockburn. Variations in photosynthetic acid metabolism in vascular plants: CAM and related phenomenon. New Phytol 101:3–24, 1985.
10. MB Wilkins. Circadian rhythms: their origin and control. New Phytol 121:347–375, 1992.
11. IP Ting, J Hann, D Sipes, A Patel, LL Walling. Expression of P-enolpyruvate carboxylase and other aspects of CAM during the development of *Peperomia camptotricha* leaves. Bot Acta 106:313–319, 1993.
12. T Amagasa. The influence of leaf age on the diurnal changes of malate and starch in the CAM plant *Kalanchoë daigremontiana* Hamet et Perr. Z Pflanzenphysiol 108:93–96, 1982.
13. DJ Von Willert, GO Kirst, S Treichel, K Von Willert. The effect of leaf age and salt stress on malate accumulation and phosphoenolpyruvate carboxylase activity in *Mesembryanthemum crystallinum*. Plant Sci Lett 7:341–346, 1976.
14. IP Ting, A Patel, S Kaur, J Hann, L Walling. Ontogenetic development of crassulacean acid metabolism as modified by water stress in *Peperomia*. In: K Winter, JAC Smith, eds. Crassulacean Acid Metabolism: Biochemistry, Ecophysiology and Evolution. Berlin: Springer-Verlag, 1996, pp 204–215.
15. K Winter, JAC Smith. Crassulacean acid metabolism: current status and perspectives. In: K Winter, JAC Smith, eds. Crassulacean Acid Metabolism: Biochemistry, Ecophysiology and Evolution. Berlin: Springer-Verlag, 1996, pp 389–426.
16. P Adams, DE Nelson, S Yamada, W Chmara, RG Jensen, HJ Bohnert, H Griffiths. Growth and development of *Mesembryanthemum crystallinum* (Aizoaceae). New Phytol 138:171–190, 1998.

17. DM Vernon, JA Ostrem, JM Schmitt, HJ Bohnert. PEPcase transcript levels in *Mesembryanthemum crystallinum* decline rapidly upon relief from salt stress. Plant Physiol 86:1002–1004, 1988.

18. HSJ Lee, H Griffiths. Induction and repression of CAM in *Sedum telephium* L. in response to photoperiod and water stress. J Exp Bot 38:187–192, 1987.

19. AM Borland, H Griffiths. Properties of phosphoenolpyruvate carboxylase and carbohydrate accumulation in the C_3-CAM intermediate *Sedum telephium* L. grown under different light and watering regimes. J Exp Bot 43:353–361, 1992.

20. U Lüttge. One morphotype, three physiotypes: sympatric species of *Clusia* with obligate C-3 photosynthesis, obligate CAM and C-3-CAM intermediate behaviour. Plant Biol 1:138–148, 1999.

21. AM Borland, LI Tecsi, RC Leegood, RP Walker. Inducibility of crassulacean acid metabolism (CAM) in *Clusia* species; physiological/biochemical characterisation and intercellular localization of carboxylation and decarboxylation processes in three species which exhibit different degrees of CAM. Planta 205:342–351, 1998.

22. Lüttge U. *Clusia*: plasticity and diversity in a genus of C3/CAM intermediate tropical trees. In: K Winter, JAC Smith, eds. Crassulacean Acid Metabolism: Biochemistry, Ecophysiology and Evolution. Berlin: Springer-Verlag, 1996, pp 296–311.

23. DM Vernon, HJ Bohnert. A novel methyl transferase induced by osmotic stress in the facultative halophyte *Mesembryanthemum crystallinum*. EMBO J 11:2077–2085, 1992.

24. M Ishitani, AL Majumder, A Bornhauser, CB Michalowski, RG Jensen, HJ Bohnert. Coordinate transcriptional induction of *myo*-inositol metabolism during environmental stress. Plant J 9:537–548, 1996.

25. BJ Barkla, L Zingarelli, E Blumwald, JAC Smith. Tonoplast Na^+/H^+ -antiport activity and its energization by the vacuolar H^+ -ATPase in the halophytic plant *Mesembryanthemum crystallinum* L. Plant Physiol 109:549–556, 1995.

26. I Struve, A Weber, U Lüttge, E Ball, JAC Smith. Increased vacuolar ATPase activity correlated with CAM induction in *Mesembryanthemum crystallinum* and *Kalanchoë blossfeldiana* cv. Tom Thumb. J Plant Physiol 117:451–468, 1985.

27. R Ratajczak, J Richter, U Lüttge. Adaptation of the tonoplast V-type H^+ -ATPase of *Mesembryanthemum crystallinum* to salt stress, C_3-CAM transition and plant age. Plant Cell Environ 17:1101–1112, 1994.

28. C Bremberger, HP Haschke, U Lüttge. Separation and purification of the tonoplast ATPase and pyrophosphatase from plants with constitutive and inducible crassulacean acid metabolism. Planta 175:465–470, 1988.

29. JE Keeley. CAM photosynthesis in submerged aquatic plants. Bot Rev 64:121–175, 1998.

30. DR Webb, MR Tattray, JMA Brown. A preliminary survey of crassulacean acid metabolism (CAM) in submerged aquatic macrophytes in New Zealand. N Z J Mar Freshwater Res 22:231–235, 1988.

31. JE Keeley. Aquatic CAM photosynthesis. In: K Winter, JAC Smith, eds. Crassulacean Acid Metabolism: Biochemistry, Ecophysiology and Evolution. Berlin: Springer-Verlag, 1996, pp 281–295.

32. JA Raven, RA Spicer. The evolution of crassulacean acid metabolism. In: K Winter,

JAC Smith, eds. Crassulacean Acid Metabolism: Biochemistry, Ecophysiology and Evolution. Berlin: Springer-Verlag, 1996, pp 360–388.

33. SC Maberly, DHN Spence. Photosynthetic inorganic carbon use by freshwater plants. J Ecol 71:705–724, 1983.

34. JR Newman, JA Raven. Photosynthetic carbon assimilation by *Crassula helmsii*. Oecologia 101:494–499, 1995.

35. K Sand-Jensen. Environmental variables and their effect on photosynthesis of aquatic plant communities. Aquat Bot 34:5–25, 1989.

36. DR Sandquist, JE Keeley. Carbon uptake characteristics in two high elevation populations of the aquatic CAM plant *Isoëtes bolanderi* (Isoetaceae). Am J Bot 77: 682–688, 1990.

37. JE Keeley. Crassulacean acid metabolism in the seasonally submerged aquatic *Isoëtes howellii*. Oecologia 58:57–62, 1983.

38. RJ Helder, MV Van Harmelen. Carbon assimilation pattern in the submerged leaves of the aquatic angiosperm *Vallisneria spiralis* L. Acta Bot Neerl 31:281–295, 1982.

39. U Lüttge, JAC Smith. Mechanism of passive malic acid efflux from vacuoles of the CAM plant *Kalanchoë daigremontiana*. J Membr Biol 81:149–158, 1984.

40. WH Kenyon, AS Holaday, CC Black. Diurnal changes in metabolite levels and crassulacean acid metabolism in *Kalanchoë daigremontiana* leaves. Plant Physiol 68:1002–1007, 1981.

41. CM Cheffings, O Pantoja, FM Ashcroft, JAC Smith. Malate transport and vacuolar ion channels in CAM plants. J Exp Bot 48:623–631, 1997.

42. H Arata, I Iwasaki, K Kusumi, M Nishimura. Thermodynamics of malate transport across the tonoplast of leaf cells of CAM plants. Plant Cell Physiol 33:873–880, 1992.

43. AM Borland, H Griffiths, C Maxwell, MSJ Broadmeadow, NM Griffiths, JD Barnes. On the ecophysiology of the Clusiaceae in Trinidad: expression of CAM in *Clusia minor* L. during the transition from wet to dry season and characterization of the endemic species. New Phytol 122:349–357, 1992.

44. CB Osmond. Crassulacean acid metabolism: a curiosity in context. Annu Rev Plant Physiol 29:379–414, 1978.

45. G Marigo, E Ball, U Lüttge, JAC Smith. Use of the DMO technique for the study of relative changes of cytoplasmic pH in leaf cells in relation to CAM. Z Pflanzenphysiol 108:223–233, 1982.

46. WH Kenyon, R Kringstad, CC Black. Diurnal changes in the malic acid content of vacuoles isolated from leaves of the crassulacean acid metabolism plant *Sedum telephium*. FEBS Lett 94:281–283, 1978.

47. C Giersch. Capacity of the malate/oxaloacetate shuttle for transfer of reducing equivalents across the envelope of leaf chloroplasts. Arch Biochem Biophys 219: 379–387, 1982.

48. HG Nimmo. The regulation of phosphoenolpyruvate carboxylase in CAM plants. Trends Plant Sci 5:75–80, 2000.

49. GA Nimmo, HG Nimmo, CA Fewson, MB Wilkins. Diurnal changes in the properties of phosphoenolpyruvate carboxylase in *Bryophyllum* leaves: a possible covalent modification. FEBS Lett 178:199–203, 1984.

50. PJ Carter, HG Nimmo, CA Fewson, MB Wilkins. Circadian rhythms in the activity of a plant protein kinase. EMBO J 10:2063–2068, 1991.

51. J Hartwell, A Gill, GA Nimmo, MB Wilkins, GI Jenkins, HG Nimmo. Phosphoenol-pyruvate carboxylase kinase is a novel protein kinase regulated at the level of expression. Plant J 20:333–342, 1999.

52. B Baur, KJ Dietz, K Winter. Regulatory protein phosphorylation of phosphoenol-pyruvate carboxylase in the crassulacean-acid-metabolism plant *Mesembryanthemum crystallinum* L. Eur J Biochem 209:95–101, 1992.

53. JC Cushman, CB Michalowski, HJ Bohnert. Developmental control of crassulacean acid metabolism inducibility by salt stress in the common ice plant. Plant Physiol 94:1137–1142, 1990.

54. W Herppich, M Herppich, DL Von Willert. The irreversible C_3 to CAM shift in well-watered and salt-stressed plants of *Mesembryanthemum crystallinum* is under strict ontogenetic control. Bot Acta 105:34–40, 1992.

55. K Winter, G Zotz, B Baur, KJ Dietz. Light and CO_2 fixation in *Clusia uvitana* and the effect of plant water status. Oecologia 91:47–51, 1992.

56. M Popp, D Kramer, H Lee, M Diaz, H Ziegler, U Lüttge. Crassulacean acid metabolism in tropical dicotyledonous trees of the genus *Clusia*. Trees 1:238–247, 1987.

57. A Haag-Kerwer, AC Franco, U Lüttge. The effect of temperature and light on gas exchange and acid accumulation in the C_3-CAM plant *Clusia minor*. J Exp Bot 43: 345–352, 1992.

58. AC Franco, E Ball, U Lüttge. Differential effects of drought and light levels on accumulation of citric and malic acids during CAM in *Clusia*. Plant Cell Environ 15:821–829, 1992.

59. E Martinoia. Malate compartmentation—responses to a complex metabolism. Annu Rev Plant Physiol Plant Mol Biol 45:447–467, 1994.

60. E Martinoia, UI Flügge, G Kaiser, U Heber, HW Heldt. Energy-dependent uptake of malate into vacuoles isolated from barley mesophyll protoplasts. Biochim Biophys Acta 806:311–319, 1985.

61. U Lüttge, JAC Smith, G Marigo, CB Osmond. Energetics of malate accumulation in the vacuoles of *Kalanchoë tubiflora* cells. FEBS Lett 126:81–84, 1981.

62. R Neff, B Blasius, F Beck, U Lüttge. Thermodynamics and energetics of the tonoplast membrane operating as a hysteresis switch in an oscillatory model of crassulacean acid metabolism. J Membr Biol 165:37–43, 1998.

63. R Ratajczak, I Kemna, U Lüttge. Characteristics, partial purification and reconstitution of the vacuolar malate transporter of the CAM plant *Kalanchoë daigremontiana* Hamet et Perrier de la Bathie. Planta 195:226–236, 1994.

64. S Steiger, T Pfeiffer, R Ratajczak, E Martinoia, U Lüttge. The vacuolar malate transporter of *Kalanchoë daigremontiana*: a 32 kDa polypeptide? J Plant Physiol 151:137–141, 1997.

65. K Lahjouji, A Carrasco, H Bouyssou, L Cazaux, G Marigo, H Canut. Identification with a photoaffinity reagent of a tonoplast protein involved in vacuolar malate transport of *Catharanthus roseus*. Plant J 9:799–808, 1996.

66. K Lahjouji, H Canut. Oligomeric state of the tonoplast proteins: structure of the putative malate transporter in *Catharanthus roseus*. Physiol Plant 105:32–38, 1999.

67. A Brune, P Gonzalez, R Goren, U Zehavi, E Echeverria. Citrate uptake into to-

noplast vesicles from acid lime (*Citrus aurantifolia*) juice cells. J Membr Biol 166: 197–203, 1998.

68. E Martinoia. Transport processes in vacuoles of higher plants. Bot Acta 105:232–245, 1992.

69. D Rentsch, J Görlach, E Vogt, N Amrhein, E Martinoia. The tonoplast-associated citrate binding protein (CBP) of *Hevea brasiliensis*. J Biol Chem 270:30525–30531, 1995.

70. C Bremberger, U Lüttge. Dynamics of tonoplast proton pumps and other tonoplast proteins of *Mesembryanthemum crystallinum* during induction of crassulacean acid metabolism. Planta 188:575–580, 1992.

71. M Forgac. Structure, function and regulation of the vacuolar (H^+)-ATPase. FEBS Lett 440:258–263, 1998.

72. KJ Dietz, B Arbinger. cDNA sequence and expression of subunit E of the vacuolar H^+-ATPase in the inducible crassulacean acid metabolism plant *Mesembryanthemum crystallinum*. Biochim Biophys Acta 1281:134–138, 1996.

73. JC Cushman, Bohnert HJ. Crassulacean acid metabolism: molecular genetics. Annu Rev Plant Physiol Plant Mol Biol 50:305–332, 1999.

74. B Rockel, R Ratajczak, A Becker, U Lüttge. Changed densities and diameters of intra-membrane tonoplast particles of *Mesembryanthemum crystallinum* in correlation with NaCl-induced CAM. J Plant Physiol 143:318–324, 1994.

75. KJ Dietz, U Heber, T Mimura. Modulation of the vacuolar H^+-ATPase by adenylates as basis for the transient CO_2 dependent acidification of the leaf vacuole upon illumination. Biochim Biophys Acta 1373:87–92, 1998.

76. P Dittrich. Nicotineamide adenine dinucleotide–specific malic enzyme in *Kalanchoë daigremontiana* and other plants exhibiting crassulacean acid metabolism. Plant Physiol 57:310–314, 1976.

77. JT Christopher, JAM Holtum. Patterns of carbon partitioning in leaves of crassulacean acid metabolism species during deacidification. Plant Physiol 112:393–399, 1996.

78. JAM Holtum, K Winter. Activities of enzymes during induction of crassulacean acid metabolism in *Mesembryanthemum crystallinum* L. Planta 155:8–16, 1982.

79. CC Black, JQ Chen, RL Dong, MN Angelov, SJS Sung. Alternative carbohydrate reserves used in the daily cycle of crassulacean acid metabolism. In: K Winter, JAC Smith, eds. Crassulacean Acid Metabolism: Biochemistry, Ecophysiology and Evolution. Berlin: Springer-Verlag, 1996, pp 31–45.

80. JC Cushman. Characterization and expression of a NADP-malic enzyme cDNA induced by salt stress from the facultative crassulacean acid metabolism plant, *Mesembryanthemum crystallinum*. Eur J Biochem 208:259–266, 1992.

81. JC Cushman. Molecular cloning and expression of chloroplast NADP-malate dehydrogenase during crassulacean acid metabolism induction by salt stress. Photosynth Res 35:15–27, 1993.

82. JA Ostrem, DM Vernon, HJ Bohnert. Increased expression of a gene coding for NAD:glyceraldehyde-3-phosphate dehydrogenase the transition from C_3 photosynthesis to crassulacean acid metabolism in *Mesembryanthemum crystallinum*. J Biol Chem 265:3497–3502, 1990.

83. R Löw, B Röckel, M Kirsch, R Ratajczak, S Hortensteiner, E Martinoia, U Lüttge,

T Rausch. Early salt stress effects on the differential expression of vacuolar H(+)-ATPase genes in roots and leaves of *Mesembryanthemum crystallinum*. Plant Physiol 110:259–265, 1996.

84. MS Tsiantis, DM Bartholomew, JA Smith. Salt regulation of transcript levels for the c subunit of a leaf vacuolar H(+)-ATPase in the halophyte *Mesembryanthemum crystallinum*. Plant J 9:729–736, 1996.

85. D Golldack, KJ Dietz. Developmental control and tissue-specificity of salt-induced expression of the vacuolar H⁺-ATPase in *Mesembryanthemum crystallinum*. Plant Physiol 125:1643–1654.

86. H Gehrig, T Taybi, M Kluge, J Brulfert. Identification of multiple PEPC isogenes in leaves of the facultative crassulacean acid metabolism (CAM) plant *Kalanchoë blossfeldiana* Poelln. cv. Tom Thumb. FEBS Lett 377:399–402, 1995.

87. H Gehrig, K Faist, M Kluge. Identification of phosphoenolpyruvate carboxylase isoforms in leaf, stem and roots of the obligate CAM plant *Vanilla planifolia* Salib. (Orchidaceae): a physiological and molecular approach. Plant Mol Biol 38:1215–1223, 1998.

88. JC Cushman, HJ Bohnert. Molecular genetics of crassulacean acid metabolism. Plant Physiol 113:667–676, 1997.

89. JC Cushman, HJ Bohnert. Transcriptional activation of CAM genes during development and environmental stress. In: K Winter, JAC Smith, eds. Crassulacean Acid Metabolism: Biochemistry, Ecophysiology and Evolution. Berlin: Springer-Verlag, 1996, pp 135–158.

90. CB Michalowski, SW Olson, M Piepenbrock, JM Schmitt, HJ Bohnert. Time course of mRNA induction elicited by salt stress in the common ice plant (*M. crystallinum*). Plant Physiol 89:811–816, 1989.

91. RP Walker, RC Leegood. Phosphorylation of phosphoenolpyruvate carboxykinase in plants. Studies in plants with C4 photosynthesis and crassulacean acid metabolism and in germinating seeds. Biochem J 317:653–658, 1996.

92. R Chollet, J Vidal, MH O'Leary. Phosphoenolpyruvate carboxylase: a ubiquitous, highly regulated enzyme in plants. Ann Rev Plant Physiol Plant Mol Biol 47:273–298, 1996.

93. Ting IP. Effects of abscisic acid on CAM in *Portulacaria afra*. Photosynth Res 2: 39–48, 1981.

94. GE Edwards, Z Dai, SH Cheng, MSB Ku. Factors affecting the induction of crassulacean acid metabolism in *Mesembryanthemum crystallinum*. In: K Winter, JAC Smith, eds. Crassulacean Acid Metabolism: Biochemistry, Ecophysiology and Evolution. Berlin: Springer-Verlag, 1996, pp 119–134.

95. JC Thomas, EF McElwain, HJ Bohnert. Convergent induction of osmotic stress-responses. Abscisic acid, cytokinin, and the effects of NaCl. Plant Physiol 100: 416–423, 1992.

96. EF McElwain, HJ Bohnert, JC Thomas. Light moderates the induction of phosphoenolpyruvate carboxylase by NaCl and abscisic acid in *Mesembryanthemum crystallinum*. Plant Physiol 99:1261–1264, 1992.

97. JM Schmitt, B Fißlthaler, A Sheriff, B. Lenz, M Bäßler, G. Meyer. Environmental control of CAM induction in *Mesembryanthemum crystallinum*—a role for cytokinin, abscisic acid and jasmonate? In: K Winter, JAC Smith, eds. Crassulacean Acid

Metabolism: Biochemistry, Ecophysiology and Evolution. Berlin: Springer-Verlag, 1996, pp 159–175.

98. JM Schmitt, M Piepenbrock. Regulation of phosphoenolpyruvate carboxylase and crassulacean acid metabolism induction in *Mesembryanthemum crystallinum* L. by cytokinin. Plant Physiol 99:1664–1669, 1992.

99. W Peters, E Beck, M Piepenbrock, B Lenz, JM Schmitt. Cytokinin as a negative effector of phosphoenolpyruvate carboxylase induction in *Mesembryanthemum crystallinum*. J Plant Physiol 151:362–367, 1997.

100. T Taybi, JC Cushman. Signaling events leading to crassulacean acid metabolism induction in the common ice plant. Plant Physiol 121:545–556, 1999.

101. Smith FA, JA Raven. Intracellular pH and its regulation. Annu Rev Plant Physiol 30:289–311, 1979.

102. DD Davis. Control of and by pH. Symp Soc Exp Biol 27:513–529, 1973.

103. DD Davis. The fine control of cytosolic pH. Physiol Plant 67:702–706, 1986.

104. K Sakano. Revision of biochemical pH-Stat: involvement of alternative pathway metabolisms. Plant Cell Physiol 39:615–619, 1998.

105. JAC Smith, J Ingram, MS Tsiantis, BJ Barkla, DM Bartholomew, M Bettey, O Pantoja, AJ Pennington. Transport across the vacuolar membrane in CAM plants. In: K Winter, JAC Smith, eds. Crassulacean Acid Metabolism: Biochemistry, Ecophysiology and Evolution. Berlin: Springer-Verlag, 1996, pp 53–71.

106. G Marigo, U Lüttge, JAC Smith. Cytoplasmic pH and the control of crassulacean acid metabolism. Z Pflanzenphysiol 109:405–413, 1983.

107. N Tavakoli, C Kluge, D Golldack, T Mimuta, KJ Dietz. Reversible redox control of the plant vacuolar H^+-ATPase activity is related to disulfide bridge formation in subunit E as well as subunit A. Plant J 28:51–59, 2001.

11

Dynamics of H⁺ Fluxes in the Plant Apoplast

Jóska Gerendás and Burkhard Sattelmacher
University of Kiel, Kiel, Germany

1 INTRODUCTION AND DEFINITIONS

Despite the fact that cell walls were the first cell structures observed by simple microscopes, the apoplast was considered "the dead excretion product of the living protoplast" and thus received much less attention until the mid 1980s [1,2]. Today, the *apoplast* is defined as all compartments beyond the plasma membrane, i.e., the interfibrillar and intermicellar space of the cell walls, the xylem as well as the gas- and water-filled intercellular space [2]. The outside boundary of the apoplast is formed by the rhizoplane and the phylloplane.

During the last two decades our knowledge about the apoplast, its composition, structures, and functions, has increased considerably. It is not simply thought of as a cell wall box around cells but is perceived as a highly complex, flexible matrix consisting of cellulose, hemicellulose, pectins, and proteins [3,4]. The chemical and physical properties of cell walls change during ontogeny, and several physical (temperature, light) and chemical (osmotic and heavy metal stress) parameters have a strong influence as well (see Sec. 2 and 4). The apoplast is of utmost importance for nutrient acquisition, as nutrients do not simply pass through the apoplast on their way to the plasma membrane but strongly interact with cell wall components, which may be of significance for both nutrient acquisition [5,6] and tolerance to toxicity [7]. Most processes during plant growth and development do involve cell walls [8], and many environmental stimuli are not received directly by the cell but via changes within the apoplast [9].

Many of the functions and processes just mentioned exhibit strong relationships to *pH*. The dissociation state of nutrients, whether essential (phosphate, borate) or beneficial (silicon), determines the uptake rate. The apoplast acts as an anion exchanger, whose capacity is affected by pH [10]. The distribution of abscisic acid (ABA) is also governed by pH profiles [11]. The activity of many enzymes shows a strong relationship to the pH of their environment, and the pH is involved in a cascade regulating cell wall extension and cell growth (see Chapter 6). Tissues involved in assimilate transport, stomata, and motor cells display acidified or alkalized apoplasts, according to their physiological functions and state (see Sec. 4).

Although many researchers agree that pH is important, this does not answer the question of how it is defined, influenced, or regulated. A simple definition of pH as the negative logarithm of the H^+ activity is not very helpful, and in view of the complexity and dynamics of the apoplastic compartment, the question is far from trivial. For example, the ion-exchange properties of cell wall components induce Donnan potentials, and the tortuosity of the apoplast in combination with the existence of unstirred layers and localized pH-affecting processes may induce substantial pH gradients (see Sec. 4). Even in ideal solutions, pH regulation is far from trivial. In recent years, the interpretation of pH control has been based on a quantitative physicochemical approach [12,13]. This approach shows that the pH of any solution is a dependent parameter. The pH and the dissociation state of weak acids and bases are governed by the independent parameters, namely the concentration of strong and weak ions in the solution and the pCO_2. Thus, experiments that measure pH changes in the apoplast require detailed information on the distribution of the independent variables before a sound conclusion on the cause of pH changes can be drawn. In Sec. 3.1 the strong ion difference (SID) concept is applied to ion relations in the xylem in particular, as a relevant biological solution of high complexity that is easily obtained.

Even though changes or differences in pH might be interpreted on a solid physicochemical basis, this does not solve the problem of how the pH of the apoplast can be measured. First of all, it is important to define the target solution pH that should be determined. Various methods have been described that either provide an indication of the bulk apoplast pH or provide an estimate of the pH in the water free space (WFS) in situ or at the membrane surface. Whereas some methods provide a general apoplast pH of a leaf, for example, which might be helpful when related to nutrient leaching, others give a reading of high spatial resolution, which is essential when looking at stomata or transfer cells (Sec. 5).

It is the objective of this chapter to review the properties and processes of the apoplast that determine the pH of this compartment, how the pH is influenced and regulated, and how this affects the function of the apoplast as a compartment for storage, transport, and reactions of nutrients and metabolites. Special attention is given to advantages and disadvantages of the various methods for investigating apoplast pH.

2 FUNCTIONS AND CHARACTERISTICS OF THE APOPLAST

2.1 The Physical and Chemical Properties of the Apoplast

The appropriate pH measurements in the apoplast and their interpretation in relation to mineral nutrition and pH regulation require brief consideration of the physicochemical properties of cell walls. More thorough reviews have been presented elsewhere [14,15]. Cell walls consist of a series of layers. The earliest layer is deposited at cell division and because the subsequent wall layers are laid down between the plasma membrane and the previous layer, the oldest cell wall is found where the cell walls of two neighboring cells adjoin, with the newest wall layer being nearest to the plasma membrane. Three clear-cut layers differing in both chemical and physical properties can be distinguished: the middle lamella, the primary cell wall, and the secondary cell wall.

The *middle lamella* of dicotyledonous plants, and to a lesser degree of monocotyledons as well, consists of pectins with different degrees of methylation. Pectins are a very heterogeneous group, with homogalacturonans and rhamnogalacturonans being just two prominent representatives. The *primary wall* consists of a network of cellulose of a relatively low degree of polymerization, hemicellulose (xylans in monocotyledons, xyloglucans in dicotyledons), and glycoproteins. The latter may represent between 5 and 10% of the cell wall dry weight [16], demonstrating that cell walls are important metabolic sites (see later). The *secondary cell wall* contains more cellulose of a relatively high degree of polymerization than the primary wall, while contents of hemicellulose and proteins are considerable lower than in the primary cell wall [14]. In both primary and secondary cell walls, the cellulose-hemicellulose network consists of interfibrillar and intermicellar spaces that differ in size between 3.5 and 5 nm [17–19], which does not represent a major diffusion barrier, even for larger molecules. However, due to friction and tortuosity, transport velocity may be hampered. For high-molecular-weight solutes such as fulvic acids, polymeric chelators, or viruses, the small pore size prevents transport.

The *conductivity* of the apoplast for ions is generally high when compared with the symplast [20], despite reduction by specific measures (incrustation with suberin, lignin, etc.). However, because cell walls normally have negative charges due to the predominance of free carboxyl groups of polygalacturonic acids in the middle lamella and the primary wall, movement of ions in cell walls is characterized by electrostatic interactions leading to an accumulation of cations in the apparent free space (AFS) in a nonmetabolic manner [21]. According to the current view of ion movement in cell walls, which is mainly based on the early work of Hope and Stevens [22] as well as Briggs and Robertson [23], the AFS is divided into the Donnan free space (DFS) and the WFS (Fig. 1). Both cation-exchange capacity [24,25] and electrical potential [26] have been used to describe the physical properties of the DFS. However, we now know that a clear separation between the DFS and the WFS may be an oversimplification because it is not possible to make any clear spatial differentiation between the two compartments [27,28].

The amount of nondiffusible cell wall anions is normally quantified by the cation-exchange capacity (CEC), which is much higher in dicotyledonous than in monocotyledonous species [29]. In most cases the CEC is determined on isolated cell wall material, but it should be noted that under in vivo conditions only part of the exchange sites is accessible to cations due to spatial limitations [21]. The CEC of a plant tissue is not a fixed characteristic but is responsive to several environmental factors. For example, salinity generally decreases the CEC [30], which is regulated by enzymes such as pectin methylesterase (PME). This enzyme itself may be affected by apoplastic polyamines [31–33] and thus by the N nutrition of the plants [34]. Knowledge of the distribution of PME in cell walls is limited, but the frequently observed accumulation of Ca^{2+} in the middle lamella of the junction zone (P. van Cutsem, Universitaires Notre-Dame de la Paix, Na-

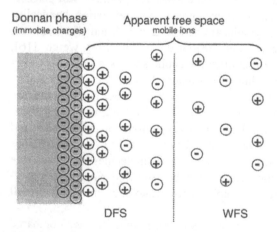

FIGURE 1 Spatial distribution of positively charged mobile ions (+) occurring in a Donnan phase containing embedded immobile negative charges (−). The apparent free space (AFS) is thus divided into the Donnan free space (DFS, surplus of positive charges) and the water free space (WFS, zero net charge).

mur, Belgium, 2000, personal communication) suggests a preferential transport of PME in this large intercellular space. Because the pK of the cell wall is in the range of 4.3 [35] or lower [36], a decreasing apoplastic pH may reduce the CEC [37]. The nondiffusible anions have a strong influence on ion movement. For example, the existence of electrical bilayers may restrict movement of anions to the larger interfibrillar spaces [38], and the velocity of cation movement (mainly Ca^{2+}) is reduced by interaction with the free carboxyl groups [21].

Unstirred layers (USLs) are defined as boundary layers of either liquids or gases in the vicinity of transport barriers. In these layers, complete mixing is not possible and thus concentration gradients are observed. USLs are of principal importance for all transport processes across barriers such as the apoplast or the plasma membrane. As a consequence, it is not only the resistance of the barrier itself that determines transport rate but also the diffusion across the USL [39]. Transport resistance across USLs depends on the mobility of each solute as well as on the thickness of the USL. As the thickness of USLs in the apoplast may be quite substantial [40,41], it may be concluded that USLs are important factors in transport processes in the apoplast.

A factor hardly considered so far is the properties of water bound to gels such as pectins. Studies determining water relaxation in a model system, however, suggest highly structured properties, quite different from those in free water [42]. Implications for the activity of enzymes and dyes frequently used in studies of ion relations in cell walls are far from being understood.

2.2 The Root Apoplast

Due to the negative charges in the root cell wall, an accumulation of cations and a repulsion of anions occur in the root apoplast [43]. This is particularly strong for di- and polyvalent ions [44]. Although accumulation in the root apoplast is not an essential step in nutrient absorption, it does explain certain well-known phenomena, such as differences in K/Ca ratio among plant species [44]. A very good example is also the preferential uptake of metals such as Zn and Cu in ionic over the chelated forms [21]. In the latter case, however, one cannot exclude a possibility that restriction of the relatively large chelate molecules by the cell wall pores is an important factor explaining the results. As a consequence of the cell wall properties of roots, ionic relations in the vicinity of the plasma membrane may vary considerably from those in the rhizosphere [10,45]. These phenomena are of fundamental importance for the understanding of processes such as ionic antagonisms [46–48] or apparent synergisms, like those between Ca^{2+} and $H_2PO_4^-$ [45].

In the apoplast of roots, the Casparian band represents the major diffusion barrier [49]. Although it is generally considered to be completely impermeable to water and ions, results [50,51] do suggest a certain degree of permeability.

Depending on species and age, the endodermis and exodermis contain cutin and suberin in different quantities [52,53]. The chemical composition of the Casparian band changes with ontogeny [53] as well as with environment [54]. Adverse ionic conditions, such as salt stress [55], accelerate formation of the endodermis, presumably to prevent bypass flow through the Casparian band. In addition, many salt-tolerant plant species have a thicker Casparian band than the less tolerant ones [56]. For the chemical analysis of the cell walls of the endodermis, the reader is referred to the review by Schreiber et al. [54].

In most plant species the hypodermis is converted into an exodermis; i.e., an outer Casparian band [57–59] is formed that in many cases contains suberin deposition [60]. The formation of the exodermis occurs later than that of the endodermis and depends largely on growing conditions [61], such as salinity stress [55]. The significance of the exodermis for water and ion uptake is discussed in the literature with some degree of controversy [62,63] and apparently depends largely on the ion under consideration [64]; however, even for mobile nutrients such as K^+, the exodermis represents a significant diffusion barrier [65]. The exodermis does not form a continuous apoplastic barrier, and development may be asynchronous to lateral root development [60]. This finding, together with the existence of passage cells [66], suggests that nutrients may diffuse into the AFS (occupying approximately 5% of the root volume [67]) in spite of the existence of the exodermis. However, for nutrients present at low concentrations in the soil solution (P, K, NH_4, micronutrients) the significance of this internal absorbing surface is limited, as the concentration of these ions is often reduced down to the critical minimal concentration (C_{min}) due to uptake by the rhizodermal cells [68].

The root apoplast is the plant compartment that first encounters adverse soil chemical conditions such as high Na^+ or high Al^{3+} concentrations, and conditions in the root apoplast determine the plant response. As has been demonstrated for numerous plant species, cessation of root growth is the first detectable symptom of Al^{3+} toxicity [69,70]. The interaction of Al^{3+} with cell wall components, such as the pectin matrix [71–73], could explain the phenomena of growth cessation. It should be borne in mind that pectins have a large influence on cell wall properties, such as hydraulic conductivity and, in connection with extensin, on wall plasticity [74]. Recent findings demonstrating a correlation between pectin methylation and Al tolerance support such a view [75]. The immediate reduction of K^+ efflux [76] as well as Ca^{2+} influx [77–79] may be interpreted as the result of an interaction of the trivalent Al cation with the plasma membrane [7,80]. However, the fact that physiological processes such as cytoplasmic streaming, which are extremely sensitive to any change in Ca^{2+} homeostasis [81], remain undisturbed by external Al^{3+} supply [82] casts doubt on the assumption that Al toxicity is causally related to the disturbance in Ca^{2+} homeostasis [78,79,83], especially in the light of new findings suggesting that Al^{3+} rather prevents an increase in cytoplasmic Ca^{2+} brought about by high external H^+ [81]. The involvement of the cell wall in the process of detoxification of Al species has been demonstrated for different plant species such as wheat [84], and it was shown that Al^{3+} sensitivity is restricted to the distal part of the transition zone [85,86]. Findings indicating the release of chelating substances such as organic acids into the apoplast in relation to Al tolerance are frequently discussed [87]. This could also explain earlier findings demonstrating that the form of N supply (NO_3^- vs. NH_4^+) exerts a strong influence on Al tolerance [88].

2.3 The Xylem

As already pointed out, the traditional scheme according to which the ions transverse the root cortex using either a symplastic or an apoplastic path is likely to be of limited significance due to the existence of the exodermis. Either way, ions acquired are subjected to selective uptake into the symplast and finally secreted into the xylem (part of the apoplast of the stele). Contrary to findings suggesting an active transport mechanism in xylem loading [89,90], more recent data demonstrate a thermodynamically passive transport via ion channels [91]. Both inward [92] and outward [93] rectifying channels have been detected in xylem parenchyma cells, contributing further to our understanding of xylem loading. The driving force is generated by H^+-ATPase, which is particularly abundant in the paratracheal parenchyma cells [94].

The cation-exchange capacity of xylem cell walls is rather high, estimated to be in the range of 1000 mmol L^{-1} for tomato [95]. Interactions of cations with these nondiffusible anions lead to a separation of ion transport from water flow. The transport of cations may be compared with that in a cation-exchange resin, while water is transported by mass flow [21,95]. The degree of retardation of ion translocation depends on the valence of the cation ($Ca^{2+} > K^+$), its own activity and surface charge, the activity of competing cations, the charge density of the nondiffusible anions, and the pH of the xylem sap [95]. Consequently, the transport rate of di- or trivalent cations is enhanced significantly by complexation [96]. Cations may be complexed by organic acids [97,98], amino acids, or peptides [97,99,100]. It should be stressed that, because of the requirement of charge balance (electroneutrality), cation transport is always tightly coupled to anion transport, whether occurring across membranes or not. This holds true irrespective of the existence of electrical potentials, as the substantial electrical potential generated even by small deviations from perfect neutrality hinders the development of unbalanced ion movements [13]. Thus, if cation transport is enhanced by the formation of soluble complexes, mobility of anions is also increased (depending on the net charge of the complexes formed).

Transfer cells present in the paratracheal parenchyma may mediate the exchange of solutes between xylem and phloem [101]. The significance of this process, especially for nitrogenous compounds, is often underestimated [102]. The processes—absorption from the xylem, release into the xylem, and transfer into the phloem—may lead to strong concentration gradients in the xylem sap, i.e., higher in the base and lower in the apical region [103]. This is often correlated with changes in the xylem sap pH [104] (see also Sec. 3).

Composition of the xylem sap is highly variable and depends, among other factors, on plant species, age [105], time of day [106,107], location of sampling [103], nutritional status, rooting medium [108], and, last but not least, nutrient cycling within the plant [109]. Mineral nutrient supply exerts a strong influence on xylem sap composition. As a rule, a positive correlation exists between ion concentration in the external solution and in the xylem sap. Contrary to variation in rhizosphere pH, the effect of nutrient uptake on xylem sap pH is not well studied and understood. Although NH_4^+ nutrition always led to an acidification of the rhizosphere due to predominant uptake of cations [21], considerable discrepancies were found in the effect on xylem sap pH (see Sec. 4.2).

2.4 The Leaf Apoplast

Upon entering the leaf, the xylem stream is predominantly directed to sites of rapid evaporation, such as leaf margins. If the leaf veins are mechanically ruptured as may occur under natural conditions, e.g., as a result of insect attack, the ruptured site is bypassed rapidly, possibly by increased transport rates in minor veins (W. Merbach, Martin-Luther-Universität, Halle-Wittenberg, Germany, 2000, personal communication).

The fate of the xylem sap in the leaf apoplast was subject of a debate between supporters of the apoplastic or symplastic route for water transport [110]. According to the current knowledge, the predominant route depends mainly on the driving force: hydrostatic pressure gradients support transport through the apoplast, whereas osmotic gradients mainly favor symplastic routes [111]. According to the Hagen-Poiseuille law, the volume flow is affected by the tube diameter to the fourth power. One would thus expect transport to be restricted mainly to the major veins. However, contrary to the number of vessels per vein, the diameter of xylem vessels is rather independent of vein size [112]. This is not true for the smallest veins where large vessels are missing and where accumulation of solutes was observed accordingly [112]. Because mass flow is difficult to imagine outside the xylem vessels and flux by diffusion is effective over very short distances [113], intercostals fields are rather small and in most cases do not exceed seven cell layers. For the particular cell wall zones in which diffusion takes place, Canny [114] has suggested the term "nanopaths."

Surprisingly little information is available on the path of apoplastic transport in the leaf tissue. Since the early work of Strugger [115], most authors have assumed transport to take place in the intercellular spaces and/or in the water films covering the outer surfaces of cell walls. This concept may not, however, be realistic because the wetability of cell walls, at least in the leaf apoplast, is thought to be low [116,117]. The formation of such an "inner cuticle" depends on both plant species and environmental conditions, and its existence would restrict nutrient transport to the "cell wall apoplast" [118]. As the water content of cell walls is rather high [119], the potential for diffusion is quite good, which has indeed been demonstrated for roots [120]. We do not yet have good information on whether such "internal cuticle" covers the entire internal leaf surface or just certain areas such as the substomatal cavity [117,121]. In any case, experiments with stable isotopes demonstrate equilibrium within rather short time periods, which suggests that the internal cuticle is not a major diffusion barrier. If this internal cuticle exists, its nature is not understood, but it is presumably the result of methylation rather than of cutin incrustation [122].

One common apoplast does not exist in leaves, at least not in C_4 plants [123]. Bundle sheath cells are connected to mesophyll cells via numerous plasmodesmata [124,125], but their apoplastic compartments are separated by a suberin lamella [118,124,126–128]. Thus, ionic conditions in the two apoplastic compartments may differ significantly, and although no direct evidence is available so far, a large difference in pH optima of K^+ channels in the two compartments supports the idea [123]. Because a similar situation has been described for wheat [129], it may be anticipated that a separation of the leaf apoplast into smaller compartments by diffusion barriers is a more common phenomenon in the plant kingdom.

3 APOPLAST pH AND ITS REGULATION

3.1 Quantitative Acid-Base Chemistry

As pointed out before, the pH of the apoplast has numerous implications for several functions and processes. The key question is, what is the pH of the apoplast? A simple definition of pH does not answer the question, and in view of quantitative approaches to understand pH regulation, the question is far from trivial [12]. The ionic composition of biological solutions, such as apoplastic fluid and xylem sap, is determined by physical (e.g., gas exchange, transmembrane transport, membrane potential) and chemical processes (e.g., metabolic conversion, dissociation). In more recent years the interpretation of changes and regulation of pH has been based on a quantitative physicochemical approach [12], particularly in the animal [130–132] and human sciences [133–135]. This approach shows that the pH of any solution is a dependent parameter. The misleading term "SID concept" (SID = strong ion difference, the surplus of fully dissociated cations, e.g., K^+, Ca^{2+}, over fully dissociated anions, e.g., Cl^-, NO_3^-) should be avoided, as it suggests that the pH is primarily dependent on the distribution of strong ions. In fact, the pH, as well as the dissociation state of weak acids and bases [ion exchangers, phosphate, borate, ammonia, ABA, metabolites, protein (hence conformation and enzyme activity)] and the concentration of carbonate species, is governed by the independent parameters, namely the concentration of strong and weak ions (partly dissociated anions, e.g., amino and carboxylic acids) and the pCO_2. The relationship between dependent and independent parameters is characterized by physicochemical principles (ion and water dissociation equilibria, solubility of CO_2, conservation of mass, and electrical neutrality). The electrical potential, as a difference between the apoplast and either the symplast (membrane potential) or the xylem sap (transroot potential), can be thought of as additional negative or positive charges. However, due to the large consequences of even small deviations from electric neutrality, the number of extra charges within the range encountered by biological solutions is minute compared with the total concentration of charges [136]. Therefore, in most cases biological solutions can be regarded as electrically neutral. The concept thus allows an integrated quantitative approach to acid-base relations.

Experiments that measure pH changes require detailed information on the distribution of the independent variables before a sound conclusion on the cause of pH changes can be drawn. The concept attracted only limited attention in the plant sciences but has been applied to ion transport processes across biological membranes [13,137,138] and has been discussed in relation to intracellular pH regulation during anoxia [139]. Recently, the concept of quantitative acid-base chemistry (QABC) has been used to analyze ion relations of xylem sap in detail [13]. Xylem sap is particularly suitable for applying quantitative acid-base chemistry as it can be obtained in relatively large quantities, permitting quantitative analysis of all major constituents. Results obtained for *Populus* plants showed that on average calculated values were in good agreement with observed data, particularly under the assumption of a 10-fold higher pCO_2, but individual data sets revealed substantial deviations (Fig. 2). The following examples illustrate the potential of QABC.

As pointed out by several investigators [104], the composition of the xylem sap is not uniform along the axis. Taking the composition of xylem sap obtained from *Ricinus communis* as a standard model [13], the selective removal of 0.1 mmol L^{-1} NO_3^- is interpreted as follows. The uptake of NO_3^- represents the selective removal of 0.1 mmol L^{-1} of anions from the xylem sap and thus represents an increase in SID. The system "xylem sap" responds to this removal of negative charges with the generation of negative charges from internal sources. This is accomplished by increasing the dissociation of CO_2, amino acids, carboxylic acids, and phosphate (Fig. 3A). As all these processes are related to the proton concentration in the solution, a concomitant pH increase is predicted (Fig. 3B). NO_3^- could also be exchanged for glutamate or glutamine. In the first case

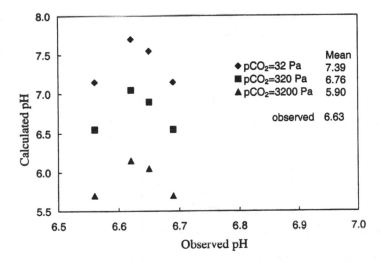

FIGURE 2 Relationship between calculated and observed pH values of xylem sap obtained from *Populus* [13].

pH changes are small, as the number of negative charges remains stable, and in the latter case pH changes are similar to those for the removal of NO_3^- alone. It should be noted that, according to this quantitative approach, an increase in pH, along with the increased dissociation of weak acids, is the consequence of a change in the independent variables, rather than its cause [12,13]. Similarly, when evaluating the "influence" of pH on enzyme activities, it should be expressed more carefully as the relationship between pH and enzyme activity because of substantial changes in other dependent variables, such as the dissociation state of metabolites and enzyme proteins, having far-reaching consequences for substrate affinity and hence enzyme activity. This quantitative approach can be applied to other systems as well, e.g., phloem sap, regulation of cytoplasmic pH, and pH gradients in chloroplasts and mitochondria [13,140,141]. A major constraint is the need for quantitative data on the composition of the system and their quantitative interaction, mainly the dissociation equilibria. For example, just an underestimation of the strong cations by 5% changes the predicted pH by 0.4 units [141]. However, the analysis of the quantitative impact of variations of independent variables on the pH (sensitivity analysis) allows the quantitative assessment of the capacity of any proposed pH-disturbing and pH-regulating mechanisms with respect to pH homeostasis, even in complex solutions.

The bicarbonate concentration of the xylem sap not only provides negative charges to the ionic balance (Fig. 3) but also may contribute substantially to the

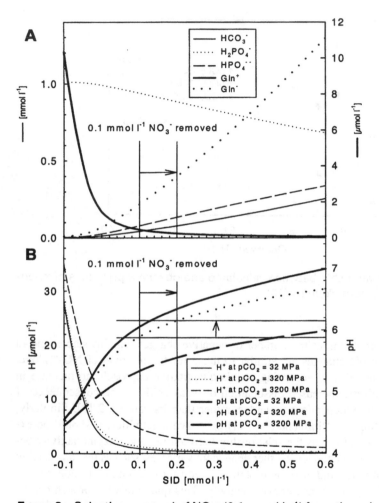

FIGURE 3 Selective removal of NO_3^- (0.1 mmol L^{-1}) from the xylem of *Ricinus communis*. (A) Changes of dissociation of weak ions (generation of negative charges). (B) Changes in proton concentration and pH. Details on data sources and calculations are given elsewhere [13].

provision of photosynthetic CO_2 fixation in woody plants [142]. High CO_2 partial pressures were found, ranging from 3000 to 9200 Pa, allowing a CO_2 supply that represented 0.5 to 7.1% of typical leaf photosynthetic rates. As expected, these values exhibited a strong relationship to sap pH.

The continuous removal of ions from the xylem sap may result in strong concentration gradients, i.e., higher in the base and lower in the apical region

[103]. This is often correlated with an increase in the xylem sap pH in the acropetal direction [104]. In addition, an inverse relation between solute concentration and xylem flow rate is often observed [107,143].

3.2 Donnan and Diffusion Potentials

The complex nature of the apoplast requires further consideration. A proton activity can be considered only in an aqueous phase, the AFS. Because pectin and other macromolecules in the cell wall have a large number of carboxyl groups (—COOH), which give these cell wall constituents a net negative charge (Fig. 4), cations are electrostatically attracted to the negatively charged cell surface. The region containing the immobile charges—such as dissociated carboxyl

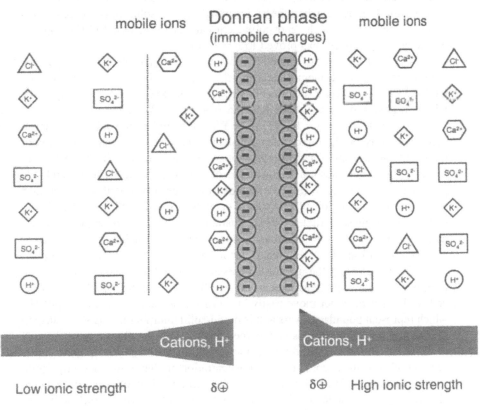

FIGURE 4 Spatial distribution of positively charged mobile ions at low or high ionic strength occurring on either side of a Donnan phase in which immobile negative charges are embedded.

groups in the case of the cell wall—is generally referred to as the *Donnan phase*. At equilibrium, a distribution of positively charged ions electrostatically attracted to these immobile charges occurs between the Donnan phase and the adjacent aqueous phase (WFS). This sets up an ion concentration gradient, so a Donnan potential is created between the center of the Donnan phase, where the attraction of positively charged species can increase the local concentration of solutes (ionic strength) to about 0.6 M [136], and the bulk of the solution next to it, which is not influenced by fixed charges and thus resembles an ideal solution of net zero charge. The Donnan potential arising from electrostatic interactions at such a solid-liquid interface can be regarded as a special type of diffusion potential (Fig. 4). The physical properties and the size of the DFS depend on the cation-exchange capacity [24,25] and electrical potential [26]. Because the influence of the fixed negative charges diminishes gradually with increasing distance from them, the extent of the DFS is not fixed [144]. For example, the size of the Donnan phase is substantially reduced when the ionic strength of the solution is increased [10] or when the activity of di- and polyvalent cations is increased [145] (Fig. 4). Nonetheless, the model has proved to be helpful.

All small ions in the immediate vicinity of Donnan phases (e.g., pectins, large proteins, plasma membrane) are affected, including H^+, thus influencing the local pH. As cations are attracted to the Donnan phase, the H^+ activity in the DFS is higher and hence the pH is lower than in the WFS.

Which pH is relevant, the pH of the WFS, the DFS, or the bulk? The answer depends largely on the topic under consideration. When the characterization of the significance of pH for the passage of weak acids through the plasma membrane is the target question, then the pH of the DFS of the plasma membrane provides key information (e.g., determined by weak acid distribution; see Sec. 5). When the dissolution of Ca oxalate crystals is considered, then the pH observed around these particles (WFS), spatially resolved, seems appropriate. Finally, when focus is on the relationship between pH and leaching from the leaf as a whole, the bulk apoplast pH appears adequate (see Sec. 5 for further discussion).

The structure of the apoplast represents a complex arrangement of cellulose, pectin, lignin, proteins, and so on (see Sec. 2). A considerable part of the cell wall water does not move easily but forms unstirred layers (USLs) [40,41], which represent boundary layers at this liquid-solid interphase. In USLs, transport of ions and neutral molecules is governed by diffusion. As the transport capacity of diffusion diminishes rapidly with increasing distance and the apoplast represents a highly tortuous compartment, concentration gradients are not easily equilibrated. Also, the diffusibility of the ions and compounds in question differs considerably [146] (Table 1). Because protons do not need to diffuse in the conventional way, their diffusibility is about 10-fold higher (Table 1). The diffusion of charged ions is controlled not only by the concentration gradient but also by the electrical potential (diffusion potential), which develops when, e.g., more mobile protons tend to diffuse faster than the less mobile mineral ions (K^+, Cl^-,

TABLE 1 Mobilities of Mineral Ions
in Water

Ion	Mobility (cm s⁻¹ volt⁻¹ cm⁻¹ at 25°C)
H^+	3.62×10^{-3}
K^+	7.12×10^{-4}
Na^+	5.20×10^{-4}
Li^+	4.57×10^{-4}
Ca^{2+}	6.16×10^{-4}
Cl^-	7.91×10^{-4}
NO_3^-	6.40×10^{-4}

SO_4^{2-}) (Fig. 5) [136]. Consequently, pH gradients within the apoplast do not diminish easily and have been observed in a number of cases (see Sec. 4).

4 STABILITY, CHANGES, AND GRADIENTS OF APOPLASTIC PH

Whereas some time ago the apoplast was regarded a dead compartment with no need for homeostasis of its components, convincing evidence has accumulated during the past decades demonstrating the homeostasis of apoplastic ionic compo-

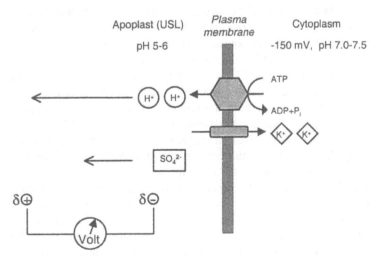

FIGURE 5 Diffusion potential resulting from selective uptake of ions and their different mobilities (Donnan potential is considered to be unchanged).

sition, including pH. At the same time it has become evident that certain responses of plants to internal (development) and external (nutrient supply, light, water stress) stimuli require specific changes of apoplast pH, sometimes as part of a signal transduction pathway. Yet, other situations have been identified in which adverse environmental conditions and other types of stress override the plant's capacity to maintain homeostasis, resulting in substantial pH changes in the apoplast. In the following paragraphs, several examples illustrating these situations are presented. They are, however, by no means comprehensive. For convenience, examples are given for (1) pH homeostasis, (2) pH changes and gradients during normal development, and (3) pH changes under adverse conditions for the root apoplast, the xylem, and the leaf apoplast, where applicable.

4.1 pH Homeostasis and Fluctuations in the Root Apoplast

Unlike the xylem and the leaf apoplast, the root apoplast is linked to the soil solution. It is therefore not as easily controlled by the plant but on the other hand represents a large reservoir, whose composition usually does not change as rapidly.

Important pH changes have been recorded in response to gravitropism [147]. Apoplastic pH in the elongation zone of corn roots was approximately 4.9. Upon gravitropic stimulation, the pH on the convex side of actively bending roots was 4.5, with lowering of the apoplastic pH by 0.4 units claimed to be sufficient to account for the increased growth on that side.

Among the external factors that influence pH of the root apoplast, the nutrient uptake by the plant is the dominant process because the net proton balance is largely determined by the cation/anion uptake ratio, which in turn depends mostly on the nature of the nitrogen source [148,149]. When plants take up more cations than anions, net proton efflux occurs. Otherwise, the net fluxes of charge would not be electrically neutral [150]. Growth with nitrate supply is more likely to remove protons from the apoplast in comparison with growth with ammonium supply [148,151]; such a situation tends to produce alkalization of this compartment with nitrate and an acidification with ammonium supply [21,152]. It needs to be stressed that this results principally from the requirement for charge balance during nutrient uptake and not from the consumption and release of protons during nitrate and ammonium assimilation, respectively.

The effect of N form on apoplast pH is superimposed on a spatial variation along the root axis that tends to ensure that the pH around the root tip is relatively alkaline [153–155]. In a recent study, a fluorescence dye bound covalently to the root apoplast allowed determination of an apoplastic pH of about 5.1 in the division zone, between pH 4.8 and 4.9 in the elongation region, and about 4.9 in the root hair zone at a bulk solution pH of 5 [156]. Addition of 1 mmol L^{-1}

NH_4^+ caused a small apoplastic pH decrease (0.05 pH unit) in all root zones. Apoplastic alkalization upon application of 6 mmol L^{-1} NO_3^- was highest (0.3 pH unit) in the zone where root hairs emerge, while in the division and early elongation zones apoplastic pH increased transiently (see also Chapter 12).

Further pH changes related to nutrient acquisition can occur if the root faces a shortage of a particular nutrient. Examples include the acidification of the rhizosphere that may occur when organic acids are released in response to a shortage of phosphorus [157] and the formation of cluster (proteoid) roots in white lupin [158,159] and Proteaceae [160–162]. Cluster roots in *Casuarina glauca* are also induced by Fe deficiency and enhanced by nitrate supply [163], indicating that a high apoplast pH promoted the formation of these acidifying structures. An acidification of the rhizosphere [164,165] and the apoplast [166] has also been observed in plants not forming cluster roots in response to Fe deficiency. The magnitude of the pH change is likely to be dependent on the conditions experienced by the plant (e.g., it is more difficult to reduce the apoplastic and rhizosphere pH in high-carbonate soils) [166,167].

With regard to the biochemical nature of the organic acid extrusion, earlier publications seem to suggest a passive efflux of free carboxylic acids driven by a concentration gradient. However, outward-conducting anion channels of broad specificity (carboxylates, Cl^-, and others) have been identified in root cells [168–170] and, because such anion efflux is electrochemically downhill, high efflux rates can be established. An efflux of carboxylic anions can be sustained only in the presence of concomitant anion uptake or cation export. In situations of apoplast acidification, proton export by the plasma membrane ATPase is the most likely candidate. However, charge may be balanced by K efflux when an alkalization of the rhizosphere occurs because of partial protonation of the carboxylic anion (weak acid) [171]. The release of carboxylates through anion channels has indeed been observed in response to toxic concentrations of Al [87] and is important in conferring resistance to aluminum [171,172]. Organic anion efflux leads to Al resistance through both complexation, as shown for malate in wheat [173] and citrate in corn [172], and increasing the pH of the apoplast and the rhizosphere [174].

4.2 pH Homeostasis and Fluctuations in the Xylem

pH homeostasis and pH fluctuations in the xylem have been studied to some extent because the xylem sap is relatively easily obtained (but see Sec. 5 for a discussion of the advantages and disadvantages of the various methods). The *buffering* capacity of xylem vessel walls for H^+ is high [175], and a strong pH regulation can be demonstrated by perfusion experiments [176]. Nonetheless, substantial variations in xylem sap pH have been observed. The effect most often described is the increase of xylem sap pH in response to *drought stress* [177,178] in relation to the distribution of ABA (a weak acid) between apoplast and symplast [11]. It can be deduced from mathematical models that pH gradients explain a two- to threefold accumulation of ABA in the xylem [179]. A whole-plant model also stresses the importance of the stele tissue and the rhizosphere ABA concentration, although root-to-shoot communication hardly depends on ABA redistribution after stress-induced pH shifts in leaves [180]. Changes in the concentrations of calcium or other xylem components, such as ABA conjugates, together with possible changes in the ability of the *Lupinus albus* leaves to degrade and/or to compartmentalize ABA, could partly explain the midday increase in the apparent stomatal sensitivity to xylem ABA [181].

Although the dynamics of pH changes (see Sec. 3.1) suggest that the pH increase is due to changes in the composition of either strong (SID) or weak ions (amino acids, carboxylates), explaining the pH increase on a quantitative physicochemical basis has not been achieved so far [11]. A good example of the consequences of inadequate ion homeostasis was reported [182], where an increase in the xylem sap pH of wheat by approximately 1 unit was observed with $Ca(NO_3)_2$ as compared with $NaNO_3$ or KNO_3; it was concluded that roots grown in the absence of monovalent cations had serious pH problems due to both a high rate of alkalinity release to the medium in exchange for NO_3^- uptake and the secretion of alkalinity into the xylem.

The *form of N* supply has large effects on the pH of the root apoplast and the rhizosphere (see Sec. 4.1). The N form also strongly influences the ionic composition of the xylem sap [183–185], but consequences for xylem sap pH are not uniform. In some studies an acidification from pH 5.37 (NO_3^-) down to pH 5.03 (NH_4^+) [186] was observed, whereas others revealed no effect of N form at all [184] or even an alkalization (V. Römheld, University of Hohenheim, Stuttgart, Germany, 2000, personal communication). Generally speaking, the influence of N form on the ionic composition of the xylem sap, mainly the reduced cation content and increased amino acid level, should affect the pH homeostasis only if it is not fully compensated for by reduced anion and carboxylate contents. Just a few percent of uncompensated charge could easily explain the observed pH differences quantitatively (see Ref. 13). To our knowledge, the data are not available to explain the pH changes in the xylem fluid induced by different N forms on a quantitative basis.

Substantial variations of xylem pH over time have been observed. *Seasonal* variations in pH, carbohydrate, and N in the xylem exudate of *Vitis vinifera* have been characterized, and a strong inverse correlation between volume flow and pH was found [187]. The pH values decreased until budburst (pH 5.0–5.5) and increased thereafter to pH 7.5–8.0. Strong seasonal variations in the xylem sap pH were also observed in poplar stems, where winter tracheal sap contained sucrose as the dominant sugar at pH 7.0–7.5, whereas in spring sap hexoses prevailed at pH 5.5 [188]. *Diurnal* pH variations were observed in tomato xylem sap (bleeding technique), strongly correlating with a diurnal cycle; lowest values were observed during the day and highest values were observed at night [189]. The changes in the pH of the xylem sap are typically buffered but may possibly influence processes localized within the shoot. Diurnal variations in xylem exudate composition were also detected using a pressure technique and discussed in relation to fluxes of ABA [107].

Variations of xylem sap pH have far-reaching consequences for the stability of *complexes* between micronutrients, heavy metals, and chelators, mostly amino acids and carboxylates. Calculating the stability of Fe, Mn, Zn, and Cu complexes in tomato and soybean showed that their free concentration was substantially affected by the xylem pH [190]. The initial xylem pH for soybean plants was 6.1 but dropped to 5.5 within 7 h. The impact of xylem pH variations was also demonstrated for Ni complexes [191] (Fig. 6).

Few investigators commented on the influence of *temperature* on the xylem sap pH. Whereas in *Lupinus albus* plants no pH change was observed in response to variations in growth temperature (1015 or 2025°C night/day) [181], other investigators detected a decrease in the xylem sap pH by 0.2 units upon transfer of *Capsicum annuum* plants from a root zone temperature of 20°C to a diurnally fluctuating ambient temperature regime (25–40°C), even though temperature had no effect on abscisic acid or nitrate concentration in the xylem [192]. The pH

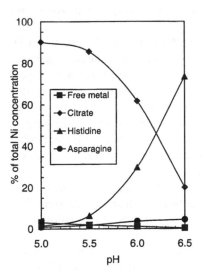

FIGURE 6 Influence of pH on Ni species in xylem sap of soybean plants at 0.1 µM Ni. Data taken from Ref. 190 for xylem exudate (0–1 h) of plants grown with adequate Zn. Details on calculations are given elsewhere [191].

changes in response to temperature are accompanied by substantial changes in the ionic composition of xylem saps, mainly mineral ions and amino acids [193].

Salinity represents another factor that strongly affects the ionic composition of the xylem sap [194,195], but to the best of our knowledge salinity-induced pH changes in the sap have not been reported so far.

4.3 pH Homeostasis and Changes in the Leaf Apoplast

Ion homeostasis in the apoplast has been suggested several times [196], partly on the basis that this compartment has a limited size, and regulation within the symplast at the expense of the apoplast would result in a very harsh internal environment for leaf cells (see also Ref. 197). Several examples of ion and pH homeostasis in the leaf apoplast have been reported. The NH_4^+ concentration in the leaf apoplast of *Brassica napus* plants was shown to be highly regulated [198], and similar observations were made for phosphate in barley leaves [199]. When leaves of *Bromus erectus* plants were exposed to NH_3, a steady-state alkalization of the apoplast of the substomatal cavity by 0.53 units was observed [200]. An apoplastic buffer capacity of 6 mM pH^{-1} unit was calculated from the initial changes in pH and $[NH_4^+]$, whereas the steady-state values yielded 2.7 mM pH^{-1}. The pH readily returned to its initial value after replacing the NH_3 with clean air; it was concluded that pH was controlled by regulatory processes involving both membrane transport and NH_4^+ assimilation. Similar observations regarding the pH dynamics in the substomatal cavity in response to ABA have been described [201].

Buffering capacities for the spinach apoplast obtained by titration with CO_2 were in the range of 3 to 5 mM pH^{-1} [202]. In a study on apoplastic pH regulation, fumigation with HCl and NH_3 caused rapid acidification and alkalization, respectively, and return to initial pH values was slow after cessation of fumigation [203]. The effects of increased CO_2 concentrations were variable, but an overshoot upon removal of CO_2 (transient alkalization) was observed. The high capacity of leaf cells to rapidly reattain pH homeostasis in the apoplast under acid or alkaline stress was related to the rapid activation and deactivation of proton pumping and corresponding ion fluxes [203].

Despite this general background of ion and pH homeostasis, numerous reports demonstrate strong spatial gradients and substantial changes over time. Possibly the best characterized is the acidification of the cell wall in relation to the acid growth theory (see Chapter 6).

Spatial gradients may be the result of several factors, such as differences in the rate of uptake, delivery by mass flow, or efflux from the symplast, and have been considered in greater detail [113]. Strong H^+ concentration gradients in the vicinity of leaf teeth have been observed [204,205].

Temporal variations in apoplastic ion relations are the results of changes in metabolic activity or environmental conditions. A dark-light transition leads to a biphasic apoplastic pH response [206,207] in which an alkalization observed immediately after the onset of illumination is considered to reflect the beginning of photosynthetic electron transport. This results in an alkalization of the stroma that is compensated by H^+ uptake from the cytosol and finally the apoplast. Measuring the apoplastic pH of the substomatal cavity of *Bromus erectus* leaves using microelectrodes revealed that leaf apoplast pH responded to exposure to darkness with an initial transient acidification and subsequent sustained alkalization of 0.2–0.3 pH, whereas exposure to light caused the opposite response [201]. These reports seem to be the first indications of the role of the leaf apoplast as a transient reservoir of protons, but it should be remembered that this requires compensatory fluxes of other ions (see Sec. 3).

Other examples of marked *spatial and temporal variations* of apoplastic ion and proton concentrations are related to swelling and shrinking of motor and guard cells [208]. Guard cells extruded protons 20 min before opening of the stoma [209]. Acidification of the apoplast spread outwards from the guard cells to the surrounding cells, and this phenomenon persisted for some time after subsequent stomatal closure. Substantial changes of apoplast pH in response to light were also observed in *Samanea* pulvini [210]. The pH of the extensor apoplast was higher than that of the flexor apoplast in darkness, and extensor cells acidified and flexor cells alkalized their environment in response to white light, with reverse changes occurring in response to darkness.

The pH changes were also discussed in relation to K^+ transport through the plasma membranes of motor cells. This is supported by similar studies on *Phaseolus* pulvini [28], where a tight coexistence of ultradian and circadian leaf movements as well as of ultradian and circadian pH changes in the water free space (WFS) of the extensor apoplast in situ was demonstrated. When extensor cells swell (upward movement of the lamina), the H^+ activity increases from approximately pH 6.7 to 5.9 and the K^+ concentration decreases from approximately 50 to 10 mmol L^{-1}. It was suggested that the cell walls of pulvinar cells serve as reservoirs for K^+ and H^+.

The supply of nutrients has far-reaching consequences for the pH of the leaf apoplast. The ionic balance and potentially the apoplast pH are influenced to the greatest extent by the *form of N supply* (NO_3^- vs. NH_4^+). However, effects on apoplastic pH in leaves have been debated. It was argued that NO_3^- nutrition leads to alkalization and NH_4^+ induces acidification [211,212]. Other reports suggest that NO_3^- nutrition may lead to alkalization in relation to the NO_3^- concentration in the xylem sap, which has been discussed in relation to the NO_3^- concentration in the nutrient solution and the NO_3^- reductase activity in the root [206]. However, NH_4^+ nutrition should normally exert little influence on apoplastic pH in leaves because NH_4^+ concentration in the leaf apoplast is unaffected by the form of N supplied, at least at low N concentration in the rooting medium. This has been questioned when relatively high NH_4^+ concentrations in the xylem sap and in the leaf apoplast were detected [213]; the latter is probably due to photorespiration and is highly regulated [198,214]. Although the form of N supply to the rooting medium has relatively little effect on apoplastic pH, this is not true with foliar uptake of NH_4^+, following either NH_4^+ deposition or application of foliar fertilizer, which leads to an immediate acidification of the leaf cell apoplast [215].

The potassium supply to the rooting medium strongly affects the ionic relations in the leaf apoplast, with the apoplastic K^+ concentration being a reflection of K^+ supply [216]. The rapid decrease in apoplastic K^+ that occurs well in advance of a decrease in total tissue K^+ demonstrates the sensitivity of this parameter. Because K^+ is by far the most abundant cation [206], it may be anticipated that a decrease in K^+ has far-reaching consequences for the ion homeostasis of the apoplastic solution, but pH changes associated with such imbalance have not been reported.

The function of the leaf apoplast as a reservoir of ions such as K^+ has been demonstrated for the vicinity of guard cells or motor cells [217,218]. The apoplast has several advantages over the vacuole with respect to storage of cations and anions [10,206]. The advantages are mainly the high CEC of the cell walls (see earlier) and the ease with which ions can be taken up from the apoplast. As discussed in detail elsewhere [10], ions are taken up easily because the nondiffusible anions can be neutralized by H^+. An H^+/cation exchange thus leads to a reduction of the negative charge and an increase in the electrochemical gradient. Because H^+ is osmotically inactive, the osmotic gradient increases simultaneously. Due to the very small dimension of the leaf apoplast, even small amounts of ions can cause significant changes in osmotic potential [197].

The apoplast is important for nutrient dynamic in leaves, and Fe is a well-characterized example. Under certain conditions leaves showing *Fe deficiency* symptoms may have higher total Fe content than control leaves [219]. This has been interpreted as the result of Fe immobilization in the leaf apoplast [220] due to high apoplastic pH [221,222]. In spite of the fact that evidence for high apoplastic pH in leaves is still missing [206], a point has been made that the Fe content in leaves should be compared only if leaf size is comparable [223], which may not have been the case in the study just mentioned. In addition, detailed studies of the mechanism of Fe uptake from the leaf apoplast revealed that varying the pH of the incubation solution in the relevant pH range of the leaf apoplast (5.0–6.0) had no significant effect on Fe^{III} reduction, which is a prerequisite for iron uptake [224]. It was concluded that Fe inactivation by high pH in the leaf apoplast is not a primary cause of Fe deficiency chlorosis induced by bicarbonate.

The leaf apoplast is strongly influenced by the aerial environment. *Brassica napus* plants may represent both a sink and a source for ammonia [225]. Exchange properties depend largely on physiological conditions in the leaf apoplast [226], such as pH in the apoplastic solution [227], as well as on stomatal conductance

[225] and the NH_3 concentration in the atmosphere [228]. When leaves of *Bromus erectus* plants were exposed to 280 nmol NH_3 mol^{-1} air, alkalization of the apoplast by 0.22 pH units occurred, followed by a slower pH increase to reach a steady-state alkalization of 0.53 units [200]. This pH shift was persistent as long as NH_3 was flushed and readily returned to its initial value after replacing NH_3 with air. It was concluded that the initial alkalization was due to rapid conversion of NH_3 to NH_4^+, whereas regulatory processes involving both membrane transport and NH_4^+ assimilation caused the slower pH increase.

5 METHODS TO DETERMINE APOPLAST pH AND TO STUDY ITS REGULATION

The pH of the apoplast is influenced by a number of parameters and, above all, difficult to measure. Various methods have been described (Table 2), which can be roughly classified into "in vitro" and "in vivo" methods and which may be subdivided according to the physical principle used in pH electrode and pH indicator methods. However, the target fluid or region in which the pH should be determined has to be defined too. Some methods provide an indication of the "bulk apoplast" pH; others provide an estimate of the pH in the WFS in situ or at the membrane surface. With respect to spatial resolution, some methods provide a general leaf apoplast pH (which might be helpful when related to nutrient leaching), and others give a reading of high spatial resolution (essential when looking at stomata, motor cells, or transfer cells with an uneven distribution of transmembrane ion fluxes). A review of this subject has been published [229].

Although the *xylem sap* is easy to access, care must be taken with the collection method and data interpretation. Obtaining bleeding sap from decapitated plants is certainly not an adequate sampling technique, as changes of ionic composition and pH over time have been documented [230]. Pressurization of the root system and collection of xylem sap on cut petioles from transpiring plants allow continuous sampling and recordings of diurnal fluctuations [231]. However, others have questioned whether unadulterated samples may be obtained (U. Zimmermann, University of Würzburg, Germany, 2000, personal communication). A rather harsh technique by which the solution contained in the xylem vessels is driven out by high compression immediately followed by decompression is supposed to yield unadulterated xylem sap [104]. Other techniques, such as suction [232] and perfusion [175,176,233,234], may be suitable for some aspects of pH regulation, but roots or root systems are detached. Whenever the xylem sap is recovered, the influence of CO_2, the concentration of which may be increased by several orders of magnitude compared with the ambient level [235] should be considered (see Sec. 3.1). Only in a few cases have true in situ techniques been tried, such as ion-sensitive field-effect transistors [236]. Preliminary in vivo measurements revealed the ability of the living xylem to buffer the

TABLE 2 Overview of Methods to Estimate Apoplast pH and Investigate Its Regulation

Method (grouped)	Characteristics	Selected references
Xylem		
"In vitro"	pH of WFS	
Bleeding sap	Simple. Plant decapitated, sap composition not stable and not representative	230, 268
Pressure chamber	Continuous collection from intact plants Root system pressurized	13, 177, 231
Pressure/decompression	No adulteration of xylem sap composition, mostly used with wood species	104, 269
Suction method	Flow rate adjustable to transpiration of reference plants. Plant decapitated	230, 232
Perfusion method	Easily performed, suitable to study pH regulation, but organ detached	175, 176, 233, 234
"In situ"		
Ion-sensitive field-effect transistors	First attempts to study xylem pH in an intact plant, difficult	236
Xylem pH probe	Allows study of xylem pH on line in an intact plant	U. Zimmermann, University of Würzburg, Germany, 2001, personal communication
Apoplastic tissue fluid		
"In vitro"		
Equilibration with external solution, sometimes small volumes (in combination with microelectrodes) pressure technique	Different degree of sophistication Suitable to follow pH changes over time Often does not represent the real apoplast pH but may be strongly correlated	28, 41, 208, 240–250
Elution	Perturbation of pH gradients. Cytoplasmic contamination possible	255, 270, 271
and		
Forced perfusion	Perturbation of pH gradients. Dilution does not allow correct pH estimate.	252
Centrifugation	Cytoplasmic contamination possible	254, 270
With previous infiltration	Widely used. Perturbation of pH gradients. Dilution does not allow correct pH estimate. Cytoplasmic contamination possible	206, 216, 258, 270, 272, 273

Without previous infiltration	Undiluted apoplastic solution (WFS) Limited volume and centrifugal force Cytoplasmic contamination possible	256, 257, 274
Weak acid distribution	Provides the pH at the membrane surface (DFS) Sulforhodamine G as tracer for water transport	275–277 205
"In situ" pH-sensitive electrodes Roots (outer cells) Substomatal cavity Guard cells	Direct determination of pH (ion activity)	200, 209, 251, 252
pH-sensitive dyes	High spatial and temporal resolution Nondestructive. Expensive equipment. Limited availability of suitable dyes (cross-sensitivity)	2, 229
Detection by fluorescence microscopy Infiltration	Different infiltration solution to manipulate apoplastic fluid. Gradients disturbed or dissipated. Either high spatial resolution or Low spatial resolution	206, 207, 262
Covalently bound fluorescence ration dyes	High spatial and temporal resolution	202, 203, 211, 278 156
Feeding: Via xylem pressure changes Via petiole (transpiration)	No disturbance of pH gradients. May take long time, high-quality ratio required	262, 279
Dyes (pH-sensitive GFP) synthesized by transgenic plants and targeted to the apoplast	No manipulation of apoplast during measurements. Very limited availability (molecular techniques, plant species, suitable dyes)	C. Plieth, University of Kiel, Germany, 2000, personal communication
Confocal laser scanning microscopy	High spatial and temporal resolution	147, 265

pH of the sap to its own characteristic value, but the sensor used had limited stability. A recent development for online in vivo recordings of xylem sap pH is the xylem pH probe, and the first results will be available soon (U. Zimmermann, University of Würzburg, Germany, 2001, personal communication).

In earlier work, *apoplastic pH* values were estimated using *pH indicators* included in agar sheets. With this approach, localized pH changes or gradients along the root could be visualized in response to, e.g., N form or Fe supply [237]. A further option is to combine colorimetric methods with (micro-)electrode measurements in which the roots are grown in an agar medium containing a pH indicator such as bromocresol purple [152,238]. Videodensitometric analysis of the pH dye provides an accurate method for quantifying the pH gradients that develop around roots [239], and there is a good agreement between the H^+ fluxes calculated from colorimetric and potentiometric data [155]. These approaches do not necessarily indicate the real apoplast pH and require free access to the apoplast (roots, abraded shoots tissue). However, the techniques are inexpensive and may be used for a variety of species and tissues.

The *pH-sensitive electrodes* have been used in a variety of ways. Tissues were equilibrated with small volumes of solution, sometimes of different initial pH, to determine the "equilibrium" pH or pH that corresponds to zero H^+ net flux [240–244]. Occasionally, microelectrodes were placed in small holes, pricked with a needle that contained a solution equilibrated with the surrounding apoplast [28,245,246]. Other groups placed the microelectrode near [208,247, 248] or on the tissue surface [41,249,250]. Only recently have attempts been successful to insert microelectrodes into the apoplast proper, which gives access to the apoplast in the immediate vicinity of injured cells [209,251,252] or in the substomatal cavity [201].

For leaves, several other in vitro methods have been suggested, usually in combination with potentiometric pH determination. Examples include elution procedures [253], the vacuum perfusion of leaf discs [254], a pressure technique [255], and different centrifugation techniques [216,256,257]. All these methods have some special advantages. For example, the infiltration centrifugation technique allows the use of solutions differing in exchange strength [206] and thus differentiation between free and adsorbed cations, contributing significantly to current knowledge of ionic relations in this plant compartment. However, as discussed in the previous sections, the pH of the apoplast is partly regulated but nonetheless highly variable; i.e., both temporal and spatial concentration gradients do exist. This is why conventional methods, leading to an average concentration that does not exist in most locations, are inadequate to describe a complex situation. Cytoplasmic contamination has frequently been considered as one factor affecting the apoplastic washing fluids. However, if the experiments are restricted to healthy nonstressed plants, cytoplasmic contamination is rather unlikely to occur. Even at high centrifugation forces, composition is relatively little affected [257,258].

One problem in any application of the infiltration-centrifugation technique is the precise determination of the air- and water-filled apoplastic spaces [259] in order to convert concentration in the washing fluid correctly into concentration in the apoplastic fluid [258]. One of the major problems with those techniques in recovering the apoplast fluid after infiltration is that the original solution is diluted, and depending on the ionic composition of the infiltrating solution, new equilibria with the apoplast matrix are established. Particularly when infiltrating with pure water, the adsorption equilibria of protons and other cations with the exchange sites in the apoplast, and hence the pH values, are affected. More so, the dilution of the apoplastic fluid itself dilutes the concentration of mineral cations proportionally, but not the protons, which are buffered by low-molecular-weight compounds dissolved in the apoplastic fluid. Consequently, it is inappropriate to calculate the pH using the dilution factor. Infiltrating the apoplast with a solution that mimics the composition of the apoplastic fluid may partly compensate for this effect. However, this would require knowledge of the composition of the apoplastic fluid, including the pH, which is the aim of the procedure and cannot be a prerequisite. Due to the dilution, precipitates may dissolve, which also affects the pH. In view of all these difficulties and constraints, the pH obtained by recovering the apoplastic fluid after infiltration can only provide a rough estimate of the apoplast pH (not to mention the elimination of pH gradients).

The use of *pH-sensitive fluorescence dyes* offers an increasingly powerful alternative to microelectrodes for the measurement of apoplastic pH values. In the early days, single-wavelength fluorescence dyes were used [209,260], which did not allow compensation for uneven dye concentration. The introduction of ratiometric fluorescence dyes improved considerably the reliability of the measurements and provided the possibility of determining ionic activity at a high temporal and spatial resolution with a minimal invasive disturbance [261]. The dextran conjugate of fluorescein isothiocyanate (FITC-dextran), which is not membrane permeable to any significant extent, has been widely used [202,203, 206,211,216,262,263] and significantly contributed to our knowledge on apoplastic pH regulation.

An important feature of this approach to pH determination is that spatially resolved fluorescence measurements can be obtained with confocal laser scanning microscopy (LSM), allowing the construction of pH maps within cells and tissues [147,229,264]. Unfortunately, optical sectioning by confocal LSM is restricted to the first two or three cell layers of a root, but much can still be achieved using appropriate plants such as *Arabidopsis*, and the method is a powerful alternative to the use of microelectrodes for probing cell-specific pH events [265].

A newly synthesized fluorescent ratiometric probe, fluorescein boronic acid

($pK_a = 5.48$), which covalently binds to the cell wall of the outer cell layers, has been developed [266]. This allowed determination of apoplastic pH in the outer root cortex of corn roots with high spatial and temporal sensitivity [156]. More reports using this powerful technique are expected.

The power of the fluorescence technique has been increased still further through the development of pH-dependent green fluorescence protein (GFP) probes that can be targeted to particular subcellular locations [267]. This approach is currently being extended to pH measurements in the apoplast (C. Plieth, University of Kiel, Kiel, Germany, 2000, personal communication).

In general, reported apoplastic pH values range from just above 4 to just above 7, reflecting partly the variety of plant material and experimental conditions. The majority of data lie between pH 5 and pH 6.5. As a rule, the pH values reported for dicotyledons are higher than for monocotyledons and higher for angiosperms than for gymnosperms [260]. A compilation of apoplast pH values has been presented [229].

In conclusion, the xylem sap pH can be adequately measured using the pressure balance method or the xylem pressure probe (monitoring possible) or the compression-decompression technique (destructive). Regarding the apoplast of leaves and roots, the in vitro techniques in which the apoplast fluid is either removed or equilibrated with the external solution cannot be used to characterize the conditions in the apoplast proper, and gradients are erased as well. However, these techniques may be adequate when comparing the effect of various treatments on the bulk apoplast pH or the effect of feeding or removal of nutrients via the leaves. When gradients or localized pH changes are to be investigated, the microelectrodes and fluorescence ratiometric dyes are the recommended methods. The dyes can be fed via the petiole to preserve pH gradients in leaves. In other techniques, (1) dyes bound covalently to the wall matrix allow sufficient equilibration after infiltration and (2) the synthetic dyes may be targeted to the wall by molecular techniques; however, these two options are difficult to implement and will be used in only a limited number of cases.

ABBREVIATIONS

AFS	apparent free space (DFS + WFS)
DFS	Donnan free space
WFS	water free space
CEC	cation-exchange capacity
PME	pectin methylesterase
QABC	quantitative acid-base chemistry
SID	strong ion difference (fully dissociated cations minus fully dissociated anions, in mol L^{-1})
USL	unstirred layer

REFERENCES

1. T Schindler. Das Neue Bild der Zellwand. Biol Unserer Zeit 23:113–120, 1993.
2. B Sattelmacher. The apoplast and its significance for plant mineral nutrition. New Phytol 149:167–192, 2001.
3. N Sakurai, DJ Nevins. Changes in physical properties and cell wall polysaccharides of tomato (*Lycopersicon esculentum*) pericarp tissues. Physiol Plant 89:681–686, 1993.
4. N Carpita, M McCann, LR Griffing. The plant extracellular matrix: news from the cell's frontier. Plant Cell 8:1451–1463, 1996.
5. B Thornton, AE Macklon. Copper uptake by ryegrass seedlings: contribution of cell wall adsorption. J Exp Bot 40:1105–1111, 1989.
6. N Ae, T Otani. The role of cell wall components from groundnut roots in solubilizing sparingly soluble phosphorus in low fertility soils. Plant Soil 196:265–270, 1997.
7. WJ Horst. The role of the apoplast in aluminium toxicity and resistance of higher plants: a review. Z Pflanzenernaehr Bodenkd 58:419–428, 1995.
8. N Sakurai. Dynamic function and regulation of apoplast in the plant body. J Plant Res 111:133–148, 1998.
9. T Hoson. Apoplast as the side of response to environmental signals. J Plant Res 111:167–177, 1998.
10. C Grignon, H Sentenac. pH and ionic conditions in the apoplast. Annu Rev Plant Physiol Plant Mol Biol 42:103–128, 1991.
11. S Wilkinson. pH as a stress signal. Plant Growth Regul 29:87–99, 1999.
12. PA Stewart. Modern quantitative acid-base chemistry. Can J Physiol Pharmacol 61:1444–1461, 1983.
13. J Gerendás, U Schurr. Physicochemical aspects of ion relations and pH regulation in plants—a quantitative approach. J Exp Bot 50:1101–1114, 1999.
14. C Brett, K Waldron. Physiology and Biochemistry of Plant Cell Walls. London: Chapman & Hall, 1996.
15. CA Peterson, E Cholewa. Structural modifications of the apoplast and their potential impact on ion uptake. Z Pflanzenernaehr Bodenkd 161:521–531, 1998.
16. GI Cassab, JE Varner. Cell wall proteins. Annu Rev Plant Physiol Plant Mol Biol 39:321–359, 1988.
17. JP Gogarten. Physical properties of the cell wall of photoautotropic suspension cells from *Chenopodium rubrum* L. Planta 174:333–339, 1988.
18. VA Shepherd, PB Gootwin. The porosity of permeabilised *Chara* cells. Aust J Plant Physiol 16:231–239, 1989.
19. A Chesson, PT Gardner, TJ Wood. Cell wall porosity and available surface area of wheat straw and wheat grain fractions. J Sci Food Agric 75:289–295, 1997.
20. P Touchard, M Demarty, C Ripoll, C Marvan, M Thellier. Estimate of ionic mobilities in flax cell walls. In: J Dainty, M De Michelis, E Marré, F Rasi-Caldogno, eds. Plant Membrane Transport: The Current Position. Amsterdam: Elsevier, 1989, pp 603–606.
21. H Marschner. Mineral Nutrition of Higher Plants. San Diego: Academic Press, 1995.

22. T Hoson. Inhibiting effect of auxin on glucosamine incorporation into cell walls of rice coleoptiles. Plant Cell Physiol 28:301–308, 1987.

23. GE Briggs, RN Robertson. Apparent free space. Annu Rev Plant Physiol 8:11–30, 1957.

24. M Demarty, C Morvan, M Thellier. Exchange properties of isolated cell walls of *Lemna minor* L. Plant Physiol 62:477–481, 1978.

25. DS Bush, JG McColl. Mass action expressions of ion exchange applied to Ca^{2+}, H^+, K^+, and Mg^{2+} sorption on isolated cell walls of leaves from *Brassica oleracea*. Plant Physiol 85:247–260, 1987.

26. RG Stout, LR Griffing. Transmural secretion of a highly-expressed cell surface antigen of oat root cap cells. Protoplasma 172:27–37, 1993.

27. KA Platt-Aloia, WW Thomson, RE Young. Ultrastructural changes in the walls of ripening avocados: transmission, scanning and freeze fracture microscopy. Bot Gaz 141:366–373, 1980.

28. N Starrach, WE Mayer. Changes in the apoplastic pH and K concentration in the *Phaseolus pulvinus* in situ in relation to rhythmic leaf movements. J Exp Bot 40: 865–873, 1989.

29. P Keller, H Deuel. Kationenaustauschkapazität und Pektingehalt von Pflanzenwurzeln. Z Pflanzenernaehr Bodenkd 79:119–131, 1957.

30. J Bigot, P Binet. Study of the cation exchange capacity and selectivity of isolated root cell walls of *Cochlearia anglica* and *Phaseolus vulgaris* grown in media of different salinities. Can J Bot 64:955–958, 1986.

31. D Charnay, J Nari, G Noat. Regulation of plant cell-wall pectin methylesterase by polyamines—interactions with the effects of metal ions. Eur J Biochem 205:711–714, 1992.

32. J Messiaen, P Cambier, P Van Cutsem. Polyamines and pectins. I. Ion exchange and selectivity. Plant Physiol 113:387–395, 1997.

33. G Berta, MM Altamura, A Fusconi, F Cerruti, F Capitani, N Bagni. The plant cell wall is altered by inhibition of polyamine biosynthesis. New Phytol 137:569–577, 1997.

34. J Gerendás, RG Ratcliffe, B Sattelmacher. Relationship between intracellular pH and N metabolism in maize (*Zea mays* L.) roots. Plant Soil 155/156:167–170, 1993.

35. EAH Baydoun, CT Brett. Properties and possible physiological significance of cell wall calcium binding in etiolated pea epicotyls. J Exp Bot 39:199–208, 1988.

36. C Richter, J Dainty. Ion behavior in plant cell walls. I. Characterization of the *Sphagnum russowii* cell wall ion exchanger. Can J Bot 67:451–459, 1989.

37. DL Allan, WM Jarrel. Proton and copper adsorption to maize and soybean root cell walls. Plant Physiol 89:823–832, 1989.

38. DT Clarkson. Root structure and sites of ion uptake. In: Y Waisel, A Eshel, U Kafkafi, eds. Plant Roots—The Hidden Half. New York: Marcel Dekker, 1991, pp 417–453.

39. U Zimmermann, J Rygol, A Balling, G Klock, A Metzler, A Haase. Radial turgor and osmotic pressure profiles in intact and excised roots of *Aster tripolium*. Plant Physiol 99:186–196, 1992.

40. AD Thompson, JM Dietschy. Experimental demonstration of the effect of unstirred

water layer on the kinetic constants of the membrane transport of D-glucose in rabbit jejunum. J Membr Biol 54:221–229, 1980.

41. RL Preston. Effects of unstirred layers on the kinetics of carrier mediated solute transport by two systems. Biochim Biophys Acta 668:422–428, 1982.

42. M Esch, VL Sukhorukov, M Kürschner, U Zimmermann. Dielectric properties of alginate beads and bound water relaxation studied by electrorotation. Biopolymers 50:227–237, 1999.

43. DT Clarkson. Roots and the delivery of solutes to the xylem. Philos Trans R Soc Lond Ser B 341:5–17, 1993.

44. B Haynes. Ion exchange properties of roots and ionic interactions within the root apoplasm: their role in ion accumulation by plants. Bot Rev 46:75–99, 1980.

45. RE Franklin. Effects of adsorbed cations on phosphorus uptake by excised roots. Plant Physiol 44:697–700, 1969.

46. GW Borst-Pauwels, PP Severens. Effect of the surface potential upon ion selectivity found in competitive inhibition of divalent cation uptake. A theoretical approach. Physiol Plant 60:86–91, 1984.

47. PW Barts, GW Borst-Pauwels. Effects of membrane potential and surface potential on the kinetics of solute transport. Biochim Biophys Acta 813:51–60, 1985.

48. KA Collier, MJ O'Donnell. Analysis of epithelial transport by measurement of K⁺, Cl⁻ and pH gradients in extracellular unstirred layers: ion secretion and reabsorption by Malpighian tubules of *Rhodnius prolixus*. J Exp Biol 200:1627–1638, 1997.

49. J Sanderson. Water uptake by different regions of burley root. Pathways of radial flow in relation to development of the endodermis. J Exp Bot 34:240–253, 1983.

50. E Steudle, M Murrmann, CA Peterson. Transport of water and solutes across maize roots modified by puncturing the endodermis. Further evidence for the composite transport model of the root. Plant Physiol 103:335–349, 1993.

51. E Steudle. Water transport across roots. Plant Soil 167:79–90, 1994.

52. L Schreiber. Chemical composition of Casparian strips isolated from *Clivia miniata* Reg. roots: evidence for lignin. Planta 199:596–601, 1996.

53. J Zeier, L Schreiber. Comparative investigation of primary and tertiary endodermal cell walls isolated from the roots of five monocotyledoneous species: chemical composition in relation to fine structure. Planta 206:349–361, 1998.

54. L Schreiber, K Hartmann, M Skrabs, J Zeier. Apoplastic barriers in roots: chemical composition of endodermal and hypodermal cell walls. J Exp Bot 50:1267–1280, 1999.

55. DH Reinhardt, TL Rost. Salinity accelerates endodermal development and induces an exodermis in cotton seedling roots. Environ Exp Bot 35:563–574, 1995.

56. A Poljakoff-Mayber. Morphological and anatomical changes in plants as a response to salinity stress. In: A Poljakoff-Mayber, J Gale, eds. Plants in Saline Environment. Berlin: Springer, 1975, pp 97–117.

57. CJ Perumalla, CA Peterson. Deposition of Casparian bands and suberin lamellae in the exodermis and endodermis of young corn and onion roots. Can J Bot 64: 1873–1878, 1986.

58. CA Peterson, CJ Perumalla. A survey of angiosperm species to detect hypodermal Casparian bands. II. Roots with a multiseriate hypodermis or epidermis. Bot J Linn Soc 103:113–125, 1990.

59. M Damus, RL Peterson, DE Enstone, CA Peterson. Modification of cortical cell walls in roots of seedless vascular plants. Bot Acta 110:190–195, 1997.

60. DE Enstone, CA Peterson. Suberin deposition and band plasmolysis in the corn (*Zea mays* L.) root exodermis. Can J Bot 75:1188–1199, 1997.

61. DE Barrowclough, CA Peterson. Effects of growing conditions and development of the underlying exodermis on the vitality of the onion root epidermis. Physiol Plant 92:343–349, 1994.

62. DT Clarkson, AW Robards, JE Stephens, M Stark. Suberin lamellae in the hypodermis of maize (*Zea mays*) roots: development and factors affecting the permeability of hypodermal layers. Plant Cell Environ 10:83–94, 1987.

63. CA Peterson. Exodermal Casparian bands: their significance for ion uptake by roots. Physiol Plant 72:204–208, 1988.

64. DE Enstone, CA Peterson. The apoplastic permeability of root apices. Can J Bot 70:1502–1512, 1992.

65. M Gierth, R Stelzer, H Lehmann. An analytical microscopical study on the role of the exodermis on apoplastic Rb^+ (K^+) transport in barley roots. Plant Soil 207:209–218, 1999.

66. CA Peterson, DE Enstone. Functions of passage cells in the endodermis and exodermis of roots. Physiol Plant 97:592–598, 1996.

67. MG Shone, AV Flood. Measurement of free space and sorption of large molecules by cereal roots. Plant Cell Environ 8:309–315, 1985.

68. A Jungk, N Claassen, R Kuchenbuch. Potassium depletion of the soil-root interface in relation to soil parameters and root properties. Adv Plant Nutr 1:250–255, 1982.

69. WJ Horst, H Goppel. Aluminium-tolerance of horse bean, yellow lupin, barley and rye. 1. Shoot and root growth as affected by aluminium supply. Z Pflanzenernaehr Bodenkd 149:83–93, 1986.

70. FP Blamey, CJ Asher, DG Edwards, GL Kerver. In vitro evidence of aluminium effects on solution movement through root cell walls. J Plant Nutr 16:555–562, 1993.

71. DT Clarkson. Interaction between aluminium and phosphorous on root surfaces and cell wall material. Plant Soil 27:347–356, 1967.

72. FP Blamey, CJ Asher, GL Kerven, DG Edwards. Factors affecting aluminium sorption by calcium pectate. Plant Soil 149:87–94, 1993.

73. H Le Van, S Kuraishi, N Sakurai. Aluminum-induced rapid root inhibition and changes in cell-wall components of squash seedlings. Plant Physiol 106:971–976, 1994.

74. LG Wilson, JC Fry. Extensin—a major cell wall glycoprotein. Plant Cell Environ 9:239–260, 1986.

75. N Schmohl, WJ Horst. Pectin methylesterase modulates aluminium sensitivity in *Zea mays* and *Solanum tuberosum* L. Physiol Plant 109:419–427, 2000.

76. WJ Horst, CJ Asher, I Cakmak, P Szulkiewicz, AH Wissemeier. Short-term responses of soybean roots to aluminium. J Plant Physiol 140:174–178, 1992.

77. JW Huang, JE Shaff, DL Grunes, LV Kochian. Aluminium effects on calcium fluxes at the root apex of aluminium-sensitive wheat cultivars. Plant Physiol 98:230–237, 1992.

78. Z Rengel. Disturbance of cell Ca^{2+} homeostasis as a primary trigger of Al toxicity syndrome. Plant Cell Environ 15:931–938, 1992.
79. Z Rengel. Role of calcium in aluminium toxicity. New Phytol 121:499–513, 1992.
80. LV Kochian. Cellular mechanisms of aluminum toxicity and resistance in plants. Annu Rev Plant Physiol Plant Mol Biol 46:237–260, 1995.
81. C Plieth, B Sattelmacher, UP Hansen, MR Knight. Low pH-mediated elevations in cytosolic calcium are inhibited by aluminium: a potential mechanism for aluminium toxicity. Plant J 18:643–650, 1999.
82. B Sattelmacher, I Heinecke, KH Mühling. Influence of minerals on cytoplasmic streaming in root hairs of intact wheat seedlings (*Triticum aestivum* L.). Plant Soil 155/156:107–110, 1993.
83. S Lindberg, H Strid. Aluminium induces rapid changes in cytosolic pH and free calcium and potassium concentrations in root protoplasts of wheat (*Triticum aestivum*). Physiol Plant 99:405–414, 1997.
84. A Maison, PM Bertsch. Aluminium speciation in the presence of wheat root cell walls: a wet chemical study. Plant Cell Environ 20:504–512, 1997.
85. M Sivaguru, WJ Horst. The distal part of the transition zone is the most aluminum-sensitive apical root zone in maize. Plant Physiol 116:155–163, 1998.
86. M Sivaguru, F Baluska, D Volkmann, HH Felle, WJ Horst. Impacts of aluminum on the cytoskeleton of the maize root apex. Short-term effects on the distal part of the transition zone. Plant Physiol 119:1073–1082, 1999.
87. M Kollmeier, P Dietrich, CS Bauer, WJ Horst, R Hedrich. Aluminum activates a citrate-permeable anion channel in the aluminum sensitive zone of the maize root apex. A comparison between an aluminum-sensitive and an aluminum-resistant cultivar. Plant Physiol 125:1–16, 2000.
88. UE Grauer, WJ Horst. Effect of pH and nitrogen source on aluminium tolerance of rye (*Secale cereale* L.) and yellow lupin (*Lupinus luteus* L.). Plant Soil 127:13–21, 1990.
89. AH De Boer, HB Prius, PE Zanstra. Biphasic composition of transroot electrical potential in roots of plantago species: involvement of spatially separated electrogenic pumps. Planta 157:259–266, 1983.
90. K Mizuno, H Kojima, K Katou, H Okamoto. The electrogenic proton pumping from parenchyma symplast into xylem—direct demonstration by xylem perfusion. Plant Cell Environ 8:525–529, 1985.
91. LH Wegner, K Raschke. Ion channels in the xylem parenchyma of barley roots. A procedure to isolate protoplasts from this tissue and a patch-clamp exploration of salt passageways into xylem vessels. Plant Physiol 105:799–813, 1994.
92. LH Wegner, AH De Boer. Properties of two outward-rectifying channels in root xylem parenchyma cells suggest a role in K^+ homeostasis and long-distance signalling. Plant Physiol 115:1707–1719, 1997.
93. LH Wegner, AH De Boer. Two inward K^+ channels in the xylem parenchyma cells of barley roots are regulated by G-protein modulators through a membrane-delimited pathway. Planta 203:506–516, 1997.
94. T Jahn, F Baluska, W Michalke, JF Harper, D Volkmann. Plasma membrane H^+-ATPase in the root apex: evidence for strong expression in xylem parenchyma and

asymmetric localization within cortical and epidermal cells. Physiol Plant 104:311–316, 1998.

95. HT Wolterbeek. Cation exchange in isolated xylem cell walls of tomato. I. Cd^{2+} and Rb^+ exchange in adsorption experiments. Plant Cell Environ 10:39–44, 1987.

96. CJ Clark, PT Holland, GS Smith. Chemical composition of bleeding xylem sap from kiwifruit vines. Ann Bot 58:353–362, 1986.

97. MHMN Senden, AJGM Van der Meer, TG Verburg, HT Wolterbeek. Effects of cadmium on the behaviour of citric acid in isolated tomato xylem cell walls. J Exp Bot 45:597–606, 1994.

98. XE Yang, VC Baligar, JC Foster, DC Martens. Accumulation and transport of nickel in relation to organic acids in ryegrass and maize grown with different nickel levels. Plant Soil 196:271–276, 1997.

99. GL Mullins, LE Sommers, TL Housley. Metal speciation in xylem and phloem exudates. Plant Soil 96:377–391, 1986.

100. UW Stephan, I Schmidke, VW Stephan, G Scholz. The nicotianamine molecule is made-to-measure for complexation of metal micronutrients in plants. Biometals 9:84–90, 1996.

101. WD Jeschke, JS Pate. Cation and chloride partitioning through xylem and phloem within the whole plant of *Ricinus communis* L. under conditions of salt stress. J Exp Bot 42:1105–1116, 1991.

102. AJE Van Bel. Xylem-phloem exchange via the rays: the undervalued route of transport. J Exp Bot 41:631–644, 1990.

103. A Berger, R Oren, ED Schulze. Element concentrations in the xylem sap of *Picea abies* (L) Karst. seedlings extracted by various methods under different environmental conditions. Tree Physiol 14:111–128, 1994.

104. V Schill, W Hartung, B Orthen, MH Weisenseel. The xylem sap of maple (*Acer platanoides*) trees—sap obtained by a novel method shows changes with season and height. J Exp Bot 47:123–133, 1996.

105. PD Prima, B Botton. Organic and inorganic compounds of xylem exudates from five woody plants at the stage of bud breaking. J Plant Physiol 153:670–676, 1998.

106. M Urrestarazu, A Sanchez, FA Lorente, M Guzman. A daily rhythmic model for pH and volume from xylem sap of tomato plants. Commun Soil Sci Plant Anal 27:1859–1874, 1995.

107. U Schurr, ED Schulze. The concentration of xylem sap constituents in root exudate, and in sap from intact, transpiring castor bean plants (*Ricinus communis* L.). Plant Cell Environ 18:409–420, 1995.

108. JC Förster, WD Jeschke. Effects of potassium withdrawal on nitrate transport and on the contribution of the root to nitrate reduction in the whole plant. J Plant Physiol 141:322–328, 1993.

109. PJ White. The regulation of K^+ influx into roots of rye (*Secale cereale* L.) seedlings by negative feedback via the K^+ flux from shoot to root in the phloem. J Exp Bot 48:2063–2073, 1997.

110. MJ Canny. What becomes of the transpiration stream? New Phytol 114:341–368, 1990.

111. ME Westgate, E Steudle. Water transport in the midrib tissue of maize leaves.

Direct measurement of the propagation of changes in cell turgor across a plant tissue. Plant Physiol 78:183–191, 1985.

112. MJ Canny. Fine veins of dicotyledon leaves as sites for enrichment of solutes of the xylem sap. New Phytol 115:511–516, 1990.

113. MJ Canny. Rates of apoplastic diffusion in wheat leaves. New Phytol 116:263–268, 1990.

114. MJ Canny. Water pathways in wheat leaves. IV. The interpretation of images of a fluorescent apoplastic tracer. Aust J Plant Physiol 15:541–555, 1988.

115. E Strugger. Die lumineszenzmikroskopische Analyse des Transpirationsstromes im Parenchym. III. Untersuchungen an Helxine soleirolii. Biol Zentralbl 29:17–24, 1939.

116. BR Ray, JR Anderson, JJ Scholz. Wetting of polymer surfaces. I. Contact angles of liquids on starch, amylose amylopectin, cellulose and polyvinyl alcohol. J Phys Chem 62:1220–1227, 1958.

117. J Schönherr, MJ Bukovac. Penetration of stomata by liquids. Dependence on surface tension, wettability, and stomatal morphology. Plant Physiol 49:813–819, 1972.

118. MJ Canny. Apoplastic water and solute movement: new rules for an old space. Annu Rev Plant Physiol Plant Mol Biol 46:215–236, 1995.

119. SP Hardegree. Derivation of plant cell wall water content by examination of the water-holding capacity of membrane-disrupted tissues. J Exp Bot 40:1099–1104, 1989.

120. C Bayliss, C Van der Weele, MJ Canny. Determinations of dye diffusivities in the cell-wall apoplast of roots by a rapid method. New Phytol 134:1–4, 1996.

121. LV Edington, CA Peterson. Systemic fungicides: theory, uptake, and translocation. In: MR Siegel, HD Sisler, eds. Antifungal Compounds. Interaction in Biological and Ecological Systems. New York: Marcel Dekker, 1977, pp 51–81.

122. P Sitte. Morphologie. In: P Sitte, H Ziegler, F Ehrendorfer, A Bresinsky, eds. Strassburger Lehrbuch der Botanik. Stuttgart: Gustav Fischer Verlag, 1991, pp 13–238.

123. M Keunecke, UP Hansen. Different pH-dependence of K⁺ channel activity in bundle sheath mesophyll cells of maize leaves. Planta 210:798–800, 1999.

124. RF Evert, W Eschrich, W Heyser. Distribution and structure of the plasmodesmata in mesophyll and bundle-sheath cells of Zea mays L. Planta 136:77–89, 1977.

125. CE Botha. Plasmodesmatal distribution, structure and frequency in relation to assimilation in C3 and C4 grasses in southern Africa. Planta 187:348–358, 1992.

126. PW Hattersley, AJ Browing. Occurrence of the suberized lamella in leaves of grasses of different photosynthetic types. I. In parenchymatous bundle sheaths and PCR ("Kranz") sheaths. Protoplasma 109:371–401, 1981.

127. CEJ Botha, RHM Cross, DM Marshall. The suberin lamella, a possible barrier to water movement from veins to the mesophyll of Themedia trianda Forsk. Protoplasma 112:1–8, 1982.

128. RF Evert, CE Botha, RJ Mierzwa. Free-space marker studies on the leaf of Zea mays L. Protoplasma 126:62–73, 1985.

129. KJ Dietz. Functions and responses of the leaf apoplast under stress. Prog Bot 58:221–254, 1997.

130. RL Pieschl, PW Toll, DE Leith, LJ Peterson, MR Fedde. Acid-base changes in the running greyhound—contributing variables. J Appl Physiol 73:2297–2304, 1992.
131. WF Brechue, KE Gropp, BT Ameredes, DM Odrobinak, WN Stainsby, JW Harvey. Metabolic and work capacity of skeletal muscle of PFK-deficient dogs studied in situ. J Appl Physiol 77:2456–2467, 1994.
132. KJ Whitehair, SC Haskins, JG Whitehair, PJ Pascoe. Clinical applications of quantitative acid-base chemistry. J Vet Intern Med 9:1–11, 1995.
133. MI Lindinger, GJF Heigenhauser, RS McKelvie, NL Jones. Blood ion regulation during repeated maximal excercise and recovery in humans. Am J Physiol 262: R126–R136, 1992.
134. V Fencl, DE Leith. Stewart's quantitative acid-base chemistry—applications in biology and medicine. Respir Physiol 91:1–16, 1993.
135. DB Jennings. The physicochemistry of [H$^+$] and respiratory control—rules of pCO$_2$, strong ions, and their hormonal regulators. Can J Physiol Pharmacol 72: 1499–1512, 1994.
136. PS Nobel. Physicochemical and Environmental Plant Physiology. San Diego: Academic Press, 1991.
137. CI Ullrich, AJ Novacky. Extra- and intracellular pH and membrane potential changes induced by K$^+$, Cl$^-$, H$_2$PO$_4{}^-$ and NO$_3{}^-$ uptake and fusicoccin in root hairs of *Limnobium stoloniferum*. Plant Physiol 94:1561–1567, 1990.
138. CI Ullrich, AJ Novacky. Recent aspects of ion-induced pH changes. In: DD Randall, R Sharp, AJ Novacky, DG Blevins, eds. Current Topics in Plant Biochemistry and Physiology. Vol 11. Columbia: University of Missouri-Columbia, 1992, pp 231–248.
139. RG Ratcliffe. In vivo NMR studies of higher plants and algae. Adv Bot Res 20: 43–123, 1994.
140. RG Ratcliffe. Intracellular pH regulation in plants under anoxia. In: S Egginton, EW Taylor, JA Raven, eds. Regulation of Acid-Base Status in Animals and Plants. Cambridge: Cambridge University Press, 1999, pp 193–213.
141. NE Good. Active transport, ion movements, and pH changes. I. The chemistry of pH changes. Photosynth Res 19:225–236, 1988.
142. PE Levy, P Meir, SJ Allen, PG Jarvis. The effect of aqueous transport of CO$_2$ in xylem sap on gas exchange in woody plants. Tree Physiol 19:53–58, 1999.
143. JS Liang, JH Zhang. Collection of xylem sap at flow rate similar to in vivo transpiration flux. Plant Cell Physiol 38:1375–1381, 1997.
144. RJ Ritchie, AW Larkum. Cation exchange properties of the cell walls of *Enteromorpha intestinalis* L. Link. (Ulvales, Chlorophyta). J Exp Bot 132:125–139, 1982.
145. J Buchert, T Tamminen, L Viikari. Impact of the Donnan effect on the action of xylanases on fibre substrates. J Biotechnol 57:217–222, 1997.
146. NA Walker, MG Pitman. Measurement of fluxes across membranes. In: A Pirson, MH Zimmermann, eds. Encyclopedia of Plant Physiology: New Series. Vol 3. Berlin: Springer, 1976, pp 93–126.
147. DP Taylor, J Slattery, AC Leopold. Apoplastic pH in corn root gravitropism: a laser scanning confocal microscopy measurement. Physiol Plant 97:35–38, 1996.
148. JA Raven. Biochemical disposal of excess H$^+$ in growing plants? New Phytol 104: 175–206, 1986.

149. RE Cleland. Fusicoccin-induced growth and hydrogen ion excretion of *Avena* coleoptiles: relation to auxin responses. Planta 128:20–26, 1976.
149. JA Raven, B Wollenweber. Temporal and spatial aspects of acid-base regulation. Curr Top Plant Biochem Physiol 11:270–294, 1992.
150. R Behl, K Raschke. Close coupling between extrusion of H^+ and uptake of K^+ by barley roots. Planta 172:531–538, 1987.
151. JA Raven, FA Smith. Nitrogen assimilation and transport in vascular land plants in relation to intracellular pH regulation. New Phytol 76:415–431, 1976.
152. H Marschner, V Römheld. In vivo measurement of root induced pH changes at the soil-root interface: effect of plant species and nitrogen source. Z Pflanzenphysiol 111:241–251, 1983.
153. WR Fischer, H Flessa, G Schaller. pH values and redox potentials in microsites of the rhizosphere. Z Pflanzenernaehr Bodenkd 152:191–195, 1989.
154. AL Miller, GN Smith, JA Raven, NAR Gow. Ion currents and the nitrogen status of roots of *Hordeum vulgare* and non-nodulated *Trifolium repens*. Plant Cell Environ 14:559–567, 1991.
155. C Plassard, M Meslem, G Souche, B Jaillard. Localization and quantification of net fluxes of H^+ along maize roots by combined use of pH indicator dye videodensitometry and H^+-selective microelectrodes. Plant Soil 21:129–139, 1999.
156. H Kosegarten, F Grolig, A Esch, KH Glüsenkamp, K Mengel. Effects of NH_4^+, NO_3^- and HCO_3^- on apoplast pH in the outer cortex of root zones of maize, as measured by the fluorescence ratio of fluorescein boronic acid. Planta 209:444–452, 1999.
157. E Hoffland. Quantitative evaluation of the role of organic acid exudation in the mobilization of rock phosphate by rape. Plant Soil 140:279–289, 1992.
158. B Dinkelaker, V Römheld, H Marschner. Citric acid excretion and precipitation of calcium citrate in the rhizosphere of white lupin (*Lupinus albus* L.). Plant Cell Environ 12:285–292, 1989.
159. P Hinsinger, RJ Gilkes. Root-induced dissolution of phosphate rock in the rhizosphere of lupins grown in alkaline soil. Aust J Soil Res 33:477–489, 1995.
160. PF Grierson. Organic acids in the rhizosphere of *Banksia integrifolia* L. Plant Soil 144:259–265, 1992.
161. A Silber, R Ganmore-Neumann, J Ben-Jaacov. Effects of nutrient addition on growth and rhizosphere pH of Leucadendron 'Safari Sunset'. Plant Soil 199:205–211, 1998.
162. J Ben Jaacov, ed. Proceedings of the Fourth International Protea Working Group Symposium, Jerusalem, Israel, 17–21 March 1996. Acta Hortic No 453, 1997.
163. M Arahou, HG Diem. Iron deficiency induces cluster (proteoid) root formation in *Casuarina glauca*. Plant Soil, 196:71–79, 1997.
164. V Römheld, C Müller, H Marschner. Localization and capacity of proton pumps in roots of intact sunflower plants. Plant Physiol 76:603–606, 1984.
165. CR De Vos, HJ Lubberding, HF Bienfait. Rhizosphere acidification as a response to iron deficiency in bean plants. Plant Physiol 81:842–846, 1986.
166. V Toulon, H Sentenac, J-B Thibaud, J-C Davidian, C Moulineau, C Grignon. Role of apoplast acidification by the H^+ pump: effect on the sensitivity to pH and CO_2 of iron reduction by roots of *Brassica napus* L. Planta 186:212–218, 1992.

167. R Hauter, K Mengel. Measurement of pH at the root surface of red clover (*Trifolium pratense*) grown in soils differing in proton buffer capacity. Biol Fertil Soils 5: 295–298, 1988.

168. LA Papernik, LV Kochian. Possible involvement of Al-induced electrical signals in Al tolerance in wheat. Plant Physiol 115:657–667, 1997.

169. PR Ryan, M Skerrett, GP Findlay, E Delhaize, SD Tyerman. Aluminum activates an anion channel in the apical cells of wheat roots. Proc Natl Acad Sci USA 94: 6547–6552, 1997.

170. M Kollmeier, HH Felle, WJ Horst. Genotypical differences in aluminium resistance of maize are expressed in the distal part of the transition zone. Is reduced basipetal auxin flow involved in inhibition of root elongation by aluminium? Plant Physiol 122:945–956, 2000.

171. PR Ryan, E Delhaize, PJ Randall. Characterisation of Al-stimulated efflux of malate from the apices of Al-tolerant wheat roots. Planta 196:103–110, 1995.

172. DM Pellet, DL Grunes, LV Kochian. Organic acid exudation as an aluminum-tolerance mechanism in maize (*Zea mays* L.). Planta 196:788–795, 1995.

173. DM Pellet, LA Papernik, LV Kochian. Multiple aluminum-resistance mechanisms in wheat. Roles of root apical phosphate and malate exudation. Plant Physiol 112: 591–597, 1996.

174. J Degenhardt, PB Larsen, SH Howell, LV Kochian. Aluminum resistance in the *Arabidopsis* mutant *alr*-104 is caused by an aluminum-induced increase in rhizosphere pH. Plant Physiol 117:19–27, 1998.

175. A Mizuno, K Katou. The characteristics of the adjustment in the pH of the xylem exudate of segments of *Vigna* hypocotyls during xylem perfusion. Plant Cell Physiol 32:403–408, 1991.

176. DT Clarkson, JB Hanson. Proton fluxes and the activity of a stelar proton pump in onion roots. J Exp Bot 37:1136–1150, 1986.

177. U Schurr, ED Schulze. Effects of drought on nutrient and ABA transport in *Ricinus communis*. Plant Cell Environ 19:665–674, 1996.

178. S Wilkinson, WJ Davies. Xylem sap pH increase: a drought signal received at the apoplastic face of the guard cell that involves the suppression of saturable abscisic acid uptake by the epidermal symplast. Plant Physiol 113:559–573, 1997.

179. W Daeter, S Slovik, W Hartung. The pH gradients in the root system and the abscisic acid concentration in xylem and apoplastic saps. Philos Trans R Soc Lond Ser B 341:49–56, 1993.

180. S Slovik, W Daeter, W Hartung. Compartmental redistribution and long-distance transport of abscisic acid (ABA) in plants as influenced by environmental changes in the rhizosphere—a biomathematical model. J Exp Bot 46:881–894, 1995.

181. MJ Correia, ML Rodrigues, ML Osorio, MM Chaves. Effects of growth temperature on the response of lupin stomata to drought and abscisic acid. Aust J Plant Physiol 26:549–559, 1999.

182. AJ Barneix, H Breteler. Effect of cations on uptake, translocation and reduction of nitrate in wheat seedlings. New Phytol 99:367–379, 1985.

183. AD Peuke, J Glaab, WM Kaiser, WD Jeschke. The uptake and flow of C, N ions between roots and shoots in *Ricinus communis* L. J Exp Bot 47:377–385, 1996.

184. P Zornoza, O Carpena. Study on ammonium tolerance of cucumber plants. J Plant Nutr 15:2417–2426, 1992.

185. S Allen, JA Raven, JI Sprent. The role of long-distance transport in intracellular pH regulation in *Phaseolus vulgaris* grown with ammonium or nitrate as nitrogen source, or nodulated. J Exp Bot 39:513–528, 1988.

186. S Allen, JA Raven. Intracellular pH regulation in *Ricinus communis* grown with ammonium or nitrate as N source: the role of long distance transport. J Exp Bot 38:580–596, 1987.

187. JA Campbell, S Strother. Seasonal variation in pH, carbohydrate and nitrogen of xylem exudation of *Vitis vinifera*. Aust J Plant Physiol 23:115–118, 1996.

188. JJ Sauter. Seasonal changes in the efflux of sugars from parenchyma cells into the apoplast in poplar stems (*Populus X canadensis* 'robusta'). Trees Struct Funct 2: 242–249, 1988.

189. M Urrestarazu, A Sanchez, FA Lorente, M Gurman. A daily rhythmic model for pH and volume from xylem sap of tomato plants. Commun Soil Sci Plant Anal 27:1859–1874, 1996.

190. MC White, FD Baker, RL Chaney, AM Decker. Metal complexation in xylem fluid. II. Theoretical equilibrium model and computational computer program. Plant Physiol 67:301–310, 1981.

191. J Gerendás, J Polacco, SK Freyermuth, B Sattelmacher. Significance of nickel for plant growth and metabolism. Z Pflanzenernaehr Bodenkd 162:241–256, 1999.

192. IC Dodd, J He, CGN Turnbull, SK Lee, C Critchley. The influence of supra-optimal root-zone temperatures on growth and stomatal conductance in *Capsicum annuum* L. J Exp Bot 51:239–248, 2000.

193. P Brunet, B Sarrobert, N Paris-Pireyre, AM Risterucci. Composition chimique de sèves xylemiques du genre *Lycopersicon* (Solanaceae) en relation avec l'environnement. I. Effet de la température. Can J Bot 68:1942–1947, 1990.

194. R Munns. Effect of high external NaCl concentrations on ion transport within the shoot of *Lupinus albus*. I. Ions in xylem sap. Plant Cell Environ 11:283–289, 1988.

195. B Sarrobert, P Brunet, N Paris-Pireyre, AM Risterucci. Composition chimique de sèves xylemiques du genre *Lycopersicon* (Solanaceae) en relation avec l'environnement. II. Effet de la salinité. Can J Bot 68:1948–1952, 1990.

196. KJ Dietz. Functions and responses of the leaf apoplast under stress. Prog Bot 58: 221–254, 1997.

197. M Blatt. Extracellular potassium activity in attached leaves and its relation to stomatal function. J Exp Bot 36:240–251, 1985.

198. KH Nielsen, JK Schjoerring. Regulation of apoplastic NH_4^+ concentration in leaves of oilseed rape. Plant Physiol 118:1361–1368, 1998.

199. T Mimura, ZH Yin, E Wirth, KJ Dietz. Phosphate transport and apoplastic phosphate homeostasis in barley leaves. Plant Cell Physiol 33:563–568, 1992.

200. S Hanstein, HH Felle. The influence of atmospheric NH_3 on the apoplastic pH of green leaves: a non-invasive approach with pH sensitive microelectrodes. New Phytol 143:333–338, 1999.

201. HH Felle, S Hanstein, R Steinmeyer, R Hedrich. Continuous analysis of dark- and abscisic acid–induced changes in K^+, Cl^- and H^+ activities in the substomatal cavity of *Vicia faba* leaves. Plant J 24:297–304, 2000.

202. V Oja, G Savchenko, B Jakob, U Heber. pH and buffer capacities of apoplastic and cytoplasmic cell compartments in leaves. Planta 209:239–249, 1999.
203. G Savchenko, C Wiese, S Neimanis, R Hedrich, U Heber. pH regulation in apoplastic and cytoplasmic cell compartments of leaves. Planta 211:246–255, 1999.
204. MJ Canny. Locating active proton extrusion pumps in leaves. Plant Cell Environ 10:271–274, 1987.
205. TP Wilson, MJ Canny, ME McCully. Leaf teeth, transpiration and the retrieval of apoplastic solutes in balsam poplar. Physiol Plant 83:225–232, 1991.
206. KH Mühling, B Sattelmacher. Apoplastic ion concentration of intact leaves of field bean (Vicia faba) as influenced by ammonium and nitrate nutrition. J Plant Physiol 147:81–86, 1995.
207. KH Mühling, C Plieth, UP Hansen, B Sattelmacher. Apoplastic pH of intact leaves of Vicia faba as influenced by light. J Exp Bot 46:377–382, 1995.
208. DJF Bowling, A Edwards. Gradients in the stomatal complex of Tradescantia virginiana. J Exp Bot 35:1641–1645, 1984.
209. MC Edwards, GN Smith, DJF Bowling. Guard cells extrude protons prior to stomatal opening—a study using fluorescence microscopy and pH microelectrodes. J Exp Bot 39:1541–1547, 1988.
210. Y Lee, RL Satter. Effects of white, blue and red light and darkness on pH of the apoplast in the Samanea pulvinus. Planta 178:31–40, 1989.
211. B Hoffmann, R Plänker, K Mengel. Measurements of pH in the apoplast of sunflower leaves by means of fluorescence. Physiol Plant 84:146–153, 1992.
212. K Mengel, R Plänker, B Hoffmann. Relationship between leaf apoplast pH and iron chlorosis of sunflower (Helianthus annuus L.). J Plant Nutr 17:1053–1065, 1994.
213. J Finnemann, JK Schjoerring. Translocation of NH_4^+ in oilseed rape plants in relation to glutamine synthetase isogene expression and activity. Physiol Plant 105: 469–477, 1999.
214. HJ Kronzucker, JK Schjoerring, Y Erner, GJD Kirk, MY Siddiqi, ADM Glass. Dynamic interactions between root NH_4^+ influx and long-distance N translocation in rice: insights into feedback processes. Plant Cell Physiol 39:1287–1293, 1998.
215. AD Peuke, WD Jeschke, KJ Dietz, L Schreiber, W Hartung. Foliar application of nitrate or ammonium as sole nitrogen supply in Ricinus communis. I. Carbon and nitrogen uptake and inflows. New Phytol 138:675–687, 1998.
216. KH Mühling, B Sattelmacher. Determination of apoplastic K^+ in intact leaves by ratio imaging of PBFI fluorescence. J Exp Bot 8:337–337, 1997.
217. DJ Bowling. Measurement of the apoplastic activity of K^+ and Cl^- in the leaf epidermis of Commelina communis in relation to stomatal activity. J Exp Bot 38: 1351–1355, 1987.
218. C Freudling, N Starrach, D Flach, D Gradmann, WE Mayer. Cell walls as reservoirs of potassium ions for reversible volume changes of pulvinar motor cells during rhythmic leaf movements. Planta 175:193–203, 1988.
219. K Mengel, W Bübl, HW Scherer. Iron distribution in vine leaves with HCO_3-induced chlorosis. J Plant Nutr 7:715–724, 1984.
220. K Mengel, G Geurtzen. Relationship between iron chlorosis and alkalinity in Zea mays. Physiol Plant 72:460–465, 1988.

221. MJ Hodson, AG Sangster. Silica deposition in the inflorescence bracts of wheat (*Triticum aestivum*). I. Scanning electron microscopy and light microscopy. Can J Bot 66:829–838, 1988.
222. AJ Smolders, RJ Hendriks, HM Campschreur, JG Roelofs. Nitrate induced iron deficiency chlorosis in *Juncus acutiflorus*. Plant Soil 196:37–45, 1997.
223. M Häussling, V Römheld, H Marschner. Beziehungen zwischen Chlorosegrad, Eisengehalten und Blattwachstum von Weinreben auf verschiedenen Standorten. Vitis 24:158–168, 1985.
224. M Nikolic, V Römheld. Mechanism of Fe uptake by the leaf symplast: is Fe inactivation in leaf a cause of Fe deficiency chlorosis? Plant Soil 215:229–237, 1999.
225. S Husted, JK Schjoerring. Ammonia flux between oilseed rape plants and the atmosphere in response to changes in leaf temperature, light intensity, and air humidity —interactions with leaf conductance and apoplastic NH_4^+ and H^+ concentrations. Plant Physiol 112:67–74, 1996.
226. M Mattsson, S Husted, JK Schjoerring. Influence of nitrogen nutrition and metabolism on ammonia volatilization in plants. Nutr Cycl Agroecos 51:35–40, 1998.
227. S Husted, JK Schjoerring. Apoplastic pH and ammonium concentration in leaves of *Brassica napus* L. Plant Physiol 109:1453–1460, 1995.
228. S Hanstein, M Mattsson, HJ Jaeger, JK Schjoerring. Uptake and utilization of atmospheric ammonia in three native Poaceae species: leaf conductances, composition of apoplastic solution and interactions with root nitrogen supply. New Phytol 141: 71–83, 1999.
229. Q Yu, C Tang, J Kuo. A critical review on methods to measure apoplastic pH in plants. Plant Soil 219:29–40, 2000.
230. MJ Canny, ME McCully. The xylem sap of maize roots: its collection, composition and formation. Aust J Plant Physiol 154:557–566, 1988.
231. U Schurr. Xylem sap sampling—new approaches to an old topic. Trends Plant Sci 3:293–298, 1998.
232. C Gil de Carrasco, M Guzman, FA Lorente, M Urrestarazu. Xylem sap extraction: a method. Commun Soil Sci Plant Anal 25:1829–1839, 1994.
233. DT Clarkson, L Williams, JB Hanson. Perfusion of onion root xylem vessels: a method and some evidence of control of the pH of the xylem sap. Planta 162:361–369, 1984.
234. H Okamoto, A Mizuno, K Katou, Y Ono, Y Matsumura, H Kojima. A new method in growth-electrophysiology: pressurized intra-organ perfusion. Plant Cell Environ 7:139–147, 1984.
235. L Eklund. Endogenous levels of oxygen, carbon dioxide and ethylene in stems of Norway spruce trees during one growing season. Trees 4:150–154, 1990.
236. V Herrmann, M Tesche. In vivo pH measurement in the xylem of broad-leaved trees using ion-sensitive field-effect transistors. Trees 6:13–18, 1992.
237. H Marschner, V Römheld, H Ossenberg-Neuhaus. Rapid method for measuring changes in pH and reducing processes along roots of intact plants. Z Pflanzenphysiol 105:407–416, 1982.
238. HT Gollany, TE Schumacher. Combined use of colorimetric and microelectrode methods for evaluating rhizosphere pH. Plant Soil 154:151–159, 1993.

239. B Jaillard, L Ruiz, JC Arvieu. pH mapping in transparent gel using color indicator videodensitometry. Plant Soil 183:1–11, 1996.
240. JA Raven, OD Farquhar. In: J Dainty, M De Michelis, E Marré, F Rasi-Caldogno, eds. Plant Membrane Transport: The Current Position. Amsterdam: Elsevier, 1989, pp 607–610.
241. MJ Vesper. Use of a pH-response curve for growth to predict apparent wall pH in elongating segments of maize coleoptiles and sunflower hypocotyls. Planta 166:96–104, 1985.
242. K Kumon, S Soda. Changes in extracellular pH of the motor cells of *Mimosa pudica* L. during movement. Plant Cell Physiol 26:375–377, 1985.
243. WS Peters, C Lommel, HH Felle. IAA breakdown and its effect on auxin-induced cell wall acidification in maize coleoptile segments. Physiol Plant 100:415–422, 1997.
244. WS Peters, H Lüthen, M Böttger, HH Felle. The temporal correlation of changes in apoplast pH and growth rate in maize coleoptile segments. Aust J Plant Physiol 25:21–25, 1998.
245. Y Lee, RL Satter. Effects of temperature on H^+ uptake and release during circadian rhythmic movements of excised *Samanea* motor organs. Plant Physiol 86:352–354, 1988.
246. A Ballarin-Denti, D Antoniotti. An experimental approach to pH measurement in the intercellular free space of higher plant tissues. Experientia 47:478–482, 1991.
247. LV Kochian, JE Shaff, WJ Lucas. High affinity K^+ uptake in maize roots. A lack of coupling with H^+ efflux. Plant Physiol 91:1202–1211, 1989.
248. S Shabala, I Newman. Light-induced changes in hydrogen, calcium, potassium, and chloride ion fluxes and concentrations from the mesophyll and epidermal tissues of bean leaves. Understanding the ionic basis of light-induced bioelectrogenesis. Plant Physiol 119:1115–1124, 1999.
250. B Aloni, J Daie, RE Wvse. Regulation of apoplastic pH in source leaves of *Vicia faba* by gibberellic acid. Plant Physiol 88:367–369, 1988.
251. H Felle. Ion-selective microelectrodes: their use and importance in modern plant cell biology. Bot Acta 106:5–12, 1993.
252. H Felle. The apoplastic pH of the *Zea mays* root cortex as measured with pH-sensitive microelecrodes: aspects of regulation. J Exp Bot 49:987–995, 1998.
253. JM Long, IE Widders. Quantification of apoplastic potassium content by elution analysis of leaf lamina tissue from pea (*Pisum sativum* L. cv. Argenteum). Plant Physiol 94:1040–1047, 1990.
254. L Bernstein. Methods for determining solutes in the cell walls of leaves. Plant Physiol 47:361–365, 1971.
255. JJ Jachetta, AP Appleby, L Boersma. Use of the pressure vessel to measure concentrations of solutes in apoplastic and membrane-filtered symplastic sap in sunflower leaves. Plant Physiol 82:995–999, 1986.
256. F Dannel, H Pfeffer, H Marschner. Isolation of apoplasmic fluid from sunflower leaves and its use for studies on influence of nitrogen supply on apoplasmic pH. J Plant Physiol 146:273–278, 1995.

257. Q Yu, C Tang, Z Chen, J Kuo. Extraction of apoplastic sap from plant roots by centrifugation. New Phytol 143:299–304, 1999.

258. G Lohaus, K Pennewiss, B Sattelmacher, M Hussmann, KH Mühling. Is the infiltration centrifugation technique appropriate for the isolation of apoplastic fluids? A critical evaluation with different plant species. Physiol Plant 111:457–465, 2001.

259. K Leidreiter, A Kruse, D Heineke, D Robinson, H Heldt. Subcellular volumes and metabolite concentrations in potato (*Solanum tuberosum* cv. Desiree) leaves. Bot Acta 108:439–444, 1995.

260. H Pfanz, KJ Dietz. A fluorescent method for the determination of the apoplastic proton concentration in intact leaf tissues. J Plant Physiol 29:41–48, 1987.

261. GR Bright, GW Fisher, J Rogowska, DL Taylor. Fluorescence ratio imaging microscopy. Methods Cell Biol 30:157–192, 1989.

262. B Hoffmann, H Kosegarten. FITC-dextran for measuring apoplast pH and apoplastic pH gradients between various cell types in sunflower leaves. Physiol Plant 95:327–335, 1995.

263. KH Mühling, M Wimmer, H Goldbach. Apoplastic and membrane-associated Ca²⁺ in leaves and roots as affected by boron deficiency. Physiol Plant 102:179–184, 1998.

264. BC Gibbon, DL Kropf. Cytosolic pH gradients associated with tip growth. Science 263:1419–1421, 1994.

265. TN Bibikova, T Jacob, I Dahse, S Gilroy. Localised changes in apoplastic and cytoplasmic pH are associated with root hair development in *Arabidopsis thaliana*. Development 125:2925–2934, 1998.

266. KH Glüsenkamp, H Kosegarten, K Mengel, F Grolig, A Esch, HE Goldbach. A fluorescein boronic acid conjugate as a marker for borate binding sites in the apoplast of growing roots of *Zea mays* L. and *Helianthus annus* L. In: RW Bell, B Rerkasem, eds. Boron in Soils and Plants. Dordrecht: Kluwer Academic Press, 1997, pp 229–235.

267. J Llopis, JM McCaffery, A Miyawaki, MG Farquhar, RY Tsien. Measurement of cytosolic, mitochondrial, and Golgi pH in single living cells with green fluorescent proteins. Proc Natl Acad Sci USA 95:6803–6808, 1998.

268. E Peterlunger, B Marangoni, R Testolin, G Vizzotto, G Costa. Carbohydrates, organic acids and mineral elements in xylem sap bleeding from kiwifruit canes. Acta Hortic 282:273–282, 1990.

269. H Schneider, N Wistuba, R Reich, H-J Wagner, LH Wegner, U Zimmermann. Minimal- and noninvasive characterization of the flow-force pattern of higher plants. In: M Terazawa, ed. Tree Sap II. Hokkaido: Hokkaido University Press, 2000, pp 77–91.

270. DJ Cosgrove, RE Cleland. Solutes in the free space of growing stem tissues. Plant Physiol 72:326–331, 1983.

271. W Hartung, JW Radin, DL Hendrix. Abscisic acid movement into the apoplastic solution of water-stressed cotton leaves. Role of apoplastic pH. Plant Physiol 86:908–913, 1988.

272. J Aked, JL Hall. Effect of powdery mildew infection on concentrations of apoplastic sugars in pea leaves. New Phytol 123:283–288, 1993.

273. ML Pinedo, C Segarra, RD Conde. Occurrence of two endoproteinases in wheat leaf intercellular washing fluid. Physiol Plant 88:287–293, 1993.

274. FC Meinzer, PH Moore. Effect of apoplastic solutes on water potential in elongating sugarcane leaves. Plant Physiol 86:873–879, 1988.

275. NA Walker, FA Smith. Intracellular pH in *Chara corallina* measured by DMO distribution. Plant Sci Lett 4:125–132, 1975.

276. H Sentenac, C Grignon. Effect of H^+ excretion on the surface pH of corn root cells evaluated by using weak acid influx as a pH probe. Plant Physiol 84:1367–1372, 1987.

277. JB Thibaud, JC Davidian, H Sentenac, A Soler, C Grignon. H^+ cotransports in roots as related to the surface pH shift induced by H^+ excretion. Plant Physiol 88: 1469–1473, 1988.

278. H Kosegarten, G Englisch. Effect of various nitrogen forms on the pH in leaf apoplast and on iron chlorosis of *Glycine max* L. Z Pflanzenernaehr Bodenkd 157: 401–405, 1994.

279. U Heckenberger, U Schurr, ED Schulze. Stomatal response to abscisic acid fed into the xylem of intact *Helianthus annuus* (L.) plants. J Exp Bot 49:1405–1412, 1996.

12

H⁺ Currents around Plant Roots

Miguel A. Piñeros and Leon V. Kochian
Cornell University, Ithaca, New York

1 INTRODUCTION

Roots have the ability to induce pH changes at the root-soil interface, thereby lowering or raising the pH of alkaline or acidic surroundings, respectively. Consequently, the soil pH at the root surface may differ considerably from that in the soil a few millimeters away from the root surface. Because pH governs many of the parameters involved in plant metabolism and mineral nutrition (i.e., ion uptake), the root's ability to interact and actively modify its immediate surroundings (i.e., rhizosphere) has a direct effect on the uptake of both mineral nutrients and phytotoxic metals. For example, root-induced acidification of the rhizosphere mobilizes limiting macro- and micronutrients such as phosphorus, zinc, and iron. Consequently, plant nutrition relies on the plant's ability to interact with the surrounding soil medium, recognizing, binding, and transporting an array of substances into the root.

A few models have been put forward to predict plant-induced pH fields in the rhizosphere [1,2]. These models vary in complexity: one model assumes a uniform spatial distribution of ion fluxes along the rhizosphere of nongrowing parts of the root [1], and others take into account the root's varying growth rates and spatially varying fluxes along the growing root [2]. Nevertheless, these models provide powerful tools to explain how soil properties are affected by H⁺ fluxes associated with root metabolism. Because many of the transport processes across the plasma membrane of root cells depend directly or indirectly on proton transport, there is a need for the understanding of both endogenous and environmental factors influencing root-induced pH patterns in the rhizosphere.

Since the late 19th century, researchers have been aware of the occurrence of self-generated endogenous electric fields by diverse organisms, including growing roots of intact plants [3]. This chapter summarizes the progress in characterizing and understanding these ionic currents surrounding roots. Researchers have put forward a large number of hypotheses associating these electric fields and ion currents with differentiation and growth processes, as well as nutrient uptake and root environmental responses. Unfortunately, although a significant literature exists on the net ionic currents surrounding roots, the exact role or contribution of protons to these currents is less well documented. The contribution of H^+ fluxes to net currents has generally been assumed from indirect rather than direct measurements. Thus, although the present chapter emphasizes H^+ fluxes around roots, much of the literature review focuses on net ionic currents, under the assumption that protons play a significant role (if not in the magnitude at least in the direction) as a component of these net currents. Rather than quantitatively describing currents, a qualitative overview on the directions of the currents (inward or outward) and spatial patterns along roots is given.

2 TECHNIQUES

Different techniques have been used to estimate and characterize fluxes of macro- and micronutrient ions in higher plant cells. Early studies of intact tissue ion transport processes in higher plants relied on methodologies involving analysis of bulk solutions as the means to gain general understanding of the mechanism and regulation of root ion transport processes. Analysis of radioisotope fluxes, quantification of the ion content of bulk root tissue following ion absorption, and studies monitoring the depletion of a particular ion in the solution surrounding the root have commonly been used [e.g., 4,5]. Such techniques are limited with regard to spatial and temporal resolution, as the experimental conditions usually yield fluxes that are an integration of ion uptake over the entire tissue surface. Furthermore, these measurements provide an averaged measurement over a period of time (ranging from several minutes up to hours), limiting the study of root ion transport to excised roots or entire root systems over fairly long time periods. Thus, these techniques cannot be used to study transient responses or transport events associated with a small number of cells at the root surface.

This section provides an overview of some recent technologies that have allowed researchers to overcome some limitations of the earlier techniques, thus improving our understanding of some of the mechanisms underlying root ion transport processes. The implementation of voltage- and ion-sensing microelectrodes in studying ion fluxes has improved the temporal and spatial resolution with which we can study root ion transport. Although these techniques vary in their sensitivity and time resolution, they provide a noninvasive approach that allows multiple measurements of ion fluxes without damage to or destruction of the roots. Because of the small microelectrode tip diameter (about 2–10 μm), such systems allow a high degree of spatial resolution. Likewise, because ion flux measurements can be performed quite rapidly, these systems can achieve a high degree of temporal resolution, allowing measurements of fairly rapid changes in fluxes. Finally, implementation of these technologies has allowed measurements not only of one type of ion flux but also of simultaneous ion fluxes (e.g., K^+ and H^+), enlightening studies relating associations among different ion fluxes.

2.1 Static and Vibrating-Microelectrode Systems

In addition to the diffusion of ions either toward or away from the root, the transport of ions across the plasma membrane of cells at the root surface results in the generation of steady-state ion gradients in the unstirred layer surrounding the root. Researchers have taken advantage of this phenomenon to measure and map ionic activity gradients that develop around intact roots. Based on the measurement of radial ion activity gradients in the unstirred layer at the root surface, it is possible to quantify and map net ionic fluxes or currents along the length of roots and even study fluxes associated with a small number of or individual root epidermal cells.

Regardless of the type of microelectrode used, the basic vibrating electrode setup generally consists of at least one piezoelectric pusher element that allows the discrete movement of the electrode between two given positions (i.e., vibration amplitude). Additional piezoelectric elements can also be stacked in orthogonal directions to create a microstage, introducing experimental flexibility as the electrode can then be vibrated in several dimensions. Regardless of the arrangement, the piezoelectric elements are in turn held by manually or electrically driven translation stages that allow the coarse positioning of the microelectrode. This system is generally mounted on the stage of an inverted microscope. The piezoelectric pusher is driven by a damped squared voltage wave at low frequency that causes the microelectrode vibration. The vibration's amplitude and frequency need to be optimized for a given ion or voltage gradient. Computer software controls the manner in which the microelectrode is moved between the two preset positions (i.e., vibration amplitude) to minimize mixing of the gradients. The voltage or ion activity difference is calculated by measuring the microelectrode output at each extreme position of the vibration excursion (with a predetermined sampling rate). Any DC voltage difference between the two end points of vibration is transformed into a sinusoidal AC voltage that is measured with the aid of a lock-in amplifier. Readings from the extreme of the vibration amplitude are pooled into two separate data buffers representing the two extremes of vibration, and then subtraction of the averaged data of one data buffer from the other provides the difference in potential between the two measuring points.

The vibrating microelectrodes either can be constructed to measure voltages and thus net ion currents in the unstirred layer next to root or can provide the means for measuring specific ion currents when the microelectrode is made with ion-selective liquid membranes. The former allows measurements of net currents, and the latter allows the researcher to link the measured electric fields to the flux of a particular ion. The electrode used for measuring net currents generally consists of a small (about 30 μm in diameter) platinum-black ball electroplated onto gold-coated tips of steel needles. A concentric reference electrode, made of platinum black is located about 5 mm behind the tip. Although this type of electrode provides good time and spatial resolution, its limitation is that it provides information only on the total electric field set up by unknown ion currents.

The construction of liquid-membrane ion-selective microelectrodes and their use in identifying specific ion fluxes that make up total ion currents have been described in detail elsewhere [6–8]. Briefly, this type of microelectrode is constructed from commercially available glass capillaries. The glass is heated and pulled using either a horizontal or vertical electrode puller, resulting in a glass microelectrode with a tip diameter of approximately 1–5 μm, which is then silanized in order to increase the hydrophobicity of the surface of the internal glass walls. The microelectrode is then back-filled with an electrolyte buffer (e.g., 40 mM KH_2PO_4). The microelectrode's tip is front-filled with the desired ion-selective liquid membrane, creating a short column adjacent to the back-filling solution. Proton-selective cocktails are readily available commercially. The electrical connection between the back-filling solution of the microelectrode and amplifier is made through an Ag:AgCl wire. The ion activity at the tip of the microelectrode is actually translated at the microelectrode tip to an associated voltage (i.e., the voltage drop across the ion-selective membrane). Measured voltages can in turn be converted back to an ion activity based on calibration curves.

The microelectrodes based on a liquid membrane have a very high resistance, in the range of 10^9–10^{10} Ω. The "electrical noise" associated with this inherent resistance and noise associated with the amplifying electronics and electrochemical instabilities at the liquid membrane, bathing and back-filling solutions, and Ag:AgCl wire interfaces determine the lower limit of the detectable ion activity gradient. A detailed comparison of static and vibrating ion-selective microelectrode techniques has been conducted, indicating that the vibrating electrodes are approximately 50 times more sensitive than static microelectrodes [9]. Although static ion-selective microelectrodes have proved to be quite useful in studying root ion fluxes, the vibrating electrodes have allowed researchers to overcome many of the limitations associated with the high resistance of these microelectrodes, thus greatly improving the sensitivity. These features have made the vibrating electrodes a technique of choice when studying root ion fluxes, especially when dealing with the small ion fluxes associated with single cells.

Regardless of the type of electrode (static vs. vibrating) used to measure H⁺ fluxes that make up root ion currents, the measured H⁺ activity gradients can be expressed as net H⁺ fluxes. The ion fluxes flowing from the root surface out into the unstirred layer measured at the two extremes of electrode vibration (vibration amplitude equals ΔX) are driven by two forces: the concentration gradient (ΔC) and the electric field (ΔE). The electrical gradient can usually be neglected in the case of H⁺ currents, as it is generally much smaller than ΔC. Thus, net H⁺ fluxes at the root surface can be determined from the H⁺ activities measured at the two extremes of electrode vibration in the unstirred layer next to the root surface. Subsequently, the current density can be converted into corresponding ion fluxes. The calculations involved in measuring H⁺ proton fluxes have been outlined previously [10]. Briefly, the net fluxes for a particular ion i at the root surface are calculated based on Fick's first law of diffusion:

$$J_i = 2\pi D_i(C_1 - C_2)/\ln(R_1/R_2)$$

where J_i is the net flux of the ion i per unit length of root (generally expressed as μmol cm^{-1} s^{-1}), D_i is the diffusion coefficient for i (in cm^2 s^{-1}), C_1 and C_2 are the ion activities at the two positions, and R_1 and R_2 are the respective distances from the positions where the ion activities are measured to the center of the root.

The problems and limitations associated with measuring and interpreting H⁺ fluxes have been discussed elsewhere [11]. Primarily, it is not possible to measure directly the unidirectional H⁺ flux associated with the plasma membrane H⁺-ATPase. Ion-selective microelectrodes allow only quantification of the net H⁺ extrusion (or consumption). Thus, the reported values of net H⁺ fluxes, reflecting to some extent H⁺-ATPase activity, should be interpreted cautiously. Carbon dioxide resulting from respiratory processes, H⁺ binding by the cell wall, and H⁺ influx associated with H⁺-driven symport and antiport mechanisms can all contribute to net H⁺ fluxes.

2.2 pH Dye Indicators in Agar Media

When measuring net ionic currents, it is desirable to establish the identity of the ions carrying these currents. Prior to the development of ion-selective electrodes, the involvement of H⁺ fluxes in the net currents recorded along the root was generally inferred from indirect measurements. The net current direction and magnitude changes were often correlated with the color changes that took place around the root grown in medium that contained some type of pH indicator dye, such as bromocresol purple. This type of correlation generally suggested that H^+ ions were probably the major component of the net current recorded [12–16]. Although several improvements have been made to the original methodology [11,17], pH dye indicators are limited in their spatial and temporal resolution and can provide only qualitative information on the pH changes taking place along the root surface. Models developed to predict the plant-induced pH fields around the rhizosphere of growing roots have indicated that the agar-contact technique to measure pH may not be appropriate for measuring gradients in a rhizosphere that extends just a few millimeters from the root [2]. Consequently, given the limitations of pH dye indicators, their full utility may be exploited only as an additional visual tool when used in conjunction with some type of temporally and spatially sensitive technique, such as H^+-selective electrodes [11,17].

3 ELECTRICAL FIELDS, ION CURRENTS, AND H^+ FLUXES ALONG ROOTS

3.1 General Patterns

Growing roots are constantly generating and driving ionic currents through themselves as well as their surroundings. The recognition of the existence of these natural electrical fields surrounding roots dates back to research published at the end of the 19th century and beginning of the 20th century [3,18]. Subsequently, in the 1960s, techniques using static electrodes placed at the surface of vertically growing bean roots were refined to gain further insights into these endogenously generated electrical fields [19–21]. Since then, a significant number of studies have established that the magnitude and patterns of these endogenous currents are similar in roots of many different plant species (Table 1). In early reports, the measurements of the direction and magnitude of electrical fields around roots showed a net current flow emerging from the nongrowing root zones and entering the root at its actively growing regions. Recent techniques have allowed more detailed characterization of the spatial distribution of root ion currents. The basic pattern (see Fig. 1) describes a current that flows out of the mature root regions just adjacent to the elongation zone as well as the proximal (basal) elongation zones. This current then enters the root apex, including the root cap, meristem, and apical and central part of the elongation zone. The mean current density increases basipetally from the root cap and reaches a maximum at the apical elongation zone.

Some more detailed studies have also indicated that the currents at the root apex can fluctuate between inward and outward currents. Other studies have suggested that although there is a general inward current for the entire root apex, within the root cap there can be an inward current at the very tip of the cap and an outward current from the cap base [24,36]. Despite these differences in the

TABLE 1 Examples of Plant Species in Which
Endogenous Currents along Roots Have Been Reported

Plant species (common name)	Reference
Allium cepa (onion)	18
Arabidopsis thaliana	22
Biota orientalis	3
Cucumis sativus (cucumber)	23
Cucurbita pepo (pumpkin)	23
Hordeum vulgare (barley)	13
Lepidium sativum (cress)	12, 24
Limnobium stoloniferum	25
Nicotiana tobacum (tobacco)	26
Phaseolus vulgaris (bean)	19–21, 27
Phleum pratense (timothy)	28
Pisum sativum (pea)	23, 29
Raphanus sativus (radish)	30
Trifolium repens (white clover)	31
Triticum aestivum (wheat)	32, 33
Vicia faba (broad bean)	19, 23
Zea mays (maize)	3, 9, 10, 23, 34–46

fine details for the inward current entering the root apex, the overall pattern of net current leaving the older regions of the root and entering the younger, apical regions is conserved among different plant species. As an example, in a study on corn roots [36], the researchers measured an average inward current of 15 to 17 mA m⁻² entering the root cap and meristem regions. At the distal part of the elongation zone, the highest inward current density of 27 mA m⁻² was recorded. This current reversed to an outward current, which ranged from 1 to 15 mA m⁻² along the main elongation zone and more mature root regions. These electrical fields are symmetrical in vertically grown roots. However, as discussed below, these patterns vary under diverse endogenous and/or environmental stimuli.

The primary goal in the understanding of these electrical fields is to link the net current flow with the flux of a particular ion. Prior to the development of ion-selective electrodes, the ionic nature of these currents was inferred from studies in which changes in current direction and magnitude were associated with changes in the ionic composition of the surrounding medium. Changes in the mineral ion composition generally did not lead to changes in root-associated currents. However, increasing the pH of the medium led to small but rapid changes of current density and in some cases also caused a substantial shift of the main

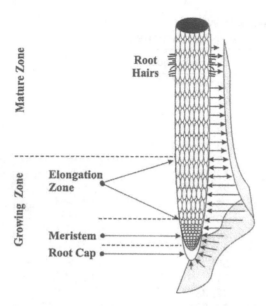

Figure 1 The endogenous current flux profile along the main root. The main and persistent current sink (i.e., a site of inward current) is located at the meristem and the apical elongating zone, while the main current source (i.e., the site of outward current) is localized in the basal elongating zone and root hair zone and probably extends up to the seed in young seedlings. The position and the magnitude of the fluxes are indicated by arrows, such that arrows directed toward the root indicate influx and arrows directed away from the root denote efflux. The scale of the flux magnitude is arbitrary.

site of inward current toward the root apex [24]. In contrast to the effect on the inward currents, the outward current in the root hair zone did not respond to pH changes of the external solution. In other studies, lowering the H^+ concentration of the medium 30-fold (from pH 5.6 to 7.1) reduced the inward current density by 50%. These types of observations led researchers to suggest an involvement of H^+ ions in the root-associated net currents described before. Experimental methods other than the vibrating voltage probe, such as the use of pH dye indicators used in conjunction with the electrophysiological techniques, also helped support the hypothesis that the natural currents in growing roots consist mainly of H^+ ions [e.g., 13].

The development of ion-specific microelectrodes for use in stationary and vibrating systems, in connection with diffusion analysis of the steady-state ion gradients, allowed further characterization of the role of certain ion fluxes (including H^+) in the endogenous currents mentioned earlier. Net K^+, H^+, Ca^{2+}, NH_4^+,

NO_3^-, and Cd^{2+} fluxes have been associated with localized regions of the root surface of several species [9,10,25,34,35,38,47–50], and the effects of other ions that influence root growth (e.g., Al^{3+}) on endogenous currents and various root ion-transport systems have been reported [25,32,33,51].

In pioneering studies, vibrating H^+-selective microelectrodes were used to measure H^+ fluxes into both intact and excised maize roots [9,10,34,38]. A large H^+ influx could readily be observed close to the root tip, and a strong H^+ efflux was observed along more mature root regions. Although the net H^+ fluxes fluctuated significantly with time, these observations suggested that H^+ fluxes could account, although not entirely, for the magnitude and direction of the net currents described before. Under the assumption that net H^+ efflux is due mainly to the activity of the plasma membrane H^+-ATPase, it might be expected that the net H^+ efflux could be variable because this transport system is involved in a number of different physiological processes (secondary transport systems, pH regulation, signal transduction, etc.).

3.2 Role of Endogenous Currents and H+ Fluxes in Root Growth, Development, and Tropisms

As mentioned earlier, plant roots exhibit a defined electric field pattern when grown in a vertical position. Currents flow symmetrically out of both sides of the root from the basal elongation and mature root zones and enter the more apical root regions (see Fig. 1). In an effort to elucidate the role of these endogenous currents in plant responses to the environment, some researchers have studied the deviations occurring from this current pattern upon application of a diverse range of stimuli. Figure 2 illustrates the net current patterns observed in some of the cases outlined in the following section, which summarizes the literature with regard to the role of H^+ and net ion currents in root function in response to environmental stimuli.

3.2.1 Cell Growth

Endogenous currents have been reported in several tip-growing model systems, such as pollen tubes, root hairs, algal rhizoids, and fungal hyphae [52–62]. All of these tip-growing cells exhibit an inward current localized to the growing tip that ranges from 10 mA m^{-2} in zygotes to 40 mA m^{-2} in pollen tubes. The presence of the localized inward current entering the tip region and outward current leaving the cells in regions behind the tip has led researchers to suggest that these endogenous currents may play a fundamental role in the growth process. Spatial and/or operational separation of membrane transport systems is generally believed to be the main cause of these transcellular endogenous currents.

When an external pH gradient is applied to algae, the zygote polarizes and the polar rhizoid growth is oriented along the pH gradient such that the rhizoid

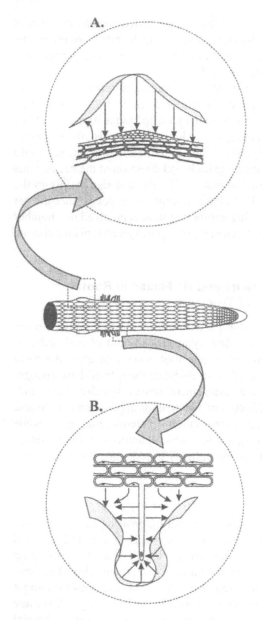

FIGURE 2 The endogenous current patterns along (A) emerging lateral roots and (B) root hairs. The position and the magnitude of the fluxes are indicated by arrows, such that arrows directed toward the root indicate influx and arrows directed away from the root denote efflux. The scale of the flux magnitude is arbitrary.

grows toward an acid source [62]. This response appears to be due to the regional cytoplasmic differentiation resulting from the generation of longitudinal cytoplasmic pH gradients (apical cytosol being slightly more acidic than that at the base of the cell) within the growing rhizoid cells. These types of studies have often tempted researchers to include the growing root in the extensive list of tip-growing organs and cells that generate endogenous currents. However, root elongation is not a true example of tip growth. In addition to the root being an elongating multicellular organ, root growth is a function of cells behind the tip. Nevertheless, roots do share some common features with tip-growing cells such as polarized elongation and defined spatial ion current patterns. Because the ion current studies with roots suggested that H^+ influx was associated with root growth, other studies [e.g., 25] have focused on measuring the magnitude and spatial localization of H^+ fluxes in growing and nongrowing root hairs, as a model for root cells. In root hairs, H^+ influx is localized almost exclusively at the growing tips and steadily decreases further back from the tip. At the root hair base, the H^+ influx turns into a large H^+ efflux. This pattern is similar to that described earlier for elongating roots, i.e., inward current at the growth zone and outward current at the nongrowing, subapical region. These observations for root hairs have led some researchers to hypothesize that as in true tip-growth systems, rhizosphere pH and H^+ currents may be involved in directing root growth. The extensive and controversial literature on the acid growth theory has been reviewed [63] and is beyond the scope of the present chapter (but see Chapter 6). However, some indirect evidence appears to support an association between root growth and endogenous H^+ currents. For example, root growth–stimulating treatments associated with decreases in rhizosphere pH (e.g., low pH or fusicoccin) lead to an increase in the apical inward current (at the root tip and the elongation zone), whereas root growth–inhibitory treatments [high pH or indoleacetic acid (IAA)] lead to a reduction in the magnitude of this current in corn roots [42]. Likewise, omission of the growth hormone IAA from the growth medium stimulated both the growth rate and the density of the basal (outward) and apical (inward) currents measured in young lateral roots of radish [30].

The correlation between the net current changes that take place at the root apex and simultaneous changes in root growth rates has led researchers to look more closely at the possibility that H^+ fluxes may play an active role in the process of cell elongation [37,40,64]. Ryan et al. [33] found a significant correlation between root growth and the magnitude of the H^+ fluxes along wheat roots. Exposing the root to extracellular aluminum (Al) inhibited net H^+ currents as well as the root growth rate. In about half of the roots examined, Al inhibition of H^+ influx occurred concurrently with root growth inhibition. However, in about 25% of the roots, Al-induced inhibition of root growth occurred before any inhibition of H^+ influx was apparent. Consequently, given that the correlation was not observed in every root examined, the authors concluded that there were sufficient grounds to reject the hypothesis that growth is directly dependent on the magnitude of current, at least in wheat roots.

As there is no evidence to disprove the possibility that H^+ fluxes are simply a response to tip growth or the result of enhanced metabolic activity in this region, the relation between H^+ fluxes and root growth is still poorly understood. Nevertheless, it has been repeatedly suggested that these endogenous currents might provide a polar axis around which differentiation and developmental processes could be linked.

3.2.2 Root Development

Specific endogenous currents associated with developing young roots as well as emerging lateral roots have also been studied in several plant species [26,29,30, 36,65]. It is worth noting that prior to any indication of lateral root development, the mature root regions from which the lateral root eventually emerges are associated with outwardly directed ion currents. Emerging lateral roots exhibit a different and distinct pattern of currents compared with that found in the mature root region from which laterals emerge. Outward currents have been reported to be associated with the small bulges on the corn root epidermis (prior to rupturing of tissue and emergence of the lateral root) [36]. In this study by Meyer and Weisenseel [36], the highest current density (about 21 mA m^{-2}) was observed above the bulge, immediately before the lateral root breaks through the surface of the primary root. At the tip of the emerging lateral roots that have just pushed their way through the primary root epidermis, a large inward current starts to develop. Thereafter, the inward current is associated with the root tip as the lateral root elongates. However, as the lateral root length increases further, the current density at its tip decreases (to about 5 mA m^{-2}). Subsequently, an outward current is observed associated with the basal region of the young lateral root, while an inward current starts developing and entering the elongation zone, meristem, and root cap, just as described before for primary roots. However, in other studies [e.g., 26,30], the currents associated with the small bulges on the epidermis were inward. For example, a peak inward current of 2.3 mA m^{-2} was recorded for this region in tobacco roots prior to the emergence of the lateral root. However, in all of these studies on lateral roots, after emergence and development they exhibit the same current pattern found in the primary root with the current density found in lateral roots generally being only about 20% of that found in primary roots.

3.2.3 Gravitropism

A significant portion of the research on root endogenous electrical fields and H^+ currents has focused on their role in gravitropic responses [12,24,28,44,45,66]. In these studies, the changes in ion current are followed from the currents measured on both sides of the vertically grown root and those measured on the upper and lower sides after the roots are positioned horizontally. Although in some cases the identity of the net current was not established, the correlation between the currents and the changes in the color of a pH indicator dye in the agar medium surrounding the root suggests that H^+ ions are likely to be one of the major ionic components of the current [12,13,24]. Within a short period of time (about 3 min) after horizontal orientation, the magnitude and direction of the currents along the root surface start to change [e.g., 12,24]. The most distinct feature is the change in current direction that takes place on the upper side of the root cap and meristematic regions of the root; in these regions the current changes from an inward current (in vertically growing roots) to an outward current (in horizontal roots). Furthermore, the current in these regions becomes asymmetrical, as the currents on the lower side of the root remain unchanged. The change in current direction is sometimes accompanied by an increase in current density. However, along the root elongation zone and the mature root, the magnitude and direction of the currents on both sides of the root remain unchanged. Within 30 min (for the root cap) and 2 h (for the root meristem), the currents at the upper side of the horizontally positioned root start to return to their original direction observed in vertically oriented roots. This reorientation of inward currents signals the end of the gravistimulated response. Comparisons of the electrical events with the time course of root curvature indicate that the changes in currents occur before a visible curvature. This suggests a possible role or function of the current in the events that lead to the gravitropic root growth response.

Other observations indicate that, in addition to H^+, other ions significantly contribute to the net current changes in gravistimulated roots. The gravitropic response in root is dependent on external Ca^{2+} [67,68]. Taking advantage of the fact that low external Ca^{2+} concentrations abolish the gravitropic response [69], endogenous currents were investigated along vertically and horizontally placed roots in which the gravitropic response was abolished by adding a Ca^{2+} chelator, EGTA, to the medium [24]. The current density for vertically growing roots in the medium without Ca^{2+} was significantly larger and the current pattern was drastically altered compared with those seen for roots in the Ca^{2+}-containing medium. While inward currents were measured at the root cap and in the basal elongating zone, outward currents dominated in the apical and central elongation zones. This pattern was opposite to the one observed for the Ca^{2+}-containing medium, where the main sink (i.e., inward current) was located along the root cap, meristem, and apical elongating zones while the sources of current (i.e., outward current) were associated with the basal elongating and mature root zones. Upon orienting the roots horizontally, the currents recorded in the medium without Ca^{2+} did not change. Given that only net currents were measured, it was not possible to determine the contribution of particular ions to the total current changes recorded. Consequently, the authors were unable to elucidate the cause of the changes in the endogenous current density and direction that take place at low Ca^{2+} concentrations. The authors suggest (based on the assumption that

H^+ ions are the main source of the currents recorded) that the H^+-ATPase is responding differently in different cells of the root surface to Ca^{2+} gradients along the root. Under such an assumption, differences in H^+ extrusion rates would result in altered patterns of net current. However, direct evidence is still missing and will be provided only by selectively measuring the fluxes for particular ions upon gravistimulation (i.e., using ion-selective microelectrodes).

3.2.4   Oscillatory Patterns—Root pH Homeostasis, Cell Growth, and Nutations

Although the preceding observations indicate that a characteristic "fingerprint" pattern exists for electrical fields around roots, several studies have indicated that at specific root locations, large variations in the magnitude and direction of root H^+ fluxes can be observed over time [19,27,37,39]. In some cases, H^+ fluxes at any given point along the roots surface can vary from a moderately large efflux to a zero net flux and often even to a moderate H^+ influx. Early studies established that these changes in net current magnitude and direction could occur rhythmically with time with a fairly constant period [19,27]. More recent studies have provided evidence for discrete oscillations in H^+ fluxes measured along the root surface [39,40]. The strong correlation between the changes in amplitude (i.e., magnitude/direction) of the current and the root growth rate and the fact the largest oscillations take place at the elongation zone have led some researchers to suggest an active role for these net ion currents in the cell elongation process [70,71]. The role of H^+ fluxes as part of the ionic basis for the oscillations of net current has also been established [37].

The oscillatory behavior of inward H^+ fluxes along the elongation zone of decapped roots was shown to consist of two superimposed components: a slow (period of about 90 min) and a fast (7 min) oscillation. Both oscillatory components are strongly dependent on the pH [37] and osmolarity [39] of the solution bathing the root. Although the physiological significance of these oscillations in current is not known, it is likely that they incorporate H^+ fluxes involved in several functional associations such as symport and/or antiport transport mechanisms coupled to the H^+-extruding ATPase, such that the resulting H^+ current loops could contribute to the dynamic and non–steady-state transport behavior of the root cell plasma membrane [10,26]. Thus, the oscillatory components just described may reflect different biophysical and biochemical mechanisms of root pH homeostasis: the fast and slow components could reflect H^+ fluxes associated with active and passive transport mechanisms, respectively. Consequently, the osmotic inhibition of the fast oscillatory component would be attributed to the suppression of active H^+ pumping.

There have also been some efforts to link the oscillatory patterns of current with different modes and sites of cell growth [24]. Weisenseel et al. [24] proposed that root growth processes dominated by increases in cytoplasmic volume require uptake of organic nutrients, such as sugars and amino acids, whose intake across the plasma membrane is coupled to H^+ influx. In contrast, growth processes dominated by turgor-driven cell elongation may require uptake of ions important for osmotic adjustment, such as K^+ and Na^+, and may thus be associated with the H^+ extrusion that drives uptake of these ions.

In addition to the well-defined tropisms, plant roots show another type of movement, the so-called circumnutations and micronutations (here abbreviated as CNs). These vertical (upward/downward) and lateral (sideways) movements are endogenously generated. Although CNs are largely independent of the root's gravitropic response, gravity does modify the qualitative parameters of CNs. Thus, CNs movements have been studied in decapped roots in an effort to remove the gravitropic response [40]. The meristem and the elongation zone of decapped roots show typical CN movements. However, the oscillation patterns of ion fluxes (see previous paragraphs) associated with the CNs are different for each of these two regions. Although there is a strong correlation between H^+ flux oscillations and root CNs found in the elongation region, there is no correlation between the ion flux behavior and the CNs observed along the meristematic region. Regardless of the novelty of these observations, it remains clear that there is still a lack of direct evidence for the understanding of the cause and role of these oscillations in currents along roots.

3.2.5 Nutrition: Ions, Sugars, and Amino Acid Uptake

The preceding observations have assumed that net H^+ fluxes measured along roots reflect almost entirely the plasma membrane H^+-ATPase activity and H^+ extrusion from the cytosol. However, these actively generated H^+ fluxes can also be functionally associated with other secondary transport systems. Active H^+ extrusion via the H^+-ATPase generates a large electrochemical H^+ gradient (the so-called proton motive force, pmf). As this pmf will be directed back across the plasma membrane into the cytosol, it can be used to drive the active transport of many other solutes against their own electrochemical gradient via H^+-coupled transport systems. These transporters catalyze ion or solute flux in the same (i.e., symporters or cotransporters) or opposite (antiporters) direction as the passive H^+ flux across the membrane.

Different experimental approaches can be used to characterize these H^+-coupled transport mechanisms. Although biochemical and membrane isolation techniques have been used to characterize uptake of radioisotope-labeled solutes, the following section will focus on transport mechanisms characterized in conjunction with net H^+ flux measurements. Electrophysiological approaches such as the use of intracellular microelectrodes to monitor changes in membrane potentials in the presence and absence of the substrate of interest have been pivotal for our understanding of some of these transporters. The quick depolarization of the membrane potential in the presence of the substrate of interest suggests the flow of positive (e.g., H^+) electrical charge into the cell. Such experimental approaches have been successfully used to characterize plasma membrane H^+/sugar [72–76] and H^+/amino acid [77–79] cotransporters. Membrane potential measurements have also suggested the existences of Cl^-/H^+ [80] and NO_3^-/H^+ [46] cotransport mechanisms. The use of ion-selective microelectrodes enabled researchers to investigate and establish an association between H^+ and NO_3^- influx in intact roots [35,41,81]. For these proton/anion cotransporters, an H^+/anion flux stoichiometry greater than one has been suggested, which would explain the depolarizing nature of these transporters. The development and implementation of molecular approaches have allowed researchers to clone the genes encoding many of these transporters and the expression of these transporters in heterologous systems. Electrophysiological research with these systems has confirmed the existence of H^+-coupled symporters for uptake of sugars, amino acids, and some mineral anions [82].

The literature on the role of H^+-coupled symport systems in K^+ uptake in roots is somewhat more controversial (for a recent review see Ref. 83). A well-characterized low-affinity K^+ channel operates at external K^+ concentrations higher than about 1 mM, allowing the flux of K^+ down its electrochemical gradients. However, the mechanistic nature of a coexisting high-affinity K^+ transporter is still unclear. Based on similarities to an H^+/K^+ symporter in *Neurospora* [84], researchers have suggested the existence of a similar mechanism in higher plants [34]. Studies with intact roots where H^+ and K^+ fluxes were measured simultaneously have not provided clear evidence for a high-affinity H^+-K^+ symporter in roots [9,10,34,38]. These studies showed that high-affinity root K^+ uptake was insensitive to changes in the pH of the external medium and exhibited extremely variable flux stoichiometries for K^+ and H^+, with K^+ uptake often being well in excess of H^+ efflux. Furthermore, when both fluxes (i.e., H^+ and K^+) were measured at specific locations along the root with pH and K^+ microelectrodes, the H^+ efflux showed a significant time-dependent variability over time, while the K^+ influx was fairly constant. These whole-root observations have suggested a lack of coupling between the high-affinity K^+ uptake and H^+ efflux, reinforcing the concept of a lack of such a cotransport mechanism in roots from higher plants.

Molecular studies of the root high-affinity K^+ transporter have also provided equivocal results. The original electrophysiological characterization of *HKT1*, a gene encoding a putative high-affinity K^+ transporter in wheat roots, suggested that this system was an H^+-K^+ cotransporter [85]. However, in a later report, the same authors suggested this system was a K^+-Na^+ cotransporter similar to that found in algae that grow in brackish water containing high levels of Na^+ [86]. Thus, this is a currently unresolved research topic.

3.3 Environmental Responses

3.3.1 Aluminum Tolerance

The rhizotoxic aluminum species, Al^{3+}, solubilizes from aluminosilicate clays into the soil solution at acidic pH values (pH 5 or below), accumulating to levels that inhibit root growth and crop production [87]. If the pH of the solution surrounding roots is increased, the solubility and toxicity of Al are decreased. This observation has encouraged conjecture on root Al tolerance mechanisms based on roots altering the rhizosphere pH, leading to an alkalization process that reduces the concentration of Al^{3+} in the rhizosphere. Several efforts have been made to elucidate the possible role of H^+ fluxes as part of such mechanisms [22,32,33,88]. Jones et al. [25] measured the effect of Al on the influx of H^+ localized at the tip of the giant root hairs from Limnobium stoloniferum, as a model system for studying effects of Al on root growth. An addition of external Al did not significantly affect either the magnitude or the spatial pattern of H^+ fluxes along root hairs over a short period of time. However, over longer periods of time, an apparent increase in H^+ influx was seen at the root tip, although not at more basal regions of the root hair. The magnitude of the H^+ influx increased with time. This Al stimulation of H^+ influx correlated with a swelling at the root hair tip, which eventually resulted in bursting of the root hair and loss of cellular content. No other morphological or H^+ flux changes were seen at any other region of the root hair. In contrast, corn roots treated with Al exhibited an increased H^+ efflux near the root apex and root cap compared with roots growing in the absence of Al [88].

In a detailed comparative study of Al-tolerant and Al-sensitive wheat cultivars, Miyasaka et al. [32] measured the rhizosphere pH and H^+ fluxes in the rhizosphere of apical and mature root zones using pH microelectrodes in the absence and presence of Al. These authors found no significant correlations between root Al tolerance, rhizosphere pH, and H^+ fluxes that could account for the exhibited Al tolerance. Ryan et al. [33] also compared the net ionic currents (based on measuring electric fields around roots) and the ionic composition of these currents (using H^+, K^+, Cl^-, and Ca^{2+} microelectrodes) for Al-tolerant and Al-sensitive wheat cultivars. The study focused on the effect of Al on the fluxes at the root apex, the site where Al toxicity takes place and where root endogenous currents have generally been linked to cell growth. A net inward current of 10 to 40 mA m^{-2} was associated with the apical root region, while a smaller net outward current was observed along the mature root regions. Given that in the absence of Al the growth rate of the root was strongly correlated with the magnitude of the net currents, the authors investigated the relation between the changes in these currents and the initiation of the inhibition of root growth induced by Al. Whereas external Al inhibited root elongation and the net current associated with it in the Al-sensitive cultivar, neither of these two parameters was altered in the Al-tolerant cultivars. Aluminum seemed to inhibit both the net currents and the growth rates of the sensitive cultivar in a similar manner. Further characterization of the ionic nature of the currents showed that H^+ influx was responsible for most of the net inward current entering the wheat root apex. A significant correlation, similar to the one described for the net currents, was found between root growth and the magnitude of the H^+ flux. As mentioned earlier (see Sec. 3.2.1), the Al inhibition of the H^+ fluxes and root growth was simultaneous in only 75% of the cases, leading the authors to reject the hypothesis that root growth is directly dependent upon the magnitude of current around the root. The authors suggested that the Al-induced decrease in ionic current is not the direct cause of the Al-induced inhibition of growth in roots.

Degenhardt et al. [22] performed similar comparative studies using an Al-tolerant *Arabidopsis thaliana* mutant. This *alr* mutant was shown to exclude Al from entering the root apex to a greater extent than the wild-type *Arabidopsis* seedlings. The authors compared the patterns and magnitude of the H^+ fluxes along the root surface of both the *alr* and wild-type seedlings in the absence and presence of external Al. In the absence of Al, the H^+ flux patterns for the mutant and wild types were similar, resembling those reported in other species (see Sec. 3.1), showing a strong H^+ influx into the root apical region, with the maximum influx occurring acropetally from the meristem region. The root cap and the more mature regions maintained a smaller net H^+ influx. When the roots were exposed to Al, the surface pH for the root apex of *alr* increased, while the root surface pH of the wild type remained unchanged. This pH increase resulted from the Al-inducible stimulation in H^+ influx at the root tip of *alr*, which was not found in the case of the wild type. The magnitude of the resulting increase in root apical rhizosphere pH was large enough to account for the Al tolerance observed in the *alr* mutant. The study did not investigate the mechanistic basis for this Al-induced change in root apical H^+ fluxes. Regardless of the exact mechanism underlying these differences, this was the first direct evidence of an Al tolerance mechanism mediated by alteration in root H^+ fluxes.

3.3.2 Root Wounding and Infection by Microorganisms

Although lateral root emergence could be considered a "self-induced" wounding of the mature root regions from which the laterals emerge, some attention has also been focused on determining the effects of mechanical wounding on the endogenous electrical fields and ion fluxes around roots [26,29,36]. In these studies, wounding was induced by excising wedge-shaped pieces of root tissue from the proximal elongation zone or by piercing the epidermis and cortex of the mature root region with a needle, taking care not to damage the stelar tissue. Prior to wounding, the proximal elongation zone and mature root regions exhibited the characteristic inward and outward currents described previously (see Sec. 3.2.2). Upon wounding, a large inward current (approximately 600 mA m^{-2}: about seven times larger than that measured prior to wounding) was observed associated with the wound site. These current densities were always significantly higher on the wounded side of the root than at the opposite side. The currents associated with other zones along the injured root remained unchanged. In some of these studies, ion-selective microelectrodes were used to characterize further the nature of the induced net current. The use of pH microelectrodes indicated a decrease in proton concentration near the wound site, suggesting an inwardly directed flow of protons at the site of the wound. Interestingly, H^+, K^+, Ca^{2+}, and Cl^- fluxes contributed only about 10% of the total wound-induced current. The inward H^+ current decreased with time after the wounding. Regardless of the nature and the direction of the endogenous field prior to wounding, it seems that mechanical wounding caused the stimulation of a large inward current. Although the exact nature of these wound-induced currents remains unknown, it has been suggested that a large part of the wound-induced current must be carried by a large mass flow of negatively charged molecules. This wound-induced current is clearly different from that observed in the natural self-induced wounding of the root cortex (i.e., development of lateral roots).

The fact that roots are electrically active and generate endogenous electrical fields has led researchers to attribute a relationship between the root's endogenous electrical fields and root infections by pathogenic microorganisms. For example, the pathogenic *Phytophthora* zoospores are electro- and chemotactic; that is, they respond to electric fields and chemical gradients. When a current is applied to a medium containing these zoospores, they migrate to the cathode [90]. In roots, these zoospores are also preferentially attracted to root zones with endogenous inwardly directed currents, such as root tips, emerging lateral roots, and root regions that are physically damaged [89,90]. Taken together, these observations have led some authors [26,31] to propose that electrotaxis may play a role in the attraction and subsequent invasion of roots by these microorganisms.

4 CONCLUSIONS

It has long been known that roots can dramatically alter the chemical makeup of their rhizosphere environment. One of the most important ways in which this can be accomplished is via changes in rhizosphere pH that are mediated by root H^+ transport processes. These transport processes include H^+ extrusion via the plasma membrane H^+-ATPase, and H^+ influx via coupled solute transport systems. Roots generate electric fields that are the product of ionic currents traversing root cell mambranes. Research conducted at the intersection of electrical fields and root ion transport processes has indicated that H^+ fluxes constitute a significant fraction of the ion currents that are the driving force for root-associated electric fields. There is considerable circumstantial evidence in the literature implicating these H^+ currents and their associated electrical fields in a number of root functions, including growth and development, gravitropism, root responses to abiotic and biotic stress, and mineral acquisition. However, the underlying mechanisms that link H^+ currents to these various root functions and processes still remain to be elucidated. A better understanding of the roles of these electrical phenomena and H^+ currents in root function will go a long way toward helping researchers understand how roots respond to their soil environment and facilitate changes in root-associated processes that allow plants to interact positively with the environment.

REFERENCES

1. PH Nye. Changes of pH across the rhizosphere induced by roots. Plant Soil 61:7–26, 1981.
2. TK Kim, WK Silk, AY Cheer. A mathematical model for pH patterns in the rhizospheres of growth zones. Plant Cell Environ 22:1–12, 1999.
3. J Muller-Hettlingen. Ueber galvanische erscheinungen an keimenden samen. Pfluegers Arch Gesamte Physiol Menschen Tiere 31:193–214, 1883.
4. D Cataldo, TR Garland, RE Wildung. Cadmium uptake in intact soybean plants. Plant Physiol 73:844–848, 1983.
5. GL Mullins, LE Sommers. Cadmium and zinc influx characteristics by intact corn (*Zea mays* L.) seedlings. Plant Soil 96:153–164, 1986.
6. WJ Lucas, LV Kochian. Ion transport processes in corn roots: an approach utilizing microelectrode techniques. In: WG Gensler, ed. Advanced Agricultural Instrumentation. Design and Use, NATO ASI series No 111. Dordrecht: Martinus Nijhoff, 1986, pp 402–425.
7. WM Kühtreiber, LF Jaffe. Detection of extracellular calcium gradients with a calcium-specific vibrating electrode. J Cell Biol 110:1565–1573, 1990.
8. PJS Smith, RH Sanger, LF Jaffe. The vibrating Ca^{2+} electrode: a new technique for detecting plasma membrane regions of Ca^{2+} influx and efflux. Meth Cell Biol 40:115–134, 1994.
9. LV Kochian, JE Shaff, WM Kühtreiber, LF Jaffe, WJ Lucas. Use of an extracellular, ion-selective, vibrating microelectrode system for the quantification of K^+, H^+, and Ca^{2+} fluxes in maize roots and maize suspension cells. Planta 188:601–610, 1992.
10. IA Newman, LV Kochian, MA Grusak, WJ Lucas. Fluxes of H^+ and K^+ in corn roots. Plant Physiol 84:1177–1184, 1987.
11. HT Gollany, TE Schumacher. Combined use of colorimetric and microelectrode methods for evaluating rhizosphere pH. Plant Soil 154:151–159, 1993.
12. HM Behrens, MH Weisenseel, A Sievers. Rapid changes in the pattern of electric current around the root tip of *Lepidium sativum* L. following gravistimulation. Plant Physiol 70:1079–1083, 1982.
13. MH Weisenseel, A Dorn, LF Jaffe. Natural H^+ currents traverse growing roots and root hairs of barley (*Hordeum vulgare* L.). Plant Physiol 64:512–518, 1979.
14. M Häussling, E Leisen, H Marschner, V Römheld. An improved method for non-destructive measurements of the pH at the root-soil interface (rhizosphere). J Plant Physiol 117:371–375, 1985.
15. AJ Gijsman. Rhizosphere pH along different root zones of Douglas-fir (*Pseudotsuga menziesii*), as affected by source of nitrogen. Plant Soil 124:161–167, 1990.

16. JWM Pijnenborg, TA Lie, AJB Zehnder. Simplified measurements of soil pH using an agar-contact technique. Plant Soil 126:155–160, 1990.
17. C Plassard, M Meslem, G Souche, B Jaillard. Localization and quantification of net fluxes of H⁺ along maize roots by combined use of pH-indicator dye videodensitometry and H⁺-selective microelectrodes. Plant Soil 211:29–39, 1999.
18. EJ Lund, WA Kenyon. Relations between continuous bioelectric currents and cell respiration. I. Electrical correlation potentials in growing root tips. J Exp Zool 48: 333–357, 1927.
19. BIH Scott, AL McAulay, P Jeyes. Correlation between the electrical current generated by a bean root growing in water and the elongation of the root. Aust J Biol Sci 8:36–46, 1955.
20. BIH Scott, DW Martin. Bioelectric fields of bean roots and their relation to salt accumulation. Aust J Biol Sci 15:83–100, 1962.
21. BIH Scott. Electric fields in plants. Annu Rev Plant Physiol 18:409–418, 1967.
22. J Degenhardt, PB Larsen, SH Howell, LV Kochian. Aluminum resistance in the *Arabidopsis* mutant alr-104 is caused by an aluminum-induced increase in rhizosphere pH. Plant Physiol 117:19–27, 1998.
23. SN Shabala, IA Newman. H⁺ flux kinetics around plant roots after short-term exposure to low temperature: identifying critical temperatures for plant chilling tolerance. Plant Cell Environ 20:1401–1410, 1997.
24. MH Weisenseel, HF Bectker, JG Ehlgötz. Growth, gravitropism, and endogenous ion currents of cress roots (*Lepidium sativum* L.). Plant Physiol 100:16–25, 1992.
25. DL Jones, JE Shaff, LV Kochian. Role of calcium and other ions in directing root hair tip growth in *Limnobium stoloniferum*. I. Inhibition of tip growth by aluminum. Planta 197:672–680, 1995.
26. AL Miller, E Shand, NAR Gow. Ion currents associated with root tips, emerging laterals and induced wound sites in *Nicotiana tabacum*: spatial relationship proposed between resulting electrical fields and phytophthoran zoospore infection. Plant Cell Environ 11:21–25, 1988.
27. BIH Scott. Feedback-induced oscillations of five-minute period in the electric field of the bean root. Ann N Y Acad Sci 98:890–900, 1962.
28. HE Zieschang, K Köhler, A Sievers. Changing proton concentrations at the surfaces of gravistimulated *Phleum* roots. Planta 190:546–554, 1993.
29. JM Hush, IA Newman, RL Overall. Utilization of the vibrating probe and ion-selective microelectrode techniques to investigate electrophysiological responses to wounding in pea roots. J Exp Bot 43:1251–1257, 1992.
30. KS Rathore, KB Hotary, KR Robinson, A two-dimensional vibrating probe study of currents around lateral roots of *Raphanus sativus* developing in culture. Plant Physiol 92:543–546, 1990.
31. AL Miller, JA Raven, JI Sprent, MH Weisenseel. Endogenous ion currents traverse growing roots and root hairs of *Trifolium repens*. Plant Cell Environ 9:79–83, 1986.
32. SC Miyasaka, LV Kochian, JE Shaff, CD Foy. Mechanisms of aluminum tolerance in wheat. Plant Physiol 91:1188–1196, 1989.
33. PR Ryan, JE Shaff, LV Kochian. Aluminum toxicity in roots. Plant Physiol 99: 1193–1200, 1992.

34. LV Kochian, JE Shaff, WJ Lucas. High affinity K^+ uptake in maize roots. Plant Physiol 91:1202–1211, 1989.
35. PR McClure, LV Kochian, RM Spanswick, JE Shaff. Evidence for cotransport of nitrate and protons in maize roots. II. Measurement of NO_3^- and H^+ fluxes with ion-selective microelectrodes. Plant Physiol 93:290–294, 1990.
36. AJ Meyer, MH Weisenseel. Wound-induced changes of membrane voltage, endogenous currents, and ion fluxes in primary roots of maize. Plant Physiol 114:989–998, 1997.
37. SN Shabala, IA Newman, J Morris. Oscillations in H^+ and Ca^{2+} ion fluxes around the elongation region of corn roots and effects of external pH. Plant Physiol 113: 111–118, 1997.
38. PR Ryan, IA Newman, B Shields. Ion fluxes in corn roots measured by microelectrodes with ion-specific liquid membranes. J Membr Sci 53:59–69, 1990.
39. SN Shabala, IA Newman. Osmotic sensitivity of Ca^{2+} and H^+ transporters in corn roots: effect on fluxes and their oscillations in the elongation region. J Membr Biol 161:45–54, 1998.
40. SN Shabala, IA Newman. Proton and calcium flux oscillations in the elongation region correlate with root nutation. Physiol Plant 100:917–926, 1997.
41. AR Taylor, AJ Bloom. Ammonium, nitrate, and proton fluxes along the maize root. Plant Cell Environ 21:1255–1263, 1998.
42. AL Miller, NAR Gow. Correlation between root-generated ionic currents, pH, fusicoccin, indoleacetic acid, and growth of the primary root of *Zea mays*. Plant Physiol 89:1198–1206, 1989.
43. AL Miller, NAR Gow. Correlation between profile of ion-current circulation and root development. Physiol Plant 75:102–108, 1989.
44. JM Versel, PE Pilet. Distribution of growth and proton efflux in gravireactive roots of maize (*Zea mays* L.). Planta 167:26–29, 1986.
45. DA Collings, RG White, RL Overall. Ionic current changes associated with the gravity-induced bending response in roots of *Zea mays* L. Plant Physiol 100:1417–1426, 1992.
46. PR McClure, LV Kochian, RM Spanswick, JE Shaff. Evidence for cotransport of nitrate and protons in maize roots. I. Effects of nitrate on the membrane potential. Plant Physiol 93:281–289, 1989.
47. M Piñeros, J Shaff, LV Kochian. Development, characterization, and application of a cadmium-selective microelectrode for the measurement of cadmium fluxes in roots of *Thlaspi* species and wheat. Plant Physiol 116:1393–1401, 1998.
48. G Henriksen, DR Raman, LP Walker, RM Spanswick. Measurement of net fluxes of ammonium and nitrate at the surface of barley roots using ion specific microelectrodes. II. Patterns of uptake along the root axis and evaluation of the microelectrode flux estimation technique. Plant Physiol 99:734–747, 1992.
49. JW Huang, DL Grunes, LV Kochian. Aluminum effects on the kinetics of calcium uptake into cells of the wheat root apex. Planta 188:414–421, 1992.
50. PR Ryan, LV Kochian. Interaction between aluminum toxicity and calcium uptake at the root apex of near-isogenic lines of wheat (*Triticum aestivum* L.) differing in aluminum tolerance. Plant Physiol 102:975–982, 1993.
51. JW Huang, DL Grunes, LV Kochian. Aluminum effects on calcium fluxes at the

root apex of aluminum-tolerant and aluminum-sensitive wheat cultivars. Plant Physiol 98:230–237, 1992.

52. R Nuccitelli, LF Jaffe. Spontaneous current pulses through developing fucoid eggs. Proc Natl Acad Sci USA 71:4855–4859, 1974.

53. KR Robinson, LF Jaffe. Polarizing fucoid eggs drive a calcium current through themselves. Science 187:70–72, 1975.

54. MH Weisenseel, LF Jaffe. The major growth current through lily pollen tubes enters as K^+ and leaves as H^+. Planta 133:1–7, 1976.

55. MH Weisenseel, R Nuccitelli, LF Jaffe. Large electrical currents traverse growing pollen tubes. J Cell Biol 66:556–567, 1975.

56. LF Jaffe. Electrical currents through the developing *Fucus* egg. Proc Natl Acad Sci USA 56:1102–1109, 1966.

57. R Nuccitelli. Ooplasmic segregation and secretion in the *Pelvetia* egg is accompanied by a membrane generated electrical current. Dev Biol 62:13–33, 1978.

58. RF Stump, KR Robinson, RL Harold, FM Harold. Endogenous electrical currents in the water mold *Blastocladiella emersonii* during growth and sporulation. Proc Natl Acad Sci USA 77:6673–6677, 1982.

59. BL Armbruster, MH Weisenseel. Ionic currents traverse growing hyphae and sporangia of the mycelial water mold *Achlya debaryana*. Protoplasma 115:65–69, 1983.

60. NAR Gow, DL Kropf, FM Harold. Growing hyphae of *Achlya bisexualis* generate a longitudinal pH gradient in the surrounding medium. J Gen Microbiol 130:1–8, 1984.

61. A Dorn, MH Weisenseel. Growth and the current pattern around internodal cells of *Nitella flexilis* L. J Exp Bot 35:373–383, 1984.

62. BC Gibbon, DL Kropf. Cytosolic pH gradients associated with tip growth. Science 263:1419–1421, 1994.

63. H Lüthen, M Bigdon, M Böttger. Reexamination of the acid growth theory of auxin action. Plant Physiol 93:931–939, 1990.

64. WS Peters, HH Felle. The correlation of profiles of surface pH and elongation growth in maize roots. Plant Physiol 121:905–912, 1999.

65. S Hamada, S Ezaki, K Hayashi, K Toko, K Yamafuji. Electric current precedes emergence of a lateral root in higher plants. Plant Physiol 100:614–619, 1992.

66. A Iwabuchi, M Yano, H Shimizu. Development of extracellular electric pattern around *Lepidium* roots: its possible role in root growth and gravitropism. Protoplasma 148:94–100, 1989.

67. M Friedmann, BW Poovaiah. Calcium and protein phosphorylation in the transduction of gravity signal in corn roots. Plant Cell Physiol 32:299–302, 1991.

68. R Moore, IL Cameron, NKR Smith. Movement of endogenous calcium in the elongation zone of graviresponding roots of *Zea mays*. Ann Bot 63:589–593, 1989.

69. A Miyazaki, K Kobayashi, S Ishizaka, T Fujii. Redistribution of phosphorus, sulfur, potassium and calcium in relation to light-induced gravitropic curvature in *Zea* roots. Plant Cell Physiol 27:693–700, 1986.

70. K Toko, M Souda, T Matsuno, K Yamafuji. Oscillations of electrical potential along a root of a higher plant. Biophys J 57:269–279, 1990.

71. B Hecks, Z Hejnowicz, A Sievers. Spontaneous oscillations of extracellular electrical potentials measured on *Lepidium sativum* L. roots. Plant Cell Environ 15:115–121, 1992.

72. JH Xia, P Saglio. H^+ efflux and hexose transport under imposed energy status in maize root tips. Plant Physiol 93:453–459, 1990.

73. DR Bush. Proton-coupled sugar and amino acid transporters in plants. Annu Rev Plant Physiol Plant Mol Biol 44:513–542, 1993.

74. F Lichtner, RM Spanswick. Electrogenic sucrose transport in developing soybean cotyledons. Plant Physiol 67:869–874, 1981.

75. H Felle, FW Bentrup. Hexose transport and membrane depolarization in *Riccia fluitans*. Planta 147:471–476, 1980.

76. H Felle, JP Gogarten, FW Bentrup. Phlorizin inhibits hexose transport across the plasmalemma of *Ricca fluitans*. Planta 157:267–270, 1983.

77. B Etherton, B Rubinstein. Evidence for amino acid–H^+ co-transport in oat coleoptiles. Plant Physiol 61:933–937, 1978.

78. H Felle. Stereospecificity and electrogenicity of amino acid transport in *Riccia fluitans*. Planta 147:471–476, 1980.

79. D Sanders, CL Slayman, ML Pall. Stoichiometry of H^+/amino acid co-transport in *Neurospora crassa* revealed by current-voltage analysis. Biochim Biophys Acta 735: 67–76, 1983.

80. H Felle. The H^+/Cl^- symporter in root-hair cells of *Sinapis alba*. An electrophysiological study using ion-selective microelectrodes. Plant Physiol 106:1131–1136, 1994.

81. AL Miller, GN Smith, JA Raven, NAR Gow. Ion currents and the nitrogen status of roots of *Hordeum vulgare* and non-nodulated *Trifolium repens*. Plant Cell Environ 14:559–567, 1991.

82. B Michelet, M Boutry. The plasma membrane H^+-ATPase: a highly regulated enzyme with multiple physiological functions. Plant Physiol 108:1–6, 1995.

83. LV Kochian. Molecular physiology of mineral nutrition acquisition, transport and utilization. In: B Buchanan, W Gruissem, R Jones, eds. Biochemistry & Molecular Biology of Plants. Rockville, MD: American Society of Plant Physiologists, 2000, pp 1207–1222.

84. A Rodriguez-Navarro, MR Blatt, CL Slayman. A potassium-proton symport in *Neurospora crassa*. J Gen Physiol 87:649–674, 1986.

85. DP Schachtman, JI Schroeder. Structure and transport mechanism of a high-affinity potassium uptake transporter from higher plants. Nature 370:655–658, 1994.

86. F Rubio, W Gassmann, JI Schroeder. Sodium-driven potassium uptake by the plant potassium transporter HKT1 and mutations conferring salt tolerance. Science 270: 1660–1663, 1995.

87. LV Kochian. Cellular mechanisms of aluminum toxicity and resistance in plants. Annu Rev Plant Physiol Plant Mol Biol 46:237–260, 1995.

88. RJ Bennet, CM Breen, MV Fey. The effects of aluminum on root cap function and root development in *Zea mays* L. Environ Exp Bot 27:91–104, 1987.

89. JL Troutman, WH Wills. Electrotaxis of *Phytophthora parasitica* zoospores and its possible role in infection of tobacco by the fungus. Phytopathology 54:225–228, 1984.

90. PD Dukes, JL Apple. Chemotaxis of zoospores of *Phytophthora parasitica* var. *nicotianae* by plant roots and certain chemical solutions. Phytopathology 51:195–197, 1961.

13

Role of pH in Availability of Ions in Soil

Zdenko Rengel
University of Western Australia, Perth, Western Australia, Australia

1 INTRODUCTION

Roots growing in soil are immersed in the soil solution, a dynamic and complex ionic environment from which they draw water and nutrients but may encounter toxic ions as well. Nonessential ions (e.g., Al, As, Cd, Cr, Hg, and Pb) as well as micronutrients and beneficial elements [cations: Fe, Cu, Zn, Mn, and Ni; anions: $B(OH)_4^-$, selenate SeO_4^-, and selenite SeO_3^-] may become toxic to plants when present in sufficiently high concentrations or activities in the soil solution.

The soil-plant system is open to inputs (fertilizers, pollutants, etc.) and losses (erosion, leaching, etc.). Metal ions are released into soil solution directly through the weathering of soil minerals and the decomposition of organic matter as well as from the ion-exchange processes. Concentrations or activities of metal ions in the soil-plant system depend on soil chemical and physical processes, relevant aspects of which are described here.

2 FUNDAMENTAL ASPECTS OF SOIL CHEMISTRY

2.1 The Concept of Ion Availability and Ion Activity

Available ions are those that roots are able to take up in a given time. Strictly speaking, only ions in soil solution would be considered available; however, due to a dynamic equilibrium that exists between the soil solution and other ion pools from which ready transfer into the soil solution can occur, ions adsorbed onto exchange sites can also be considered available (or at least controlling the available soil solution fraction).

Although a definition of ion availability is relatively unambiguous, there is a great deal of uncertainty in the analytical methods available for measuring available nutrients. Chemical extractants used for the purpose frequently do not yield results that are comparable to those of bioassays done with live plants; therefore the term "extractable" rather than "available" should be used in connection with amounts of ions extracted with a particular chemical extractant. Similarly, although solubility and availability are clearly related (ions have to be in soil solution to be taken up by roots), the term "soluble ions" should be reserved for situations where soil solution is being studied and the term "available ions" for cases in which actual uptake of ions by plant roots is being studied.

The assumption that plants respond to concentrations rather than activities of ions in the soil solution may be a relatively reasonable assumption for the mass flow–supplied ions [1,2] or for those for which roots exhibit neither preference nor selectivity. However, for diffusion-supplied ions or those taken up into plant roots by some sort of carrier mechanism (most nutrient and toxic ions would fall into this category), activities rather than concentrations are the realistic expression of chemical potential. Even though accurate determination of ion activities in a medium as complex as a soil solution is practically impossible [3], the concept of ion activities in soil solution has gained wide support and yielded useful approaches to the plant-soil system, mainly due to the widespread use of computer speciation models (e.g., GEOCHEM [4]) capable of simultaneous and comprehensive estimation of the equilibrium states for most ionic species occurring in a given soil solution. This chapter therefore emphasizes ion activities rather than concentrations as a measure of soluble ions in the soil solution and thus of availability of a particular ionic species. An attempt has been made here to elucidate some of the processes that govern ion desorption and solubility and influence availability of ions, both nutrient and toxic, the distinction between which frequently relies only on activity of a particular ion in the soil solution.

2.2 Intensity and Capacity

Factors affecting ion supply to plant roots are ion activity in the soil solution (usually referred to as intensity) and the degree and the rate of replenishment of the soil solution ion pool from other pools (ions adsorbed on solid soil particles or labile organic compounds and ions present in other readily soluble compounds), usually referred to as capacity. The capacity factor therefore determines buffer power for a particular ion [5]. The relationship between capacity and intensity factors for each ion is heavily dependent on pH.

2.3 Cation-Exchange Capacity

Cation-exchange capacity (CEC) refers to the sum of negative sites on the soil solid phase that can bind cations and therefore protect them from being leached but keep them in a state that will allow replenishment of cations depleted from the soil solution. The CEC can be measured in a $BaCl_2$ solution buffered at pH 8.2 or at field pH; the latter case yields effective CEC (ECEC) (frequently determined as a sum of extractable cations). Because the value of CEC is dependent on the pH of the measurement solution [6], it may be difficult to interpret the CEC values if there is a large difference between the field pH and the pH of the measuring solution.

Clay particles are negatively charged and therefore surrounded by a swarm of positively charged ions. Clay minerals owe a part of their negative charge to isomorphous substitution (cations of higher charge, such as Al^{3+}, are substituted by those of lower charge, such as Mg^{2+}), thus leaving a surplus of (non-neutralized) negative charges that are balanced by adsorbed, exchangeable cations [7,8]. In addition, soil colloids (e.g., humus, hydrous oxides) that exhibit protonated complexing functional groups ($-OH$, $-COOH$) contribute to the soil CEC.

$$M_1^+ + M_2X \leftrightarrow M_2^+ + MX \tag{1}$$

(solution) (solid) $\leftrightarrow$ (solid) (solution)

Because dissociation of $-OH$ and $-COOH$ groups (especially those on organic matter) is pH dependent, CEC increases with an increase in pH [9–12]. Such an increase can be substantial; an increase between 5 and 90 mmol (+) kg^{-1} was measured for an increase of one pH unit in a range of soils in Queensland, Australia [13], and an increase of 50 mmol (+) kg^{-1} was reported in the Oxisols [12]; these increases may represent about doubling of the ECEC capacity in some soils. With an increased CEC, metal cations are attracted to these negative sites on solid soil particles, which may result in a greater capacity to replenish metal cations in the soil solution and thus positively influence plant growth (e.g., increased retention of K resulted in increased corn yield because leaching of K was minimized [12]). In addition, increased CEC may cause increased binding of toxic cations onto negative charges on the solid soil particles and decrease availability (and thus toxicity) of these cations in the soil solution, e.g., Cd [14,15]. In contrast, with increased acidity, there is a decrease in CEC [16] and a decrease in the capacity of solid soil particles to adsorb cations. As a consequence, greater amounts of cations (especially basic cations such as Ca^{2+} and Mg^{2+}) are present in soil solution and are therefore prone to leaching into deeper layers of the soil profile [16,17] and eventually out of reach of roots. With increasing acidity, clay minerals are irreversibly weathered, and Mn and Al progressively mobilized, with Al occupying up to 70% of the cation-exchange complex in some situations [16]. In contrast, liming of acidic soils results in an increase in ECEC, with extra cation-exchange sites generally neutralized by Ca^{2+} binding [11].

It should be borne in mind that plant roots possess cation-exchange capacity as well [18]. Therefore, one cation exchanger (root) is immersed into the other (soil), resulting in competition for cations based on physicochemical principles. Plant capacity to take up cations is at least partly influenced by the root CEC [19] because the ionic environment at the plasma membrane surface, where uptake of all ions takes place, is influenced by the strength of the electrical field developed in the interaction of two charged surfaces (root and soil particles). However, the capacity of plant roots to take up some ions across the plasma membrane more than others would influence ion concentrations at the exchanger surfaces and would thus influence physicochemical interactions happening at the plasma membrane surface. These interactions are important not just for nutrient uptake but also for ion toxicity [20–23].

2.4 Retention of Cations in Soils

Cations are held more strongly (i.e., less reversibly) on the soil solid particles when pH increases from 5 to 7; in other words, Cu, Zn, Ni, Cd, and other metals become significantly less soluble and less exchangeable when pH increases from 5 to 7. Retention of metals in soil can occur through several processes [7,8]; (1) cation exchange (nonspecific adsorption), (2) specific adsorption, (3) organic complexation, and (4) coprecipitation.

An adsorption by cation exchange represents electrostatic binding through the formation of outer-sphere complexes with the surface functional groups. An outer-sphere complex means that at least one molecule of a solvent comes between the functional group and the ion.

Specific adsorption is pH dependent and related to the hydrolysis of the metal ion. In the specific adsorption, partly covalent bonds are formed with the ions within the crystal lattice of the clay minerals. The partly covalent bonds are inherently stronger than the electrostatic binding involved in the nonspecific cation exchange (e.g., Zn can be sorbed on Fe and Al oxides 7 and 26 times more strongly, respectively, than their corresponding cation-exchange affinities at the pH 7.6 would imply [7]). The metals most able to form hydroxy complexes are specifically adsorbed to the greatest extent, e.g., Hg > Pb > Cu $\gg$ Zn > Co > Ni > Cd. Specific adsorption may also include diffusion of metals into mineral interlayer spaces [24] and their fixation there. Such diffusion increases with an increase in pH [25].

Organic matter may either increase or decrease the availability of micronutrients, Al, and heavy metals [11,14,26–28]. Reduced availability is due to complexation with humic acids, lignin, and other organic compounds of high molecular weight (insoluble precipitates are thus formed). On the other side, increased ion availability may result from solubilization and thus mobilization of metals by the low-molecular-weight organic ligands (e.g., short-chain organic acids, amino acids, and other organic compounds).

Coprecipitation represents formation of mixed solids by simultaneous precipitation, i.e., Fe and Mn oxides [7,8,29]. These coprecipitation reactions can be facilitated by microbes [30].

To maintain electroneutrality, negative charges on the soil solid particles (soil colloids) are balanced by cations; an exchange refers to the exchange between ions neutralizing the surface negative charge on the soil colloids and the ions that are in soil solution. Such an exchange is reversible, stoichiometric, diffusion controlled, and with a certain degree of selectivity of the adsorbent. The higher the valence of an ion, the greater its replacing power (H^+ is an exception in having a large replacing power, thus behaving like a polyvalent cation); the greater the hydration, the lower its replacing power (other things being equal) because water molecules provide a sort of insulating shield [7,8].

3 SOIL ACIDIFICATION

Soil acidification is a slow, continuous natural process resulting in acid soils being common in areas where soil development continued for long, geological periods of time and under climatic conditions in which rainfall exceeds evapotranspiration. The process of soil acidification is aided by water leaching base cations to lower (subsoil) horizons; acid soils are therefore common in wet tropics [31,32]. Also, internal production of organic acids through decomposition of organic litter can acidify soils, especially forest soils of the temperate zones of the earth. There are other soil-related acidifying processes: nitrification (nitrate-producing) reactions, oxidation of reduced S compounds, leaching of NO_3^-, etc. as well as those related to plant activity (greater uptake of cations than anions, etc.). Human activities may intensify and speed up the acidification process [32]. Soil acidification resulting from either continuous use of ammonia- and amide-containing fertilizers or the nitrogen fixation by legumes worsens Al toxicity and contributes to an increase in the soil area affected. Acid rain can also initiate and/or speed up the soil acidification processes. For more detailed accounts of plant and soil mechanisms underlying soil acidification, see reviews [33,34].

The widely accepted ameliorative measure for acid soils is liming [32]. Spreading of lime or other Ca-containing material and incorporating it into the topsoil effectively increases soil pH. The situation is clearly more complicated with the subsoil because incorporation of lime into deeper horizons is technically difficult and economically infeasible. Alternatively, the surface-applied liming material may in some cases allow sufficient leaching of Ca^{2+} down the profile to provide effective amelioration of subsoil acidity [35]; this approach works effectively only in light-textured, sandy soils.

4   THE pH IN THE SOIL-PLANT SYSTEM

The pH represents a measure of the H^+ activity in the soil solution that is in a dynamic equilibrium with a negatively charged solid phase. The H^+ ions are strongly attracted to these negative sites and have power to replace other cations from them. The diffuse layer in the vicinity of a negatively charged surface has higher H^+ activity than the bulk soil solution. However, the pH value most relevant to soil and plant chemical processes is the pH of the soil solution. For example, the pH of the soil solution is a better predictor of corn yields than the soil pH [36].

The measurement of pH of the soil solution shows the most immediate impact that pH can have on ion availability and plant growth, even though it cannot predict the extent of buffering of soil solution pH. It is important that the soil solution pH is measured immediately upon displacement of the soil solution [37] because equilibration with external CO_2 (which is at lower partial pressure in the air than in soils) results in changes in pH. In contrast, formation of carbonic acid as a consequence of respiration by roots and soil microflora and fauna plays only a minor role in determining soil pH [38].

Despite the obvious importance of soil solution pH, the pH of soil is usually measured in a slurry of soil and extractant (water, $CaCl_2$, or KCl) because extraction of soil solution is a tedious process. The soil pH measured in the soil-extractant mixture, even though a useful index, should be regarded as purely empirical.

Soil pH is a dynamic parameter, with significant spatial [39] and temporal differences [40]. Diurnal fluctuations of as much as one pH unit may occur, as well as season-to-season variations. During seasons with low to moderate rainfall (evapotranspiration exceeds precipitation), salts are not being removed by deep percolation; increased salts tend to reduce pH by forcing more of the exchangeable H^+ ions into the soil solution. Conversely, during wet seasons, salts are removed from the topsoil and the pH goes up. This season-to-season fluctuation in the total salt content should not be confused with a long-term effect occurring over decades and centuries when increased rainfall leaches basic cations from the soil, causing acidification.

Greater spatial differences in pH exist in the soil under vegetation than in bare soil [41], mainly because of root capacity to change the soil pH. Storing soil, even for a relatively short time (e.g., for only 20 days), can result in a significant decrease in the pH of alkaline soils [42], probably because of acid generation arising from organic matter decomposition.

4.1   pH Effects on Distribution of Plant Species in Ecosystems

Soil acidity is one of the major factors regulating the species composition in the natural ecosystems as well as limiting crop production in agriculture. Only a few plant species grow equally well in both the acidic (pH $\leq$ 5) and the alkaline range (pH > 7). In addition to pH, soluble Al concentration or activity is another important factor in determining species distribution. In general, calcicole species (e.g., *Mycelis muralis* and *Geum urbanum*) are sensitive and calcifuge species insensitive to Al (e.g., dicots: *Calluna vulgaris*, *Digitalis purpurea*, and *Galium harcynicum*; monocots: *Avenella flexuosa* and *Juncus squarrosus*) [43].

Lucerne, barley, beans, and sugar beet are acid sensitive and need to be grown on neutral or mildly acid soils. Tolerance to acid soils would be greater in wheat, red clover, peas, and vetches and even more so in white clover, many grasses, oats, rye, lupins, and potatoes. The most tolerant to acid soils are soybean, millets, sorghum, Sudan grass, and sweet potatoes [38]. However, it should be borne in mind that genotypes within each crop species differ in tolerance to acid soils and Al; the preceding listing should therefore be taken only as a rough general guide.

For a long time it was thought that differential growth of plant species in acid and alkaline soils was due to differential plant sensitivity to H^+ ion. However, solution culture experiments have shown that plants can grow vigorously between pH 4 and 8. Only if the pH is as low as 3 are roots definitely injured (strong inhibition of root growth occurs at pH 3.8–4.0 for dicots and 3.0–3.5 for monocots in solution containing 1 mM Ca); if the pH is 9, root growth is repressed and plants cannot take up any phosphate.

It is interesting to note that root growth is restricted at both low and high pH (corresponding to high and low activities of H^+). Therefore, H^+ may be considered a beneficial ion because its relative deficiency (e.g., pH increased from an optimum of 5.5 to 6.0–7.5) caused reduction of root growth, disintegration of the root surface, and impairment of the root hair formation in sensitive species (e.g., lupins [44]).

4.2 The Relationship between pH and Ion Availability

Soil pH has a dominant effect on the solubility and therefore availability and potential phytotoxicity of ions. Whereas low pH favors free metal cations and protonated anions, higher pH favors carbonate or hydroxyl complexes. Therefore, availability of the micronutrient and toxic ions present in soil solution as cations increases with increasing soil acidity, and availability of those present as anions [MoO_4^{2-}, CrO_4^{2-}, SeO_4^-, SeO_3^-, and $B(OH)_4^-$] increases with increasing alkalinity.

Increased availability of Al^{3+} and Fe^{2+} and lower availability of the essential base cations Ca^{2+}, Mg^{2+}, and K^+ cause growth problems in acid soils. In addition, phosphate availability in acid soils is low because of the formation of sparingly soluble Al and Fe phosphates.

The concentration of metal cations in soil solution increases with an increase in acidity for two reasons [45]:

1. Adsorbed ions (e.g., Zn) are released by H^+ from exchangeable sites into the soil solution

$$\text{Soil-Zn} + 2H^+ \leftrightarrow Zn^{2+} + \text{Soil-2H} \tag{2}$$

$$\log\{Zn^{2+}\} = 5.8* - 2pH \tag{3}$$

2. Soil minerals containing mineral ions (Mn^{2+}, Al^{3+}, Fe^{2+}, . . . , M^{n+}) are being dissolved

$$MnO_2 + 4H^+ + 2e^- \leftrightarrow Mn^{2+} + 2H_2O \tag{4}$$

$$Al(OH)_3 + 3H^+ \leftrightarrow Al^{3+} + 3H_2O \tag{5}$$

$$\text{Soil-}M^{n+} + nH^+ \leftrightarrow M^{n+} + \text{soil-}H_n^+ \tag{6}$$

With decreased pH there is increased leaching of metal cations (because an increased portion of the total amount is present in the soil solution) and even deficiency may occur (e.g., Zn deficiency on acid soils) [46]. Sometimes, the solubility of metals at high pH may be greater than expected from the simple pH relationships of solubility of relevant soil minerals because formation of the metal–organic ligand complex may result in higher availability of metal than would be expected simply because of pH (e.g., Zn [45,47]). However, the amount of Zn extracted by various extractants to determine "available" Zn may not show any relationship with the soil pH [48].

Although metal cation availability generally decreases with an increase in pH (e.g., Cu [49,50]; Mn [51], Zn [50,51], Ni [50], Cd [50,52,53]; Pb [54,55]), the diethylenetrinitrilopentaacetic acid (DTPA)-extractable measure of available Zn, Cu, Cd, and Pb actually increased with an increase in pH [56]. Such an anomaly may be due to a relatively large amount of organic matter.

Soil solution concentrations of Cd and Zn decreased dramatically when pH increased above 5.5 and 5.1, respectively [57]. However, at neutral and alkaline pH chloro-Cd complexes dominated Cd speciation and were related positively to Cd uptake by potato tubers [58]. Thus, liming, with its pH-increasing effects and the general fixation of Cd on the soil cation-exchange complex [53], is not a guarantee at all of decreased Cd content in potato tubers [59]. Therefore, factors other than pH have to be taken into account when determining ion availability because some complexes may have greater availability than free ions.

Increasing soil pH caused a decline in availability of P and K [51,60,61]. Conversely, decreasing soil pH resulted in a marked increase in K and P leaching losses from the sandy soil [62]. However, if there are substantial amounts of Fe

---

* The empirically determined thermodynamic stability constant for reaction (3) is 5.8. It is clear from the equation that the solubility of Zn increases 100-fold with each unit decrease in pH.

and Al oxides in the soil, acidification causes a decrease in available P, but binding of P to the oxide surfaces increases rather than decreases P leaching [63].

Availability of ammonium and conversion to ammonia, and thus possible ammonia toxicity to plants, are influenced by the soil pH. Similarly, increasing the pH in the nutrient solution increased the risk of ammonia toxicity to cereal plants due to an increase in the ammonia partial pressure in the growing medium [64]. Soil pH could increase in the vicinity of the fertilizer band in a slightly acidic soil [65,66]; such a pH increase could lead to a high ammonia pressure in soil.

The process of hydrolysis of urea in slightly acidic soil is [67]

$$(NH_2)_2CO + 2H_2O = (NH_4)_2CO_3 \tag{7}$$

The enzyme urease can break urea in a similar manner; it has maximal activity at pH 7.0, increasing to pH 8.0 upon binding of the enzyme to hydroxyapatite, a common Ca-P mineral in alkaline soils [68].

Volatilization of ammonia occurs at soil pH above neutral. Because addition of urea frequently results in an increase in soil pH, ammonia volatilization from urea-amended soil may be significant [69]. However, not only pH but also soil CEC can determine the extent of ammonia loss through volatilization, as an increase in CEC resulted in a decrease in the net loss of ammonia, even at the relatively high soil pH of 8.0 [70].

4.3 Rhizosphere

Plant growth is dependent on availability of water and nutrients in the rhizosphere, the soil-root interface consisting of a soil layer varying in thickness from 0.1 to a few millimeters depending on the length of root hairs. Availability of nutrients in the rhizosphere is controlled by the combined effects of soil properties, plant characteristics, and the interactions of plant roots with microorganisms and the surrounding soil [71–73].

Chemical conditions in the rhizosphere are usually very much different from those in the bulk soil farther away from roots. Root-induced changes in the rhizosphere pH are a result of the balance between H^+ and HCO_3^- (OH^-) excretion, the evolution of CO_2 by respiration, and the excretion of various organic compounds collectively known as root exudates [72]. Corn plants fed ammonium-N acidified the rhizosphere soil, whereas those fed nitrate-N had a rhizosphere with an increased pH [74,75]. Excretion of OH^- or HCO_3^- as a consequence of relatively large amounts of nitrate taken up can result in a pH increase in the rhizosphere [76].

The balance between excretion of H^+ and HCO_3^- (OH^-) depends on the cation/anion uptake ratio. Greater net excretion of H^+ that accompanies greater absorption of cations than anions results in rhizosphere acidification; the reverse occurs when uptake of anions exceeds that of cations, and therefore net excretion of HCO_3^- exceeds that of H^+ [77]. The form of nitrogen is most influential in determining the cation/anion ratio: ammonium-fed plants take up more cations than anions, and they usually have a more acidic rhizosphere than the bulk soil, whereas nitrate-fed plants take up more anions than cations and show the opposite relationship between the rhizosphere and the bulk soil pH [78–82]. The difference between the rhizosphere pH of the ammonium-fed and nitrate-fed plants was as large as 1.9 pH units with an initial soil pH of 5.2 and as small as 0.2 unit when the soil pH prior to N application was 7.8 [83]. It should, however, be borne in mind that the effects on rhizosphere pH are also dependent on plant species as well as the genotype of the particular species [73,84]. There was no significant acidification in the rhizosphere of ammonium-fed wheat [85] and barley [86] in calcareous soil (80% $CaCO_3$ and pH 8.5). However, decreases of up to 2–3 pH units in the rhizosphere of ammonium-fed peach trees [*Prunus persica* (L.) Batsch] growing in alkaline calcareous soils were recorded [87]. In addition, the pH changes in the rhizosphere depend on the buffering capacity of soils, bulk soil pH, nitrogen sources, and other factors [88,89].

The rhizosphere soil, being generally more acidic than the bulk soil, can significantly increase the solubility and mobility of many metal ions, thus increasing their activities in the soil solution to the toxic range. Aluminum solubility increased several orders of magnitude in the rhizosphere soil due to increased acidity (e.g., pH of rhizosphere dropped to 4.4), while the bulk soil remained at the almost neutral value (pH = 6.8) and therefore free from any measurable quantities of soluble Al [90]. Also, rhizosphere acidification has an important role in mobilizing soil Mn [79,86,91,92].

Exudation of H^+ and organic acids from roots can lower the pH of alkaline soil and thus increase micronutrient availability [93–95]. Availability of inorganic P also increased upon acidification of the rhizosphere [96–98], but no effect was observed on the solubility and availability of organic P [96]. Other authors also showed that organic ligands exuded by plant roots have a differential effect on the availability of different ions [99], depending on the chemical characteristics of the metal-ligand complex.

Because availability of Mn is low at neutral to alkaline pH, acidification of the rhizosphere has an important role in mobilizing soil Mn [79]. The importance of organic acids (e.g., malic and citric) in mobilizing Mn [92] remains unclear because their effectiveness in forming stable complexes with micronutrients is low at high pH [100,101], where Mn deficiency usually occurs.

Dicot plants under Fe deficiency acidify the rhizosphere [102,103] and thus increase the availability of Fe and other micronutrient cations. Such acidification is due to nonspecific H^+/cation antiport mediated by the plasma membrane H^+-ATPase [103]. Availability of micronutrients can also be increased by ammonium-N nutrition due to rhizosphere acidification [87].

With an increase in pH, more Zn is adsorbed onto solid soil particles [95,104], making Zn unavailable to plants. However, at relatively higher soil pH where Zn and other metallic cations would be adsorbed, organic ligands (citrate, oxalate, and other organic acid anions) decrease adsorption of cations onto solid soil particles. Conversely, when pH is low and the metallic cations would be free in soil solution and thus either toxic to plants or subject to leaching losses, organic ligands increase adsorption of cations onto soil particles, e.g., Zn [105]. Breakdown of organic ligands by soil microflora may minimize the impact of these ligands on the adsorption of metallic cations [106].

Various plant species influence chemical and biochemical properties in the rhizosphere in a different way when grown in the same soil [107], indicating the specificity of the plant-rhizosphere interactions. Plant species that are capable of exuding the "right" ligand in a given situation (to lower potential toxicity or to alleviate deficiency of a given ion) show better adaptation and have, in the long run, evolutionary advantages over the other species [108].

Because plant roots change chemical and biological characteristics of the rhizosphere soils, and the availability of ions in the rhizosphere soil directly determines the accessibility of these ions to plants, bulk-soil measurements of soil characteristics cannot provide better than just a rough ballpark figure regarding the ion availability to plants [73]. This is true regardless of whether the extractants used really do mimic well the capacity of plants to extract ions from soil.

5 SOIL pH AND ION TOXICITIES

Decreasing pH influences ion availability [43]: at pH ≤ 5.5 Mn oxides solubilize and release Mn^{2+} into soil solution; at pH < 4.2 Al ions may become dominant ions in the soil solution; at pH < 3.8 Fe becomes the dominant ion, with H^+, Fe, and Al as the main exchangeable ions; and at pH < 3.2 H^+ and Fe ions are the main exchangeable ions.

The pH dependence of metal toxicity is a complex phenomenon: increased soil acidity releases more metal ions into soil solution, i.e., increases availability and therefore potential phytotoxicity of these metal ions; however, increased activity of H^+ ions (i.e., lower pH) decreases adsorption onto and absorption into plant roots of metal cations by noncompetitive inhibition, thus effectively reducing their accumulation and toxicity. As a result, an increase in accumulation of metal ions in soil-grown plants upon acidification is always less than could be expected considering the increased availability of a particular metal alone. Strictly speaking, such an effect can be called amelioration of metal toxicity by H^+ and has especially been studied in the case of Al.

At low pH, the relatively high activity of H^+ ions results in competition with other cationic metals for uptake, e.g., with Mg [20,109] and Zn [110]. As a consequence, it has been reported frequently that uptake of metal cations decreases with increasing acidity. The situation is a bit more complicated in the soil because of increased solubilization of metallic complexes with decreasing pH, so the competition between H^+ and other metal cationic species is not so prominent. Nevertheless, it has been generally accepted that decreasing pH (i.e., increasing H^+ activity) results in partial alleviation of Al^{3+} toxicity by increasing competition between H^+ and Al^{3+} ions [111].

5.1 Aluminum

Aluminum is the most abundant metal and the third most abundant element in the earth's crust (total Al is approximately 8% of the earth's crust). Most of the Al, however, is incorporated in aluminosilicate soil minerals; very small quantities appear in soluble forms capable of influencing living organisms. Solubilization of Al-containing minerals is enhanced in acidic environments, making Al the most yield-limiting factor in many acid soils throughout the world, possibly affecting about 40% and perhaps up to 70% of the world's arable land that is potentially usable for food and biomass production. Due to influences related to industrial and agricultural development, soil acidity and therefore toxicity of Al are likely to expand in time and space [112].

In acidic sandy soils of many European countries (e.g., Belgium, Denmark, The Netherlands, and northwestern Germany), base saturation is low and the rate of basic cation weathering does not increase with the rate of acid input. Any additional strong acid deposition in these soils is almost exclusively neutralized by Al solubilization. The source of mobilized Al is mostly free non–silicate-bound Al [43]. In addition to hydrolysis of Al, which provides most of the buffering capacity in soils, other factors (e.g., deprotonation of organic groups and variable charge minerals, i.e., CEC) are involved [113].

Different soil solution Al concentrations exist for the same pH in different soils (depending on the parent material). In general, Al toxicity is particularly severe at pH < 5 but may occur in kaolinitic soils with pH as high as 5.5 [114].

5.1.1 Ionic Speciation of Al

The crystal (ionic) radius of Al^{3+} is 0.051 nm, considerably smaller than that of commonly encountered trivalent metal ions (e.g., Fe 0.064 nm). High ionic charge and small crystalline radius of Al^{3+} combine to yield a level of chemical activity unmatched by any other solute found in soil solution. The distribution of Al among its various physicochemical forms (ionic speciation) is widely accepted to regulate the biogeochemistry of Al, i.e., mobility, transportability, bioavailability, and ultimately toxicity of Al to terrestrial and aquatic organisms. In aqueous solutions Al^{3+} forms hexahydronium ion (octahedral formation) surrounded by six water molecules. A high charge/size ratio of an ion may cause rupture of the H—O bonds in the coordinated water molecules. As pH increases, protons are removed from the coordinated water molecules, giving rise to several hydrolytic species.

$$Al^{3+}(H_2O)_6 \leftrightarrow Al(OH)^{2+}(H_2O)_5 + H^+ \leftrightarrow \tag{8}$$

$$\leftrightarrow Al(OH)_2^+(H_2O)_4 + H^+ \leftrightarrow \tag{9}$$

$$\leftrightarrow Al(OH)_3^0(H_2O)_3 + H^+ \leftrightarrow \tag{10}$$

$$\leftrightarrow Al(OH)_4^-(H_2O)_2 + H^+ \leftrightarrow \tag{11}$$

$$\leftrightarrow Al(OH)_5^{2-}(H_2O) + H^+ \tag{12}$$

Successive hydrolysis reactions occur in solutions of successively higher pH because the sink for H^+ increases with increasing pH. An addition of an acid shifts the equilibrium to the left (increasing the positive charge of the Al complexes), whereas an addition of a base shifts it to the right (decreasing the positive charge). The Al^{3+} ion (coordinated water molecules are generally omitted for clarity) is predominant below pH 4.7, $Al(OH)^{2+}$ exists only over a narrow pH range, $Al(OH)_2^+$ predominates between 4.7 and 6.5, $Al(OH)_3^0$ between pH 6.5 and 8.0, and $Al(OH)_4^-$ above pH 8.0. The $Al(OH)_5^{2-}$ occurs only at pH values above those common to soils.

Aluminum species may be monomeric (those containing only one Al atom in the complex) and polymeric (complexes or aggregations of any sort containing more than one Al atom). Polynuclear Al species are frequently formed when either pH or the concentration of total Al increases. Although techniques for differentiating between mononuclear and polynuclear Al species as well as among various monomeric Al species in a bulk solution exist [115–118], there is no experimental technique that can estimate either of these two principal groups of Al species in the root free space, i.e., in the vicinity of the plasma membrane where the multitude of Al-related effects were expected to occur [112].

There is little doubt that Al^{3+} is toxic to plants, but the extent of toxicity of other Al species and the determination of relative toxicities of various hydroxy-Al ionic species remain elusive to direct experimental procedures. A number of difficulties exist with respect to determining relative toxicities of hydroxy-Al species [112,119,120]:

1. Some Al species coexist in the solution.
2. Calculation of ionic activities of individual species is based on equilibrium data that are, at best, uncertain.
3. The possible existence of the toxic Al_{13} [$AlO_4Al_{12}(OH)_{24}(H_2O)_{12}^{7+}$] [121] as well as an ameliorative effect of H^+ per se [23,111,122] may cause significant errors in attributing toxicity to other ionic species.
4. Identity and activities of Al species at the root surface (inaccessible to direct measurements) are uncertain but are expected to be different from those in the bulk solution.

Important inorganic complexes of Al include phosphate, sulfate, fluoride, and silica complexes, none of which is considered toxic to plants [119,123–130], although fluoride-Al complexes may be toxic to *Avena sativa* and *Lycopersicon esculentum* grown in solution [131]. Organic complexes of Al are generally nontoxic [132–138], with citric and oxalic acids being stronger ligands for Al than lactic, acetic, or malic acid.

5.2 Ion Toxicities Influenced by the Redox Equilibria and pH

In addition to soil pH, redox reactions (controlled by the aqueous free-electron activity) influence the solubility and thus the availability and potential toxicity of some metal ions (Fe and Mn) [139,140], and Ag, As, Cr, Cu, Hg, and Pb can also be affected to a certain degree [8,141]. In contrast to Mn^{4+}, Mn^{3+}, and Fe^{3+} that are plant unavailable, Mn^{2+} and Fe^{2+} are plant-available forms that are more prevalent under the reduced (electron-rich) conditions that exist in most poorly aerated and ill-drained soils [8,92]. Low pH is frequently associated with such conditions.

5.2.1 Iron Toxicity

The effect of redox potential is less for Fe than for Mn, and overall solubility of Fe compounds is lower than that of Mn complexes. As a consequence, the pH needs to drop to lower values for Fe to become toxic [45]. Consequently, toxicity of Fe may occur only in very acid soils (pH < 3.2) or soils contaminated with Fe-containing waste materials. Reduced conditions as well as lowering of pH favor conversion of nonavailable (nontoxic) Fe^{3+} into plant-available Fe^{2+} ion [142,143], which is toxic if present in high ionic activity. Increasing soil pH by application of lime alleviates Fe toxicity [144].

5.2.2 Manganese Toxicity

Manganese toxicity occurs when pH is ≤5.5 if sufficient total Mn is present in the soil and also at higher pH in poorly drained soils where reducing conditions prevail [92,145]. In the case of Mn toxicity, redox potential is more important than just pH. The expression (pE + pH) (pE = −log of free-electron activity) is frequently used to characterize the soil with respect to solubilization of Mn compounds [values around 16 are associated with relatively high solubility of Mn^{2+}, the reverse applies to (pE + pH) values around 20]. It is therefore hard to predict which soils and under what conditions are going to be Mn toxic [45].

In addition to chemical reduction of Mn, which is enhanced under the pH and pE conditions just described, microbial reduction of Mn^{4+} and Mn^{3+} to the Mn^{2+} form is extremely important in soils [92,146–148] as well as the groundwater [149].

Manganese toxicity is intensified by N fertilization that lowers pH, fumigation that kills (oxidative) microorganisms, air drying, or flooding (for references see Ref. 92). Oxidation of Mn^{2+} to the Mn^{4+} form renders Mn plant unavailable and therefore nontoxic. Conversely, under reduced conditions (e.g., upon flooding), water-soluble Mn increases significantly within 1 to 2 weeks, then drops as $MnCO_3$ precipitates.

The chemistry of Mn in soils of high pH where differences in availability of Mn occur is not clear at present [92]. In aerated soils, Mn^{2+} concentration in soil solution should theoretically decrease 100-fold for every unit of pH increase [5]. However, with various organic compounds capable of complexing Mn and changing solubility equilibria, a decrease in Mn^{2+} concentration with an increase in pH is not that severe, e.g., 5- to 10-fold decrease for a 0.5-unit pH increase [150]. The complexity of the relationship between Mn concentration in soil solution and pH was illustrated by Rule and Graham [151], who used ^{54}Mn to determine soil Mn pools. With an increase in pH, soil Mn pools under *Trifolium repens* actually increased, while pools in soils planted to *Festuca elatior* decreased. So the soil supply of Mn is a complex variable that depends not only on soil chemistry but also on plant responses as well as activity of microorganisms.

Flax is so efficient in taking up Mn that it suffers from Mn toxicity even when grown on the calcareous soil with pH 8.1, which is Mn deficient for most plants [152]. An exceptional ability of flax plants to create reduced conditions and/or induce low pH in the rhizosphere is considered responsible for the high efficiency of Mn uptake.

Banksia attenuatta trees (family Proteaceae) possess a high capacity to accumulate Mn (up to 300 μg g^{-1} leaf dry matter) even when growing in Mn-deficient soils (<2 μg Mn g^{-1} soil in the DTPA extract) [153], probably because they acidify the rhizosphere or exude Mn-chelating compounds to increase Mn availability. The exact mechanisms have yet to be researched [92].

5.3 Acid Sulfate Soils

In acid sulfate soils, acidic conditions result from natural acidification processes, such as oxidation of pyritic materials [154]. These soils occur worldwide in most climatic zones (e.g., between 67,000 and 130,000 ha are present in Finland [155]) but are most prevalent in tropical regions [156,157]. The pH is generally below 4 and often as low as 2.

In tidal marshes, brackish and saline swamps, as well as mangrove swamps, acid sulfate soils may develop with formation of FeS_2 (pyrite) in a flooded (anaerobic, reduced) stage (pyrite forms from sulfate in seawater and iron in sediments) and pyrite conversion into H_2SO_4 in an oxidized stage (upon draining of such soils) [17,158]. Upon reflooding, the soil pH increases [159]. In addition to marshes and swamps, the development of acid sulfate soils is frequently hastened by flooding-drainage cycles involved in wetland rice production in southeastern Asia and west Africa [157].

The plant growth stress factors associated with acid sulfate soils are similar to those in other acid soils except for intensity: acid sulfate soils usually have lower pH. Therefore, direct toxicity of H^+ may be prominent together with Al^{3+}, Fe^{2+}, and sulfide toxicities [154,156,157].

5.4 Heavy Metal Toxicity

The term "heavy metals" is commonly used in the literature to describe all metals with atomic number > 20, generally excluding the alkali metals and alkali earths [160]. Even though very loosely defined, the term is still widely used by biologists because of lack of a more suitable alternative. Heavy metals/metalloids are Ag, As, Au, Cd, Cu, Cr, Co, Hg, Mn, Mo, Ni, Pb, Se, Sb, Tl, U, V, and Zn. Metalloids (underlined) do not form cations in solution. In contrast, heavy metals except Mo are cationic (even Cr when added as chromate anion cannot stay in such an oxidized state and thus converts to Cr^{3+}) [161].

Rocks of different types vary considerably in their mineralogical and elemental composition. The heavy metal concentrations inherited from the soil parent material are modified by pedogenic and biogeochemical processes; by natural inputs such as dust particles from soil, rocks, and volcanic ash; and by anthropogenic inputs, i.e., pollution [162,163].

The theoretical framework required for successful tackling of adsorption-desorption and solubility-precipitation of heavy metals in soils is still inadequate [162,164–166] (see also Sec. 5.1.1). The nature of heavy metals at soil surfaces is still somewhat obscure, and appropriate stability constants are mostly lacking. However, the recent intensive research in these areas should soon allow better understanding of physical, chemical, and biochemical reactions governing heavy metal solubility and mobility in a complex soil-plant-microbe system [166–169]. At present, researchers frequently resort to using theoretical models and computer simulations [164,170,171] to gain insight into processes governing solubility, mobility, and ionic speciation and, ultimately, toxicity of heavy metals.

The heavy metals or metalloids that frequently occur as environmental pollutants are Cd, Pb, Zn, Ni, Sb, and Bi. There are anthropogenic sources of heavy metal pollution (agricultural chemicals and waste products used as soil amendments, urban pollution, factory and power station emissions, in addition to natural sources such as windblown dust, volcanoes, and bushfires) [169,172,173]. Soils can be enriched in heavy metals to the point of toxicity if they are close to natural metalliferous ore outcrops or as a result of mining, smelting, or other industrial activities [163,174]. An application of municipal or other sewage sludge and other types of waste as well as polluted fumes from either waste incineration or fossil fuel burning (including emissions from car exhausts) may increase the heavy metal burden in soil. In the case of Cd, phosphorus fertilizers contain certain amounts of Cd that are therefore regularly and deliberately deposited on agricultural soils [162,172]. Similarly, some other heavy metals (e.g., Zn, Cu, Pb, and As) are components of other agricultural chemicals regularly applied to soil.

The extent of transfer of metals from soil to plants depends on the plant species, the particular metal involved, and the soil chemical characteristics [162, 166,169,175,176]. In contrast to Zn or Cu toxicity, which affects plant growth considerably [92], some metals (e.g., Cd, Pb, and Tl) can accumulate in vigorously growing, healthy-looking plants without an apparent effect [166,169,172]; however, due to such an accumulation, the concentrations of these heavy metals in plants (if in edible parts) may be above those considered desirable for human intake.

Similarly to the Al, Mn, and Fe toxicities already discussed, soil pH is the most important factor in determining heavy metal solubility and therefore availability [164] and potential toxicity to soil microflora, fauna, and plants. Soil microflora appears to be more sensitive to heavy metals than animals or crop plants [177]. Heavy metal toxicity will be prevalent on acid soils and less frequent or absent on calcareous alkaline soils or heavy metal–polluted soils that had been limed [174,178].

Cadmium adsorption by a range of soil types increased significantly with an increase in pH [53,179,180]. Cadmium reversibly adsorbed by Fe-Mn oxides increased when pH increased [181]. Therefore, acidic soils that contain less Fe and Mn oxides capable of adsorbing Cd will have greater Cd availability. Indeed, dissolution of Mn oxides resulted in the release of previously sorbed Cd, thus increasing Cd availability [165].

5.5 Mine Spoils

When mine spoils (tailings) that contain pyrite or other sulfides are exposed to air, a local acidification may occur due to oxidation of pyrite to sulfuric acid [182,183], resulting in acid mine drainage (solutions from such spoils may be as acidic as pH 2 or lower). When such drainage enters the watercourses, severe toxicity to aquatic biota may occur [184]. When the supply of reduced S is exhausted, the sulfuric acid is gradually leached away and pH reverts to higher values. However, complete S oxidation may take at least a decade and is dependent on the content of reduced S, moisture, and temperature conditions; during the time S is being oxidized and acidity produced, plant growth is inhibited and erosion can be severe, with effects outlasting adverse effects of acid production for decades [185]. However, some plant species appear to be relatively tolerant to acidic mine spoils (e.g., *Berberis repens* Lindl. [186] and woolgrass (*Scirpus cyperinus*) [187]), thus speeding up reclamation of such spoils. Moreover, when acid mine drainage solutions were treated with lime, the resulting effluent could be used for irrigation of crops, with significant increases in yield occurring in comparison with rainwater irrigation [188].

5.6 Ion Toxicities Occurring at Alkaline pH

Selenium is considered to be a beneficial element in some instances. It mimics and replaces S in biological compounds [189]. Selenate, SeO_4^-, exists in alkaline and well-oxidized environments, and selenite, SeO_3^-, occurs in neutral and mildly oxidized environments. Oxidized, high-pH environments are conducive to Se oxidation to selenite and eventually selenate [190]. In addition, increasing pH in soil results in decreased adsorption of Se [191], thus making it more available to plants. Increased accumulation of selenate in plants growing on alkaline, Se-rich soils [192] may result in direct plant toxicity [193]. In addition, livestock grazing on plants containing high levels of Se may suffer from poisoning in the form of alkali disease and blind staggers. Examples of soils high in Se (seleniferous soils) where these problems with livestock may occur can be found in the major grazing range in the central and southwestern United States.

6 CONCLUSIONS

The pH is a truly master variable in determining the solubility and availability of nutrient elements as well as toxic ions in soil. Relatively small changes in soil pH can increase availability of some chemical elements enough to result in a change from an element being an essential nutrient to the same element becoming a toxic ion. As farming expands into more marginal areas, including those with extremes of pH, greater emphasis on managing soil pH will be needed. Similarly, soil pH management is going to play a crucial role in minimizing entry of heavy metals into crops and pastures and thus into the human food chain.

REFERENCES

1. Z Rengel, DL Robinson. Modeling magnesium uptake from an acid soil. I. Net nutrient relationships at the soil-root interface. Soil Sci Soc Am J 54:785–791, 1990.
2. Z Rengel. Mechanistic simulation models of nutrient uptake: a review. Plant Soil 152:161–173, 1993.
3. G Sposito. The future of an illusion: ion activities in soil solutions. Soil Sci Soc Am J 48:531–536, 1984.
4. DR Parker, WA Norvell, RL Chaney. GEOCHEM-PC—a chemical speciation program for IBM and compatible personal computers. In: RH Loeppert, AP Schwab, S Goldberg, eds. Chemical Equilibrium and Reaction Models. Madison, WI: Soil Science Society of America, 1995, pp 253–269.

5. SA Barber. Soil Nutrient Bioavailability. A Mechanistic Approach. 2nd ed. New York: John Wiley & Sons, 1995.
6. D Curtin, HPW Rostad. Cation exchange and buffer potential of Saskatchewan soils estimated from texture, organic matter and pH. Can J Soil Sci 77:621–626, 1997.
7. HL Bohn, BL McNeal, GA O'Connor. Soil Chemistry. 2nd ed. New York: Wiley-Interscience, 1985.
8. KH Tan. Principles of Soil Chemistry. 3rd ed. New York: Marcel Dekker, 1998.
9. MR Bakker, C Nys, JF Picard. The effects of liming and gypsum applications on a sessile oak (*Quercus petraea* (M.) Liebl.) stand at La Croix-Scaille (French Ardennes). I. Site characteristics, soil chemistry and aerial biomass. Plant Soil 206: 99–108, 1998.
10. RA Falkiner, CJ Smith. Changes in soil chemistry in effluent-irrigated *Pinus radiata* and *Eucalyptus grandis* plantations. Aust J Soil Res 35:131–147, 1997.
11. Z Hochman, DC Edmeades, E White. Changes in effective cation exchange capacity and exchangeable aluminium with soil pH in lime-amended field soils. Aust J Soil Res 30:177–187, 1992.
12. DJ Hunter, LGG Yapa, NV Hue. Effects of green manure and coral lime on corn growth and chemical properties of an acid Oxisol in Western Samoa. Biol Fertil Soils 24:266–273, 1997.
13. RL Aitken, PW Moody, T Dickson. Field amelioration of acidic soils in south-east Queensland: I. Effect of amendments on soil properties. Aust J Agric Res 49:627–637, 1998.
14. K Kalbitz, R Wennrich. Mobilization of heavy metals and arsenic in polluted wetland soils and its dependence on dissolved organic matter. Sci Total Environ 209: 27–39, 1998.
15. ML Fernandes, MM Abreu, F Calouro, MC Vaz. Effect of liming and cadmium application in an acid soil on cadmium availability to sudangrass. Commun Soil Sci Plant Anal 30:1051–1062, 1999.
16. L Blake, KWT Goulding, CJB Mott, AE Johnston. Changes in soil chemistry accompanying acidification over more than 100 years under woodland and grass at Rothamsted Experimental Station, UK. Eur J Soil Sci 50:401–412, 1999.
17. NV Golez, K Kyuma. Influence of pyrite oxidation and soil acidification on some essential nutrient elements. Aquacultural Eng 16:107–124, 1997.
18. Z Rengel, DL Robinson. Determination of cation exchange capacity of ryegrass roots by summing exchangeable cations. Plant Soil 116:217–222, 1989.
19. Z Rengel. Net Mg^{2+} uptake in relation to the amount of exchangeable Mg^{2+} in the Donnan free space of ryegrass roots. Plant Soil 128:185–189, 1990.
20. Z Rengel, DL Robinson. Competitive Al^{3+} inhibition of net Mg^{2+} uptake by intact *Lolium multiflorum* roots. I. Kinetics. Plant Physiol 91:1407–1413, 1989.
21. FPC Blamey, DC Edmeades, DM Wheeler. Role of root cation-exchange capacity in differential aluminum tolerance of *Lotus* species. J Plant Nutr 13:729–744, 1990.
22. WJ Horst. The role of the apoplast in aluminium toxicity and resistance of higher plants: a review. Z Pflanzenernaehr Bodenkd 158:419–428, 1995.
23. DL Godbold, G Jentschke. Aluminum accumulation in root cell walls coincides

with inhibition of root growth but not with inhibition of magnesium uptake in Norway spruce. Physiol Plant 102:553–560, 1998.

24. LG Wesselink, JJMv Grinsven, G Grosskurth. Measuring and modeling mineral weathering in an acid forest soil, Solling, Germany. Proceedings of a symposium sponsored by Division S 5:91–110, 1992.

25. DC Bain, A Mellor, MJ Wilson. Nature and origin of an aluminous vermiculite weathering product in acid soils from upland catchments in Scotland. Clay Miner 25:467–475, 1990.

26. M Conyers. The control of aluminium solubility in some acidic Australian soils. J Soil Sci 41:147–156, 1990.

27. MTF Wong, RS Swift. Amelioration of aluminium toxicity with organic matter. In: RA Date, NJ Grundon, GE Rayment, ME Probert, eds. Plant-Soil Interactions at Low pH. Dordrecht: Kluwer Academic Publishers, 1995, pp 41–45.

28. LM Shuman, Z Li. Amelioration of zinc toxicity in cotton using lime or mushroom compost. J Soil Contam 6:425–438, 1997.

29. DC Golden, CC Chen, JB Dixon, Y Tokashiki. Pseudomorphic replacement of manganese oxides by iron oxide minerals. Geoderma 42:199–211, 1988.

30. ME Akhtar, WI Kelso. Electron microscopic characterization of iron and manganese oxide/hydroxide precipitates from agricultural field drains. Biol Fertil Soils 16:305–312, 1993.

31. HR Von Uexkull, E Mutert. Global extent, development and economic impact of acid soils. Plant Soil 171:1–15, 1995.

32. Z Rengel, ed. Handbook of Soil Acidity. New York: Marcel Dekker, 2001.

33. NS Bolan, MJ Hedley, RE White. Processes of soil acidification during nitrogen cycling with emphasis on legume based pasture. Plant Soil 134:53–63, 1991.

34. C Tang, Z Rengel. Role of plant cation/anion uptake ratio in soil acidification. In: Z Rengel, ed. Handbook of Soil Acidity. New York: Marcel Dekker, 2001, in press.

35. MG Whitten, MTF Wong, AW Rate. Amelioration of subsurface acidity in the south-west of Western Australia: downward movement and mass balance of surface-incorporated lime after 2–15 years. Aust J Soil Res 38:711–728, 2000.

36. PW Moody, RL Aitken, T Dickson. Field amelioration of acidic soils in south-east Queensland: III. Relationships of maize yield response to lime and unamended soil properties. Aust J Agric Res 49:649–656, 1998.

37. SE Lorenz, RE Hamon, SP McGrath. Differences between soil solutions obtained from rhizosphere and non-rhizosphere soils by water displacement and soil centrifugation. Eur J Soil Sci 45:431–438, 1994.

38. EJ Russell. Soil Conditions and Plant Growth. 10th ed. London: Longman, 1973.

39. A Gottlein, A Heim, E Matzner. Mobilization of aluminium in the rhizosphere soil solution of growing tree roots in an acidic soil. Plant Soil 211:41–49, 1999.

40. PJ Gregory, P Hinsinger. New approaches to studying chemical and physical changes in the rhizosphere: an overview. Plant Soil 211:1–9, 1999.

41. TC Zhang, H Pang. Applications of microelectrode techniques to measure pH and oxidation-reduction potential in rhizosphere soil. Environ Sci Technol 33:1293–1299, 1999.

42. KP Prodromou, AS Pavlatou-Ve. Changes in soil pH due to the storage of soils. Soil Use Manage 14:182–183, 1998.

43. B Ulrich, ME Sumner, eds. Soil Acidity. Berlin: Springer-Verlag, 1991.

44. C Tang, J Kuo, NE Longnecker, CJ Thomson, AD Robson. High pH causes disintegration of the root surface in *Lupinus angustifolius* L. Ann Bot 71:201–207, 1993.

45. ME Sumner, MV Fey, AD Noble. Nutrient status and toxicity problems in acid soils. In: B Ulrich, ME Sumner, eds. Soil Acidity. Berlin: Springer-Verlag, 1991, pp 149–182.

46. PN Takkar, CD Walker. The distribution and correction of zinc deficiency. In: AD Robson, ed. Zinc in Soils and Plants. Dordrecht: Kluwer Academic Publishers, 1993, pp 151–165.

47. A Fotovat, R Naidu, ME Sumner. Water:soil ratio influences aqueous phase chemistry of indigenous copper and zinc in soils. Aust J Soil Res 35:687–709, 1997.

48. E Arriechi, R Ramirez. Soil test for available zinc in acid soils of Venezuela. Commun Soil Sci Plant Anal 28:1471–1480, 1997.

49. LA Brun, J Maillet, J Richarte, P Herrmann, JC Remy. Relationships between extractable copper, soil properties and copper uptake by wild plants in vineyard soils. Environ Pollut 102:151–161, 1998.

50. IA Scotti, S Silva, C Baffi. Effects of fly ash pH on the uptake of heavy metals by chicory. Water Air Soil Pollut 109:397–406, 1999.

51. JC Nkana, A Demeyer, G Verloo. Availability of nutrients in wood ash amended tropical and soils. Environ Technol 19:1213–1221, 1998.

52. DH Han, JH Lee. Effects of liming on uptake of lead and cadmium by *Raphanus sativa*. Arch Environ Contam Toxicol 31:488–493, 1996.

53. V Ramachandran, DTJ Souza. Adsorption of cadmium by Indian soils. Water Air Soil Pollut 111:225–234, 1999.

54. DM Heil, Z Samani, AT Hanson, B Rudd. Remediation of lead contaminated soil by EDTA. I. Batch and column studies. Water Air Soil Pollut 113:77–95, 1999.

55. S-Z Lee, L Chang, H-H Yang, C-M Chen, M-C Liu. Adsorption characteristics of lead onto soils. J Hazard Mater 63:37–49, 1998.

56. KR Baldwin, JE Shelton. Availability of heavy metals in compost-amended soil. Bioresource Technol 69:1–14, 1999.

57. NT Basta, JJ Sloan. Bioavailability of heavy metals in strongly acidic soils treated with exceptional quality biosolids. J Environ Qual 28:633–638, 1999.

58. MJ McLaughlin, G Tiller, MK Smart. Speciation of cadmium in soil solutions of saline/sodic soils and relationship with cadmium concentrations in potato tubers (*Solanum tuberosum* L.). Aust J Soil Res 35:183–198, 1997.

59. MJ Mench. Cadmium availability to plants in relation to major long-term changes in agronomy systems. Agric Ecosyst Environ 67:175–187, 1998.

60. S Marlet, L Barbiero, V Valles. Soil alkalinization and irrigation in the Sahelian Zone of Niger. II: Agronomic consequences of alkalinity and sodicity. Arid Soil Res Rehabil 12:139–152, 1998.

61. PJ Smallidge, DJ Leopold. Effects of watershed liming of *Picea rubens* seedling biomass and nutrient element concentration. Water Air Soil Pollut 95:193–204, 1997.

62. ZL He, AK Alva, DV Calvert, YC Li, DJ Banks. Effects of nitrogen fertilization of grapefruit trees on soil acidification and nutrient availability in a Riviera fine sand. Plant Soil 206:11–19, 1998.

63. AM Kooijman, JCR Dopheide, J Sevink, I Takken, JM Verstraten. Nutrient limitations and their implications on the effects of atmospheric deposition in coastal dunes; lime-poor and lime-rich sites in the Netherlands. J Ecol 86:511–526, 1998.

64. GR Findenegg. A comparative study of ammonium toxicity at different constant pH of the nutrient solution. Plant Soil 103:239–243, 1987.

65. XK Zhang, Z Rengel. Role of soil pH, Ca supply and banded P fertilisers in modulating ammonia toxicity to wheat. Aust J Agric Res 51:691–699, 2000.

66. X Zhang, Z Rengel. Gradients of pH, ammonium, and phosphorus between the fertiliser band and wheat roots. Aust J Agric Res 50:365–373, 1999.

67. JM Bremner. Recent research on problems in the use of urea as a nitrogen fertilizer. Fertil Res 42:321–329, 1995.

68. C Marzadori, S Miletti, C Gessa, S Ciurli. Immobilization of jack bean urease on hydroxyapatite: urease immobilization in alkaline soils. Soil Biol Biochem 30: 1485–1490, 1998.

69. MS Zia, M Aslam, Rahmatullah, M Arshad, T Ahmed. Ammonia volatilization from nitrogen fertilizers with and without gypsum. Soil Use Manage 15:133–135, 1999.

70. TJ Purakayastha, JC Katyal. Evaluation of compacted urea fertilizers prepared with acid and non–acid producing chemical additives in three soils varying in pH and cation exchange capacity. I. NH_3 volatilization. Nutr Cycl Agroecosyst 51:107–115, 1998.

71. V Römheld. The soil-root interface in relation to mineral nutrition. Symbiosis 9: 19–27, 1990.

72. DE Crowley, Z Rengel. Biology and chemistry of nutrient availability in the rhizosphere. In: Z Rengel, ed. Mineral Nutrition of Crops: Fundamental Mechanisms and Implications. New York: Food Products Press, 1999, pp 1–40.

73. Z Rengel. Physiological mechanisms underlying differential nutrient efficiency of crop genotypes. In: Z Rengel, ed. Mineral Nutrition of Crops: Fundamental Mechanisms and Implications. New York: Food Products Press, 1999, pp 227–265.

74. TS Gahoonia, N Claassen, A Jungk. Mobilization of residual phosphate of different phosphate fertilizers in relation to pH in the rhizosphere of ryegrass. Fertil Res 33: 229–237, 1992.

75. C Hoffmann, E Ladewig, N Claassen, A Jungk. Phosphorus uptake of maize as affected by ammonium and nitrate nitrogen—measurements and model calculations. Z Pflanzenernaehr Bodenkd 157:225–232, 1994.

76. M Schottelndreier, U Falkengren-Grerup. Plant induced alteration in the rhizosphere and the utilisation of soil heterogeneity. Plant Soil 209:297–309, 1999.

77. RJ Haynes. Active ion uptake and maintenance of cation-anion balance: a critical examination of their role in regulating rhizosphere pH. Plant Soil 126:247–264, 1990.

78. SC Jarvis, AD Robson. The effects of nitrogen nutrition of plants on the development of acidity in Western Australian soils. II. Effects of differences in cation/anion balance between plant species grown under non-leaching conditions. Aust J Agric Res 34:355–365, 1983.

79. H Marschner, V Römheld, WJ Horst, P Martin. Root-induced changes in the rhizo-

sphere: importance for the mineral nutrition of plants. Z Pflanzenernaehr Bodenkd 149:441–456, 1986.

80. SP Loss, GSP Ritchie, AD Robson. Effect of lupins and pasture on soil acidification and fertility in Western Australia. Aust J Exp Agric 33:457–464, 1993.

81. C Tang, MJ Unkovich, JW Bowden. Factors affecting soil acidification under legumes. III. Acid production by N_2-fixing legumes as influenced by nitrate supply. New Phytol 143:513–521, 1999.

82. C Tang, C Raphael, Z Rengel, JW Bowden. Understanding subsoil acidification: effect of nitrogen transformation and nitrate leaching. Aust J Soil Res 38:837–849, 2000.

83. D Riley, SA Barber. Effects of ammonium and nitrate fertilization on phosphorus uptake as related to root-induced pH changes at the root-soil interface. Soil Sci Soc Am Proc 35:301–306, 1971.

84. C Tang, MJ Unkovich, JW Bowden. Pasture legume species differ in their capacity to acidify a low-buffer soil. Aust J Agric Res 49:53–58, 1997.

85. NE Marcar. Genotypic variation for manganese efficiency in cereals. PhD thesis, University of Adelaide, Adelaide, 1986.

86. Y Tong, Z Rengel, RD Graham. Interactions between nitrogen and manganese nutrition of barley genotypes differing in manganese efficiency. Ann Bot 79:53–58, 1997.

87. M Tagliavini, A Masia, M Quartieri. Bulk soil pH and rhizosphere pH of peach trees in calcareous and alkaline soils as affected by the form of nitrogen fertilizers. Plant Soil 176:263–271, 1995.

88. PH Nye. Changes of pH across the rhizosphere induced by roots. Plant Soil 61:7–26, 1981.

89. CJ Thomson, H Marschner, V Römheld. Effect of nitrogen fertilizer form on pH of the bulk soil and rhizosphere, and on the growth, phosphorus, and micronutrient uptake of bean. J Plant Nutr 16:493–506, 1993.

90. TS Gahoonia. Influence of root-induced pH on the solubility of soil aluminium in the rhizosphere. Plant Soil 149:289–291, 1993.

91. SE Petrie, TL Jackson. Effect of fertilization on soil solution pH and manganese concentration. Soil Sci Soc Am J 48:315–318, 1984.

92. Z Rengel. Uptake and transport of manganese in plants. In: A Sigel, H Sigel, eds. Metal Ions in Biological Systems. New York: Marcel Dekker, 2000, pp 57–87.

93. SC Jarvis, DJ Hatch. Rates of hydrogen ion efflux by nodulated legumes grown in flowing solution culture with continuous pH monitoring and adjustment. Ann Bot 55:41–51, 1985.

94. B Dinkelaker, C Hengeler, H Marschner. Distribution and function of proteoid roots and other root clusters. Bot Acta 108:183–200, 1995.

95. LQ Sarong, DR Bouldin, WS Reid. Total and labile zinc concentrations in water extracts of rhizosphere and bulk soils of oats and rice. Commun Soil Sci Plant Anal 20:271–290, 1989.

96. TS Gahoonia, NE Nielsen. The effects of root-induced pH changes on the depletion of inorganic and organic phosphorus in the rhizosphere. Plant Soil 143:185–191, 1992.

97. AR Gillespie, PE Pope. Rhizosphere acidification increases phosphorus recovery

of black locust. I. Induced acidification and soil response. Soil Sci Soc Am J 54: 533–537, 1990.

98.   XL Li, E George, H Marschner. Phosphorus depletion and pH decrease at the root-soil and hyphae-soil interfaces of VA mycorrhizal white clover fertilized with ammonium. New Phytol 119:397–404, 1991.

99.   M Mench, E Martin. Mobilization of cadmium and other metals from two soils by root exudates of *Zea mays* L., *Nicotiana tabacum* L. and *Nicotiana rustica* L. Plant Soil 132:187–196, 1991.

100. DL Jones, AC Edwards, K Donachie, PR Darrah. Role of proteinaceous amino acids released in root exudates in nutrient acquisition from the rhizosphere. Plant Soil 158:183–192, 1994.

101. DL Jones, PR Darrah. Role of root derived organic acids in the mobilization of nutrients from the rhizosphere. Plant Soil 166:247–257, 1994.

102. HF Bienfait, HJ Lubberding, P Heutink, L Lindner, J Visser, R Kaptein, K Dijkstra. Rhizosphere acidification by iron deficient bean plants: the role of trace amounts of divalent metal ions. A study on roots of intact plants with the use of carbon-11 and phosphorus-31 NMR. Plant Physiol 90:359–364, 1989.

103. Y Ohwaki, K Sugahara. Active extrusion of protons and exudation of carboxylic acids in response to iron deficiency by roots of chickpea (*Cicer arietinum* L.). Plant Soil 189:49–55, 1997.

104. P Chairidchai, GSP Ritchie. Zinc adsorption by a lateritic soil in the presence of organic ligands. Soil Sci Soc Am J 54:1242–1248, 1990.

105. P Chairidchai, GSP Ritchie. The effect of pH on zinc adsorption by a lateritic soil in the presence of citrate and oxalate. J Soil Sci 43:723–728, 1992.

106. P Chairidchai, GSP Ritchie. Zinc adsorption by sterilized and non-sterilized soil in the presence of citrate and catechol. Commun Soil Sci Plant Anal 24:261–275, 1993.

107. RH McKenzie, JF Dormaar, GB Schaalje, JWB Stewart. Chemical and biochemical changes in the rhizospheres of wheat and canola. Can J Soil Sci 75:439–447, 1995.

108. RFR Roelofs, Z Rengel, GR Cawthray, KW Dixon, H Lambers. Exudation of carboxylates in Australian Proteaceae: chemical composition. Plant Cell Environ 24: 891–903.

109. Z Rengel. Competitive Al^{3+} inhibition of net Mg^{2+} uptake by intact *Lolium multiflorum* roots. II. Plant age effects. Plant Physiol 93:1261–1267, 1990.

110. P Chairidchai, GSP Ritchie. The effect of citrate and pH on zinc uptake by wheat. Agron J 85:322–328, 1993.

111. UE Grauer, WJ Horst. Modeling cation amelioration of aluminum phytotoxicity. Soil Sci Soc Am J 56:166–172, 1992.

112. Z Rengel. Tansley Review No. 89: uptake of aluminium by plant cells. New Phytol 134:389–406, 1996.

113. RL Aitken. Relationships between extractable aluminum, selected soil properties, pH buffer capacity and lime requirement in some acidic Queensland soils. Aust J Soil Res 30:119–130, 1992.

114. CD Foy. Plant adaptation to acid, aluminum-toxic soils. Commun Soil Sci Plant Anal 19:959–987, 1988.

115. IR Willett. Direct determination of aluminum and its cationic fluoro-complexes by ion chromatography. Soil Sci Soc Am J 53:1385–1391, 1989.

116. T Bantan, R Milacic, B Pihlar. Possibilities for speciation of Al-citrate and other negatively charged Al complexes by anion-exchange FPLC-ICP-AES. Talanta 46: 227–235, 1998.

117. B Mitrovic, R Milacic, B Pihlar, P Simoncic. Speciation of trace amounts of aluminium in environmental samples by cation-exchange FPLC-ETAAS. Analysis 26: 381–388, 1998.

118. LM Shuman. Chelate and pH effects on aluminum determined by differential pulse polarography and plant root bioassay. J Environ Sci Health A Environ Sci Eng 29: 1423–1438, 1994.

119. TB Kinraide. Identity of the rhizotoxic aluminum species. Plant Soil 134:167–178, 1991.

120. Z Rengel. Role of calcium in aluminium toxicity. New Phytol 121:499–513, 1992.

121. DR Parker, TB Kinraide, LW Zelazny. On the phytotoxicity of polynuclear hydroxy-aluminum complexes. Soil Sci Soc Am J 53:789–796, 1989.

122. TB Kinraide, PR Ryan, LV Kochian. Interactive effects of Al^{3+}, H^+, and other cations on root elongation considered in terms of cell-surface electrical potential. Plant Physiol 99:1461–1468, 1992.

123. J Barcelo, P Guevara, C Poschenrieder. Silicon amelioration of aluminium toxicity in teosinte (*Zea mays* L. ssp. *mexicana*). Plant Soil 154:249–255, 1993.

124. PM Bertsch. Aluminum speciation: methodology and applications. In: SA Norton, SE Lindberg, AL Page, eds. Advances in Environmental Science. Vol 4. Acidic Precipitation: Soil, Aquatic Processes, and Lake Acidification. Berlin: Springer-Verlag, 1990, pp 63–105.

125. KM Cocker, DE Evans, MJ Hodson. The amelioration of aluminium toxicity by silicon in higher plants: solution chemistry or an in plants mechanism? Physiol Plant 104:608–614, 1998.

126. KE Hammond, DE Evans, MJ Hodson. Aluminium/silicon interactions in barley (*Hordeum vulgare* L.) seedlings. Plant Soil 173:89–95, 1995.

127. MJ Hodson, DE Evans. Aluminium/silicon interactions in higher plants. J Exp Bot 46:161–171, 1995.

128. TB Kinraide, DR Parker. Non-phytotoxicity of the aluminum sulfate ion, $AlSO_4^+$. Physiol Plant 71:207–212, 1987.

129. DC MacLean, KS Hansen, RE Schneider. Amelioration of aluminium toxicity in wheat by fluoride. New Phytol 121:81–88, 1992.

130. DR Parker, TB Kinraide, LW Zelazny. Aluminum speciation and phytotoxicity in dilute hydroxy-aluminum solutions. Soil Sci Soc Am J 52:438–444, 1988.

131. DP Stevens, MJ McLaughlin, AM Alston. Phytotoxicity of aluminium-fluoride complexes and their uptake from solution culture by *Avena sativa* and *Lycopersicon esculentum*. Plant Soil 192:81–93, 1997.

132. DL Jones, AM Prabowo, LV Kochian. Aluminium–organic acid interactions in acid soils. II. Influence of solid phase sorption on organic acid–Al complexation and Al rhizotoxicity. Plant Soil 182:229–237, 1996.

133. JF Ma, SJ Zheng, H Matsumoto, S Hiradate. Detoxifying aluminium with buckwheat. Nature 390:569–570, 1997.

134. JF Ma, SJ Zheng, H Matsumoto. Specific secretion of citric acid induced by Al stress in *Cassia tora* L. Plant Cell Physiol 38:1019–1025, 1997.

135. JF Ma, S Hiradate, H Matsumoto. High aluminum resistance in buckwheat. II. Oxalic acid detoxifies aluminum internally. Plant Physiol 117:753–759, 1998.

136. DM Pellet, DL Grunes, LV Kochian. Organic acid exudation as an aluminum-tolerance mechanism in maize (*Zea mays* L.). Planta 196:788–795, 1995.

137. PR Ryan, E Delhaize, PJ Randall. Malate efflux from root apices and tolerance to aluminium are highly correlated in wheat. Aust J Plant Physiol 22:531–536, 1995.

138. LM Shuman, DO Wilson, EL Ramseur. Amelioration of aluminium toxicity to sorghum seedlings by chelating agents. J Plant Nutr 14:119–128, 1991.

139. RH Abrams, K Loague. A compartmentalized solute transport model for redox zones in contaminated aquifers. 2. Field-scale simulations. Water Resour Res 36: 2015–2029, 2000.

140. S Ratering, S Schnell. Localization of iron-reducing activity in paddy soil by profile studies. Biogeochemistry 48:341–365, 2000.

141. A Carbonell-Barrachina, A Jugsujinda, RD DeLaune, WH Patrick Jr, F Burlo, S Sirisukhodom, P Anurakpongsatorn. The influence of redox chemistry and pH on chemically active forms of arsenic in sewage sludge–amended soil. Environ Int 25:613–618, 1999.

142. BP Singh, M Das, RN Prasad, M Ram. Characteristics of Fe-toxic soils and affected plants and their correction in acid Haplaquents of Meghalaya. Int Rice Res Newslett 17:18–19, 1992.

143. JG Genon, Nd Hepcee, JE Duffy, B Delvaux, PA Hennebert. Iron toxicity and other chemical soil constraints to rice in highland swamps of Burundi. Plant Soil 166: 109–115, 1994.

144. KMD Devi, CS Gopi, G Santhakumari, PV Prabhakaran. Effect of water management and lime on iron toxicity and yield of paddy. J Trop Agric 34:44–47, 1996.

145. RJ Bartlett, BR James. Redox chemistry of soils. Adv Agron 50:152–208, 1994.

146. DJ Burdige, SP Dhakar, KH Nealson. Effects of manganese oxide mineralogy on microbial and chemical manganese reduction. Geomicrobiol J 10:27–48, 1992.

147. K Posta, H Marschner, V Römheld. Manganese reduction in the rhizosphere of mycorrhizal and nonmycorrhizal maize. Mycorrhiza 5:119–124, 1994.

148. Z Rengel. Root exudation and microflora populations in rhizosphere of crop genotypes differing in tolerance to micronutrient deficiency. Plant Soil 196:255–260, 1997.

149. RA Boyd. Groundwater geochemistry in the seminole well field, Cedar Rapids, Iowa. J Am Water Resour Assoc 35:1257–1268, 1999.

150. D Neilsen, GH Neilsen, AH Sinclair, DJ Linehan. Soil phosphorus status, pH and the manganese nutrition of wheat. Plant Soil 145:45–50, 1992.

151. JH Rule, ER Graham. Soil labile pools of manganese, iron, and zinc as measured by plant uptake and DTPA equilibrium. Soil Sci Soc Am J 40:853–857, 1976.

152. JT Moraghan, TJ Freeman. Influence of FeEDDHA on growth and manganese accumulation in flax. Soil Sci Soc Am J 42:455–460, 1978.

153. W Foulds. Nutrient concentrations of foliage and soil in south-western Australia. New Phytol 125:529–546, 1993.

154. MHR Khan, T Adachi. Effects of selected natural factors on soil pH and element

dynamics studied in columns of pyritic sediments. Soil Sci Plant Nutr 45:783–793, 1999.

155. M Yli-Halla, M Puustinen, J Koskiaho. Area of cultivated acid sulfate soils in Finland. Soil Use Manage 15:62–67, 1999.

156. C Lin. Acid sulfate soils in Australia: characteristics, problems and management. Pedosphere 9:289–298, 1999.

157. PA Moore Jr, N Van Bremen, WH Patrick Jr. Effects of drainage on the chemistry of acid sulfate soils. In: Agricultural Drainage. Special Publication 54. Madison, WI: American Society of Agronomy, 1999, pp 1107–1123.

158. J Shamshuddin, I Jamilah, JA Ogunwale. Formation of hydroxy-sulfates from pyrite in coastal acid sulfate soil environments in Malaysia. Commun Soil Sci Plant Anal 26:2769–2782, 1995.

159. HR Khan. Dynamics of soil solutions of submerged acid sulfate soils under rice as influenced by various treatments. Int J Trop Agric 16:81–95, 1998.

160. KG Tiller. Heavy metals in soils and their environmental significance. Adv Soil Sci 9:113–142, 1989.

161. Z Rengel. Mechanisms of plant resistance to toxicity of aluminium and heavy metals. In: AS Basra, RK Basra, eds. Mechanisms of Environmental Stress Resistance in Plants. Amsterdam: Harwood Academic Publishers, 1997, pp 241–276.

162. R Naidu, RS Kookana, ME Sumner, RD Harter, KG Tiller. Cadmium sorption and transport in variable charge soils: a review. J Environ Qual 26:602–617, 1997.

163. H Chen, C Zheng, C Tu, Y Zhu. Heavy metal pollution in soils in China: status and countermeasures. Ambio 28:130–134, 1999.

164. HJ Percival, TW Speir, A Parshotam. Soil solution chemistry of contrasting soils amended with heavy metals. Aust J Soil Res 37:993–1004, 1999.

165. IR Phillips. Copper, lead, cadmium, and zinc sorption by waterlogged and air-dry soil. J Soil Contam 8:343–364, 1999.

166. J Pichtel, K Kuroiwa, HT Sawyerr. Distribution of Pb, Cd and Ba in soils and plants of two contaminated sites. Environ Pollut 110:171–178, 2000.

167. E Lombi, MH Gerzabek, O Horak. Mobility of heavy metals in soil and their uptake by sunflowers grown at different contamination levels. Agronomie 18:361–371, 1998.

168. M Barbafieri. The importance of nickel phytoavailable chemical species characterization in soil for phytoremediation applicability. Internat J Phytoremed 2:105–115, 2000.

169. MJ McLaughlin, RE Hamon, RG McLaren, TW Speir, SL Rogers. Review: a bioavailability-based rationale for controlling metal and metalloid contamination of agricultural land in Australia and New Zealand. Aust J Soil Res 38:1037–1086, 2000.

170. RJ Reid, Z Rengel, FA Smith. Membrane fluxes and comparative toxicities of aluminium, scandium and gallium. J Exp Bot 47:1881–1888, 1996.

171. Z Rengel. Effects of Al, rare earth elements, and other metals on net $^{45}Ca^{2+}$ uptake by *Amaranthus* protoplasts. J Plant Physiol 143:47–51, 1994.

172. MJ McLaughlin, KG Tiller, R Naidu, DP Stevens. Review: the behaviour and environmental impact of contaminants in fertilizers. Aust J Soil Res 34:1–54, 1996.

173. Z Lin, K Harsbo, M Ahlgren, U Qvarfort. The source and fate of Pb in contaminated soils at the urban area of Falun in central Sweden. Sci Total Environ 209:47–58, 1998.

174. J Derome, A Saarsalmi. The effect of liming and correction fertilization on heavy metal and macronutrient concentrations in soil solution in heavy-metal polluted Scots pine stands. Environ Pollut 104:249–259, 1999.

175. DE Salt, RD Smith, I Raskin. Phytoremediation. Annu Rev Plant Physiol Plant Mol Biol 49:643–668, 1998.

176. G Jentschke, DL Godbold. Metal toxicity and ectomycorrhizas. Physiol Plant 109: 107–116, 2000.

177. KE Giller, E Witter, SP McGrath. Toxicity of heavy metals to microorganisms and microbial processes in agricultural soils: a review. Soil Biol Biochem 30:1389–1414, 1998.

178. JE Eriksson. The influence of pH, soil type and time on adsorption and uptake by plants of Cd added to the soil. Water Air Soil Pollut 48:3–4, 1989.

179. R Naidu, NS Bolan, RS Kookana, KG Tiller. Ionic-strength and pH effects on the sorption of cadmium and the surface charge of soils. Eur J Soil Sci 45:419–429, 1994.

180. A Filius, T Streck, J Richter. Cadmium sorption and desorption in limed topsoils as influenced by pH: isotherms and simulated leaching. J Environ Qual 27:12–18, 1998.

181. X Xian, GI Shokohifard. Effect of pH on chemical forms and plant availability of cadmium, zinc, and lead in polluted soils. Water Air Soil Pollut 45:3–4, 1989.

182. VP Evangelou, YL Zhang. A review: pyrite oxidation mechanisms and acid mine drainage prevention. Crit Rev Environ Sci Technol 25:141–199, 1995.

183. T Xu, SP White, K Pruess, GH Brimhall. Modeling of pyrite oxidation in saturated and unsaturated subsurface flow systems. Transport Porous Media 39:25–56, 2000.

184. S Jooste, C Thirion. An ecological risk assessment for a South African acid mine drainage. Water Sci Technol 39:297–303, 1999.

185. NW Menzies, DR Mulligan. Vegetation dieback on clay-capped pyritic mine waste. J Environ Qual 29:437–442, 2000.

186. PJ Voeller, BA Zamora, J Harsh. Growth response of native shrubs to acid mine spoil and to proposed soil amendments. Plant Soil 198:209–217, 1998.

187. M Demchik, K Garbutt. Growth of woolgrass in acid mine drainage. J Environ Qual 28:243–249, 1999.

188. NZ Jovanovic, RO Barnard, NFG Rethman, JG Annandale. Crops can be irrigated with lime-treated acid mine drainage. Water S A (Pretoria) 24:113–122, 1998.

189. H Marschner. Mineral Nutrition of Higher Plants. 2nd ed. London: Academic Press, 1995.

190. PH Masscheleyn, RD Delaune, WH Patrick Jr. Transformations of selenium as affected by sediment oxidation-reduction potential and pH. Environ Sci Technol 24:91–96, 1990.

191. MT Pardo, ME Guadalix. Chemical factors affecting selenite sorption by allophanic soils. Geoderma 63:43–52, 1994.

192. L Johnsson. Selenium uptake by plants as a function of soil type, organic matter content and pH. Plant Soil 133:57–64, 1991.

193. RL Mikkelsen, GH Haghnia, AL Page. Effects of pH and selenium, oxidation state on the selenium accumulation and yield of alfalfa. J Plant Nutr 10:937–950, 1987.

14

Regulation of Microbial Processes by Soil pH

David E. Crowley and Samuel A. Alvey
University of California, Riverside, Riverside, California

1 INTRODUCTION

Microorganisms are an interactive component of soil systems in that they contribute to soil acidification and in turn respond to changes in soil pH. The regulation of soil biological processes by pH, as well as the effects of microorganisms on soil acidification, can be studied at different scales, ranging from microsite effects in the vicinity of plant roots to long-term effects on soil genesis and ecosystem properties. At the most fundamental level, microbial acid production leads to the weathering of rocks and minerals into secondary minerals that constitute soil [1]. Changes in soil acidity affect the leaching and bioavailability of base cations, the bioavailability of phosphorus, and the solubility and toxicity of metal cations. At the ecosystem level, biologically driven changes in soil pH influence the selection of plant and microbial species that make up the community. Soil pH also affects the primary productivity by controlling nutrient turnover rates, nitrogen chemistry, accumulation of phytotoxic compounds, and the populations of symbiotic, plant growth–promoting, and root disease–causing microorganisms that colonize the rhizosphere. In this manner, soil pH is one of the most fundamental variables in plant microbial ecology and is routinely measured and reported in virtually all studies on soil biology.

In this chapter, the role of soil pH in belowground microbial processes is reviewed with an emphasis on nutrient dynamics, plant growth, and rhizosphere ecology. The chapter starts by examining mechanisms of acid formation by microorganisms, which is followed by a review of the role of pH in microbially mediated transformations of carbon, phosphorus, nitrogen, and metals. These topics are followed by a consideration of the effect of pH on plant beneficial symbioses, and previously unpublished data are presented in which a 16S ribosomal DNA (rDNA) fingerprinting approach was used to study the effects of pH on rhizosphere bacterial communities. Lastly, the effects of soil pH on root pathogens and their interactions with microbial communities as controlled by soil pH are considered. One theme of this review is that soil microbial processes are carried out by many different bacteria and fungi, many of which have overlapping roles in terrestrial ecology. This functional redundancy provides for a continuity of ecosystem processes where there is a broad overlap of species. Conversely, processes that are carried out by few species have greater pH dependence. Most soil processes that are affected by pH are considered at the level of individual microorganisms that are studied at a particular pH in the laboratory or conversely at a very broad level in which ecosystem processes such as organic matter formation and soil respiration are considered. Nonetheless, there is still a major gap in studies on pH effects on microbial communities. As we develop better methods to characterize microbial communities in relation to various environmental factors, the study of pH effects will continue to be an important topic for research on plant nutrition and microbial ecology for many years to come.

2   ACID-FORMING PROCESSES AND THE pH TOLERANCE OF MICROORGANISMS

2.1   Contributions of Microorganisms to Soil Acidity

Acid formation by microorganisms can be accomplished by a variety of mechanisms, including respiration, release of organic acids and protons, and chemolithotrophy (transformation of inorganic substrates). The simplest mechanism involves the release of carbon dioxide by heterotrophic microorganisms that use plant-derived carbon for metabolism. As the second largest biomass component after plant roots, mycorrhizal fungi also contribute directly to carbonic acid production by their respiration and by increasing basal root respiration with concomitant release of CO_2 [2]. Mechanisms involving organic acid production are somewhat more complex. Organic acids are commonly released as components of root exudates and are also produced by microorganisms. However, under aerobic conditions, organic acids are excellent growth substrates and do not accumulate [3]. Conversely, under low-oxygen conditions that commonly occur in the plant rhizosphere and under wet soil conditions, microorganisms transform plant-derived organic acids and other compounds into various fermentation products, including acetic, lactic, and butyric acids, all of which are relatively weak acids. These acids can have a variety of effects on soil chemistry and dissolution of minerals, including the formation of metal complexes, changes in metal solubilities, and effects on the adsorption of metals and phosphorous on clay mineral surfaces.

Microbial acidification leading to direct acidification of soils by proton secretion is caused by many different processes as well as by nitrogen-involving interactions between bacteria and plant roots. The release of protons into the cell envelope is fundamental to microbial metabolism and is used to generate an electrochemical potential across the cell membrane to drive synthesis of ATP. Protons are also liberated by certain chemolithotrophic microorganisms involved in nitrogen and sulfur transformations. Nitrogen transformations further affect pH by influencing the cation balance in the rhizosphere. These processes are reviewed in detail in a later section. In brief, processes involving nitrogen start with the mineralization of ammonia from organic matter. When ammonia is taken up by plant roots, ammonia transport is accompanied by a concomitant release of protons by the plant, which is used to maintain an ionic balance across the root cell membrane. Conversion of ammonia to nitrate by nitrifying bacteria also results in the release of protons but is subsequently offset in the rhizosphere by soil alkalization when hydroxyl ions are released during plant uptake of nitrate nitrogen. In addition to nitrogen interactions, certain microorganisms such as *Azospirillum* sp. can increase proton efflux and rhizosphere acidification [4], although the relative importance of such bacteria in rhizosphere acidification is not yet well understood.

Another chemolithotrophic activity leading to acid production involves the activity of iron- and sulfur-oxidizing microorganisms, represented by species such as *Thiobacillus ferrooxidans* that produce protons as a result of the mineral oxidation. The activity of acid-forming chemolithotrophs, however, is generally of concern only in soils that have exposed pyrite or that have been amended with elemental sulfur. When large quantities of pyrite or sulfur are exposed to the atmosphere, such as during mining, they can cause extreme acidification of the soil, lowering the soil pH to values below 1.0 in soils located on mine spoils that contain exposed pyrite [5,6]. These bacteria typically thrive under extreme acid conditions but are also present in soils having a higher pH. With elemental sulfur, this mechanism can be readily exploited to lower the soil pH by addition of elemental sulfur that is then oxidized to sulfuric acid. There are also many diverse bacteria and fungi that can grow well beyond the normal pH range of soil (pH 3.5 to 9), which are beyond the scope of this review. These microorganisms, termed extremophiles, include acidophilic bacteria such as *Thiobacillus thiooxi-*

dans that can grow at pH 1.0, as well as alkaliphilic organisms such as *Bacillus alcalophilus* that can grow at pH 11. Many of these bacteria can be isolated from soils at intermediate pH but are generally present in low population numbers.

2.2 Adaptations of Microorganisms to pH

The influence of pH on soil microbial community structure is largely the result of selection for different microorganisms that have a tolerance for specific pH conditions. Despite the ability of certain microorganisms to tolerate pH extremes, all microorganisms must maintain the pH of their cytoplasm at near-neutral pH. This is accomplished by specific mechanisms that result in individual species and strains having relatively narrow pH optima. Microorganisms that are adapted to acid or alkaline pH conditions use a variety of mechanisms to maintain pH homeostasis [7]. These include changes in the cell walls, membranes, and cell physiology. Changes in the cell envelope frequently involve the use of alternative membrane lipids, alterations in the amino acid composition of membrane proteins, and the use of ion transporters that actively help to maintain pH homeostasis. Microorganisms that are adapted to extremely acid conditions typically use reverse electron transport to rid the cell of excess hydrogen ions. Certain outer membrane proteins that confer increased tolerance to acidification also can increase resistance to other environmental stresses [8,9]. Alkaliphilic bacteria use similar mechanisms to establish and maintain a neutral cytoplasmic pH. In alkaliphilic *Bacillus* species, general adaptations include the use of proteins with reduced basic amino acid content, increased use of acidic cell wall components such as teichuronic acid and teichuronopeptides, and the establishment of an electrochemical gradient by proton extrusion [10]. In some bacteria, an electrogenic Na^+/H^+ antiport may also be used to achieve net acidification of the cytoplasm relative to the outside pH. In other alkaliphilic bacteria, adaptations may involve changing the net surface charge of the wall by increased inclusion of acidic phospholipids such as cardiolipin, phosphatidylethanolamine, and phosphatidylglycerol in the membrane [11].

3 INFLUENCE OF pH ON MICROBIAL COMMUNITY STRUCTURE AND FUNCTION

3.1 Biomass Composition

Current thought on organic matter decomposition is that fungi carry out most decomposition in undisturbed ecosystems, whereas bacteria dominate under intensively managed systems such as in agricultural fields [12]. The relative proportion of bacterial biomass and fungal biomass is strongly affected by soil pH, with bacteria constituting a smaller proportion of the biomass at low soil pH. The significance of a shift from a fungal- to a bacterial-dominated biomass or vice versa is still being studied, but it has been speculated to be important in causing altered nutrient mineralization dynamics, organic matter accumulation, and soil respiration rates. As a rule, fungi assimilate 30 to 40% of the carbon in plant litter into their biomass as compared with 5 to 10% for bacteria. An immediate consequence of a shift to bacterial-dominated decomposition is that the soil respiration rate per unit biomass, or the soil metabolic quotient, is elevated as compared with respiration in fungal-dominated systems [13]. The relevance of metabolic quotients to organic matter accumulation has led to the hypothesis that fungal-dominated systems may have greater retention of soil organic matter than bacterial-dominated systems [13]. Effects of soil acidity on the soil biomass ratio of fungi and bacteria are also being studied in relation to nutrient dynamics and the impact of various land management systems [14].

Biomass values generally increase with pH over the pH range 3 to 7. At very low pH, the decline in biomass is associated with an increase in aluminum and a decrease in base cations that together cause aluminum toxicity to plants and microorganisms [15]. This toxicity can be observed as a reduction in the proportion of biomass carbon to total organic carbon. At neutral pH, the microbial biomass carbon makes up between 2 and 4% of the total soil carbon. These values decrease to 1% in strongly acidic soils, which suggests that organic matter decomposition is impaired at very low pH. In a study examining this phenomenon, detailed measurements were made in a 3-year study of acid-irrigated and limed plots containing Norway spruce [16]. Addition of lime increased the microbial biomass, respiration, and C-microbial : C-organic. Paradoxically, increases in microbial biomass that are brought about by liming also can increase the acid loading rate that is caused by the increased microbial activity. In spruce forests, mineralization of ammonia and conversion of ammonia to nitrate are estimated to contribute from 50 to 75% of the acid loading in forest soils, with both of these activities being strongly correlated with the size of the microbial biomass [17]. In the absence of liming, the production of acidity contributes to an eventual decline in nitrogen mineralization as the microbial biomass becomes diminished at low soil pH [17].

Changes in the relative biomass and activity of fungi and bacteria, as affected by soil pH, are difficult to estimate using culture-based techniques. Instead, they are determined using signature phospholipid fatty acid (PLFA) indicators (biomass) or by differential respiration methods in the presence of specific inhibitors (activity). Culture-based methods enumerate only a small percentage of the microorganisms that can be detected by direct counts and are inadequate for determination of fungal biomass, as they are biased toward spore counts and further depend on the degree of fragmentation of fungal hyphae. PLFA markers include the fatty acid 18:2w6 for fungi and a variety of marker fatty acids for bacteria.

Typical fungi-to-bacteria ratios in unmanaged meadow soils measured using PLFA methods range from 2.0 to 2.3 versus approximately 1 : 1 for managed soils [12]. These values tend to be more variable in agricultural soils and range from 1 : 1 to 3.7 : 1 following amendment with different organic matter materials. Using selective inhibition methods, fungi were shown to be responsible for 90 and 94% of the total CO_2 respired in soils planted to spruce (*Picea abies*) and beech (*Fagus sylvatica*), respectively, at pH 3 [18]. At higher soil pH levels, achieved by liming the soil to pH 6, the fungal respiration component was reduced to 74 and 84% of the total respired CO_2. Shifts in the bacterial biomass are also observed using thymidine incorporation methods specific to the measurement of bacteria [19]. Liming of forest soils from pH 4 to near-neutral pH increased both the thymidine incorporation rates and numbers of culturable bacteria fivefold.

Although these changes in biomass reflect very broad changes in the properties of the belowground system, there is very little detailed information on the effects of pH on eubacterial community structure and species composition at different values of soil pH. Nonetheless, shifts in community structure are grossly indicated by the fact that the bacterial communities that develop in acid soils are well adapted to the prevailing soil pH [20], which suggests a shift to specifically adapted species. Higher resolution techniques are now being applied to examine these shifts, and there are various methods for examining changes in species composition, including fatty acid analyses and molecular methods using 16S rDNA profiles. One of most difficult problems facing soil ecologists in the future will be to demonstrate how shifts in eubacterial community structure might be associated with altered functional properties at the ecosystem level. Answers to these questions will require diagnostic methods for correlating changes in microbial community fingerprints with changes in specific biological processes.

3.2 Soil pH Effects on Mineralization of Nitrogen

Factors affecting nitrogen mineralization are important in controlling the productivity of natural ecosystems, particularly those that do not receive nitrogen fertilizers. Soil pH has a significant effect on organic matter decomposition kinetics, microbial biomass, and microbial activity and thus is very important in determining the availability and fate of nitrogen. In a major study examining factors that control nitrogen dynamics across pH and climatic gradients, a climatic sequence in western Europe was examined in which ^{14}C- and ^{15}N-labeled substrates were used to measure the changes in carbon and nitrogen mineralization rates [21]. Because many factors control basal respiration rates and nutrient dynamics, statistical procedures involving principal components or factor analysis were required to estimate the impact of individual variables on these parameters. Eleven organic matter decomposition variables were analyzed by means of correspondence analysis, and descriptors of the soil properties were used as additional variables. Results of this study showed that ordination of all of the major organic matter decomposition variables was highly related to soil pH and that pH was the best soil-related predictor of decomposition parameters [21]. One possible explanation involves differential effects of pH on soil enzyme activities, which are substantially decreased at low soil pH [22]. In a detailed study that compared the activities of 14 enzymes involved in the C, N, S, and P cycles, 10-fold or greater increases were observed 7 years after liming an acid soil, which concomitantly increased the pH from 4.9 to 6.9 [22]. However, there is still some controversy regarding liming effects on organic matter mineralization, which provides a positive benefit only in some soils and is subject to artifacts resulting from temporary increases in soluble organic matter caused by calcium effects on soil structure and chemistry [23,24].

Organic matter decomposition dynamics can also be influenced by the addition of nitrogen to acid and alkaline soils. In experiments conducted in Greece with soils having a pH of 5.6 and 7.8, applied nitrogen increased the biomass and stimulated net mineral nitrogen and phosphorus release in the near-neutral soil [25]. In contrast, application of nitrogen to the acid soil decreased the rate of decomposition and release of mineral nitrogen and calcium. These complex effects of nitrogen additions suggest that soil pH may have complex if not unpredictable effects on nutrient dynamics in soils that are subjected to fertilization or other perturbations.

3.3 Acid Rain Effects on Microbial Decomposition Processes

The detrimental impact of acid rain has been of concern in Europe and North America, where the productivity of many forest soils with low concentrations of base cations is already impaired by low soil pH, and may have long-term effects even after the pH and base saturation levels are restored [26]. In one of the major studies on this topic, employing a simulated acid rain application over 12 growing seasons, small shifts in pH on an already highly acid forest soil caused significant declines in the total biomass as well as shifts in microbial community composition [27]. As the pH of the soil humus was decreased from 3.83 to 3.65 by acid rain application, basal respiration decreased along with the total microbial biomass, and the phospholipid fatty acid signature pattern suggested an increase in gram-positive bacteria. In a companion study combining acid rain with heavy metal additions of nickel and copper, shifts in microbial community composition were shown to be due primarily to the increased acidity as opposed to an increase in heavy metal toxicity effects for these metals [28]. Nonetheless, aluminum toxicity at low pH is still suspected as one of the most important changes associated with the decline in biomass but is difficult to separate out from other pH effects in studies of field soils.

4 SOIL pH AND METAL BIOAVAILABILITY

In addition to the problem of pH homeostasis discussed earlier, some of the most important effects of pH on soil microorganisms are those caused by changes in metal solubilities. Under alkaline conditions, essential trace metals such as iron have low solubility and require special adaptations that facilitate metal uptake. Conversely, under acid conditions that permit dissolution of metal ions from solid-phase minerals, metal toxicities can occur, which include aluminum toxicities in natural and agricultural systems and heavy metal toxicities in heavy metal–contaminated soils. In acid soils, some plants and microorganisms use adaptive mechanisms to avoid heavy metal toxicities, and metal toxicity to the roots can be influenced by both rhizosphere pH and mycorrhizal fungi [29]. In both acid and alkaline soils, microbial processes involving uptake, chelation, precipitation, methylation, and surface binding of metals can potentially affect the bioavailability of metal ions and their transfer into the food chain. This subject remains an active area of investigation with respect to environmental quality and remediation of contaminated soils.

4.1 Effects of Acidity on Metal Toxicities to Microorganisms

Under acidic soil conditions, the solubility of various metal ions is increased by a factor of 10 to 1000 for every unit decrease in pH, depending on the metal's valence and its speciation characteristics in the soil solution. Monovalent metal ions such as Cu^+ and Ni^+ increase by a factor of 10 for every unit decrease in pH, whereas divalent ions including Zn^{2+} and Mn^{2+}, and the trivalent metal ion iron increase in solubility by 100- and 1000-fold, respectively. All of these metal ions also form complexes of different valences with hydroxyl ions, phosphate, sulfate, and carbonate that follow similar solubility relationships with respect to pH. These relationships have been well modeled [30,31] and can be described using computer equilibrium models to predict metal solubilities in solutions of varying composition. Of particular concern are metal toxicities caused by Al, Zn, Cu, Pb, Mn, Ni, and Hg that have detrimental effects on soil microbial processes and plant growth.

In general, the toxic effects of heavy metals are strongly controlled by pH [32], although many studies have also emphasized the equal importance of metal loading rates. Methods to assess the bioavailability and toxicity of metals continue to be developed and involve a variety of chemical and biological assays. One complication in predicting metal bioavailability is that pH controls not only the free activity of the metal but also the speciation of other ions in the soil solution and their potential for forming ion complexes with soil organic matter and organic acids. The potential importance of different organic acids for metal complexation varies depending on their pK values and stability constants with different metals. The types of organic acids in the soil solution may also vary depending on the plant species that are present as well as on prevailing environmental conditions. Organic acids are rapidly degraded by microorganisms and are strongly sorbed by ferrihydrite and clay minerals in acid soils [3,33]. There are also complex interactions with soil organic matter, carbonates, and other ions that result in deviations from simple linear models of metal solubility with respect to pH. For the time being, the problem of modeling metal bioavailability has been studied empirically, especially with regard to land applications of sewage sludge that represent a major source of metal loading. Empirical models show that total concentrations of Pb, Zn, and Cu increase as soil pH is decreased but also indicate that there are possible threshold pH values, generally below pH 6, at which there may be significantly increased metal toxicities [34].

Relatively little is known about possible long-term adverse effects of metals on soil microorganisms that may adapt to pH-induced metal toxicities by a variety of mechanisms. Genes for metal resistance are carried on plasmids and can be amplified throughout different groups of bacteria via gene transfer. Short-term experiments generally show detrimental effects of metals on soil microorganisms, but there are very few long-term studies, and the appropriate methods for assessing metal toxicities at low pH are still being debated and developed.

Controlled field experiments with sewage sludges conducted in the United Kingdom, Sweden, Germany, and the United States show that microbial activity and populations of cyanobacteria, *Rhizobium leguminosarum*, mycorrhizae and the total microbial biomass are adversely affected by increased metal concentrations that, in some cases, are below the European Union's maximum allowable concentration limits for metals in sludge-treated soils [35]. Studies of the effects of heavy metals on populations of *Rhizobium leguminosarum* and nitrogen fixation rates by free-living diazotrophs and cyanobacteria have generated particular concern and show that adverse effects by metal concentrations can occur even below the relatively conservative levels established by the European Union [36]. However, data from these and other studies strongly suggest that in addition to heavy metal loading rates, soil pH and texture need to be considered in determining the risk of heavy metal toxicities to plants and microorganisms. Soils having a high pH and increased contents of clay and organic matter have considerably reduced metal toxicities.

4.2 Microbial Effects on Plant Metal Toxicities in Acid Soils

Soil microorganisms can have a wide variety of effects on uptake and potential toxicities of metals to plants. This may involve microbially mediated changes in metal speciation, metal complex formation, intercellular binding, metal efflux proteins, and other processes such as methylation or precipitation [37]. In this regard, there has been particular interest in certain bacteria that may help to increase the metal tolerance of plants. In one such study using soil inoculants to increase plant tolerance to acid pH, barley plants were grown in an acid soil and inoculated with various strains of *Azospirillum* and *Flavobacterium* [38]. When inoculated with the metal-resistant strains, the barley seedlings showed increased resistance to stress caused by high acidity and aluminum, which was manifested by stimulated root growth and decreased proline content. However, the response of plants to the inoculation strongly varied from positive to negative with the soil pH and barley cultivar. Another mechanism by which certain bacteria protect plants from heavy metal toxicity is by limiting the damaging effects of heavy metal–induced ethylene production [39]. This particular mechanism was shown for canola plants following inoculation with the plant growth–promoting bacterium *Kluyvera ascorbata* SUD165. Uptake of Ni by plants was not decreased, but ethylene concentrations were reduced by microbial degradation of this volatile plant hormone.

Another mechanism by which microorganisms influence heavy metal toxicities is their ability to adsorb metals passively on their cell walls. Metal ion binding to biotic surfaces is pH dependent, as is metal ion binding to soil mineral surfaces. At high pH, metal solubilities are reduced, but the surfaces of plant and microbial cells become more electronegative and have increased capacity to adsorb metals. Thus, the metal toxicity for plants and microorganisms often increases with increasing pH for water-culture experiments, whereas the opposite effect is observed for plants and microorganisms growing in soils [40]. In research with *Pseudomonas putida*, passive adsorption of six metals (Cs, Sr, Eu, Zn, Cd, and Hg) has been shown to occur, and the metal accumulation capacity was comparable to that of soil organic matter [41]. Metal binding affinity followed the order Hg > Eu > Cd = Zn = Sr > Cs. The relevance of microbial binding of metals as a means of lowering plant bioavailability has not been quantitatively examined but is an interesting question because it is possible that microorganisms may afford some protection at high population densities in the plant rhizosphere. Although serving only as a passive barrier, microorganisms proliferate at critical zones in the apoplastic pathway along cell wall junctions and provide a last site of protection. In addition to adsorption, microorganisms may have many other effects on metal bioavailability by changing the pH or by production of organic acids and CO_2. For example, increases in Cd bioavailability to wheat are associated with a low rhizosphere pH and increases in low-molecular-weight organic acid complexes of Cd [35]. Microorganisms could affect both of these processes by many mechanisms that have already been described earlier in this chapter.

4.3 Soil pH and Trace Metal Deficiencies

At neutral to alkaline pH, plants and microorganisms are subject to trace metal deficiencies of iron, zinc, and manganese, all of which become increasingly unavailable with increasing pH. Because of the relatively high requirement for iron as compared with other trace metals, iron deficiencies are one of the most common trace metal deficiencies. Many plants and microorganisms have adapted to growth under iron-deficient conditions by production and utilization of iron chelates and organic acids that form iron complexes. This phenomenon has been particularly well investigated for rhizosphere microorganisms, which produce siderophores to solubilize iron and hold it in an available form that can be transported across cell membranes. Certain siderophores produced by microorganisms can be used by plants for iron uptake, and, conversely, many microorganisms can use iron citrate or phytosiderophores that are produced by dicotyledonous and monocotyledonous plants, respectively. In general, the quantities of siderophore and phytosiderophore that are produced will depend on the growth rate and relative iron demand of plants and microorganisms [42]. Certain hydroxamate siderophores, such as ferrioxamine B, and ferrichrome are widely used by a number of microorganisms including bacteria and fungi that do not produce these compounds but can take up iron from these chelators by various mechanisms. Other siderophores, such as pyoverdin, are used by narrow groups of microorganisms, sometimes in a strain-specific manner.

Microbial siderophores produced under iron deficiency conditions at high soil pH are important in the survival and competitiveness of plant growth–promoting rhizosphere bacteria and may also have both a direct and an indirect role in the suppression of certain disease-causing microorganisms [42]. For some time there has also been interest in the relative importance of microbial iron chelators for plant iron nutrition in calcareous, alkaline soils. Microorganisms can produce very large quantities of siderophore in pure culture conditions, but production of siderophores in the rhizosphere is probably limited to microsites of high activity, and their accumulation at high concentrations necessary to support plant growth is unlikely. This question has been carefully studied with pseudomonads using different gene reporters to ascertain the growth status of these bacteria in different root zones and to estimate the potential quantities of siderophore that may be released by these bacteria. When an ice nucleation reporter system was used to assess induction of the genes for synthesis of the siderophore pyoverdin, rhizosphere populations of *Pseudomonas fluorescens* Pf-5 expressed greater ice nucleation activity in soil at pH 7 or 8 than in the same soil at pH 5.4 [43]. These results are in agreement with chemical models that predict that pH is a major factor in controlling iron availability and competition for this element.

5   EFFECT OF pH ON PLANT-MICROBIAL SYMBIOSES

Soil pH has a significant effect on plant symbioses with mycorrhizal fungi and with nitrogen-fixing bacteria in legume-rhizobium symbioses. Many plants are highly dependent on mycorrhizae for uptake of phosphorus under low nutrient conditions. Because phosphorus availability is strongly controlled by dissolution of mineral P that can constitute a considerable portion of the available P, soil pH is a major factor in determining the relative importance of mycorrhizae for phosphorus uptake. Mineral phosphorus has greatest availability at slightly acid to near-neutral pH. At low pH, phosphorus solubility is limited by the low solubility of iron and aluminum phosphates, whereas at alkaline pH phosphorus forms insoluble calcium and magnesium phosphate minerals. Numbers of vesicular-arbuscular mycorrhizal (VAM) spores are highly correlated with the availability of total soil P; thus, soil pH indirectly affects the VAM population [44]. With respect to nitrogen-fixing symbioses, an acid soil pH generally is detrimental to *Rhizobium* survival and may also decrease the population size of free-living bacteria and cyanobacteria. At alkaline pH, nitrogen fixation by *Rhizobium*-legume associations may be limited by the low availability of iron. These topics are discussed in more detail next.

5.1   Soil pH Effects on Mycorrhizal Symbioses

Mycorrhizae increase plant phosphorus and zinc uptake and may also enhance uptake of water and ammonia nitrogen. This symbiosis results in increased plant growth under nutrient-limited conditions. Many ecosystems are comprised of plants that are obligately mycorrhizal. Arbuscular mycorrhizal (AM) fungi are widely adapted to many soil environments and have significant differences in ability at the species and ecotype level to promote plant growth and nutrition. Many studies have been conducted to examine the ecological relationships between various fungi and plant species under natural conditions or when introduced to different soils. Although pH is clearly an important variable in the ecology of mycorrhizal plants, the specific effects of soil pH are confounded by interactions with other soil properties such as soil structure, texture, and organic matter content, and it is difficult to predict AM responses only with respect to changes in soil pH [45].

Mycorrhizal species vary in their efficacy in increasing plant growth on acid and alkaline soils. The response of different plant species to different AM fungi varies with the isolate, plant species, and soil type, which complicates the identification of isolates that may have application for use as soil inoculants [46]. In a study with three different AM fungi, *Glomus etunicatum*, *G. diaphanum*, and *G. intraradices*, soil pH was found to affect significantly growth and root colonization of maize (*Zea mays* L.) [47]. When grown on acid (pH 4) and alkaline (pH 8.0) soils, shoot and root dry matter, leaf area, and root length were consistently higher for mycorrhizal than for nonmycorrhizal plants. Among the three fungi compared, inoculation with *G. etunicatum* resulted in higher plant biomass than inoculation with the other two fungi for plants grown on alkaline soil. However, on acid soils the differences among the mycorrhizal fungi isolates were minor. On the acid soil, plants colonized by *G. etunicatum* took up less phosphorus than plants that were infected with the other two fungal isolates [48]. This suggests that this particular fungus was slightly less adapted to the acid soil conditions. An interesting observation in the study was that very few roots had arbuscules or vesicles in acid soil, whereas these structures were abundant for all mycorrhizal plants grown in alkaline soil. Vesicles are important for lipid storage, and arbuscules are the internal structures in the root cells that facilitate nutrient exchange between the fungus and the host plant. Presumably, the absence of these structures could be the result of altered host physiology and carbon allocation, or this may be indicative of stress for the mycorrhizal fungus.

Mycorrhizae also appear to enhance the alkalization or acidification of the rhizosphere in response to nitrogen fertilizer uptake by plants [49]. Nitrate supply results in alkalization of the rhizosphere when plants release hydroxyl ions to maintain an ion balance across the cell membrane. Conversely, ammonia uptake by plants results in acidification of the rhizosphere due to the secretion of protons. This can result in a significant change in pH in the rhizosphere of 1 to 2 pH units, depending on the plant species and soil type [50]. In studies with onion, pH changes promoted by either mycorrhizal or nonmycorrhizal roots were studied by a nondestructive technique using the pH indicator bromocresol purple. Results showed that the pH changes observed depended on the symbiotic status of the root and the nitrogen form added to the soil. Mycorrhizal roots supplied with nitrate initially promoted a more intense alkalization on their surface as compared with the control roots after 30 days. However, at the end of the experiment at 60 days, intense acidification halos were observed in the mycorrhizosphere, and acidification was almost absent in the nonmycorrhizal rhizosphere. Such observations have led to speculation about the relationship between mycorrhizae-induced pH changes in the soil and the higher efficiency in the exploitation of nitrogen in the rhizosphere by the arbuscular-mycorrhizal plants [49].

The importance of soil pH in ectomycorrhizae formation and function has also been well studied and is of particular concern in forest soils that have been negatively affected by acid rain. In studies on *Pinus massoniana*, with and without inoculation by the ectomycorrhizal fungus *Pisolithus tinctorius*, plants subject to artificial acid rain showed declines in the levels of important enzymes including acid phosphatase, alkaline phosphatase, nitrate reductase, mannitol dehydrogenase, and trehalase throughout the roots, stems, and leaves [51]. Part of this affect can be attributed to increased toxicity of aluminum to the plants that occurs in conjunction with acid rain. This aluminum toxicity can be ameliorated by the addition of calcium. However, one of the concomitant problems associated with acid rain is the leaching of base cations from the soil. In this study, mycorrhizal trees were better able to withstand the stress imposed by acid rain and aluminum. Thus, the decline in mycorrhizal infection leads to a spiral effect by which the effects of acid rain and aluminum toxicity become more severe with the loss of the mycorrhizal roots.

The effects of liming on mycorrhizae formation and function also have been studied with oak under natural conditions. In a long-term study by Bakker and coworkers [52], ectomycorrhizae infection was compared on *Quercus petraea* and *Q. robur* at 10 sites in plots that were not limed or that had received lime at various times ranging from 1 to 27 years before the sampling. At all of the sites, liming slightly but significantly increased the total fine root length, which was associated with an increase in the total number of mycorrhizal root tips. Liming also caused a significant shift in the types of mycorrhizae fungal symbionts, with increased numbers of hairy types as opposed to smooth types. The hairy-type mycorrhizae have an increased absorption surface and thus are presumably more efficient in nutrient acquisition. Shifts also occurred in the species of smooth-type mycorrhizae. Although soil pH explained part of the shift in the smooth-type mycorrhizae, it did not in the hairy types. There are many possible consequences of these mycorrhizal species shifts. In low-pH soils, some ectomycorrhizae can protect against the effects of heavy metals by sequestering metals in their cell walls or on the cell surface [53]. For example, it has been shown that ectomycorrhizal fungi differ considerably in their effect on Pb accumulation in spruce, which is a function of the binding capacity of the extramatrical mycelium [54]. Shifts in mycorrhizal species may also influence resource allocation to the roots and the extent of the soil that is explored by the hyphae, depending on whether the fungi form long running rhizomorphs or localized fans. The importance of many of these factors in ecosystem function remains to be examined.

5.2 Soil pH Effects on Nitrogen Fixation

Considerable research has been conducted on the effects of soil pH on nitrogen fixation. In general, the population size and persistence of *Rhizobium* inoculants are greatest in soils with near-neutral to alkaline pH. In acid soils, even relatively small increases in pH achieved by soil liming can cause significant increases in nitrogen fixation. Up to 50% increases in N fixation have been observed for pea plants on increasing the soil pH from 3.9 to 4.5 [55]. Extremes of pH affect nodulation by reducing the population size of the *Rhizobium* as well as the ability to form nodules. Highly acidic soils (pH less than 4.0) frequently have low levels of phosphorus, calcium, and molybdenum and high concentrations of aluminum and manganese, which are toxic for both the plant and the rhizobia. In general, nodulation is more affected than host plant growth and nitrogen fixation. Highly alkaline soils with a pH greater than 8.0 tend to be high in sodium chloride, bicarbonate, and borate and are often associated with high salinity, all of which reduce nitrogen fixation [56].

Rhizobium and Bradyrhizobium strains vary substantially in their acid tolerance, and strains that are indigenous to acid soils may have greater nodulation capability in these soils than introduced strains [57]. Because of the need for effective isolates that can be used on acid soils, there has been considerable work aimed at isolating acid-tolerant Rhizobium strains in enrichment cultures. Screening of isolates on low-pH agar media has been one of the standard techniques, but this method does not necessarily select for strains that perform well when screened in vivo with plants grown under acid soil conditions. This suggests that many factors, other than pH, are involved in adaptation and the ability to form nodules under acid conditions [58]. In vitro experiments suggest that calcium interactions may be particularly important in determining the growth of acid-tolerant and acid-sensitive strains of Rhizobium. In studies with selected strains of Rhizobium meliloti and Tn5-induced acid-sensitive mutant strains, increases in the calcium concentration had strong effects on growth and survival of all of the strains that were tested [59]. Among the mutant strains, four of the five tested would grow even at pH 5.5 if high concentrations of calcium were added to the medium.

Differences in the acid tolerance of different strains of Rhizobium appear to involve changes in the cell membrane lipid composition. In experiments with two strains of Mesorhizobium loti differing in their tolerance to pH, both acid-tolerant and acid-sensitive strains showed shifts in lipid composition when grown at pH 5.5 as compared with growth at pH 7.0 [60]. In the acid-sensitive strain, membrane phosphatidylglycerol decreased and phosphatidylcholine increased, compared with cells grown at pH 7.0. Shifts in membrane components were also observed in the acid-tolerant strain in which phosphatidylglycerol, phosphatidylethanolamine, and lysophospholipid decreased 25%, 39%, and 51%, respectively, while phosphatidyl-N-methylethanolamine and cardiolipin increased 26% and 65%, respectively, compared with cells grown at pH 7.0. Also, certain long-chain fatty acids (19:0 cy and 20:0) increased in both strains at pH 5.5, which were even further elevated in the acid-tolerant strain at pH 4.0.

In addition to influencing the activity and population size of various bacteria, such as Rhizobium, that are introduced as soil inoculants, soil pH may affect the passive transport of the inoculum through the soil and active taxis of bacteria to the plant roots. As shown for Azospirillum, the sorption of bacteria to soil is significantly increased with decreasing pH. This is due to an increase in the density of positive charges on humus and on the edges of clay particles [61]. This effect is likely to be most severe for bacteria that are poor competitors in bulk soil because low pH may contribute to decreased native inoculation potential of acid soils following fallow periods in the absence of a plant host.

6 SOIL pH EFFECTS ON MICROBIAL TRANSFORMATIONS OF NITROGEN

Microorganisms are involved in transformations of nitrogen, sulfur, and phosphorus, which continually undergo biogeochemical cycling. Because the various microorganisms that carry out these processes all have different pH optima, soil pH has a profound effect on the transformation of these mineral nutrients and can affect both the chemical form of these elements in the soil and their availability to plants.

6.1 Nitrogen Transformations

The effects of soil pH on nitrogen transformations have been particularly well examined over the past 40 years. Numerous studies have clearly established the direct relationship between pH and the rates of nitrification and denitrification by soil bacteria. The first step in nitrogen mineralization is ammonification, in which ammonia nitrogen is enzymatically released from the organic matter. Ammonification is generally insensitive to pH and can be measured at significant rates over a broad pH range [62]. Nitrification, which is the oxidation of ammonia to nitrite and nitrate, is strongly affected by soil pH. This process can be carried out by both heterotrophic microorganisms and chemolithotrophs. Heterotrophic conversion of ammonia to nitrate appears to be a fortuitous process of no benefit to microorganisms and occurs at a low rate in acid soils [63]. Chemolithotrophic conversions of nitrogen by ammonia-oxidizing and nitrite-oxidizing bacteria are carried out using ammonia and nitrite as electron donors for metabolism and occur at the fastest rate in slightly alkaline pH soils [64]. Based on observations of several hundred soils in the United States, three general patterns were suggested for nitrogen transformation [65]. At pH 6.9 to 7.8, ammonia is rapidly oxidized to nitrite that accumulates before conversion to nitrate. Between pH 5.0 and 6.4, ammonia is oxidized to nitrate with no accumulation of nitrite. Below pH 5.4, ammonia is only very slowly oxidized to nitrate.

The nitrifying bacteria include the prototypic genera *Nitrosomonas* and *Nitrobacter* but also include a number of other genera and species. These bacteria have a rather narrow pH optimum, and ammonia conversion to nitrate occurs most rapidly in the pH range 7 to 8. Population sizes of ammonia-oxidizing bacteria in acidic forest soil or humus are below detection at pH 4 and increase with increasing pH to 10^3 to 10^5 per gram soil, depending on the availability of ammonia [66]. Mathematical models of nitrification rates have been used at the regional scale to estimate the effects of pH on nitrate formation. Data collected in the 12-state corn belt region of the United States during a 4-year period following a uniform protocol showed that the production of nitrate N in the surface 0 to 30 cm of soil during the preplant to V-5 growth stage of corn reached a maximum at a pH of about 6.7 and was negligible at pH <5.0 or >8.5 [67].

Denitrification occurs when nitrate is used as an alternative electron acceptor for respiration under low-oxygen conditions and is carried out by a wide range of bacteria. This process is governed by the availability of NO_3, NO_2, and organic substrates. In addition to these factors, oxygen availability, soil moisture, temperature, and pH are important secondary factors [68,69]. As this process is mediated by bacteria, denitrification occurs most rapidly at neutral to alkaline pH conditions that favor bacterial growth [69]. Based on isolation of heterotrophs, it is estimated that approximately 5% of the bacteria in soil are capable of denitrification, although soils have been found in which one third of the bacteria were capable of carrying out denitrification [70]. Denitrifying bacteria are present in soils having a pH as low as 4 but may be present at an insufficient population density to carry out significant denitrification. An exception to this rule is soils with high organic matter under water-saturated conditions, in which addition of nitrate can still result in significant denitrification rates at a pH as low as 4.7 [71]. The pH optimum for different denitrifying bacteria varies considerably, and this is one example where functional redundancy allows this process to occur over a relatively wide pH range. In studies by Valera and Alexander [70] in the 1960s, *Micrococcus* and *Achromobacter* were shown to be effective denitrifiers at alkaline pH, whereas *Bacillus lichenformis* had an optimum at neutral pH and *Pseudomonas denitrificans* at slightly acid pH.

When examined over a wide pH range, 30-fold increases in denitrification rates have been observed over the transition from pH 4.8 to 5.4 [72]. However, some caution should be used in defining a precise optimum for pH and denitrification rates, as soil microbial communities may adapt to the prevailing soil pH [73]. Because rates of denitrification depend on both denitrifying enzyme activity and nitrate availability as well as other environmental factors, this process can be characterized by a parameter termed "denitrification potential" in which standardized conditions are imposed in a laboratory assay. This assay shows that there is a simple relationship between denitrifying enzyme activity and denitrification potential with respect to pH [69]. Soil pH also affects the form of nitrogen that is volatilized because both N_2O and NO gas can be produced during denitrification. Net fluxes of N_2O decrease with decreases in soil pH, whereas NO fluxes show little dependence on pH, with the highest production levels occurring around pH 6 [73].

7 SOIL pH AND PHOSPHATE SOLUBILIZATION BY MICROORGANISMS

Phosphorus-solubilizing microorganisms are important in the dissolution of mineral phosphate, and many different bacteria and fungi are capable of solubilizing rock phosphate and calcium phosphates. A general estimate is that approximately

20% of the microorganisms isolated from soil are capable of solubilizing inorganic phosphate [74]. The mechanisms by which microorganisms enhance phosphorus solubilization need to be considered in relation to the predominant forms of phosphorus in a given soil. Many soils are dominated by organic phosphates, which are mineralized as a result of acid and alkaline phosphatase enzymes that are produced by both plants and microorganisms [75]. In alkaline soils containing calcium phosphates, or in soils that have been fertilized with rock phosphates, acid production and pH lowering are the primary mechanisms for P solubilization. Conversely, in acid soils dominated by iron phosphates, microbially produced organic acids may form iron complexes that result in dissolution of the iron phosphate minerals and can displace ligand-bound P from clay minerals. Iron and aluminum phosphates can also be solubilized by pH increases and by addition of organic acids that form complexes with iron and cause the dissolution of iron phosphate minerals.

The relative importance of phosphate-solubilizing bacteria for increasing P availability to plants is still being actively investigated and remains controversial. Results of soil inoculation experiments are inconsistent but in certain cases have suggested that the phosphate nutrition of crops can be improved through seed inoculation with these bacteria [76]. Phosphorus is one of the most limiting nutrients in many soils, and low-cost methods for increasing its availability and the efficient use of phosphate fertilizer additions to soil are of great importance for low-input agriculture. Isolation and identification of microorganisms that solubilize phosphate are accomplished using various agar-based screening media or growth in liquid media supplemented with solid-phase phosphate minerals. When screened in this manner, it is possible to isolate hundreds of microorganisms, including both bacteria and fungi, that are capable of acidifying their medium and causing phosphate solubilization [77]. Several studies have focused in particular on rhizosphere isolates that have been examined in more detail [78], including both plant growth–promoting bacteria and free-living diazotrophs [79]. The relative importance of the pH lowering effects in solubilization of P by bacteria is controversial, as it is difficult to determine the actual pH lowering effect that may occur around bacterial colonies in situ. Bacteria can solubilize rock phosphate and dicalcium phosphate on unbuffered media but not on a pH-buffered agar medium [80]. This suggest that the pH lowering effect may be most important in unbuffered soils not containing large amounts of calcium carbonate. Recently, the ability of organic acids to displace sorbed P has received more attention as a possible mechanism for increasing the bioavailability of phosphate. When rock phosphates are added to soil, negatively charged organic acids may slow the sorption of P to clay mineral surfaces and prolong its bioavailability to plants. Also, surface-bound P is largely unavailable to plants but appears to be more accessible to microorganisms that over time can increase the $NaHCO_3$-extractable or plant-available P [81,82].

8 RHIZOSPHERE BACTERIAL COMMUNITIES IN RELATION TO SOIL pH

The effects of soil pH on the soil rhizosphere microflora have not been well investigated despite the importance of root-associated microorganisms in plant nutrition and disease. One reason for this has been the lack of culture-independent methods to examine changes in the rhizosphere microflora. Plate culture methods that involve isolations of hundreds of bacteria, followed by chemical or genetic identification of each isolate, are too labor intensive to use for examining changes in rhizosphere communities with respect to pH. Moreover, such methods may miss significant changes, as recent studies using 16S rDNA profile methods suggest that culturable microorganisms isolated from plant roots may constitute only a small proportion of the bacteria in the rhizosphere [83]. To better examine changes in eubacterial communities with respect to pH, we have examined the effect of altering soil pH on the rhizosphere bacterial communities that develop in association with the root tips of sorghum and cowpea seedlings (S.A. Alvey and D.E. Crowley, unpublished data).

The soil used in this experiment was a Handford loamy sand with a baseline pH of 8.1. Prior to the experiment, batches of the soil were amended with 0.1 N HCl to pH values of 4, 5, 6, 7, and 8 by placing them into beakers in a 1:1 soil-water suspension. The pH was adjusted daily by addition of dilute HCl (with mixing) for a period of 10 days until the pH had stabilized in all soil samples. The soils were then drained, air dried, and 100-g aliquots were used to fill 95-mL plant growth container tubes. The soils were amended with minimal salts medium to provide nitrogen, phosphorus, and trace elements and were then planted with pregerminated seedling of sorghum and cowpea using three replicate plants of each species at each pH level. After 18 days of growth, the plants were removed from the containers and the soil pH was measured again to determine the final values that were used for the statistical analyses. Ten replicate root tips per plant sample with adhering soil were placed into bead beater tubes for extraction of DNA using the Fastprep system following the procedures described by Yang and Crowley [83]. The bacterial 16S rDNA was then amplified and separated by polymerase chain reaction, denaturing gradient gel electrophoresis (PCR-DGGE) to generate a bacterial community profile. The denaturing gradient gels were stained, photographed, and subjected to image analysis to generate line plots of each root sample. The line plots were then analyzed using peak-fitting software to resolve the individual peaks, and these data were then analyzed using canonical correspondence analysis. This type of analysis ordinates the sample data with respect to their similarities based on the first two factors that describe most of the variation in the data set.

As shown in Figures 1 and 2, the line plot profiles of the rhizosphere communities from sorghum and cowpea contained complex communities with a few

Sorghum

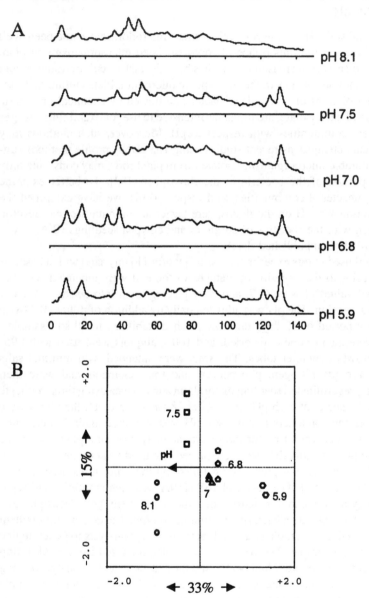

predominant bands that were common to all samples. Examining visually, it can also be seen that the community species composition and the relative predominance of different bacterial groups shift in response to pH. Statistical analysis of the community shifts is shown in Fig. 1B and 2B, with an ordination diagram for the communities that were produced at each pH for the two plant species. In the plot for sorghum, 22 and 20% or a total of 42% of the variation in the data set was described by the first two factors, which are plotted on the x and y axes, respectively. The statistical significance of pH for causing the observed shifts in the community structures was determined by Monte Carlo analysis and was found to be highly significant ($P < .01$) for both plant species. The arrow on the x axis indicates the effect of increasing soil pH as determined when pH was included in the analysis as an environmental variable and used to constrain the first factor to this variable.

Results of this analysis showed a clear separation of the communities that formed in the rhizosphere of sorghum at pH 8.1 and 7.5, with the community at 8.1 occurring at the bottom left of the diagram along the x axis. At the intermediate pH values of 7 and 6.8, the communities were indistinguishable, whereas at pH 5.9, the three samples for these bacterial communities clustered together to the far right on the x axis. In the rhizosphere of cowpea, the community that developed at pH 7.5 was clearly separated and clustered to the right along the x axis. As was observed with sorghum, communities that developed at the intermediate pH values near neutrality were highly similar with respect to factor 1 but separated into two clusters along the second factor plotted on the y axis. The top group comprised communities that developed at pH 7 but also contained points representing communities that grew at pH 6.7. Overall, the effect of lowering the pH was more variable for cowpea than for sorghum.

FIGURE 1   Influence of soil pH on bacterial community structure in the rhizosphere of sorghum as determined by 16S rDNA profiles generated by PCR-DGGE. (A) Line profiles showing 16S rDNA peaks obtained by image analysis of DGGE gel for communities associated with sorghum in a sandy loam soil adjusted to different pH levels. (B) Ordination diagram constructed using correspondence analysis statistical procedures showing similarities in bacterial communities for three replicate plants at pH 5.9, 6.8, 7, 7.5, and 8.1. Thirty-three percent and 15% of the variation in the data set are explained by the first two factors plotted on the x and y axes, respectively. The relative difference in community structure for any two samples is represented by the distance separating the points corresponding to the individual communities. Arrow on the x axis shows the effect of pH in separating the communities, when included in the data set as an environmental factor. In this case, pH explained 100% of the variation associated with factor 1, plotted on the x axis.

Cowpea

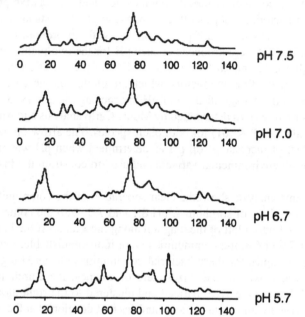

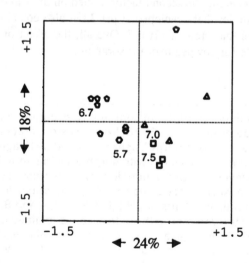

These data clearly show that soil pH can significantly affect the species composition and relative predominance of different bacterial groups in the plant rhizosphere. More detailed analysis of the community shifts may be accomplished by excising bands of interest that can then be cloned and sequenced to identify the predominant bacteria that are represented by individual bands. Alternatively, DNA probes may be used to identify changes in selected species that are represented by their 16S rDNA signatures. Although there is still much to be learned about the actual species that were affected by pH change, this type of analysis may eventually be useful for examining pH-related effects on the functional properties of these communities with respect to plant growth promotion, biodegradation of xenobiotics, and plant health.

The present difficulty in correlating shifts in community structure and species composition with enzyme activities and nutrient dynamics probably reflects the fact that many different possible community structures can carry out the same biological functions. However, so far, data on microbial communities obtained using high-resolution techniques such as DNA profiling of species composition, bacterial enumerations on agar media, and detailed measurements of specific enzyme activities have proved highly variable with respect to soil type, water, and climate. New methods to describe this variability and the relationship between community structure and function are now being applied. These procedures include neural networks and discriminate analyses in which computer models are trained to recognize patterns. Modeling of microbial communities might also eventually entail the use of fractal mathematics, in which environmental factors such as pH operate as so-called strange attractors that influence possible subsets of community structure and species composition. As we gather more information of the functional properties of various sets of communities, it should eventually be possible to examine microbial community shifts at the level of fatty acid or DNA signatures in relation to important ecosystem traits.

FIGURE 2 Influence of soil pH on bacterial community structure in the rhizosphere of cowpea as determined by 16S rDNA profiles generated by PCR-DGGE. (A) Line profiles showing 16S rDNA peaks obtained by image analysis of DGGE gel for communities associated with cowpea in a sandy loam soil adjusted to different pH levels. (B) Ordination diagram constructed using correspondence analysis statistical procedures showing similarities in bacterial communities for three replicate plants at pH 5.7, 6.7, 7.0, and 7.5. Twenty-four percent and 18% of the variation in the data set are explained by the first two factors plotted on the x and y axes, respectively. The relative difference in community structure for any two samples is represented by the distance separating the points corresponding to the individual communities.

9 EFFECTS OF pH ON BIODEGRADATION OF XENOBIOTICS

Soil pH effects on biodegradation of xenobiotics are dependent on the chemical characteristics of the target compound and the numbers and types of microorganisms that are able to degrade it. For compounds that are degraded by many different microorganisms, there is greater functional redundancy, which increases the likelihood that the compound will be degraded over a wide pH range. Different degrader species will be selected depending on the pH and other prevailing environmental conditions. On the other hand, if only one or a few species of degraders are capable of metabolizing the chemical, greater attention must be given to the pH optimum for enhancing degradation activity. Soil pH may affect not only the size of the degrader population but also its activity. An experiment clearly demonstrated this principle, in which degradation of the fungicide chlorothalonil, 2,4,5,6-tetrachloroisophthalonitrile, was compared among four soils subjected to different types of fertilizer application [84]. Biodegradation was enhanced by the addition of organic fertilizers that provided alternative carbon sources for growth. Degradation of chlorothalonil was further increased in the organic matter–amended soils by adjustment of the soil pH to a neutral value, although the most probable number of degrading microorganisms remained constant. Thus, enhanced microbial degradation of chlorothalonil was due to the increase in the degrading capacity by the maintenance of a near-neutral pH value in the soil rather than an increase in the number of degrading microorganisms.

For certain compounds, xenobiotic degraders can be isolated from soils under even extreme pH conditions. Enrichment cultures from an alkali lake contaminated with chloroaromatic compounds have degrader organisms with an optimal pH range from 9.0 to 9.4 [85]. One strain isolated in this particular research was a *Nocardioides* and was classified as a slightly halophilic alkaliphile with an optimal sodium ion concentration of 0.2 to 4 M for growth. Conversely, under acidic conditions, biodegradation of aromatic hydrocarbons has been evaluated in soil samples along gradients of both contaminant levels and pH [86]. The soil pH values in areas affected by runoff from the storage basin described in this report were as low as pH 2.0. However, even at this low pH, the indigenous microbial community was metabolically active and mineralized more than 40% of the parent hydrocarbons, naphthalene and toluene. Results of this experiment suggested that the degradation of aromatic compounds involved a microbial consortium instead of individual acidophilic bacteria. No defined mixed cultures were able to mineralize radiolabeled hydrocarbons, but an undefined mixed culture consisting of a fungus, a yeast, and several bacteria successfully metabolized approximately 27% of supplied naphthalene after 1 week. Data such as these emphasize the need for a community-based approach to study microorganisms and consortia that degrade organic compounds at different pH and soil conditions.

Techniques such as the use of PCR-DGGE of 16S rDNA and other community profiling techniques should prove to be particularly useful for this application.

10 PLANT DISEASE AND SOIL pH

Soil pH has a significant effect on root pathogens, which are mediated by a number of different mechanisms. First, soil pH may affect the growth and survival of the pathogen, which like all microorganisms is competitive only within a certain pH optimum. This may vary for different pathogens depending on the range of pH over which they can grow. A classic example in which pH changes can be used to control pathogens is the use of soil acidification to control *Streptomyces scabies*, the causative agent of potato root rot. Other effects of soil pH include the influence of pH on growth of competitors and disease-suppressive microorganisms that antagonize the pathogen and the effects of soil-borne vectors, such as nematodes and soil insects. Lastly, mineral nutrient interactions with pH may alter the ability of plants to resist the pathogen and may play a role in limiting the pathogen's ability to acquire essential nutrients such as iron.

Many plant growth–promoting bacteria have multiple roles in disease suppression, including plant growth promotion through the production of hormones, removal of toxicants such as hydrogen cyanide, suppression of pathogen growth through the production of siderophores, and direct antagonistic effects involving secretion of antibiotics or lytic enzymes. The relationships between bacterial antagonism of pathogenic fungi and pH are further moderated by the soil clay content, which may affect microsite pH at the surface of clay minerals [87]. Given all of these possible effects of pH, it may not be possible to consider only one type of interaction, and multivariate analyses should probably be considered in determining the effects and role of pH in plant-disease interactions. Disease-suppressive bacteria that have been identified for possible use in biocontrol include many *Bacillus* sp. and certain *Pseudomonas* sp. strains. In addition, some actinomycetes are disease suppressive against pathogenic fungi. Actinomycetes have been found to grow well at pH 6.5 to 8 but generally fail to grow at pH 5.5 [88]. There are also significant effects of pH on pseudomonads, which may alter their production of siderophores, antibiotics, and plant growth hormones at different pH levels [89]. Antibiotic production by *P. fluorescens* strains that produce diacetylphloroglucinol is significantly greater at pH 6.4 as compared with soil at the lower pH values of 5.4 and 4.4 [90]. Soil pH may also affect the relative efficacy of pseudomonad-produced phenazine antibiotics with some 10-fold higher antibiotic activity occurring at neutral pH than under acid soil conditions [91]. Because of the idiosyncrasy of different organisms and their interactions, the effects of pH will probably need to be determined for individual diseases in different soils. The same will also hold true for use of disease-suppressive bacteria in which pH is likely to have a significant effect on the efficacy of biocontrol. Presently, soil factors such as differences in pH, clay type and content, or organic matter content are seldom considered in the ecology of disease-suppressive microorganisms [92].

In general, acid soils are conducive to many diseases caused by fungi, and disease suppression of many fungi can be enhanced by liming the soil to a neutral pH. *Pythium* stem rot of wheat and vascular wilt caused by *Cephalosporium graminearum* are both enhanced by low pH and are inhibited in limed soils [93,94]. Cavity spot of carrots, which is also caused by *Pythium*, has been suppressed by the addition of lime to increase the soil pH from 5.1 to 6.9 [95]. This may involve a variety of pH-related changes in the soil microflora. In the case of cavity spot of carrot, suppression of the disease was associated with a general increase in the total numbers of aerobic bacteria, fluorescent pseudomonads, and actinomycetes and a concomitant decline in the colony-forming units of filamentous fungi, yeast, and the population size of the fungal pathogen [95]. In addition to competition for nutrients, pH might affect direct antagonism by elevating the population size of specific microorganisms that directly parasitize *Pythium*. Different methods are used to separate out these effects, but one of the most powerful methods is to use fungal baits to isolate microorganisms that parasitize the hyphae of these and other root pathogens and then study their ecology in relation to pH.

Soil pH also controls disease pathogens through nutritional effects involving metals. For example, liming soils to increase pH is thought to control *Fusarium* wilts by increasing competition for both carbon and iron [96]. Populations of nonpathogenic *Fusarium oxysporum* and fluorescent *Pseudomonas* sp. are at least partly responsible for competition for carbon and iron, respectively, along with other possible factors that may contribute to their biocontrol activity. In addition to competition for iron, several fungal diseases are influenced by manganese (Mn), which has increased solubility as the Mn(II) ion at low pH and thus a greater inhibitory effect in low-pH soil [97]. The specific effects of Mn vary with the disease-causing organism. *Streptomyces scabies* has inhibited vegetative growth in the presence of high concentrations of manganese [98]. *Geaumannomyces graminis*, which causes take-all decline of wheat and barley, has a growth optimum at pH 7 and has inhibited growth at a pH below 6.8. Severity of the disease increases with increasing pH over the pH range 5.5 to 8.5 [99]. Manganese does not affect growth of the fungus but increases the resistance of the plant to invasion by the fungus. When manganese is oxidized, it becomes less available, which is a mechanism used by *G. graminis* to avoid manganese toxicity. Conversely, certain manganese-reducing pseudomonads that thrive at high pH are able to solubilize Mn and also may be antagonistic to the fungus by producing various antibiotic substances. In this manner, soil pH has myriad effects on disease development, including altered plant resistance to the pathogen, metal toxicities, competition for iron, and changes in microbial community structure that influence competition for carbon and the growth of microorganisms that directly antagonize the disease-causing microorganism.

11 CONCLUSIONS

Soil pH is a master variable in soil microbial processes, with effects on organic matter decomposition, nutrient cycling, transformations of nitrogen, xenobiotic degradation, and plant-disease interactions. Much of the research on this topic so far has either involved studies on the ecology of individual strains and species or examined very broad-level ecosystem processes. For many processes, the overlapping abilities of microorganisms to mineralize organic matter and degrade various substances provide a functional redundancy that allows terrestrial ecosystems to function over a broad range of pH and soil conditions. However, for certain processes, such as nitrification, that are carried out by only a few species of microorganisms, the impact of pH is much greater, and significant changes in nutrient transformation rates are observed over relatively small pH increments.

Considerable progress has been made on the effects of soil pH at the process level, for example, in causing shifts between bacterial- and fungal-dominated decomposition. Still, there is interest in developing a better understanding of the role of microbial diversity and community in ecosystem function, where our knowledge of the effects of pH remains very limited. New molecular techniques using 16S rDNA community profile and biochemical methods using fatty acid analyses provide powerful tools for assessing the effects of pH and other environmental factors on soil microbial communities. As agricultural biotechnology increasingly moves toward methods to manipulate rhizosphere microbial communities for improved plant nutrition and health, a new consideration of the effects of soil pH on microbial communities will undoubtedly be important for making future advances in plant nutrition and soil ecology.

REFERENCES

1. J Banfield, WW Barker, S Welch, A Taunton. Biological impact on mineral dissolution: application of the lichen model to understanding mineral weathering in the rhizosphere. Proc Natl Acad Sci USA 96:3404–3411, 1999.
2. L Rigou, E Mignard. Factors of acidification of the rhizosphere of mycorrhizal plants: measurement of pCO_2 in the rhizosphere. Acta Bot Gall 141:533–539, 1994.
3. DL Jones, PR Darrah. Role of root derived organic acids in the mobilization of nutrients from the rhizosphere. Plant Soil 166:247–257, 1994.
4. Y Bashan, H Levanony. Alterations in membrane potential and proton efflux in plant roots colonized by *Azospirillum brasilense*. Plant Soil 137:99–103, 1991.
5. DB Johnson, S McGinness, MA Ghauri. Biogeochemical cycling of iron and sulfur in leaching environments. FEMS Microbiol Rev 11:63–70, 1993.
6. K Edwards, BM Goebel, T Rodgers, M Schrenk, T Gihring, M Cardona, B Hu, M McGuire, R Hamers, N Pace, J Banfield. Geomicrobiology of pyrite (FeS_2) dissolution: case study at Iron Mountain, California. Geomicrobiol J 16:156–179, 1999.

7. H Kobayashi. Bacterial adaptation to change in environmental pH. Jpn J Bacteriol 51:745–753, 1996.

8. T Koga, F Sakamoto, A Yamoto, K Takumi. Acid adaptation induces cross-protection against some environmental stresses in *Vibrio parahaemolyticus*. J Gen Appl Microbiol 45:155–161, 1999.

9. GL Lorca, RR Raya, MP Taranto, GF De Valdez. Adaptive acid tolerance response in *Lactobacillus acidophilus*. Biotechnol Lett 20:239–241, 1998.

10. TA Krulwich, M Ito, R Gilmour, AA Guffanti. Mechanisms of cytoplasmic pH regulation in alkaliphilic strains of *Bacillus*. Extremophiles 1:163–169, 1997.

11. K Enomoto, N Koyama. Effect of growth pH on the phospholipid contents of the membranes from alkaliphilic bacteria. Curr Microbiol 39:270–273, 1999.

12. R Bardgett, E McAlister. Measurement of fungal:bacterial ratios as an indicator of ecosystem self regulation in temperate meadow grasslands. Biol Fertil Soils 29:282–290, 1999.

13. K Sakamoto, Y Oba. Effects of fungal to bacterial biomass ratio on the relationship between CO_2 evolution and total soil microbial biomass. Biol Fertil Soils 17:39–44, 1994.

14. JPE Anderson, KH Domsch. The metabolic quotient for CO_2 qCO_2 as a specific activity parameter to assess the effect of environmental conditions, such as pH, on the microbial biomass of forest soil. Soil Biol Biochem 25:393–395, 1993.

15. M Raubuch, F Beese. Patterns of microbial indicators in forest soils along an European transect. Biol Fertil Soils 19:362–368, 1995.

16. T Anderson. The influence of acid irrigation and liming on the soil microbial biomass in a Norway spruce (*Picea abies* (L.) K.) stand. Plant Soil 199:117–122, 1998.

17. M Raubuch, F Beese. Interaction between microbial biomass and activity and the soil chemical conditions and the processes of acid load in coniferous forest soils. Z Pflanzenernaehr Bodenkd 161:59–66, 1998.

18. E Blagodatskaya, T Anderson. Interactive effects of pH and substrate quality on the fungal-to-bacterial ratio and qCO_2 of microbial communities in forest soils. Soil Biol Biochem 30:1269–1274, 1998.

19. E Baath, K Arnebrant. Growth rate and response of bacterial communities to pH in limed and ash treated forest soils. Soil Biol Biochem 26:995–1001, 1994.

20. E Baath. Adaptation of soil bacterial communities to prevailing pH in different soils. FEMS Microbiol Ecol 19:227–237, 1996.

21. P Bottner, F Austrui, J Cortez, G Billes, M Couteaux. Decomposition of [14]C- and ^{15}N-labelled plant material, under controlled conditions,in coniferous forest soils from a north-south climatic sequence in western Europe. Soil Biol Biochem 30:597–610, 1998.

22. V Acosta-Martinez, M Tabatabai. Enzyme activities in limed agricultural soil. Biol Fertil Soils 31:85–91, 2000.

23. D Clay, C Clapp, R Dowdy, J Molina. Mineralization of nitrogen in fertilizer acidified lime-amended soils. Biol Fertil Soils 15:249–252, 1993.

24. M Nyborg, PB Hoyt. Effect of soil acidity and liming on mineralization of soil nitrogen. Can J Soil Sci 58:331–338, 1978.

25. KL Kalburtji, AP Mamolos, S Kostopoulou. Nutrient release from decomposing *Lotus corniculatus* residues in relation to soil pH and nitrogen levels. Agric Ecosyst Environ 65:107–112, 1997.

26. T Persson, A Wiren. Effects of experimental acidification on C and N mineralization in forest soils. Agric Ecosyst Environ 47:159–174, 1993.

27. T Pennanen, H Fritze, P Vanhala, O Kiikkila, S Neuvonen, E Baath. Structure of a microbial community in soil after prolonged addition of low levels of simulated acid rain. Appl Environ Microbiol 64:2173–2180, 1998.

28. T Pennanen, J Perkiomaki, O Kiikkila, P Vanhala, S Neuvonen, H Fritze. Prolonged, simulated acid rain and heavy metal deposition: separated and combined effects on forest soil microbial community structure. FEMS Microbiol Ecol 27:291–300, 1998.

29. H Kahle. Response of roots of trees to heavy metals. Environ Exp Bot 33:99–119, 1993.

30. WL Lindsay. Chemical Equilbria in Soils. New York: Wiley, 1979.

31. DL Parker, RL Chaney, WA Norvell. Chemical equilibrium models: applications to plant nutrition research. In: RH Loeppert, ed. Chemical Equilibrium and Reaction Models. Madison, WI: Soil Science Society of America, 1995, pp 163–195.

32. AM Ibekwe, JS Angle, RL Chaney, P Van Berkum. Enumeration and N_2 fixation potential of *Rhizobium leguminosarum* biovar *trifolii* grown in soil with varying pH values and heavy metal concentrations. Agric Ecosyst Environ 61:1679–1685, 1997.

33. DL Jones, DS Brassington. Sorption of organic acids in acid soils and its implications in the rhizosphere. Eur J Soil Sci 49:447–455, 1998.

34. CE Martinez, HL Motto. Solubility of lead, zinc and copper added to mineral soils. Environ Pollut 107:153–158, 2000.

35. SP McGrath, AM Chaudri, KE Giller. Long-term effects of metals in sewage sludge on soils, microorganisms and plants. J Ind Microbiol 14:94–104, 1995.

36. S Dahlin, E Witter, A Martensson, A Turner, E Baath. Where's the limit? Changes in the microbiological properties of agricultural soils at low levels of metal contamination. Soil Biol Biochem 29:1405–1415, 1997.

37. DJ Kushner. Effects of speciation of toxic metals on their biological activity. Water Pollut Res J Can 28:111–128, 1993.

38. A Belimov, A Kunakova, E Gruzdeva. Influence of soil pH on the interaction of associative bacteria with barley. Mikrobiologiya 67:561–568, 1998.

39. G Burd, D Dixon, B Glick. A plant growth-promoting bacterium that decreases nickel toxicity in seedlings. Appl Environ Microbiol 64:3663–3668, 1998.

40. ACC Plette, MM Nederlof, EJM Temminghoff, WH Van Riemsdijk. Bioavailability of heavy metals in terrestrial and aquatic systems: a quantitative approach. Environ Toxicol Chem 18:1882–1890, 1999.

41. M Ledin, K Pedersen, B Allard. Effects of pH and ionic strength on the adsorption of Cs, Sr, Eu, Zn, Cd and Hg by *Pseudomonas putida*. Water Air Soil Pollut 93:367–381, 1997.

42. P Marschner, DE Crowley. Phytosiderophores decrease iron stress and pyoverdine production of *Pseudomonas fluorescens* Pf-5 (PVD-INA Z. Soil Biol Biochem 30:1275–1280, 1998.

43. JE Loper, MD Henkels. Availability of iron to *Pseudomonas fluorescens* in rhizosphere and bulk soil evaluated with an ice nucleation reporter gene. Appl Environ Microbiol 63:99–105, 1997.

44. S Bhardwaj, S Dudeja, AL Khurana. Distribution of vesicular-arbuscular mycorrhizal fungi in the natural ecosystem. Folia Microbiol 42:589–594, 1997.

45. SD Frey, ET Elliott, K Paustian. Bacterial and fungal abundance and biomass in conventional and no-tillage agroecosystems along two climatic gradients. Soil Biol Biochem 31:573–585, 1999.

46. DM Sylvia, DO Wilson, JH Graham, JJ Maddox, P Millner, JB Morton, HD Skipper, SF Wright, AG Jarstfer. Evaluation of vesicular-arbuscular mycorrhizal fungi in diverse plants and soils. Soil Biol Biochem 25:705–713, 1993.

47. R Clark, S Zeto. Growth and root colonization of mycorrhizal maize growth on acid and alkaline soil. Soil Biol Biochem 28:1505–1511, 1996.

48. R Clark, S Zeto. Mineral acquisition by mycorrhizal maize grown on acid and alkaline soil. Soil Biol Biochem 28:1495–1503, 1996.

49. B Bago, C Azcon-Aguilar. Changes in the rhizospheric pH induced by arbuscular mycorrhiza formation in onion (*Allium cepa* L.). Z Pflanzenernaehr Bodenkd 160: 333–339, 1997.

50. H Marschner. Role of root growth, arbuscular mycorrhiza, and root exudates for the efficiency in nutrient acquisition. Field Crops Res 56:203–207, 1998.

51. FX Kong, Y Liu, W Hu, PP Shen, CL Zhou, LS Wang. Biochemical responses of the mycorrhizae in *Pinus massoniana* to combined effects of Al, Ca and low pH. Chemosphere 40:311–318, 2000.

52. M Bakker, J Garbaye, C Nys. Effect of liming on the ectomycorrhizal status of oak. Forest Ecol Manage 126:121–131, 2000.

53. DA Wilkens. The influence of sheathing ectomycorrhizae of trees on the uptake and toxicity of metals. Agric Ecosyst Environ 35:245–260, 1991.

54. P Marschner, DL Godbold, G Jentschke. Dynamics of lead accumulation in mycorrhizal and non-mycorrhizal Norway spruce (*Picea abies* (L.) Karst). Plant Soil 178: 239–245, 1996.

55. J Evans, P Chalk, G O'Connor. Potential for increasing N_2 fixation of field pea through soil management and genotype. Biol Agric Hortic 12:97–112, 1995.

56. L Bordeleau, D Prevost. Nodulation and nitrogen fixation in extreme environments. Plant Soil 161:115–125, 1994.

57. D Van Rossum, A Muyotcha, BM De Hoop, HW Van Verseveld, AH Stouthamer, FC Boogerd. Soil acidity in relation to groundnut-*Bradyrhizobium* symbiotic performance. Plant Soil 163:165–175, 1994.

58. LG Gemell, RJ Roughley. Field evaluation in acid soils of strains of *Rhizobium leguminosarum* bv. *trifolii* selected for their tolerance or sensitivity to acid soil factors in agar medium. Soil Biol Biochem 25:1447–1452, 1993.

59. WG Reeve, RP Tiwari, MJ Dilworth, AR Glenn. Calcium affects the growth and survival of *Rhizobium meliloti*. Soil Biol Biochem 25:581–586, 1993.

60. O Correa, E Rivas, A Barneix. Cellular envelopes and tolerance to acid pH in *Mesorhizobium loti* Curr Microbiol 38:329–334, 1999.

61. Y Bashan. Interactions of *Azospirillum* spp. in soils: a review. Biol Fertil Soils 29: 246–256, 1999.

62. W Dancer, A Peterson, G Chester. Ammonification and nitrification of N as influenced by soil pH and previous N treatments. Soil Sci Soc Am J 37:67–69, 1973.

63. JP Schimel, MK Firestone, KS Kilham. Identification of heterotrophic nitrification in a Sierra forest soil. Appl Environ Microbiol 48:802–806, 1984.

64. D Focht, W Verstrate. Biochemical ecology of nitrifiers and denitrifiers. Adv Microb Ecol 1:135–214, 1977.

65. L Morrill, J Dawson. Patterns observed for the oxidation of ammonia to nitrate by soil organisms. Soil Sci Soc Am Proc 31, 1967.

66. L Klemedtsson, Q Jiang, AK Klemedtsson, L Bakken. Autotrophic ammonium-oxidising bacteria in Swedish mor humus. Soil Biol Biochem 31:839–847, 1999.

67. A Olness. A description of the general effect of pH on formation of nitrate in soils. J Plant Nutr Soil Sci 162:549–556, 1999.

68. J Bandibas, A Vermoesen, C De Groot, A Van Cleemput. The effect of different moisture regimes and soil characteristics on nitrous oxide emission and consumption by different soils. Soil Sci 158:106–114, 1994.

69. M Simek, DW Hopkins. Regulation of potential denitrification by soil pH in long-term fertilized arable soils. Biol Fertil Soils 30:41–47, 1999.

70. C Valera, M Alexander. Nutrition and physiology of denitrifying bacteria. Plant Soil 15:268–280, 1961.

71. D Ekpete, A Cornfield. Effect of pH and addition of organic matter on denitrification losses from soil. Nature 280:1200, 1965.

72. DW Hopkins, AG O'Donnell, RS Sheil. Effect of fertilizer on soil nitrifier activity in experimentally limed plots. Biol Fertil Soils 5:344–399, 1988.

73. S Yamulki, RM Harrison, KWT Goulding, CP Webster. N_2O, NO and NO_2 fluxes from a grassland: effect of soil pH. Soil Biol Biochem 29:1199–1208, 1997.

74. KMN Kucey, HH Janzen, ME Leggett. Microbially mediated increase in plant available phosphorus. Adv Agron 42:119–228, 1989.

75. A Jungk, B Seeling, J Gerke. Mobilization of different phosphate fractions in the rhizosphere. Plant Soil 155 156:91–94, 1993.

76. SS Pal. Interactions of an acid tolerant strain of phosphate solubilizing bacteria with a few acid tolerant crops. Plant Soil 198:169–177, 1998.

77. M Choi, J Chung, T Sa, S Lim, S Kang. Solubilization of insoluble phosphates by *Penicillium* sp. GL-101 isolated from soil. Agric Chem Biotech 40:329–333, 1997.

78. CS Nautiyal, S Bhadauria, P Kumar, H Lal, R Mondal, D Verma. Stress induced phosphate solubilization in bacteria isolated from alkaline soils. FEMS Microbiol Lett 182:291–296, 2000.

79. KY Kim, D Jordan, GA McDonald. Effect of phosphate-solubilizing bacteria and vesicular-arbuscular mycorrhizae on tomato growth and soil microbial activity. Biol Fertil Soils 26:79–87, 1998.

80. P Gyaneshwar, GN Kumar, LJ Parekh. Effect of buffering on the phosphate-solubilizing ability of microorganisms. World J Microbiol Biotechnol 14:669–673, 1998.

81. P Illmer, F Schinner. Solubilization of inorganic calcium phosphate: solubilization mechanisms. Soil Biol Biochem 27:257–264, 1994.

82. Z He, J Zhu. Microbial utilization and transformation of phosphate adsorbed by variable charge minerals. Soil Biol Biochem 30:917–923, 1998.

83. CH Yang, DE Crowley. Rhizosphere microbial community structure in relation to root location and plant iron nutritional status. Appl Environ Microbiol 66:345–351, 2000.

84. T Mori, K Fujie, S Kuwatsuka, A Katayama. Accelerated microbial degradation of chlorothalonil in soils amended with farmyard manure. Soil Sci Plant Nutr 42:315–322, 1996.

85. O Maltseva, P Oriel. Monitoring of an alkaline 2,4,6-trichlorophenol–degrading enrichment culture by DNA fingerprinting methods and isolation of the responsible

organism, haloalkaliphilic *Nocardioides* sp. strains M6. Appl Environ Microbiol 63: 4145–4149, 1997.

86. RD Stapleton, DC Savage, GS Sayler, G Stacey. Biodegradation of aromatic hydrocarbons in an extremely acidic environment. Appl Environ Microbiol 64:4180–4184, 1998.

87. WD Rosenzweig, G Stotzky. Influence of environmental factors on antagonism of fungi by bacteria in soil: clay minerals and pH. Appl Environ Microbiol 38:1120–1126, 1979.

88. D Crawford, J Lynch, J Whipps, M Ousley. Isolation and characterization of actinomycete antagonists of a fungal root pathogen. Appl Environ Microbiol 59:3899–3905, 1993.

89. V Leinhos. Effects of pH and glucose on auxin production of phosphate-solubilizing rhizobacteria in vitro. Microbiol Res 149:135–138, 1994.

90. DC Naseby, JM Lynch. Effects of *Pseudomonas fluorescens* F113 on ecological functions in the pea rhizosphere are dependent on pH. Microb Ecol 37:248–256, 1999.

91. T Chin-A-Woeng, G Bloemberg, A Van Der Bij, K Van Der Drift, J Schripsema, B Kroon, R Scheffer, C Keel, P Bakker, H Tichy, F De Bruijn, J Thomas-Oates, B Lugtenberg. Biocontrol by phenazine-1-carboxamide–producing *Pseudomonas chlororaphis* PCL1391 of tomato root rot caused by *Fusarium oxysporum* f. sp. *radicis lycopersici*. Mol Plant Microbe Interact 11:1069–1077, 1998.

92. H Hoper, C Alabouvette. Importance of physical and chemical soil properties in the suppressiveness of soils to plant diseases. Eur J Soil Biol 32:41–58, 1996.

93. G Bruehl, P Lai. The probable significance of saprophytic colonization of wheat straw in the field by *Cephalsoporium graminearum*. Phytopathology 58:464–466, 1968.

94. R Cook, J Sitton, W Haglund. Increased growth and yield responses of wheat to reduction in the *Pythium* populations by soil treatments. Phytopathology 77:1192–1198, 1987.

95. K El-Tarabily, G Hardy, K Sivasithamparam, I Kurtboke. Microbiological differences between limed and unlimed soils and their relationship with cavity spot disease of carrots (*Daucus carota* L.) caused by *Pythium coloratum* in Western Australia. Plant Soil 183:279–290, 1996.

96. C Alabouvette. *Fusarium* wilt suppressive soils: an example of disease-suppressive soils. Aust Plant Pathol 28:57–64, 1999.

97. DM Huber, NS Wilhelm. The role of manganese in resistance to plant diseases. In: RD Graham, RJ Hannam, NC Uren, eds. Manganese in Soils and Plants. Dordrecht: Kluwer Academic, 1988, pp 155–173.

98. JJ Mortvedt, KC Berger, HM Darling. Effects of manganese and copper on the growth of *Streptomyces* and the incidence of potato scab. Am Potato J 40:96–102, 1963.

99. EM Ries, RJ Cook, BJ McNeal. Elevated pH and associated reduced trace nutrient availability as factors contributing to take-all upon soil liming. Phytopathology 73:411–413, 1983.

15

The Role of Acid pH in Symbiosis between Plants and Soil Organisms

Karen G. Ballen
Augsburg College, Minneapolis, Minnesota

Peter H. Graham
University of Minnesota, St Paul, Minnesota

1 INTRODUCTION

Acid soils constrain agricultural production on perhaps 25% of the world's crop-lands [1], including 101 million ha in the semiarid tropics and 262 million ha in the subhumid tropics [2]. Acid soils cover more than one third of sub-Saharan Africa [3], and Latin America alone includes more than 500 million ha with a soil pH less than 5.0 [4]. Stress under acid soil conditions may be due to hydrogen ion concentration per se or to deficiencies or toxicities of minerals that result from low pH. This chapter focuses on hydrogen ion effects and specifically addresses constraints that acid soils place on symbiotic relationships between plants and soil-dwelling microorganisms. These relationships are important to agriculture because they provide the plant with necessary nutrients that are often deficient in soil.

Symbiotic organisms are particularly vulnerable to soil acidity as the host, the microorganism, and the establishment of symbiosis itself are each exposed to acid soil stress. In the specific case of the legumes and their root-nodule symbiosis, a consequence is that nodulation will commonly fail in soils of pH 5.0 or less. (Note: In this chapter pH values refer to water pH.) In this chapter we examine the effects of soil acidity on symbiotic nitrogen fixation in legumes, actinorrhizal species, and *Azolla* where the microsymbionts are all bacteria, and on ecto- and endomycorrhizal symbioses involving fungal species.

2   pH AND THE LEGUME-*RHIZOBIUM* SYMBIOSIS

Symbiotic N_2-fixing bacteria, collectively known as rhizobia, infect leguminous plants, forming stem or root nodules. The bacteria derive energy from this process for growth, respiration, and N_2 fixation and are protected from the external environment; the host accesses a form of N it could not otherwise utilize. Rates of N_2 fixation vary but are commonly in the range of 50 to 200 kg N_2 fixed ha^{-1} annum^{-1} [5], reducing the plant's need for soil or fertilizer N. Globally, more than 50% of all the N used in agriculture may arise from this symbiosis [6], with consequent soil acidification a cause for considerable concern. Studies that illustrate the impact of soil acidity on different phases of the legume-rhizobial symbiosis include those of Staley and Wright [7], Tang and Thomson [8], and Slattery and Coventry [9].

2.1   pH and the Survival of Rhizobia in Soil and Culture

Not all soils contain rhizobia capable of nodulating the legume to be seeded. This is common when a legume species is introduced to a new area and usually requires seed or soil inoculation with a suitable strain of rhizobia [10]. Numbers of these introduced rhizobia on seed or in soil can be decimated by low soil pH; Carter et al. [11] noted that only three of eight strains of *Rhizobium leguminosarum* bv *viciae* maintained soil numbers in excess of 100 g^{-1} of soil 2 years after introduction. With few exceptions, rhizobia sown into acid soils must run this pH gauntlet every season. Although large numbers of rhizobia are released back into soil when nodules senesce [12], survival until the reintroduction of a suitable host can be problematic. Thus, Brockwell et al. [13] found 8.9×10^4 *Sinorhizobium meliloti* g^{-1} in soils of pH 7.0 and above but only 37 of these organisms g^{-1} in soils of pH less than 6.0.

   Species of rhizobia differ in their ability to survive and persist at low pH. Strains of *Sinorhizobium meliloti* are particularly acid sensitive and rarely grow below pH 5.5. At the other extreme, only strains of *Mesorhizobium loti*, *Rhizobium tropici*, and some *Bradyrhizobium* spp. are able to grow at pH less than 4.5 [14,15]. There are several points for discussion in this statement.

   pH tolerance has been evaluated using a number of different procedures. For acid tolerance in nutrient media, we contend that only growth of isolated colonies on a medium set to the desired pH after autoclaving but before the medium has been poured and set is a true indication of tolerance. A number of buffers have been used in pH studies; too often they are used outside the pH range for which they are effective.

Within species, variation in acid pH tolerance is often minimal, and within strains, selection for enhanced tolerance to acidity is usually ineffective. Where marked variation in pH tolerance is evident within a species, caution in reaching conclusions is advocated. The classic example is that of the bean rhizobia. A small group of strains initially distinguished on the basis of their superior acid pH tolerance [16] were subsequently shown to differ from traditional bean rhizobia in host range and other traits and were eventually reassigned to *R. tropici* [17]. In a similar vein, it was assumed for many years that acid soil pH was correlated with a reduction in the efficiency of native rhizobia [18,19]. It is more likely that pH eliminated the homologous rhizobia and allowed colonization by acid-tolerant but ineffective heterologous strains. Current examples of variation in acid pH tolerance within species that warrant careful taxonomic scrutiny include the two separate groups of relatively acid-tolerant strains currently grouped within *S. meliloti* [20,21] and perhaps some bradyrhizobia. For the latter organisms, it appears that promiscuous varieties of soybean in Africa can nodulate with an array of fast- and slow-growing rhizobia [22] and that these organisms can be dramatically different in tolerance to environmental constraints. Similarly, Ozawa et al. [23] reported marked differences between soybean isolates from Japan and Indonesia in the frequency with which pH-tolerant isolates are recovered. Date and Halliday [24] and Lesueur et al. [25] have even found some bradyrhizobia to grow better at acid pH than in a medium of pH 6.8, with R. Sylvester Bradley and colleagues (personal communication, 1985) recommending the use of pH 5.5 medium in the initial isolation of bradyrhizobia from some tropical legumes.

The correlation observed between acid pH tolerance measured in culture medium and strain survival, persistence, and contribution in soil is not necessarily very high [26]. Because of this, Howieson et al. [27,28] and Watkin et al. [29] selected for pH-tolerant strains on the basis of their ability to colonize and survive soils of acid pH.

Soil acidity may also have a marked affect on strain diversity in the soil. Anyango et al. [30] and Hungria et al. [31] found significant differences in the diversity of rhizobia associated with beans in soils of acid versus more neutral pH. Strains of *R. tropici* normally compete poorly for sites of nodule formation on beans [32,33] but are competitive with *R. etli* under acid soil conditions. Soil pH may also have temporal and spatial consequences for *Rhizobium* distribution in soil. Thus, Richardson and Simpson [34,35] isolated numerous acid-sensitive

R. trifolii from acid soil. These bacteria were presumably restricted to microenvironments of higher pH, as they did not survive well when the soil was cultivated. Soil aggregates may be intensively colonized by rhizobia [36] and are one obvious habitat for acid-sensitive organisms in soil. Acid-tolerant inoculant strains have been identified for both beans and alfalfa [20,37,38] and in each case have been major factors in extending the area over which their specific host could be grown. Hungria and Vargas [37] note, however, that the combination of soil acidity and high temperature in Brazil makes annual inoculation desirable.

Following is a description of acid tolerance mechanisms found in rhizobia. A consideration of aluminum and manganese toxicities is beyond the scope of this chapter. Readers are referred to Flis et al. [39] and Graham [40].

2.2 Mechanisms of Acid Tolerance in Rhizobia

Although low soil pH affects both the number and diversity of rhizobia populating the soil, some rhizobia do survive quite difficult soil conditions. A number of mechanisms may be invoked, including that of escape, as referred to before. The most important mechanisms are described here.

2.2.1 Internal pH Homeostasis

Acid-tolerant rhizobia can generally maintain cytoplasmic pH (pH_i) near neutrality in spite of markedly lower external pH (pH_o), permitting continued function of cell enzymes in an adverse soil environment [41]. This ability is much less evident in acid-sensitive strains. Thus, the acid-tolerant *R. tropici* UMR1899 was able to maintain a constant cytoplasmic pH as much as 1.7 units higher than the cell surroundings, whereas the pH_i of the acid-sensitive UMR1632 fell with decreasing pH_o [14]. Similarly, acid-tolerant strains of *S. meliloti* were able to maintain an alkaline cytoplasm at a pH_o as low as 5.6, whereas in acid-sensitive strains the pH_i began to decline when the external pH fell below pH 6.0 [42]. Mutants of both *S. meliloti* and *R. tropici* with reduced ability to tolerate low pH have usually shown reduced ability to regulate cytoplasmic pH as well [14,43].

Several mechanisms contributing to the regulation of cytoplasmic pH in *Rhizobium* have been recognized. These include enhanced proton-translocating ATPase activity coupled to uptake of K^+ through a K^+/H^+ antiport system that maintains charge neutrality across the cell membrane [44]. Aarons and Graham [45] noted an increase in K^+ content in acid-tolerant UMR1899 as a result of exposure to low pH; this phenomenon was not observed in acid-sensitive UMR1632 under similar conditions. For the same strains, proton-dependent ATPase activity was greater in UMR1899 than in UMR1632 at all pH levels and even greater in UMR1899 upon exposure to acid pH.

Cytoplasmic buffers may also compensate for acidification of the external medium. Graham et al. [14] noted an increase in cellular glutamate in the acid-tolerant *R. tropici* UMR1899 in response to low pH_o. Rowbury et al. [46] also noted a glutamate-induced enhancement of acid tolerance in *E. coli* and correlated this with the presence in the culture medium of specific proteins associated with the acid tolerance response. Similarly, acid-tolerant bacteria may produce homospermidine, an alkaline compound that may buffer the pH_i at low pH_o. Thus, the homospermidine content of the relatively acid-tolerant *S. fredii* P220 almost doubled on exposure to acid pH [47]. In contrast, Graham et al. [14] noted greater polyamine production in acid-sensitive UMR1632 than in acid-tolerant UMR1899 but concluded that polyamine production in these strains was a result of acid stress rather than a mechanism of recovery. Glutathione production has also been shown to be essential for growth under stress conditions, including acid pH [48].

We speculate that N_2 fixation leading to ammonia production might also be an acid stress response, made necessary by the acid environment within the nodule [49] and *Rhizobium* cell wall changes within the symbiosome [50–52]. This might be one explanation for the absence of N_2 fixation in vitro.

In a number of studies, calcium has enhanced the ability of rhizobia to tolerate low pH. Thus, the mean generation time of the acid-sensitive *S. meliloti* CC169 at pH 5.7 dropped from 60.1 h in medium with 200 μM Ca to only 12.2 h when the medium contained 2000 μM Ca [53]. The generation time at pH 7.3 was rapid regardless of Ca content. Similarly, Dilworth et al. [54] have reported a decrease in the death rate of *S. meliloti* WSM419 at pH 4.0 with increasing Ca concentration, with both *R. etli* [55] and *R. leguminosarum* [15] showing a similar response to Ca level at low pH. This Ca requirement is often (but not always) greatest for the more acid-sensitive strains. Thus, Tn5 acid-sensitive mutants of *R. meliloti* WSM419 required higher Ca levels to maintain a constant pH_i in acid medium than did the wild-type strains [40].

Acid-shock proteins—transient proteins produced in response to a rapid drop in ambient pH—have been identified in a number of bacteria, including the rhizobia [45,56]. Peick et al. [56] identified five such shock proteins produced in cells of acid-tolerant UMR1899 exposed to acid pH; four other polypeptides were decreased in synthesis following a pH shift. Parallel changes occurred in acid-sensitive UMR1632, but the polypeptides involved were different from those identified in UMR1899. Some of these polypeptides are likely to be involved in changes needed for survival at acid pH, for the repair of cell damage, or for the protection of critical cell proteins or DNA, whereas others are more likely the consequence of exposing the cell to acid pH (rather than a countermeasure to it). Functions for very few of these proteins have been established to date.

2.2.2 Cell Surface Stability

Acid-tolerant strains must have a cell surface/membrane chemistry that adapts to ambient pH but provides stability and maintains structure [41]. Cell surface lipopolysaccharides (LPSs), membrane lipids, outer membrane proteins (OMPs), and extracellular polysaccharides may all vary in composition according to external pH and play a significant role in cell protection. Howieson et al. [57] suggested that the differences in surface ionogenic properties of *R. meliloti* strains WSM540 and CC169 may contribute to differences in their acid soil tolerance.

Bhat and Carlson [58] showed that cells of *R. leguminosarum* bv. *phaseoli* CE3 grown at pH 4.8 were different in LPS composition from cells cultured at pH 7.2, as determined by composition analysis and differences in cross-reactivity with monoclonal antibodies. Similar changes have been reported by Graham et al. [14] and Ballen et al. [55]. In *R. galegae*, acid pH tolerance is associated with the presence of long O-chain lipopolysaccharide and abundant exopolysaccharide production [59].

Changes in membrane composition also influence the ability of substances to enter the cell. Hydrophilic substances of less than 700 daltons enter the cell via porins, whereas hydrophobic substances enter by dissolving across the hydrocarbon layer. Ballen et al. [55] have shown changes in fatty acid methyl ester (FAME) profiles in rhizobial cells as a function of external pH, with the percent 19:0 cyclopropane in the cell membrane of acid-sensitive UMR 1632 declining with decreased pH. In contrast, the percent 19:0 cyclopropane in the membrane of acid-tolerant UMR1899 stayed relatively constant with a change in ambient pH. Similarly, Correa et al. [60] found that membrane phosphatidylglycerol, phosphatidylethanolamine, and lysophospholipid content in the acid-tolerant *M. loti* LL56 decreased 25, 39, and 51%, respectively, when pH was changed from 7.0 to 5.5, whereas phosphatidyl-N-methylethanolamine and cardiolipin contents were increased 26 and 65%.

Calcium is vital for the maintenance of the cell envelope. Calcium ions bridge adjacent molecules of LPS and protein [61,62], anchor the outer membrane to the peptidoglycan [61], and are essential in maintaining cell wall rigidity [63]. Cells of *R. trifolii* grown under Ca deficiency differ serologically from those grown with adequate Ca, indicating a Ca effect on the LPS [64]. Increased Ca concentration stabilized the LPS structure and porin function of the acid-sensitive *R. etli* UMR 1632 at low pH [55].

Extracellular polysaccharides (EPSs) may have a localized buffering effect in the region immediately surrounding the cell, reducing the need for pumping protons or other internal changes. Cunningham and Munns [65] found a correlation between EPS production and acid tolerance in *Rhizobium*. However, neither Chen et al. [66] with *Rhizobium leguminosarum* biovar *trifolii* nor Graham et al. [16] with *R. tropici* UMR 1899 found a role for EPS in acid pH tolerance.

2.2.3 Regulation of Copper and Iron Content in Acid-Stressed Cells

Changes in cell pH_i can affect the availability and thus the toxicity of heavy metals. Ability to regulate intracellular concentrations of heavy metals appears to be related to acid tolerance. Copper- and pH-sensitive mutants have been identified in both *S. meliloti* [67,68] and *R. leguminosarum* bv *viciae* [68]. These mutants lack a Cu-transporting ATPase but can grow at acid pH, provided copper is removed from the medium. *Gus*A fusions into *act*P show gene induction by both Cu and low pH but not by addition of Cd, Hg, Ag, or Zn. A two-component sensor (*act*S) and regulatory system (*act*R) has also been described [69].

Deluca et al. [70] have noted in *R. leguminosarum* the presence of the ferric uptake regulatory (*fur*) gene that appeared essential for growth. In *Salmonella*, *fur* mutants are acid sensitive and have altered expression of several acid-shock proteins [71].

2.3 pH and Nodulation

Acidity has long been known to inhibit nodulation. Munns [72] found that low pH hinders nodulation of alfalfa (*Medicago sativa* L.) roots by *S. meliloti* at the stage of root hair infection by the bacteria. By moving plants between solutions of different pH, he was able to show that exposure to acid pH had the same effect on nodulation as did delaying inoculation for that period. In contrast, transfer to a favorable pH for as little as 48 h allowed nodulation to occur even though plants were later returned to an acid environment.

2.3.1 Steps in Nodule Formation

The process of nodule formation and the specificity inherent in this process are reviewed by Hirsch [73] and Broughton et al. [74]. The latter authors describe this process in terms of a mating ritual in which a sophisticated signaling process determines compatibility of host and microsymbiont. This interaction occurs at both a visible and a molecular level [75]. In the majority of legumes, physical evidence of impending infection is first evident from the attachment of rhizobia to the newly emerged root hairs of a susceptible host. Presence of the rhizobia leads to localized hydrolysis of the root hair cell wall and invagination of the host plasma membrane. The enzymes involved in this step appear cell bound and difficult to quantify, and several appear different from those normally associated with plant infection. Rhizobia then enter the root hair, with additional plant cell material deposited around them as they infect. Rhizobial penetration results in cessation of root hair growth at the point of infection and leads to curling of the root hairs to form so-called shepherd's crooks [73], first visible some 6–18 h after inoculation. Rhizobia remain enclosed within a host-derived infection thread but progressively move down the root hair to the cortex of their host, where cell multiplication and enlargement of the host and progressive infection of host cells by the rhizobia lead to the formation of a nodule.

At the molecular level, conversation begins with the production of flavonoid compounds by the legume host [76,77]. Different legumes produce different flavonoids, but with a compatible host and rhizobia, the result is the expression of all the nodulation genes in *Rhizobium* needed for infection. Early studies with the nodulation genes suggested some that were common to all rhizobia and thus could compensate for mutations in another organism; others were specific and varied with the particular host species nodulated. Over 50 nodulation genes have now been identified in the rhizobia, with most involved in the synthesis and export of substituted lipooligosaccharides—also known as Nod factors [78]. These molecules all have the same core structure encoded by the "common" nodulation genes but vary in the side chains each carries, affecting host range [79–81]. Nod factors act as powerful morphogens, which at concentrations as low as 10^{-9} to 10^{-11} M can deform root hairs and initiate the cortical cell division typical of nodule formation [82].

Interaction of host and *Rhizobium* is accompanied by the production of nodule-specific proteins or nodulins [83]needed for nodule production and function. Nodulin expression can vary both spatially and temporally, with "early" nodulins produced within 6 h of inoculation and involved in infection and nodule development.

2.3.2   The Effect of Acidity on the Nodulation Process

In addition to its effect on growth and persistence of rhizobia in soil, acid pH can limit *Rhizobium* attachment to host plants and hinder signaling between host and rhizobia. Exposure to acidity prior to infection affects the microsymbiont more than the host [84]. Thus, preincubation of *S. meliloti* L5-30 at a low pH inhibited subsequent attachment to alfalfa roots at a neutral pH, whereas preincubation of alfalfa roots at the same pH had a limited effect. Acidity during infection leads to destabilization of binding and detachment of rhizobia from the root hair. Exposing the cells to high Ca concentrations prior to inoculation improved later attachment at low Ca levels. Likewise, optimal "cap formation" (the aggregation of bacteria on the tip of a root hair to form a caplike structure, a step that must occur for nodule formation) of *R. leguminosarum* 248 on pea root hair tips occurred at pH 7.5, whereas little or no cap formation was observed at pH 6.0 [85]. The authors suggested that inhibition of nodulation at low pH might be attributed to this reduced cap formation. Calcium concentration may again be a factor. Increased Ca concentrations led to an increase in nodule number at low pH [86]. Howieson et al. [57] observed that addition of Ca and P increased attachment to alfalfa roots of the acid-sensitive *R. meliloti* CC169 more dramatically than that of the acid-tolerant WSM540.

Acid inhibition of signaling between host and *Rhizobium* can occur at several levels. Initial reports suggested that acid-tolerant cultivars of *Medicago* [88] and *Trifolium* [87] produced root exudates at low pH that led to better *nod* gene expression than did exudates from acid-sensitive cultivars. A more recent report [89] found no difference in *nod* gene expression by rhizobia exposed to exudates from acid-grown cultivars that differed in pH tolerance. The acid-sensitive *R.*

meliloti CC169 exhibited less *nod* gene induction in response to host root exudates at low pH than did the acid-tolerant *R. meliloti* WSM40. This could have been due either to decreased uptake of the inducer or to decreased *nod* gene expression at low pH. The availability of Ca appears to moderate the effect of low pH on flavonoid release by acid-sensitive legumes. Increased Ca concentration led to increased *nod* gene–inducing activity of exudates collected from the acid-sensitive *M. trunculata* but did not affect exudates produced by the acid-tolerant *M. murex*. Calcium spiking, the depolarization of root hair membrane potential in the presence of appropriate rhizobia or nod factor [90,91], might also be affected by external pH. Heo et al. [92] reported that specific calmodulin isoforms are involved in plant disease response. A promoter-tagging program in *Lotus japonicus* [93] has identified a gene with strong homology to those encoding calcium-modulating proteins and shown it to be regulated by the presence of infective rhizobia. Using acinar cells from rat pancreas, Tsunoda [94] showed such calcium oscillation to be sensitive to intracellular pH and completely inhibited by a drop in pH_i from 7.4 to 6.5, but no parallel studies appear to have been attempted with legumes.

Studies undertaken with *Phaseolus vulgaris* [95] give results similar to those with *Medicago*. Thus, the acid-tolerant bean cultivar Capixaba Precoce nodulated well at acid pH, irrespective of strain inoculated, whereas the acid-sensitive cultivar Negro Argel nodulated only when inoculated with the acid-tolerant *R. tropici* UMR1899. Lowered pH affected the release of flavonoid *nod* gene inducers in both bean and soybean; partial alleviation of this problem was accomplished by supplementation with *nod* gene inducers [96]. The effect of soil acidity on the relative efficiency in nodule formation of isolates of *R. etli* and *R. tropici* has already been mentioned.

2.4 Overcoming the Effects of Acid pH

While aluminum and manganese toxicities and induced deficiencies of Mo or P have profound effects on N_2 fixation [37], pH per se influences rhizobial survival in soil or rhizosphere and the process of nodulation itself. As suggested earlier, problems in rhizobial strain survival in soil and nodulation problems in a range of legumes can be expected when soil pH falls below pH 5.2. When rhizobia have gained access to their host, they can be protected from many of the effects of extreme acidity. Note, however, that the pH within the symbiosome of *S. meliloti* may be only 5.5 to 6.0 and that the combination of low pH and acetate is bacteriostatic [49]. Soil liming, use of acid-tolerant hosts and rhizobia, and lime pelleting of the seed are all common practices. The latter technique is used predominantly in the United States for preinoculation of legume seed but, properly employed, can protect the rhizobia from intimate contact with the seed surface and generate a less acid microclimate around the seed during seed germination and nodulation. Pelleting requires an adhesive such as gum arabic, with the rhizobia first mixed with the adhesive and applied to the seed, then rolled in finely ground rock phosphate or limestone [97].

3 OTHER SYMBIOSES AFFECTED BY pH

All symbioses between plants and microorganisms can be affected by pH; however, the pH at which this occurs and the relative effects of pH on microbe and symbiosis will vary. The available literature on pH and actinorhizal, *Azolla-Anabaena*, and mycorrhizal associations is limited, but inferences from information gained in the legume-rhizobial symbiosis can be helpful.

3.1 Frankia and Actinorhizal Symbiosis

Actinorhizal plants are defined by their ability to form N_2-fixing root nodule symbioses with the soil actinomycete *Frankia* [98]. Eight families of angiosperms are included; most are woody trees or shrubs with only limited commercial value. Together with the legumes they fall into a single phylogenetic clade predisposed to nodule formation [99]. *Frankia* were not isolated in pure culture until 1978 and in some cases still present isolation problems. Because of this, many recent studies emphasize molecular approaches to the ecology of this organism in soil [100–102], bypassing the need for isolation and the study of growth constraints. The relatively limited number of isolates available, coupled with the difficulty of isolating *Frankia* from particular hosts [98], could also mean that results to date are biased toward the more saprophytically competent members of this genus. This notwithstanding, it appears that the range of pH tolerance in *Frankia* is similar to that already reported for the rhizobia. Faure Raynaud et al. [103] found all 20 strains studied unable to grow at pH 4.2, though 6 isolates grew at pH 4.6 in culture media. Three of these strains survived better in acid soil than did acid-sensitive *Frankia* isolates. Burggraaf and Shipton [104] also report strains of *Frankia* able to grow at pH 4.6 and above; Griffiths and McCormick [105] found decreased viability below pH 4.5. Some *Frankia* strains are spore producing and generally considered not as effective in N_2 fixation; this fact is significant in that Markham and Chanway [106] reported a slight increase in the proportion of nodules containing spore producers as soil pH decreased.

Infection by *Frankia* has interesting similarities to and differences from that found in legume nodulation. For example:

Infection in some species is by root hair penetration, although the actual process bears greater similarity to rhizobial nodulation of the nonlegume *Parasponia* [107] than to what happens in legumes.
Flavonoid-containing preparations from seed washes of *Alnus rubra* appear to influence nodulation by *Frankia* [108].

Root hair penetration is also the mechanism of infection for some *Frankia*-host combinations, with nod factors analogous to those produced in the rhizobia reported in some *Frankia* strains [109]. These nod factors, however, have properties suggestive of differences in structure compared with those from *Rhizobium*.

Infection threads are produced in the *Comptonia, Casuarina, Alnus, Datsica*, and *Myrica* symbioses [110], but *Frankia* are not released into cells of their host [111]. Instead, host cells fill with branched infection threads.

Regulation of nodulation in some symbioses with *Frankia* parallels that found with *Rhizobium*. In *Discaria*, for example, the first-formed nodules cluster around the position of the root tip at the time of infection [112]. Culture age and inoculum density influence nodule distribution, and existing nodules inhibit further nodule development.

"Nodulins" have also been reported in *Frankia* nodule development. They include a serine protease expressed early in symbiosis [113] and a nodule-specific hemoglobin [114].

A number of actinorhizal species are considered plant pioneers [98] and likely to face low pH in some revegetation areas. Their success or failure in this role has often been studied without reference to nodulation status. Nodulation at pH 4.5 has been reported [105,115], although the optimal pH for nodulation, like that in legumes, is from pH 5.5 to 7.2. Red alder plants grown under nursery conditions actually nodulated better at pH 4.5 than at 5.6 or 7.2 [123]. Coincidentally, calcium treatments used in this study gave rise to plants with increased nodulation.

Frankia inoculation may be by cultured cells, crushed nodules, or infective soil, with inoculation procedures much less reliable than for legumes. Best results are with inoculation of nursery stock, with questions of the saprophytic competence of the different available cultures in the field still to be resolved.

3.2 The *Azolla-Anabaena* Symbiosis

The aquatic fern *Azolla* is a widely used biofertilizer and green manure in rice production. Its high productivity and ability to symbiose with *Anabaena* can lead to a doubling of cell mass in only 3 to 5 days and to N accumulations under ideal conditions of 15 to 20 kg N ha^{-1} week^{-1} [116].

In this symbiosis *Anabaena* are housed within cavities in the leaf frond and exhibit high heterocyst frequency. They can be isolated from the fern [117], although reintroduction of the microsymbiont is extremely difficult. For this reason, the information on acid pH effects considers only the symbiotic state. Optimal pH for the symbiosis is from 4.5 to 7.0 [118], with the combination of low pH and high ammonium concentration leading to selective lysis of the microsymbiont

[126]. Changes in the pH of floodwaters as a result of *Azolla* cultivation have been reported [119,120].

3.3 Ectomycorrhizal and Endomycorrhizal Symbiosis

Mycorrhizal fungi represent the underground absorbing organs of most plants in nature [121]. Their exploration of the soil allows uptake of P, N, and micronutrients (most recently reviewed by Clark and Zeto [122]); they can also enhance water supply to their hosts and protect the host from soil pathogens [123]. Mycorrhizal links between plants in natural ecosystems allow the transfer of nutrients [124] and support wider species biodiversity [125]. Several different groups of plants and fungi form mycorrhizal symbioses. In ectomycorrhizal symbiosis as occurs in pines, eucalypts, and other angiosperms, fungal hyphae are intercellular and form a dense mantle over the lateral roots. By contrast, more than 80% of plant species form endomycorrhizal symbiosis, with intraradical mycelial growth and intracellular penetration. Gianinazzi-Pearson [121] detailed the infection process in endomycorrhizal symbiosis; some structural and functional similarities to the infection process in legumes have been reported [126,127].

As with the other symbioses discussed, endomycorrhizal and ectomycorrhizal associations can be affected by soil pH. Unfortunately, although their natural habitat is the soil, we know little of the ecology, saprophytic competence, and biodiversity of these organisms and how these factors are affected by soil stresses. This is in part because they are perceived as obligate symbionts and are most studied in association with their hosts. Also, the more important *Glomalean* (endomycorrhizal) fungi are yet to be subcultured. Molecular techniques are only now beginning to be applied to the study of these organisms in soil.

Clark [128] noted that information on arbuscular mycorrhizal fungi and host plant response at low pH is limited. He reported that *Acaulospora* and *Gigaspora* spores are more common in acid soils than those of *Glomus* spp., with only *Glomus manihotis* tolerant of acid soil constraints. Shepherd et al. [129] found spore counts in soils of pH 5.0 to 6.7 to range from 44 to 126 live spores per 100 g soil, with *Scutellospora* and *Acaulospora* dominant. A difficulty in a number of pH studies has been the confounding effects of soil liming on pH, calcium supply, and phosphate availability. Thus, Habte and Soedarjo [130] attributed increased mycorrhizal colonization and response in limed soils to increased concentrations of calcium, whereas Raznikiewicz et al. [131] explained a significant decline in arbuscular mycorrhizal spores in soil limed from pH 5.1 to 6.3 by enhanced P availability.

Plant roots can be colonized with mycorrhizae at pH values as low as pH 2.7, the critical pH for 95% maximum colonization of cassava roots varying with species from 4.4 to 4.8 [132]. In corn, the optimal pH for colonization by *Gigaspora margarita* was pH 5.3 [133], whereas decreases in colonization were noted at pH 6.3. Habte and Soedarjo [134]found colonization of *Acacia mangium* roots increased from pH 4.3 to 5.0, with no further benefit above this pH. Medeiros et al. [135] observed that isolates differed in root colonization ability at low pH, although this was not necessarily correlated to yield response. When Clark et al. [136] grew *Panicum virgatum* in acid soil inoculated with eight isolates of *Glomus, Gigaspora*, and *Acaulospora*, plant P concentration varied by as much as 6.2-fold at pH 4 but only 2.9-fold at pH 5. Differences in nitrogen accumulation were relatively small. Strain biodiversity and how specific strains differ in colonizing ability and nutrient uptake as a function of pH clearly need additional study. In vitro propagation of mycorrhizal isolates on transformed carrot roots grown under controlled conditions [137] offers opportunities for more detailed studies of host-isolate interaction as a function of pH.

4 CONCLUSIONS

In general, the microsymbiont is more strongly affected than the host plant by soil acidity. Acidity restricts both total number of microsymbionts present in soil and also biodiversity, as acid-sensitive strains—which may be very effective microsymbionts—are selected against. The result for the host plant may be that it cannot establish a symbiosis because compatible microbes are not present or that the only symbiosis that it can establish is with a microorganism that is acid tolerant but is less effective in supplying nutrients than acid-sensitive microbes. A second way in which low pH disrupts symbiotic relationships is by inhibiting the establishment of a symbiosis in spite of the presence of acid-tolerant microbes. Again, the plant is left unable to obtain necessary nutrients, resulting in reduced crop vigor and yield.

This situation may be approached from a number of different directions. One "quick fix" is to lime the soil surrounding the plants, thus creating a suitable microenvironment for the establishment of symbioses. This solution is not always practical because of the large quantity (and expense) of lime required. Small amounts of calcium may be added to rhizobia used to inoculate legumes, as calcium has proved to enhance nodulation; perhaps calcium would also increase the establishment of other symbioses. Certainly, more effort should be directed to the search for acid-tolerant microbes that are very effective symbionts and can be used to inoculate crops. Identifying crop plants that exhibit a preference for these strains (over the acid-tolerant but potentially ineffective microbes present in the soil) may improve crop yield even more.

REFERENCES

1. DN Munns. Acid soils tolerance in legumes and rhizobia. Adv Plant Nutr 2:63–91, 1986.

2. JH Sanders, JC Garcia. The economics of stress and technology development in the Sahel and the "Cerrados" of Brazil. In: J Maranville et al., eds. Adaption of Plants to Soil Stresses. University of Nebraska, Lincoln: INTSORMIL 94-2, 1993, pp 28–47.

3. EF Depauw. The management of acid soils in Africa. Outlook Agric 23:11–16, 1994.

4. TT Cochrane. An ongoing appraisal of the savanna ecosystems of tropical America for beef cattle production. In: PA Sanchez and LE Tergas, eds. Pasture Production in Acid Soils of the Tropics. Cali, Colombia: CIAT, 1979, pp 1–12.

5. MK Conyers, NC Uren, KR Helyar, GJ Poile, BR Cullis. Temporal variation in soil acidity. Aust J Agric Res 35:1115–1129, 1997.

6. PH Graham, CP Vance. Nitrogen fixation in perspective: an overview of research and extension needs. Field Crops Res 65:93–106, 2000.

7. TE Staley, RJ Wright. Inoculation responses of perennial forage legumes grown in fresh hill-land ultisols as affected by soil acidity related factors. J Plant Nutr 14: 599–612, 1991.

8. C Tang, BD Thomson. Effects of solution pH and bicarbonate on the growth and nodulation of a range of grain legume species. Plant Soil 186:321–330, 1996.

9. JF Slattery, DR Coventry. Persistence of introduced strains of *Rhizobium leguminosarum* by *trifolii* in acidic soils of north eastern Victoria. Aust J Exp Res 39:829–837, 1999.

10. PH Graham. Biological dinitrogen fixation: symbiotic. In: DM Sylvia, JJ Fuhrmann, PG Hartel, DA Zuberer, eds. Principles and Applications of Soil Microbiology. Upper Saddle River, NJ: Prentice Hall, 1998, pp 322–345.

11. JM Carter, JS Tieman, AH Gibson. Competitiveness and persistence of strains of rhizobia for faba bean in acid and alkaline soils. Soil Biol Biochem 27:617–623, 1995.

12. TR McDermott, PH Graham, DH Brandwein. Viability of *Bradyrhizobium japonicum* bacteroids. Arch Microbiol 148:100–106, 1987.

13. J Brockwell, A Pilka, RA Holliday. Soil pH is a major determinant of the numbers of naturally occurring *Rhizobium meliloti* in non-cultivated soils in central New South Wales, Australia. Aust J Exp Agric 31:211–220, 1991.

14. PH Graham, KJ Draeger, ML Ferrey, MJ Conroy, BE Hammer, E Martinez, SR Aarons, C Quinto. Acid pH tolerance in strains of *Rhizobium* and *Bradyrhizobium*, and initial studies on the basis for pH tolerance of *Rhizobium tropici* UMR 1899. Can J Microbiol 40:198–207, 1994.

15. ELJ Watkin, GW O'Hara, AR Glenn. Calcium and acid stress interact to affect the growth of *Rhizobium leguminosarum* by *trifolii*. Soil Biol Biochem 29:1427–1432, 1997.

16. PH Graham, SE Viteri, F Mackie, AT Vargas, A Palacios. Variation in acid pH tolerance among strains of *Rhizobium phaseoli*. Field Crops Res 5:121–128, 1982.

17. E Martinez-Romero, L Segovia, FM Mercante, AA Franco, P Graham, MA Pardo. *Rhizobium tropici*, a novel species nodulating *Phaseolus vulgaris* L. beans and *Leucaena* sp trees. Int J Syst Bacteriol 41:417–426, 1991.

18. AJ Holding, JF Lowe. Some effects of acidity and heavy metals on the *Rhizobium*–leguminous plant association. Plant Soil Special Volume:153–166, 1971.

19. AM Ibekwe, JS Angle, RL Chaney, P Van Berkum. Differentiation of clover *Rhizo-*

bium from biosolids-amended soils with varying pH. Soil Sci Soc Am J 61:1679–1685, 1997.

20. JG Howieson, MA Ewing, MF D'Antuono. Selection for acid tolerance in *Rhizobium meliloti*. Plant Soil 105:179–188, 1988.

21. MF Del Papa, LJ Balague, SC Sowinski, C Wegener, E Segundo, FM Abarca, N Toro, K Niehaus, A Puhler, OM Aguilar, G Martinez Drets, A Lagares. Isolation and characterization of alfalfa-nodulating rhizobia present in acidic soils of Central Argentina and Uruguay. Appl Environ Microbiol 65:1420–1427, 1999.

22. S Mpepereki, F Javaheri, P Davis, KE Giller. Soyabeans and sustainable agriculture. Promiscuous soyabeans in southern Africa. Field Crops Res 65:137–149, 2000.

23. T Ozawa, Y Imay, HI Sukiman, H Karsono, D Ariani, S Saono. Low pH and aluminum tolerance of *Bradyrhizobium* strains isolated from acid soils in Indonesia. Soil Sci Plant Nutr 45:987–992, 1999.

24. RA Date, J Halliday. Selecting *Rhizobium* for acid, infertile soils in the tropics. Nature 277:62–64, 1979.

25. D Lesueur, HG Diem, M Dianda, C LeRoux. Selection of *Bradyrhizobium* strains and provenances of *Acacia mangium* and *Faidherbia albida* relationship with their tolerance to acidity and aluminum. Plant Soil 149:159–166, 1993.

26. LG Gemell, RJ Roughley. Field evaluation in acid soils of strains of *Rhizobium leguminosarum* bv *trifolii* selected for their tolerance or sensitivity to acid soil factors in agar medium. Soil Biol Biochem 25:1447–1452, 1993.

27. JG Howieson, J Malden, RJ Yates, GW O'Hara. Techniques for the selection and development of elite inoculant strains of *Rhizobium leguminosarum* in Southern Australia. Symbiosis 28:33–48, 2000.

28. JG Howieson, GW O'Hara, SJ Carr. Changing roles for legumes in Mediterranean agriculture: developments from an Australian perspective. Field Crops Res 65:107–122, 2000.

29. ELJ Watkin, GW O'Hara, JG Howieson, AR Glenn. Identification of tolerance to soil acidity in inoculant strains of *Rhizobium leguminosarum* bv *trifolii*. Soil Biol Biochem 32:1393–1403, 2000.

30. B Anyango, JK Wilson, JL Beynon, KE Giller. Diversity of rhizobia nodulating *Phaseolus vulgaris* in two Kenyan soils with contrasting pHs. Appl Environ Microbiol 61:416–421, 1995.

31. M Hungria, DD Andrade, A Colozzi, EL Balota. Interactions among soil organisms and bean and maize grown in monoculture or monocropped. Pesqui Agropecu Bras 32:807–818, 1997.

32. E Martinez, M Rosenblueth. Increased bean (*Phaseolus vulgaris* L.) nodulation competitiveness of genetically modified *Rhizobium* strains. Appl Environ Microbiol 56:2384–2388, 1990.

33. MH Chaverra, PH Graham. Cultivar variation in traits affecting early nodulation of common bean. Crop Sci 32:1432–1436, 1992.

34. AE Richardson, RJ Simpson. Enumeration and distribution of *Rhizobium trifolii* under a subterraneum clover based pasture growing in an acid soil. Soil Biol Biochem 20:431–438, 1988.

35. AE Richardson, RJ Simpson. Acid tolerance and symbiotic effectiveness of *Rhizo-*

bium trifolii associated with a *Trifolium subterraneum* L. based pasture growing in an acid soil. Soil Biol Biochem 21:87–95, 1989.

36. IC Mendes, PJ Bottomley. Distribution of a population of *Rhizobium leguminosarum* bv *trifolii* among different size classes of soil aggregates. Appl Environ Microbiol 64:970–975, 1998.

37. M Hungria, MAT Vargas. Environmental factors affecting N_2 fixation in grain legumes in the tropics, with an emphasis on Brazil. Field Crops Res 65:151–164, 2000.

38. MA Ewing, JG Howieson. The development of *Medicago polymorpha* L. as an important pasture species for Southern Australia. Proceedings XVI International Grasslands Congress, Nice, 1989, pp 197–198.

39. SE Flis, AR Glenn, MJ Dilworth. The interaction between aluminum and root nodule bacteria. Soil Biol Biochem 25:403–417, 1993.

40. PH Graham. Stress tolerance in *Rhizobium* and *Bradyrhizobium*, and nodulation under adverse soil conditions. Can J Microbiol 38:475–484, 1992.

41. JG Howieson. Characteristics of an ideotype acid tolerant pasture legume symbiosis in Mediterranean agriculture. Plant Soil 171:71–76, 1995.

42. GW O'Hara, TJ Goss, MJ Dilworth, AR Glenn. Maintenance of intracellular pH and acid tolerance in *Rhizobium meliloti*. Appl Environ Microbiol 55:1870–1876, 1989.

43. TJ Goss, GW O'Hara, MJ Dilworth, AR Glenn. Cloning, characterization, and complementation of lesions causing acid sensitivity in Tn5-induced mutants of *Rhizobium meliloti* WSM419. J Bacteriol 172:5173–5179, 1990.

44. IR Booth. Regulation of cytoplasmic pH in bacteria. Microbiol Rev 49:359–378, 1985.

45. SR Aarons, PH Graham. Response of *Rhizobium leguminosarum* bv *phaseoli* to acidity. Plant Soil 134:145–151, 1991.

46. RJ Rowbury, TJ Humphrey, M Goodson. Properties of an L-glutamate induced acid tolerance response which involves the functioning of extracellular induction components. J Appl Microbiol 86:325–330, 1999.

47. S Fujihara, T Yoneyama. Effects of pH and osmotic stress on cellular polyamine contents in the soybean rhizobia *Rhizobium fredii* P220 and *Bradyrhizobium japonicum* A1017. Appl Environ Microbiol 59:1104–1109, 1993.

48. PM Riccillo, CI Muglia, FJ De Bruijn, AJ Roe, IR Booth, OM Aguilar. Glutathione is involved in environmental stress responses in *Rhizobium tropici*, including acid tolerance. J Bacteriol 182:1748–1753, 2000.

49. R Perez-Galdona, ML Kahn. Effects of organic acids and low pH on *Rhizobium meliloti* 104A14. Microbiology 140:1231–1235, 1994.

50. AAN Van Brussel, K Planque, A Quispel. The wall of *Rhizobium leguminosarum* in bacteroid and free-living forms. J Gen Microbiol 101:51–56, 1977.

51. RE Tully, ME Terry. Decreased exopolysaccharide synthesis by anaerobic and symbiotic cells of *Bradyrhizobium japonicum*. Plant Physiol 79:445–450, 1985.

52. RA DeMaagd, RD Rijk, IHM Mulders, BJJ Lugtenberg. Immunological characterization of *Rhizobium leguminosarum* outer membrane antigens by use of polyclonal and monoclonal antibodies. J Bacteriol 171:1136–1142, 1989.

53. JG Howieson, AD Robson, LK Abbott. Calcium modifies pH effects on the growth

of acid-tolerant and acid-sensitive *Rhizobium meliloti*. Aust J Agric Res 43:765–772, 1992.

54. MJ Dilworth, FG Rynne, JM Castelli, AI Vivas-Marfisi, AR Glenn. Survival and exopolysaccharide production in *Sinorhizobium meliloti* WSM419 are affected by calcium and low pH. Microbiology 145:1585–1593, 1999.

55. KG Ballen, PH Graham, RK Jones, JH Bowers. Acidity and calcium interaction affecting cell envelope stability in *Rhizobium*. Can J Microbiol 44:582–587, 1998.

56. B Peick, P Graumann, R Schmid, M Marahiel, D Werner. Differential pH-induced proteins in *Rhizobium tropici* CIAT 899 and *R. etli* CIAT611. Soil Biol Biochem 31:189–194, 1999.

57. JG Howieson, AD Robson, MA Ewing. External phosphate and calcium concentrations, and pH, but not the products of rhizobial nodulation genes, affect the attachment of *Rhizobium meliloti* to roots of annual medics. Soil Biol Biochem 25:567–573, 1993.

58. UR Bhat, RW Carlson. Chemical characterization of pH-dependent structural isotopes of lipopolysaccharides from *Rhizobium leguminosarum* biovar *phaseoli*. J Bacteriol 174:2230–2235, 1992.

59. LA Rasanen, K Lindstrom. Stability of short and long O-chain lipopolysaccharide types in *Rhizobium galegae*, and their correlation with symbiotic properties and growth conditions, tolerance of low pH, aluminum and salt in the growth medium. FEMS Microbiol Lett 155:17–22, 1997.

60. OS Correa, EA Rivas, AJ Barneix. Cellular envelopes and tolerance to acid pH in *Mesorhizobium loti*. Curr Microbiol 38:329–334, 1999.

61. MJ Osborn, HCP Wu. Proteins of the outer membrane of gram negative bacteria. Annu Rev Microbiol 34:369–422, 1980.

62. AJ Wicken, KW Knox. Bacterial cell surface amphiles. Biochem Biophys Acta 604:1–26, 1980.

63. JM Vincent. Influence of calcium and magnesium on the growth of *Rhizobium*. J Gen Microbiol 28:653–663, 1962.

64. JM Vincent, BA Humphrey. Modification of the antigenic surface of *Rhizobium* by a deficiency of calcium. J Gen Microbiol 54:397–405, 1968.

65. SD Cunningham, DN Munns. The correlation between extracellular polysaccharide production and acid tolerance in *Rhizobium*. Soil Sci Soc Am J 48:1273–1276, 1984.

66. H Chen, AE Richardson, BG Rolfe. Studies on the physiological and genetic basis of acid tolerance in *Rhizobium leguminosarum* biovar *trifolii*. Appl Environ Microbiol 59:1798–1804, 1993.

67. RP Tiwari, WG Reeve, MJ Dilworth, AR Glenn. An essential role for actA in acid tolerance of *Rhizobium meliloti*. Microbiology 142:601–610, 1996.

68. MJ Dilworth, RP Tiwari, WG Reeve, AR Glenn. Heavy metal involvement in the acid tolerance of *Sinorhizobium meliloti* and *Rhizobium leguminosarum* bv *viciae*. Proceedings 17th North American Conference on Symbiotic Nitrogen Fixation, Quebec, 2000, p 26.

69. RP Tiwari, WG Reeve, MJ Dilworth, AR Glenn. Acid tolerance in *Rhizobium meliloti* strain WSM419 involves a two component sensor-regulator system. Microbiology 142:1693–1704, 1996.

70. NG Deluca, M Wexler, MJ Pereira, KH Yeoman, AWB Johnston. Is the *fur* gene of *Rhizobium leguminosarum* essential? FEMS Microbiol Lett 168:289–295, 1998.

71. HK Hall, JW Foster. The role of *fur* in the acid tolerance response of *Salmonella typhimurium* is physiologically and genetically separable from its role in iron acquisition. J Bacteriol 178:5683–5691, 1996.

72. DN Munns. Nodulation of *Medicago sativa* in solution culture. I. Acid sensitive steps. Plant Soil 28:129–146, 1968.

73. AM Hirsch. Developmental biology of legume nodulation. New Phytol 122:211–237, 1992.

74. WJ Broughton, S Jabbouri, X Perret. Keys to symbiotic harmony. J Bacteriol 182: 5641–5652, 2000.

75. PH Graham. Nodule formation in legumes. In: J Lederberg, ed. Encyclopedia of Microbiology. Vol 3. San Diego: Academic Press, 2000, pp 407–417.

76. HA Stafford. Roles of flavonoids in symbiotic and defense reactions in legume roots. Bot Rev 63:27–39, 1997.

77. D Werner. Organic signals between plants and organisms. In: R Pinton, Z Varanini, G Nannipieri, eds. The Rhizosphere: Biochemistry and Organic Substances at the Soil-Plant Interface. New York: Marcel Dekker, 2000, pp 197–222.

78. JA Downie. Functions of rhizobial nodulation genes. In: HP Spaink, A Kondorosi, PJJ Hooykaas, eds. The Rhizobiaceae: Molecular Biology of Model Plant-Associated Bacteria. Dordrecht: Kluwer Academic Publishers, 1998, pp 387–402.

79. P Mergaert, M Van Montagu, M Hoisters. Molecular mechanisms of nod factor diversity. Mol Microbiol 25:811–817, 1997.

80. SR Long. *Rhizobium* symbiosis: nod factors in perspective. Plant Cell 8:1885–1898, 1996.

81. J Cohn, RB Day, G Stacey. Legume nodule organogenesis. Trends Plant Sci 3: 105–110, 1998.

82. Az-Eddine Hadri, T Bisseling. Responses of the plant to nod factor. In: HP Spaink, A Kondorosi, PJJ Hooykaas, eds. The Rhizobiaceae: Molecular Biology of Model Plant-Associated Bacteria. Dordrecht: Kluwer Academic Publishers, 1998, pp 403–416.

83. AT Trese, SG Pueppke. Modification of host gene expression during initiation and early growth of nodules in cowpea, *Vigna unguiculata* (L.) Walp. Plant Physiol 92:946–953, 1990.

84. G Caetano-Anolles, A Lagares, G Favelukes. Adsorption of *Rhizobium meliloti* to alfalfa roots: dependence on divalent cations and pH. Plant Soil 117:67–74, 1989.

85. G Smit, JW Kijne, BJJ Lugtenberg. Correlation between extracellular fibrils and attachment of *Rhizobium leguminosarum* to pea root hair tips. J Bacteriol 168:821–827, 1986.

86. DN Munns. Nodulation of *Medicago sativa* in solution culture. V. Calcium and pH requirements during infection. Plant Soil 32:90–102, 1970.

87. AE Richardson, RJ Simpson, MA Djordjevic, BG Rolfe. Expression of nodulation genes in *Rhizobium leguminosarum* biovar *trifolii* is affected by low pH and by Ca and Al ions. Appl Environ Microbiol 54: 2541–2548, 1988.

88. JG Howieson, AD Robson, LK Abbott. Acid-tolerant species of *Medicago* produce

root exudates at low pH which induce the expression of nodulation genes in *Rhizobium meliloti*. Aust J Plant Physiol 19:287–296, 1992.

89. ELJ Watkin, JM Castelli, WG Best, GW O'Hara, JG Howieson. The effect of acidity on the production and recognition of signal molecules in nodulation of *Medicago* by *Sinorhizobium*. Proceedings 17th North American Conference on Symbiotic Nitrogen Fixation. Quebec, 2000, p 68.

90. DW Ehrhardt, EM Atkinson, SR Long. Depolarization of alfalfa root hair membrane potential by *Rhizobium meliloti* nod factors. Science 256:998–1000, 1992.

91. DW Ehrhardt, R Wais, SR Long. Calcium spiking in plant root hairs responding to *Rhizobium* nodulation signals. Cell 85:673–681, 1996.

92. W Heo, SH Lee, MC Kim, JC Kim, WS Chung, HJ Chun, KJ Lee, CY Park, HC Park, JY Choi, MJ Cho. Involvement of specific calmodulin isoforms in salicylic acid independent activation of plant disease resistance responses. Proc Natl Acad Sci USA 96:766–771, 1999.

93. KJ Webb, L Skot, MN Nicholson, B Jorgensen, S Mizen. *Mesorhizobium loti* increases root-specific expression of a calcium-binding protein homologue identified by promoter tagging in *Lotus japonicus*. Mol Plant Microbe Interact 13:606–616, 2000.

94. Y Tsunoda. Cytosolic free calcium spiking affected by intracellular pH change. Exp Cell Res 188:294–301, 1990.

95. AAT Vargas, PH Graham. *Phaseolus vulgaris* cultivar and *Rhizobium* strain variation in acid pH tolerance and nodulation under acid soil conditions. Field Crops Res 19:91–101, 1988.

96. M Hungria, G Stacey. Molecular signals exchanged between host plants and rhizobia; basic aspects and potential applications in agriculture. Soil Biol Biochem 29:519–530, 1997.

97. P Somasegaran, HJ Hoben. Handbook for Rhizobia. New York: Springer Verlag, 1994, p 450.

98. JI Sprent, R Parsons. Nitrogen fixation in legume and non-legume trees. Field Crops Res 65:183–196, 2000.

99. DE Soltis, PS Soltis, DR Morgan, SM Swensen, BC Mullin, JM Dowd, PG Martin. Chloroplast gene sequence data suggest a single origin of the predisposition for symbiotic nitrogen fixation in angiosperms. Proc Natl Acad Sci USA 92:2646–2651, 1995.

100. D Hahn, A Nickel, J Dawson. Assessing *Frankia* populations in plant and soil using molecular methods. FEMS Microbiol Ecol 29:215–227, 1999.

101. E Navarro, T Jaffre, D Gauthier, F Gourbiere, G Rinaudo, P Simonet, P Normand. Distribution of *Gymnostoma* spp microsymbiotic *Frankia* strains in New Caledonia is related to soil type and to host plant species. Mol Ecol 8:1781–1788, 1999.

102. P Simonet, E Navarro, C Rouvrier, P Reddell, J Zimpfer, Y Dommergues, R Bardin, P Combarro, J Hamelin, AM Domenach, F Goubiere, Y Prin, JO Dawson, P Normand. Coevolution between *Frankia* populations and host plants in the family Casuarinaceae and consequent patterns of global dispersal. Environ Microbiol 1:525–533, 1999.

103. M Faure Raynaud, MA Bonnefoy Poirier, A Moiroud. Low pH influence on *Frankia* strains viability. Plant Soil 96:347–358, 1986.

104. AJP Burggraaf, WA Shipton. Estimates of *Frankia* growth under various pH and temperature regimes. Plant Soil 69:135–148, 1982.
105. AP Griffiths, LH McCormick. Effects of soil acidity on nodulation of *Alnus glutinosa* and viability of *Frankia*. Plant Soil 79:429–434, 1984.
106. J Markham, CP Chanway. *Alnus rubra* (Bong.) nodule spore type distribution in south-western British Columbia. Plant Ecol 135:197–205, 1998.
107. AE Hadri, HP Spaink, T Bisseling, NJ Brewin. Diversity of root nodulation and rhizobial infection process. In: HP Spaink, A Kondorosi, PJJ Hooykaas, eds. The Rhizobiaceae: Molecular Biology of Model Plant-Associated Bacteria. Dordrecht: Kluwer Academic Publishers, 1998, pp 347–360.
108. LF Benoit, AM Berry. Flavonoid-like compounds from seeds of red alder (*Alnus rubra*) influence host nodulation by *Frankia* (Actinomycetales). Physiol Plant 99: 588–593, 1997.
109. H Ceremonie, F Debelle, MP Fernandez. Structural and functional comparison of *Frankia* root hair deforming factor and rhizobia nod factor. Can J Bot 77:1293–1301, 1999.
110. RH Berg. *Frankia* form infection threads. Can J Bot 77:1327–1333, 1999.
111. K Pawlowski, T Bisseling. Rhizobial and actinorhizal symbioses: what are the shared features? Plant Cell 8:1899–1913, 1996.
112. C Valverde, LG Wall. Time course of nodule development in the *Discaria trinervis* (Rhamnaceae)–*Frankia* symbiosis. New Phytol 141:345–354, 1999.
113. L Laplaze, E Duhoux, C Franche, T Frutz, S Svistoonoff, T Bisseling, D Bogusz, K Pawlowski. *Casuarina gluaca* prenodule cells display the same differentiation as the corresponding nodule cells. Mol Plant Microbe Interact 13:107–112, 2000.
114. A Smolander, C Van Dijk, V Sundman. Survival of *Frankia* strains introduced into soil. Plant Soil 106:65–72, 1988.
115. WK Crannell, Y Tanaka, DD Myrold. Calcium and pH interaction on root nodulation of nursery grown red alder (*Alnus rubra* Bong.) seedlings by *Frankia*. Soil Biol Biochem 26:607–614, 1994.
116. GM Wagner. *Azolla*—a review of its biology and utilization. Bot Rev 63:1–26, 1997.
117. GA Peters. Studies on the *Azolla–Anabaena azollae* symbiosis. In: WE Newton, CJ Nyman, eds. Proceedings First International Symposium on Nitrogen Fixation. Pullman: Washington State University Press, 1976, pp 592–610.
118. E Uheda, S Kitoh. High concentrations of ammonium ions at low pH decrease the number of cyanobionts in apical portions of *Azolla*. Plant Cell Physiol 33:205–208, 1992.
119. VV Vu, HWWF Sang, JW Kijne, K Planque, R Kraayenhoef. Effect of temperature, pH and bound nitrogen on photosynthesis and nitrogen fixation of *Azolla pinnata* and *Azolla filiculoides*. Photosynthetica 20:67–73, 1986.
120. T Kroeck, J Alkaemper, I Watanabe. Effect of an *Azolla* cover on the conditions in floodwater. J Agron Crop Sci 161:185–189, 1988.
121. V Gianinazzi-Pearson. Plant cell responses to arbuscular mycorrhizal fungi; getting to the roots of symbiosis. Plant Cell 8:1871–1883, 1996.
122. RB Clark, SK Zeto. Mineral acquisition by arbuscular mycorrhizal plants. J Plant Nutr 23:867–902, 2000.

123. JM Barea. Rhizosphere and mycorrhiza of field crops. Proceedings of Biological Resource Management: Connecting Science and Policy, New York, 2000, pp 81–92.

124. AM Martensson, I Rydberg, M Vestberg. Potential to improve transfer of N in intercropped systems by optimizing host-endophyte combinations. Plant Soil 205: 57–66, 1998.

125. R Francis, DJ Read. The contributions of mycorrhizal fungi to the determination of plant community structure. Plant Soil 159:11–25, 1994.

126. E Overholt, G Engqvist, P Lindblad, A Martensson, I Rydberg, E Zagal. Pea-rhizobial and mycorrhizal symbiotic systems—a review of their commonalities with other plant-microbe systems. Symbiosis 21:95–113, 1996.

127. AM Hirsch, Y Kapulnik. Signal transduction pathways in mycorrhizal associations: comparisons with the *Rhizobium*-legume symbiosis. Fungal Gen Biol 23:205–212, 1998.

128. RB Clark. Arbuscular mycorrhizal adaptation, spore germination, root colonization and host plant growth and mineral acquisition at low pH. Plant Soil 192:15–22, 1997.

129. KD Shepherd, J Jefwa, J Wilson, JK Ndufa, K Ingleby, KW Mbuthia. Infection potential of farm soils as mycorrhizal inocula for *Leucaena leucocephala*. Biol Fertil Soils 22:16–21, 1996.

130. M Habte, M Soedarjo. Mycorrhizal inoculation effect in *Acacia mangium* grown in an acid oxisol amended with gypsum. J Plant Nutr 18:2059–2073, 1995.

131. H Raznikiewicz, K Curlgren, A Martensson. Impact of phosphorus fertilization and liming on the presence of arbuscular mycorrhizal spores in a Swedish long-term field experiment. Swed J Agric Res 24:157–164, 1994.

132. RH Howeler, E Sieverding, SR Saif. Practical aspects of mycorrhizal technology in some tropical crops and pastures. Plant Soil 100:249–283, 1987.

133. JO Siqueira, AW Mahmud, DH Hubbell. Comportamento diferenciado de fungos formadores de micorrizas vesicular-arbusculares em relacao a acidez do solo. R Bras Cienc Solo 10:11–16, 1986.

134. M Habte, M Soedarjo. Response of *Acacia mangium* to vesicular arbuscular mycorrhizal inoculation, soil pH and P concentration in an oxisol. Can J Bot 74:155–162, 1996.

135. CAB Medeiros, RB Clark, JR Ellis. Growth and nutrient uptake of sorghum cultivated with vesicular-arbuscular mycorrhiza isolates at varying pH. Mycorrhiza 4: 185–191, 1994.

136. RB Clark, RW Zobel, SK Zeto. Effects of mycorrhizal fungus isolates on mineral acquisition by *Panicum virgatum* in soil. Mycorrhiza 9:167–176, 1999.

137. TE Pawlowska, DD Douds, I Charvat. In vitro propagation and life cycle of the arbuscular mycorrhizal fungus *Glomus etunicatum*. Mycol Res 103:1549–1556, 1999.

16

Distribution of Plant Species in Relation to pH of Soil and Water

Jacqueline Baar* and Jan G. M. Roelofs
University of Nijmegen, Nijmegen, The Netherlands

1 INTRODUCTION

Environmental variables, such as soil pH, are known to be related to the distribution and the diversity of plant species within terrestrial ecosystems. In fact, soil acidity is a determining factor of the availability of nutrients to the plants. Soil pH influences microbial processes (e.g., mineralization, nitrification, and decomposition of organic matter), resulting in differences in the form and availability of inorganic nitrogen. Furthermore, soil pH affects the availability of phosphorus, potassium, and other macronutrients. The concentrations of potential phytotoxic metals (e.g., aluminum, zinc, iron, and cadmium) are also dependent on the soil pH [1–4].

Environmental variables, the pH in particular, determine the species composition of the plant communities in (semi-)aquatic environments. For instance, the plant community in naturally acidified waters is significantly different from the plant community in alkaline waters.

In this chapter, processes and plant distribution are discussed for both terrestrial and (semi-)aquatic ecosystems. There is also a strong focus on acidifica-

---

*Current affiliation: Wageningen University and Research Center, Horst, The Netherlands.

tion, an environmental problem that has affected strongly the composition of plant communities over the last decades, particularly in Europe.

2 BUFFER CAPACITY AND pH RANGES IN TERRESTRIAL ECOSYSTEMS

Chemical reactions occurring in soils can result in products that are neither acidic nor alkaline, leaving the soil pH unaffected. This process is indicated as pH buffering or acid neutralizing capacity (ANC) [1–3]. Calcareous soils with a soil pH ranging from 6 to 9 have a high buffer capacity, and hardly any changes in pH are observed when acidifying compounds are added. Under acidifying conditions, soluble calcium carbonate leaches from these soils by dissolving to bicarbonate and calcium. Subsequently, bicarbonate dissolves into carbon dioxide and water (Fig. 1):

$$CaCO_3 + H^+ \rightarrow HCO_3^- + Ca^{2+}$$
$$HCO_3^- + H^+ \rightarrow H_2O + CO_2$$

This process is relatively fast and the pH remains stable until almost all calcium carbonate has been depleted [5]. After depletion of all calcium carbonate together with magnesium carbonate, silicates are degraded (Fig. 1). This process starts at soil pH 6.5 and is a relatively slow process with low ecological importance. Primary silicates are transformed into secondary silicates, for example:

$$CaAl_2Si_2O_8 + 2H^+ \rightarrow Ca^{2+} + Al_2Si_2O_6(OH)_2$$

In soils that are in the pH range 4.0–5.5, protons are exchanged with potassium, calcium, and magnesium of the adsorption complex. The exchange capacity of this mechanism is commonly referred to as cation-exchange capacity (CEC). Be-

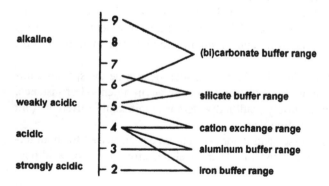

FIGURE 1 Buffering processes in the soil.

cause of the limited buffer capacity, the pH of these soils decreases rapidly under continuing acidifying conditions. At a pH below 4.5, aluminum is mobilized [6,7] (Fig. 1):

$$3AlO_2H\text{-polymer} + 9H^+ \rightarrow 3Al^{3+} + 6H_2O$$

This results in enhanced Al:Ca ratios in the soil, which can have large impact on the vegetation [3,8–10]. After mobilization of all aluminum in the soil, hydrous oxides of iron are dissolved according to [8]:

$$FeO_2H\text{-polymer} + 3H^+ \rightarrow Fe^{3+} + 2H_2O$$

The acidity of the soils where these processes occur is very high and the pH ranges from 2.0 to 4.0 (Fig. 1, see also Ref. 7). Few plant species can develop at such low soil pH values.

3 SPECIFIC ADAPTATIONS OF TERRESTRIAL PLANTS TO VARIOUS pH RANGES

The inorganic nitrogen pool in soils consists of ammonium and nitrate. In dry heathlands on acidic soils (pH < 4.2), ammonium is the dominant inorganic nitrogen form because the activity of nitrifying bacteria is low [11,12]. In forest soils, ammonium is often the dominant form too [13,14], although Falkengren-Grerup et al. [15] noted that in Swedish forests on neutral soils nitrate dominated. However, the availability of ammonium and nitrate in forest soils depends on different nitrification rates. Kriebitzsch [16] developed a classification for forest soils based on differences in nitrification in acidic forest soils in Germany. Four groups were distinguished, ranging from soils without nitrification to soils with a high nitrification rate. Similar differences in nitrification were observed for acidic soils in *Pinus sylvestris* forests in The Netherlands [17].

Differences in the availability of inorganic nitrogen are likely to have a great impact on the distribution of plant species because many plant species prefer one inorganic nitrogen source to another [18–22]. Several studies with characteristic plant species of heathlands on acidic soils (e.g., *Calluna vulgaris, Deschampsia flexuosa, Juncus squarrosus,* and *Nardus stricta*) showed that these plant species generally took up more ammonium than nitrate, whereas plant species on less acidic and more calcareous soils favor nitrate or a combination of ammonium and nitrate (e.g., *Scabiosa columbaria, Sesleria albicans* [20,21,23–26].

Rorison [27] studied nitrogen uptake in more detail and reported that plant species restricted to acidic soils, such as *Deschampsia flexuosa,* are able to take up both ammonium and nitrate, but only when the pH of the root medium is acidic. At high pH levels (>7), uptake of nitrate is inhibited, resulting in reduced growth and increased calcium concentrations in the shoots [23,27]. This indicates a lack of regulatory mechanisms to control calcium uptake as observed for plant species characteristic of calcareous soils. Plant species characteristic of acidic soils have no problem with potassium uptake, whereas plant species typical of less acidic soils frequently suffer from potassium deficiency. This is attributed to an antagonism between ammonium and potassium in the nutrient uptake process [27].

Most plant species from acidic soils can tolerate ammonium, provided they have sufficient amounts of the enzyme nitrate reductase to utilize nitrate that may be present [23,27]. However, the plant species belonging to the Ericaceae are an exception. These plant species have a poor capacity to take up nitrate [28,29] because they lack nitrate reductase activity in shoots [21,30].

Ammonium preference was also reported for a range of tree species. Keltjens and Van Loenen [31] showed that five coniferous and deciduous tree species preferred taking up ammonium over nitrate when supplied with ammonium nitrate solution. The results of another greenhouse experiment indicated that *Pinus sylvestris* seedlings preferred ammonium as the main source of nitrogen [32]. Pine trees are commonly found on acidic soils where ammonium is the dominant nitrogen form, although nitrates are present in low concentrations [33]. Arnold and Van Diest [18] explained the preferential ammonium uptake by coniferous and deciduous tree species through the absence of nitrate reductase.

High concentrations of aluminum can occur in soils of acid heathlands and forests (pH < 4.2), which has an effect on the composition of the plant communities. Species that are capable of tolerating such high concentrations of aluminum have an advantage over other species. Several grass species, such as *Deschampsia flexuosa* and *Holcus lanatus*, are relatively tolerant to low soil pH and high concentrations of aluminum [34,35]. De Graaf et al. [10] observed no detrimental effects of aluminum on the growth of *Calluna vulgaris*, indicating that this species is tolerant to acid soil conditions. In contrast, *Cirsium dissectum*, a species from less acidic soils, was sensitive to aluminum.

The presence of mycorrhizal fungi probably broadens the range of habitats where plant species can grow [36]. Ectomycorrhizal fungi are the predominant associates of coniferous and deciduous trees, and numerous herbs, grasses, and ferns develop association with arbuscular mycorrhizal fungi. Ericoid plant species associate predominantly with ericoid mycorrhizal fungi [37,38]. The major functions of mycorrhizal fungi are uptake of water and nutrients in exchange for carbohydrates provided by the host plants [38]. Mycorrhizal fungi differ in their functioning; for instance, ectomycorrhizal fungi take up mostly inorganic nitrogen (ammonium) [38–40], and arbuscular mycorrhizal fungi are of major importance for phosphorus uptake [38].

A gradient in the occurrence of different mycorrhizal fungi can be distinguished [41]. Arbuscular mycorrhizal fungi develop in litter and humus of the mull type in the pH range from 5.4 to 6.5. Ectomycorrhizal fungi occur mostly on trees growing in moder and mor soils with a pH ranging from 4.2 to 5.5. In the most acidic soils (pH < 4.2), ericoid mycorrhizal fungi can be found in litter and humus of the lowest resource quality [41].

4   ALKALINITY IN AQUATIC ECOSYSTEMS

Several studies revealed that the occurrence of the majority of aquatic plant species is correlated with pH and alkalinity. The pH in waters is dependent on the alkalinity and buffer capacity, which is mostly determined by the bicarbonate concentration [42,43]. In fact, in most waters the alkalinity is similar to the bicarbonate concentration.

Wiegleb [44] classified aquatic plant communities on the basis of alkalinity and distinguished three different groups: (1) plant species that mainly occur at a water alkalinity of less than 0.5 meq L^{-1}, (2) indifferent plant species, and (3) species that occur in waters with alkalinity higher than 0.5 meq L^{-1}. Bloemendaal and Roelofs [45] changed this classification for aquatic water plants in The Netherlands into a more detailed classification with seven groups of different alkalinity.

The distribution of aquatic plant species is generally correlated more with alkalinity than with pH, which can be explained by daily variations in the pH. These variations are highest in dense aquatic vegetation and in waters with algal bloom. During the day, plants and algae take up carbon dioxide for photosynthesis, resulting in enhancement of the carbonate concentrations and an increase in pH. Numerous plants and algae can also use bicarbonate as the carbon source, a process that is accompanied by the release of OH^- and a further increase in pH. In some waters with algal bloom, pH can raise to as high as 11. During the night, the pH decreases because carbon dioxide is released in respiration.

The pH of most nonacidic fresh waters ranges from 6 to 9 [45]. Multiple sampling of a large number of waters over time can level out the variations in pH, making it useful in relating the water pH to the distribution of plant species.

5   ADAPTATIONS OF (SEMI-)AQUATIC PLANT SPECIES
TO VARIOUS ALKALINITY RANGES

(Semi-)aquatic plant communities that vary in alkalinity and pH also differ in species composition. Evidence is growing that many plant species can occur only at a certain level of alkalinity. The bicarbonate concentration seems to be an important factor, and a number of aquatic plant species are well adapted to certain bicarbonate concentrations.

Naturally acidified waters contain very low concentrations of carbon dioxide, while bicarbonate is absent. There are only a few aquatic plant species that are able to develop under such conditions, e.g., *Utricularia minor* [46]. Because the carbon dioxide concentration in the water is too low, *Utricularia minor* derives carbon from the microfauna and from dead organic material. *Urticularia minor* is capable of catching and digesting microfauna such as bacteria and water fleas by special adapted leaves that look like small bladders.

Besides plant species, mosses can grow in naturally acidified waters. Abundant development of different *Sphagnum* species can be found along the edges and in naturally acidified waters. Characteristic of the *Sphagnum* species is that these mosses develop under acidic conditions as long as they can obtain carbon dioxide from the atmosphere. Abundant growth of *Sphagnum* species can lead to the formation of raised bogs [45].

The carbon dioxide concentration in extremely weakly buffered oligotrophic waters is also very low. In contrast, water in the sediment pores in the soft water bodies usually contains levels of carbon dioxide 10–100 times higher than in the water layers above the sediment due to decomposition processes [47–49]. Only plant species that take up carbon dioxide from water in the sediment pores can grow in the extremely weakly buffered waters, such as isoetid plant species belonging to the *Littorelletea* plant communities [49–51]. Extensive lacunal systems in the roots of isoetid plants facilitate carbon dioxide transport from roots to shoots and enable recapturing of photorespired carbon dioxide, resulting in more efficient assimilation [52]. Furthermore, some isoetid plants have developed a diurnal acidification-deacidification cycle similar to crassulacean acid metabolism (CAM) as an adaptation to environments poor in carbon [53,54]. The aquatic plants with CAM utilize an internal carbon dioxide source during the day when the external carbon dioxide content is limited. Beside carbon, the availability of other nutrients such as phosphorus and nitrogen is very low in weakly buffered oligotrophic waters. Several studies reported the presence of arbuscular mycorrhizal fungi on isoetid plants [52,55–57]. One of the major functions contributed by arbuscular mycorrhizal fungi is the uptake of phosphorus [38].

In alkaline waters containing low concentrations of carbon dioxide, most carbon is available as (bi)carbonate. Plant species that are capable of utilizing bicarbonate, e.g., *Elodea canadensis*, have an advantage in these waters.

6 DISTRIBUTION OF TERRESTRIAL PLANT SPECIES IN RELATION TO SOIL pH

Differences in plant species distribution along a soil pH gradient are often referred to as the calcifuge-calcicole gradient [58–60]. The calcifuge species are found on acidic soils, whereas the calcicole species are restricted to soils with higher pH. Falkengren-Grerup et al. [61] noted strong negative correlations between the calcium concentrations in the soil and the growth of plant species typical for acidic soils, e.g., *Carex pilulifera* and *Deschampsia flexuosa*, whereas positive correlations were found for plants characteristic of more calcareous soils, such as *Milium effusum*, *Poa nemoralis*, and *Viola reichenbachiana*. Also, germination and survival of plant species may be affected negatively by aluminum toxicity or calcium deficiency in acidic soils, which may explain the growth of some plant species on alkaline soils only [60]. However, it is questionable to what extent this distinction in plant species distribution is simply a matter of pH or whether other environmental factors are involved too [4,27,62,63].

Roelofs et al. [49] distinguished three different groups of plants species within the characteristic vegetation of wet heathlands on oligotrophic, weakly buffered soils in The Netherlands. This differentiation was based on several environmental factors, such as soil pH. The first group consisted of plant species such as *Erica tetralix*, *Molinia caerulea*, *Gentiana pneumonanthe*, *Lycopodium inundatum*, *Drosera*, and two *Rhynchospora* species. The pH of the soil where these plant species were growing ranged from 4.5 to 5.0. The second group consisted of plants species occurring most commonly on soils with pH ranging from 5.0 to 5.5, e.g., *Carex nigra*, *Carex panicea*, *Cirsium dissectum*, and *Narthecium ossifragum*. The third distinguishable group consisted of species such as *Dacty-lorhiza maculata*, *Juncus acutiflorus*, and *Parnassia palustris*, found mostly on soils with pH from 5.5 to 6.0.

A similar differentiation was made for dry heathlands in The Netherlands. On soils with pH ranging from 3.7 to 4.5, characteristic plant species for dry heathlands were *Calluna vulgaris*, *Lycopodium clavatum*, *Genista pilosa*, and *Genista anglica*. A slightly increasing soil pH (4.2–4.9) resulted in a different vegetation type with, for example, *Nardus stricta*, *Viola canina*, *Arnica montana*, and *Antennaria dioica* [49].

Lammerts et al. [64] studied the basiphilous pioneer plant communities in sand dunes along the Dutch coast. Basiphilous plant species such as *Juncus balticus*, *Sagina nodosa*, *Samolus valerandi*, and *Schoenus nigricans* build up typical basiphilous pioneer communities that can be found on almost bare sandy beach plains or blown-out calcareous slacks. A high soil pH (>6) and low nutrient availability are required [65].

Remarkable is the development of pygmy forests on the oldest and most acidic soils along a narrow coastal band in northern California [66]. This plant community consists of dwarf (<5 m tall) trees of *Cupressus pygmaea*, *Pinus contorta*, and *Pinus muricata*. The pygmy forests can be found only on extremely acidic and infertile soils that have formed from ancient marine terraces. The pH ranges from 3.0 to 5.0 in the upper mineral soil and from 2.0 to 4.0 in the litter layers where fine roots are concentrated. Abrupt changes in the plant communities can be seen at the boundaries of the pygmy forests where the pH of the adjacent soils is much higher [66].

For a unique pine forest ecosystem that occurs in the mountains, it was found that the vegetation was strongly determined by soil pH. *Pinus aristata* forests in the Rocky Mountains in Colorado belong to the oldest forests in the world; some trees are over 2000 years old [67]. Soil pH values ranged from 4.5

to 6.9, decreasing with elevation. A possible explanation is that high-elevation *Pinus aristata* forests are often underlain by granite, a parental material containing low amounts of base cations. The understory vegetation seemed to be directly influenced by the soil pH; *Vaccinium myrtilus* was found only on acidic soils (pH < 5.1) [67].

A study in arctic tundra in Alaska provided further evidence that plant species diversity and composition are directly related to soil pH [63]. Numbers of relatively sparse plant species increased on nonacidic sites, while the dominant plant species on acidic soils were still present but less abundant [63,68,69]. A similar pattern of higher plant species richness with increasing soil pH was also observed across other tundra types throughout the Arctic [68,70,71]. This suggests that the soil pH is an important factor affecting the composition of the plant communities within Arctic regions, although other factors might be of importance at the local scale [63].

7   DISTRIBUTION OF PLANT SPECIES IN SEMIAQUATIC WETLAND ECOSYSTEMS

Ecosystems that accumulate peat (mires) can be divided into ombrotrophic mires, often characterized by bogs, and with mineroptrophic mires, for which the term fens is often used. Bogs receive their water almost exclusively from precipitation, and fens receive a significant amount of water from the surrounding grounds [72]. Particularly in fens, there is much variation in water chemistry and floristic composition. Therefore, several classifications have been developed. The first classification was made by Du Rietz [73], who distinguished rich and poor fens on the basis of the vegetation composition. Rich fen types in Europe belong to the Caricion davallianae, whereas poor fens can be classified as Caricion nigrae. Extremely poor fens form a transition to bogs and are assigned to the Oxycocco-Ericion. To date, rich fens in Europe occur mostly in the subboreal regions and in the mountainous areas. Small numbers of fens are also found in the lowlands of northwestern Europe, often in areas of former peat diggings. After peat removal, the terrestrialization process started. Diggings became revegetated, usually as a floating mire, i.e., a raft of low-density peat covered with vegetation [72].

Raised bogs are characterized by vegetation that is strongly dominated by mosses, predominantly *Sphagnum* species. A small number of plant species are able to grow in this wet acidic environment, but only at low densities. Among these are members of the Cyperaceae and Ericaceae and insectivorous genera such as *Drosera* and *Sarracenia*. Raised bogs may show a distinct depth profile for bicarbonate alkalinity and pH. Lamers et al. [74] observed that the water in the upper layer of a floating peat mat in a raised bog in The Netherlands was acidic (pH 4.3) and unbuffered. The peat mat was dominated by *Sphagnum magellanicum* and *Rynchospora alba* vegetations. The water between the mat and the sediment was more alkaline (pH 6.3) and was moderately buffered because of higher calcium and magnesium bicarbonate concentrations. At sites without a floating raft, the surface water was slightly buffered and the pH was 4.9. The vegetation was characterized by the plant species *Potamogeton polygonifolius* and *Potentilla palustris*.

8 DISTRIBUTION OF PLANT SPECIES IN PERMANENT AQUATIC ECOSYSTEMS

In The Netherlands, plant communities of aquatic ecosystems can be divided into four groups. The first group consists of aquatic plant species such as *Urticularia minor*, *Juncus bulbosus*, and *Ranunculus ololeucos* that occur mostly in acid waters (pH from 4.1 to 4.9). However, the latter two species are not characteristic of acid waters but are mainly observed in waters whose layers have recently become acidified. The soil layers of these waters are still buffered.

The second group is characterized by plant species typical of slightly acidic waters (pH 5.6–5.8), e.g., *Isoetes lacustris*, *Littorella uniflora*, and *Luronium natans*, members of the plant community Littorelletea uniflorae [51,75,76]. The soft water lakes where these plant species occur are poorly buffered waters containing low concentrations of calcium and are usually fully mixed water bodies with periodically fluctuating water levels. They are mostly rainwater fed and thus oligotrophic. In western Europe, the undisturbed soft water lakes have become very rare and are almost all found in nature reserves.

The third group comprises plant species that mainly occur in slightly acid to neutral waters (pH 6.2–7.2), e.g., *Hottonia palustris*, *Myriophyllum alterniflorum*, and *Potamogeton natans*.

The largest number of aquatic plant species is found in neutral to alkaline waters (pH 7.3–8.4) and among the species in this group are *Utricularia vulgaris*, *Potamogeton*, and *Ranunculus* species. The group of plant species with their optimum at very high pH (8.5–9.1) completely consists of aquatic plant species that are often found in brackish waters: *Ceratophyllum submersum*, *Ranunculus baudotii*, and *Zannichellia palustris*.

9 ACIDIFICATION OF TERRESTRIAL ECOSYSTEMS

Over the last decades atmospheric deposition has occurred throughout Europe and parts of Northern America, and it is still an ongoing process. The main components are ammonia (NH_4^+), nitrogen oxides (NO_x), and airborne sulfur (SO_x). Ammonia originates from intensive livestock industry, and nitrogen and sulfur oxides are emitted by burning of fossil fuels [77,78].

The deposition of atmospheric SO_x and NO_x results in direct acidification of the soil. In the atmosphere, SO_x and NO_x react to SO_4^{2-} and NO_3^-, which are directly deposited at the soil. The deposition of atmospheric ammonium causes eutrophication and indirect acidification of the soil, particularly in nutrient-poor soils with low buffer capacity [79,80]. Nitrification of ammonium by nitrifying bacteria results in the release of protons [3,11] according to

$$NH_4^+ + 2O_2 \rightarrow NO_3^- + 2H^+ + H_2O$$

Indirect soil acidification is also caused by the vegetation. During uptake of ammonium, protons are extruded from plant roots, resulting in acidification of the rhizosphere [26,81–83]. The latter process occurs only when the extent of ammonium uptake by plant roots is such that anion uptake insufficiently balances the influx of positive ions. Gijsman [84] reported that Douglas fir trees acidify their rhizosphere when the contribution of nitrate to total nitrogen uptake falls below 65%. Arnold [83] estimated that *Pinus sylvestris* trees exhibit an acidifying uptake pattern as long as the contribution of nitrate to the nitrogen nutrition is lower than 70%. An additional effect of the acidification of the rhizosphere is leaching of cations, which results in reduced uptake of cations by the roots [2].

Acidification due to nitrogen and sulfur deposition was studied by Dueck and Elderson [85] by competition experiments with *Arnica montana*, *Viola canina*, and the grass species *Agrostis canina* under different nitrogen and sulfur deposition levels. The results of these experiments suggested that acidification was the primary cause of the decline of the plant species. Studies by Van Dam et al. [86], Houdijk [17], and Roelofs et al. [49] also indicated that plant species characteristic of heathland ecosystems have declined due to soil acidification, e.g., *Arnica montana*, *Cirsium dissectum*, *Polygala serpyllifolia*, and *Thymus serpyllum*. Generally, these plant species grow on weakly to moderately acidic soils [pH(H_2O) > 4.4] with relatively high base cation concentrations and relatively low Al:Ca ratios. On more acidic soils [pH(H_2O) < 4.4] with higher Al:Ca ratios the composition of the plant community has shifted to a community that predominantly consists of the grass species *Deschampsia flexuosa* and *Molinia caerulea*. Hydroculture experiments with seedlings of *Arnica montana* and *Cirsium dissectum* grown on different aluminum concentrations and different Al:Ca ratios showed that these species are sensitive to high Al:Ca ratios. When the Al:Ca ratio was greater than 1, the growth of the *Arnica montana* plants was more reduced than on medium without aluminum. *Cirsium dissectum* showed the same pattern when the Al:Ca ratio was greater than 2.

The aluminum concentration affected the plant development too. The *Arnica montana* and *Cirsium dissectum* plants showed more reduced development at 50 than at 100 µmol Al L^{-1}, regardless of the Al:Ca ratio (Fig. 2). A possible explanation for the reduced root development of plants grown at high aluminum is inhibition of cell development [6]. Other plant physiological effects of aluminum are reduction of internal cation concentrations, inhibition of nitrate uptake, and disturbance of several metabolic processes [87–90].

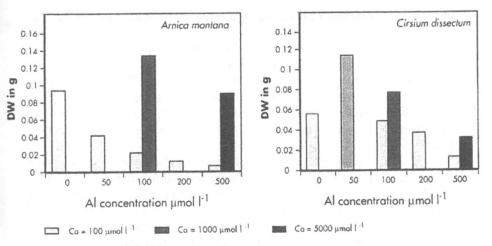

FIGURE 2 Average biomass (g dry weight) of *Arnica montana* and *Cirsium dissectum* grown in media with different concentrations of aluminum and calcium.

Fennema [91] observed relatively high aluminum concentrations in soils from which *Arnica montana* had recently disappeared. In contrast, Pegtel [92] noted that *Arnica montana* is hardly sensitive to aluminum and that the decline of this plant species in The Netherlands cannot be explained by aluminum intolerance. The results of a competition experiment with *Arnica montana* and *Deschampsia flexuosa* also suggested that extra supply of aluminum and manganese under acidifying conditions did not explain the decline of *Arnica montana* [93]. A more likely explanation is that *Arnica montana* has a different growth strategy than *Deschampsia flexuosa*, resulting in lower competitive strength. *Arnica montana* uses nutrients efficiently and grows more slowly than *Deschampsia flexuosa* as shown for fertilized humus podzol. Furthermore, *Deschampsia flexuosa* is capable of taking up a wide range of nutrients, thus avoiding nutritional imbalances and maintaining a rather constant growth rate.

Grasslands belonging to the Cirsio dissecti-Molinietum are sensitive to disturbances such as acidification and desiccation. The calcium buffering is just sufficient to buffer acidification due to natural deposition of hydrocarbonic acid and natural acidification by *Sphagnum* species. Lowering of the ground water level can rapidly result in acidification because the calcium and magnesium buffer capacity of the soil is insufficient. The reduction of the buffer capacity can be accompanied by lowering of the pH. In addition, atmospheric deposition of acidifying compounds enhances the acidification of the grasslands. The consequences for the vegetation are that various *Sphagnum* species outcompete characteristic plant species, including *Carex panicea*, *Cirsium dissectum*, and *Parnassia palustris* [4].

For calcareous grasslands, the effects of ammonium nitrate and ammonium sulfate additions were studied in The Netherlands and the United Kingdom. Soil pH values decreased from 6.8 to 6.3–6.2 after addition of 70 or 140 kg ammonium nitrate ha^{-1} year^{-1} for 7 years. Addition of 140 kg ammonium sulfate ha^{-1} year^{-1} lowered the soil pH to 5.2. The severe soil acidification could be explained by high nitrification rates resulting in a significant decline of species diversity [94,95].

A high number of *Pinus sylvestris* forests in The Netherlands have been affected by atmospheric deposition of nitrogen and sulfur during the last decades, and the soil has been strongly acidified. For instance, Baar [96] observed that the pH of the mineral soil ranged from 3.2 to 4.0, while the pH of the litter and humus layers was even lower (from 2.9 to 3.6). The understory vegetation of these *Pinus sylvestris* forests is dominated by the grass species *Deschampsia flexuosa*, which can grow well on the acidified soils [97,98]. Apparently, *Deschampsia flexuosa* has outcompeted plant species characteristic of the understory vegetation in unpolluted *Pinus sylvestris* forests, e.g., *Calluna vulgaris*, *Empetrum nigrum*, and *Vaccinium myrtillus* [99,100]. The competitive advantage of *Deschampsia flexuosa* may be mediated by the belowground mycorrhizal community [96]. Arbuscular mycorrhizal fungi associated with *Deschampsia flexuosa* may give this grass species a competitive advantage over the ericoid species in the nitrogen-enriched *Pinus sylvestris* forests on acidified soils where phosphorus has become limited while excess of inorganic nitrogen is available. The major function attributed to arbuscular mycorrhizal fungi is uptake of phosphorus, and ericoid mycorrhizal fungi have the ability to utilize nitrogen from organic nutrient compounds [38].

Acidification of temperate deciduous forests during recent decades has altered the soil chemistry, resulting in a decrease of the pH and exchangeable cations [20]. In forests in south Sweden, the exchangeable base cations were replaced by H$^+$ and/or Al^{3+} ions at the rate of about 1% year^{-1} over the last decades. The abundance and frequency of several plant species (*Melica uniflora*, *Vaccinium reichenbachiana*) in the understory were reduced, which was probably related to the altered soil chemistry [20].

10 ACIDIFICATION OF SEMIAQUATIC WETLAND ECOSYSTEMS

Freshwater wetlands occur worldwide but are threatened by acidification processes too. An important cause of the acidification of fresh water is atmospheric input of SO$_x$ [101]. Also, desiccation can lead to acidification due to proton production by chemical processes such as chemical oxidation of reduced iron and sulfide and increase of oxygen-dependent microbiological processes such as nitrification [102]. In a mesocosm study, the vegetation of a mesotrophic wetland meadow dominated with *Carex nigra* was studied to evaluate the combined effects of preceding accumulation of sulfur and desiccation [102]. Treatment with solutions containing sulfate followed by desiccation resulted in a significant drop of pH from 6.5 to 4.5, which can be explained by sulfide oxidation. Base cations and phytotoxic metals such as Al were mobilized because of acidification and the cation-exchange processes [102]. Brouwer and Roelofs [103] observed in a mesocosm experiment that the pH dropped significantly from 7.3 to 3.8 after reflooding of sediments of a desiccated soft water lake in the southern part of The Netherlands. High concentrations of sulfate were measured, while iron had precipitated [103].

Field observations revealed that floating base-rich fens in wet dune slacks and peatland areas with plant communities belonging to the Caricion davallianae are affected by acidification and eutrophication [104,105]. The main source of acidification is the surplus of nonbuffered rainwater that forms a lens above the calcareous groundwater. This results in a shift in the plant community from a highly diverse vegetation with, for instance, *Juncus subnodulosus*, *Menyanthes trifoliata*, and *Urticularia minor* to a dominance of *Sphagnum squarrosum*, *Sphagnum fallax*, and *Polytrichum commune* [105]. Boeye and Verheyen [104] concluded from a study of a rich fen vegetation in Belgium that the species of the Caricion davallianae disappear if the water supply of alkaline water ceases.

11 ACIDIFICATION OF PERMANENT AQUATIC ECOSYSTEMS

Large-scale acidification has significantly affected freshwater ecosystems through atmospheric deposition. In The Netherlands, a large proportion of the hydrologically isolated soft water lakes has become acidified, resulting in a decreased pH of the water layer and raised concentrations of sulfate [75,106]. Furthermore, nitrogen- and sulfur-enriched atmospheric deposition can lead to enhanced concentrations of carbon dioxide in the water layers and dominance of ammonium over nitrate [47,107,108]. Apart from direct acidification of the water layers, atmospheric deposition causes acceleration of weathering of the catchment soils, resulting in relatively high concentrations of aluminum, cadmium, and manganese [106]. As a result of acidification, thick organic sapropelium layers can develop on the bottom, and the water transparency increases. This results in a shift in the plant species composition; most freshwater macrophytes rarely grow in waters with pH below 5 [101,109]. However, nymphaeid plant species such as *Nympha alba* are relatively insensitive to acidification because they can occur along a wide pH range [47,75].

Observations in 70 shallow soft water lakes in The Netherlands and Germany revealed that isoetid species have disappeared from 53 of these lakes since 1950 [51,110]. In 41 of these waters, Roelofs [47] observed that *Juncus bulbosus* and *Sphagnum cuspidatum* occurred abundantly when the pH was low (3.9). Luxurious growth of *Juncus bulbosus* and concomitant suppression of isoetid species is often observed in acidified, shallow soft water bodies in western Europe [111]. Addition of different artificial rain solutions in an aquatic mesocosm study resulted in acidification; in particular, ammonium sulfate reduced the pH to 3.5 in both water and sediment. Plants characteristic of soft waters declined and were overgrown by *Sphagnum* species and *Juncus bulbosus* [112]. Increased development of these species was also seen in acidified Swedish lakes, although isoetid plant species remained abundant [113,114]. Experimental studies showed that addition of acid rain did not necessarily affect the naturally developed vegetation and could allow diverse isoetid vegetation to remain [115]. Isoetid plants such as *Littorella uniflora* are known to be acid resistant, and therefore the decline of isoetid plants is not a direct result of acidification [48,109]. However, enhancement of ammonium concentrations under acidic conditions stimulated development of *Juncus bulbosus*, submerged mosses such as *Sphagnum* species, and the grass species *Agrostis canina* [47,48]. Apparently, the increased levels of ammonium in these waters stimulated the growth of plants such as *Juncus bulbosus*, while the surplus ammonium was nitrified, resulting in an increase in H^+ ions and hence an increase in acidity. The isoetid macrophytes were outcompeted because they take up predominantly nitrate [116].

In the remaining 12 of the 53 waters, Roelofs [47] observed that eutrophic plant species such as *Lemna minor*, *Riccia fluitans*, *Myriophyllum alterniflorum*, and *Ranunculus peltatus* had replaced the isoetid species. In these waters, a relatively high alkalinity and relatively high concentrations of phosphate were measured.

12 CONCLUSIONS

The pH is an environmental variable that is important for the distribution of plant species in terrestrial and (semi-)aquatic ecosystems. In terrestrial ecosystems, the composition of the plant communities on soils with a lower pH differs from that on soils with a higher pH. This differentiation has been observed for a wide range of ecosystems such as wet and dry heathlands, sand dunes, pine forests, and arctic tundra. (Semi-)aquatic plant communities are also pH dependent and range from plant communities characteristic of acid waters to plant communities that have their optimum in alkaline waters.

Disturbances such as acidification, desiccation, and eutrophication generally affect the soil pH and/or water pH, resulting in a shift in the composition of plant communities. For terrestrial ecosystems, effects on the soil pH due to such disturbances have been observed for various ecosystems, such as heathland, grassland, and forest ecosystems. Acidification is the major disturbance factor in (semi-)aquatic ecosystems, particularly in Europe, and has caused major shifts in the composition of the plant communities.

ACKNOWLEDGMENTS

We thank A.W. Boxman, L.P.M. Lamers and H.B.M. Tomassen for detailed comments aimed at improving this review.

REFERENCES

1. B Ulrich. An ecosystem approach to soil acidification. In: B Ulrich, ME Summer, eds. Soil Acidity. Berlin: Springer Verlag, 1991, pp 28–79.
2. B Ulrich. Interactions of forest canopies with atmospheric constituents: SO_2, alkali and earth alkali cations and chloride. In: B Ulrich, J Pankrath, eds. Effects of Accumulation of Air Pollutants in Forest Ecosystems. Dordrecht: Reidel, 1983, pp 33–45.
3. N Van Breemen, J Mulder, CT Driscoll. Acidification and alkalinization of soils. Plant Soil 75:283–308, 1982.
4. MCC De Graaf. The calcifuge-calcicole gradient in dry heathlands. PhD dissertation, University of Nijmegen, The Netherlands, 2000.
5. J Rozema, P Haan, R Broekman, WHO Ernst, AJ Appelo. On lime transition and decalcification in the coastal dunes of the province of North Holland and the isle of Schiermonnikoog. Acta Bot Neerl 34:393–411, 1985.
6. H Marschner. Mechanisms of adaptations of plants to acid soils. Plant Soil 134:1–20, 1991.
7. W De Vries, M Posch, J Kämäri. Simulation of the long-term soil response to acid deposition in various buffer ranges. Water Air Soil Pollut 48:349–390, 1989.
8. W Ernst. Chemical soil factors determining plant growth. In: AHJ Freysen, JW Woldendorp, eds. Structure and Functioning of Plant Populations. Amsterdam: North Holland, 1978, pp 155–187.
9. ALFM Houdijk, PJM Verbeek, HFG Van Dijk, JGM Roelofs. Distribution and decline of endangered herbaceous heathland species in relation to the chemical composition of the soil. Plant Soil 148:137–143, 1993.
10. MCC De Graaf, R Bobbink, PJM Verbeek, JGM Roelofs. Aluminum toxicity and tolerance in three heathland species. Water Air Soil Pollut 98:229–239, 1997.
11. W De Boer. Nitrification in Dutch heathland soils. PhD dissertation, Agricultural University Wageningen, The Netherlands, 1989.
12. SR Troelstra, R Wagenaar, W De Boer. Nitrification in Dutch heathland soils: I. General soil characteristics and nitrification in undisturbed soils. Plant Soil 127:179–192, 1990.
13. DW Cole. Nitrogen uptake and translocation by forest ecosystems. In: FE Clark, T Rosswall, eds. Terrestrial Nitrogen Cycles. Stockholm, Ecol Bull, 1981.
14. T Nasholm. Qualitative and quantitative changes in plant nitrogen acquisition induced by anthropogenic nitrogen deposition. New Phytol 139:87–90, 1998.

15. U Falkengren-Grerup, J Brunet, M Diekman. Nitrogen mineralisation in deciduous forest soils in south Sweden in gradients of soil acidity and deposition. Environ Pollut 102:415–420, 1998.

16. WU Kriebitzsch. Stikstoffnachlieferung in sauren Waldböden Nordwestdeutschlands. Scripta Geobot 14:1–66, 1978.

17. ALFM Houdijk. Atmospheric ammonium deposition and the nutritional balance of terrestrial ecosystems. PhD dissertation, University of Nijmegen, The Netherlands, 1993.

18. G Arnold, A Van Diest. Nitrogen supply, tree growth and soil acidification. Fertil Res 27:29–38, 1991.

19. U Falkengren-Grerup, H Lakkenborg-Kristensen. Importance of ammonium and nitrate to the performance of herb-layer species from deciduous forests in southern Sweden. Environ Exp Bot 34:31–38, 1994.

20. U Falkengren-Grerup. Interspecies differences in the preference of ammonium and nitrate in vascular plants. Oecologia 102:305–311, 1995.

21. SR Troelstra, R Wagenaar, W Smant. Nitrogen utilization by plant species from acid heathlands. I. Comparison between nitrate and ammonium nutrition at constant low pH. J Exp Bot 46:1103–1112, 1995.

22. HJ Kronzucker, MY Siddique, ADM Glass. Conifer root discrimination against soil nitrate and the ecology of forest succession. Nature 385:59–61, 1997.

23. A Gigon, IH Rorison. The response of some ecologically distinct plant species to nitrate- and ammonium-nitrogen. J Ecol 60:93–102, 1972.

24. CJ Atkinson. Nitrogen acquisition in four co-existing species from an upland acidic grassland. Physiol Plant 63:375–387, 1985.

25. LA Peterson, EJ Stang, MN Dana. Blueberry response to NH_4-N and NO_3-N. J Am Soc Hortic Sci 113:9–12, 1988.

26. MCC De Graaf, R Bobbink, JGM Roelofs, PJM Verbeek. Differential effects of ammonium and nitrate on three heathland species. Plant Ecol 135:186–196, 1997.

27. IH Rorison. The response of plants to acid soils. Experientia 42:357–362, 1986.

28. G Gebauer, H Rehder, B Wollenweber. Nitrate, nitrate reduction and organic nitrogen in plants from different ecological and taxonomic groups of central Europe. Oecologia 75:371–385, 1988.

29. P Högberg, C Johannisson, H Nicklasson, L Högbom. Shoot nitrate reductase activities of field-layer species in different forest types. I. Preliminary surveys in northern Sweden. Scand J For Res 5:449–456, 1990.

30. JA Lee, GR Stewart. Ecological aspects of nitrogen assimilation. Adv Bot Res 6: 2–43, 1978.

31. WG Keltjens, E Van Loenen. Effects of aluminum and mineral nutrition on growth and chemical composition of hydroponically grown seedlings of five different forest tree species. Plant Soil 119:39–50, 1989.

32. AW Boxman, JGM Roelofs. Some effects of nitrate versus ammonium nutrition on the nutrient fluxes in *Pinus sylvestris* seedlings. Effects of mycorrhizal infection. Can J Bot 66:1091–1097, 1988.

33. JGM Roelofs, AW Boxman. The effect of airborne ammonium sulphate deposition on pine forests. In: H-D Gregor, M Hübner, E Niegengerd, L Ries, M Schneider,

W Schröder, eds. Neue Ursachenhypothesen. Berlin: Umwelt Bundes Amt, 1986, pp 415–422.

34. IH Rorison. Nitrogen source and the tolerance of *Deschampsia flexuosa*, *Holcus lanatus*, and *Bromus erectus* to aluminum during seedling growth. J Ecol 73:83–90, 1985.

35. AM Balsberg-Pahlsson. Growth, radicle and root hair development of *Deschampsia flexuosa* (L.) Trin. seedlings in relation to soil acidity. Plant Soil 175:125–132, 1995.

36. HJ Peat, AH Fitter. The distribution of arbuscular mycorrhizas in the British flora. New Phytol 125:845–854, 1993.

37. JL Harley, EL Harley. A check-list of mycorrhiza in the British flora. New Phytol Suppl 105:1–102, 1987.

38. SE Smith, DJ Read. Mycorrhizal Symbiosis. 3rd ed. San Diego: Academic Press, 1997.

39. RD Finlay, H Ek, G Odham, B Söderström. Mycelial uptake, translocation and assimilation of nitrogen from ^{15}N-labelled ammonium by *Pinus sylvestris* plants infected with four different ectomycorrhizal fungi. New Phytol 110:59–66, 1988.

40. G Keller. Utilization of inorganic and organic nitrogen sources by high-subalpine ectomycorrhizal fungi of *Pinus cembra* in pure culture. Mycol Res 100:989–998, 1996.

41. DJ Read. Mycorrhizas in ecosystems. Nature's response to the "law of the minimum " In: DL Hawksworth, ed. Frontiers in Mycology. Wallingford: CAB International, 1991, pp 101–130.

42. CB Hellquist. Correlation of alkalinity and the distribution of *Potamogeton* in New England. Rhodora 82:331–344, 1980.

43. Y Kadono. Photosynthetic carbon sources in some *Potamogeton* species. Bot Mag 93:185–194, 1980.

44. G Wiegleb. Untersuchungen über den Zusammenhang zwischen hydrochemischen Umweltfaktoren und Makrophytenvegetation in stehenden Gewässern. Arch Hydrobiol 84:443–484, 1980.

45. FHJL Bloemendaal, JGM Roelofs. Waterplanten en waterkwaliteit. Utrecht Stichting Uitgeverij Koninklijke Nederlanse Natuurhistorische Vereniging, 1988.

46. EJ Weeda, R Westra, C Westra, T Westra. Nederlandse oecologische flora. Wilde planten en hun relaties 3. Amsterdam: IVN, 1988.

47. JGM Roelofs. Impact of acidification and eutrophication on macrophyte communities in soft waters in The Netherlands. I. Field observations. Aquat Bot 17:139–155, 1983.

48. JGM Roelofs, JAAR Schuurkes, AJM Smits. Impact of acidification and eutrophication on macrophyte communities in soft water in The Netherlands. II. Experimental studies. Aquat Bot 18:389–411, 1984.

49. JGM Roelofs, R Bobbink, E Brouwer, MCC De Graaf. Restoration ecology of aquatic and terrestrial vegetation on non-calcareous sandy soils in The Netherlands. Acta Bot Neerl 45:517–541, 1996.

50. C Den Hartog, S Segal, A new classification of the water-plant communities. Acta Bot Neerl 13:367–393, 1964.

51. R Wittig. Verbreitung der Littorelletea-Arten in der Westfalischen Bucht. Decheniana 135:14–21, 1982.

52. M Sondergaard, S Laegaard. Vesicular-arbuscular mycorrhizae in some aquatic vascular plants. Nature 268:232–233, 1977.

53. JE Keeley. Crassulacean acid metabolism in *Isoetes bolanderi* in high elevation oligotrophic lakes. Oecologia 58:63–69, 1983.

54. JE Keeley. Distribution of diurnal acid metabolism in the genus *Isoetes*. Am J Bot 69:254–257, 1982.

55. AM Farmer. The occurrence of vesicular-arbuscular mycorrhiza in isoetid-type submerged aquatic macrophytes under naturally varying conditions. Aquat Bot 21: 245–249, 1985.

56. KK Christensen, C Wigand. Formation of root plaques and their influence on tissue phosphorus content in *Lobelia dormanna*. Aquat Bot 61:111–122, 1998.

57. C Wigand, FO Andersen, KK Christensen, M Holmer, HS Jensen. Endomycorrhizae of isoetids along a biogeochemical gradient. Limnol Oceanogr 43:508–515, 1998.

58. OE Balme. Edaphic and vegetational zoning on the carboniferous limestone of the Derbyshire Dales. J Ecol 41:331–344, 1956.

59. AA Hayati, MCF Proctor. Plant distribution in relation to mineral nutrient availability and uptake on a wet-heath site in South-west England. J Ecol 78:134–151, 1990.

60. A Zohlen, G Tyler. Immobilization of tissue iron on calcareous soil: differences between calcicole and calcifuge plants. Oikos 89:95–106, 2000.

61. U Falkengren-Grerup, J Brunet, ME Quist, G Tyler. Is the Ca:Al ratio superior to pH, Ca or Al concentrations of soils accounting for the distribution of plants in deciduous forests? Plant Soil 177:21–31, 1995.

62. W Seidling, M-S Rohner. Zusammenhänge zwischen Reaktions-Zeigerwerten und bodenchemischen Parametern am Beispiel von Waldbodenvegetation. Phytocoenologia 23:301–317, 1993.

63. L Gough, GR Shaver, J Carroll, DL Royer, JA Laundre. Vascular plant species richness in Alaskan arctic tundra: the importance of soil pH. J Ecol 88:54–66, 2000.

64. EJ Lammerts, DM Pegtel, AP Grootjans, A Van der Veen. Nutrient limitation and vegetation changes in a coastal dune slack. J Veg Sci 10:111–122, 1999.

65. AP Grootjans, FP Sival, PJ Stuyfzand. Hydrogeochemical analysis of a degraded dune slack. Vegetatio 126:27–38, 1996.

66. RR Northup, RA Dahlgren, Z Yu. Intraspecific variation of conifer phenolic concentration on a marine terrace soil acidity gradient: a new interpretation. Plant Soil 171:255–262, 1995.

67. BM Ranne, WL Baker, T Andrews, MG Ryan. Natural variability of vegetation, soils, and physiography in the bristlecone pine forests of the Rocky Mountains. Great Basin Nat 57:21–37, 1997.

68. MD Walker, DA Walker, NA Auerbach. Plant communities of a tussock tundra landscape in the Brooks Range Foothills, Alaska J Veg Sci 5:843–866, 1994.

69. GR Shaver, JA Laundre, AE Giblin, KJ Nadelhoffer. Changes in live plant biomass, primary production, and species composition along a riverside toposequence in Arctic Alaska, U.S.A Arct Alp Res 28:363–379, 1996.

70. KP Timoney, GH La Roi, MRT Dale. Subarctic forest-tundra vegetation gradients: the sigmoid wave hypothesis. J Veg Sci 4:387–394, 1993.

71. R-K Heikkinen. Predicting patterns of vascular plant species richness with composite variables: a meso-scale study in Finnish Lapland. Vegetatio 126:151–165, 1996.

72. R Van Diggelen, WJ Molenaar, AM Kooijman. Vegetation succession in a floating mire in relation to management and hydrology. J Veg Sci 7:809–820, 1996.

73. GE Du Rietz. Die Mineralbodenwasserseigergrenze als Grundlage einer natürlichen Zweigliederung der nord und mitteleuropäischen Moore. Vegetatio 5–6:571–585, 1954.

74. LPM Lamers, C Farhoush, JM Van Groenendael, JGM Roelofs. Calcareous groundwater raises bogs; the concept of ombrotrophy revisited. J Ecol 87:639–648, 1999.

75. GHP Arts, G Van der Velde, JGM Roelofs, CAM Van Swaay. Successional changes in the softwater macrophyte vegetation of (sub)atlantic, sandy, lowland regions during this century. Freshwater Biol 24:287–294, 1990.

76. JHJ Schaminee, EJ Weeda, V Westhoff. De Vegetatie van Nederland. Deel 2. Plantengemeenschappen van wateren, moerassen en natte heiden. Uppsala: Opulus Press, 1995.

77. JW Erisman. Acid deposition to nature area's in The Netherlands: part 1. Methods and results. Water Air Soil Pollut 371:51–80, 1993.

78. WAH Asman, MA Sutton, JK Schjorring. Ammonia: emission, atmospheric transport and deposition. New Phytol 139:27–48, 1998.

79. GJ Heij, T Schneider. Acidification Research in The Netherlands: Final Report of the Dutch Priority Programme on Acidification. Amsterdam: Elsevier, 1991.

80. R Bobbink, M Hornung, JGM Roelofs. The effects of air-borne nitrogen pollutants on species diversity in natural and semi-natural European vegetation. J Ecol 86:717–738, 1998.

81. GR Findenegg. A comparative study of ammonium toxicity at different constant pH of the nutrient solution. Plant Soil 103:239–243, 1987.

82. L Salsac, S Chaillon, JF Morot-Gaudry, C Lessaint, E Jolivet. Nitrate and ammonium nutrition in plants. Plant Physiol Biochem 25:805–812, 1987.

83. G Arnold. Soil acidification as caused by the nitrogen uptake pattern of Scots pine (*Pinus sylvestris*). Plant Soil 142:41–51, 1992.

84. A Gijsman. Nitrogen nutrition of Douglas fir (*Pseudotsuga menziesii*) on acid sandy soils. I. Growth, nutrient uptake and ionic balance. Plant Soil 126:53–61, 1990.

85. TA Dueck, J Elderson. Influence of NH_3 and SO_2 on the growth and competitive ability of *Arnica montana* L. and *Viola canina* L. New Phytol 122:507–514, 1992.

86. D Van Dam, HF Van Dobben, CFJ Ter Braak, T De Wit. Air pollution as a possible cause for the decline of some phanerogamic species in The Netherlands. Vegetatio 65:47–52, 1986.

87. L Galvez, RB Clark. Nitrate and ammonium uptake and solution pH changes for Al-tolerant and Al-sensitive sorghum (*Sorghum bicolor*) genotypes grown with and without aluminum. Plant Soil 134:179–188, 1991.

88. WG Keltjens, K Tan. Interactions between aluminum, magnesium and calcium with different monocotyledonous plant species. Plant Soil 155/156:485–488, 1993.

89. E Delhaize, PR Ryan. Aluminum toxicity and tolerance in plants. Plant Physiol 107:315–321, 1995.

90. RP Durieux, RJ Bartlett, FR Magdoff. Separate mechanisms of aluminum toxicity for nitrate uptake and root elongation. Plant Soil 172:229–234, 1995.
91. F Fennema. SO$_2$ and NH$_3$ deposition as possible cause for the extinction of *Arnica montana* L. Water Air Soil Pollut 62:325–336, 1992.
92. DM Pegtel. Effect of ionic Al in culture solutions on the growth of *Arnica montana* L. and *Deschampsia flexuosa* (L.) Trin. Plant Soil 102:85–95, 1987.
93. DM Pegtel. Habitat characteristics and the effects of various nutrient solutions on growth and mineral nutrition of *Arnica montana* L. grown on natural soil. Vegetatio 144:109–121, 1994.
94. JA Lee, SJM Caporn. Ecological effects of atmospheric reactive nitrogen deposition on semi-natural terrestrial ecosystems. New Phytol 139:127–134, 1998.
95. R Bobbink. Impacts of tropospheric ozone and airborne nitrogenous pollutants on natural and semi-natural ecosystems: a commentary. New Phytol 139:161–168, 1998.
96. J Baar. Ectomycorrhizal fungi as affected by litter and humus. PhD Dissertation, Agricultural University Wageningen, The Netherlands, 1995.
97. J Baar, CJF Ter Braak. Ectomycorrhizal sporocarp occurrence as affected by manipulation of litter and humus layers in Scots pine stands of different age. Appl Soil Ecol 6:61–73, 1996.
98. EJ Weeda, R Westra, C Westra, T Westra. Nederlandse oecologische flora. Wilde planten en hun relaties 5. Amsterdam: IVN, 1994.
99. J Fanta. Spontane Waldentwicklung auf diluvialen Sandböden und ihre Bedeutung für den Naturstudie. Ein Fallstudie aus der Veluwe, Niederlande. NNA Ber 5:23–27, 1992.
100. BWL De Vries, E Jansen, HF Van Dobben, ThW Kuyper. Partial restoration of fungal and plant species diversity by removal of litter and humus layers in stands of Scots pine in the Netherlands. Biodiv Conserv 4:156–164, 1995.
101. MJS Bellemakers. Reversibility of the effects of acidification and eutrophication of shallow surface water perspective for restoration. PhD dissertation, University of Nijmegen, The Netherlands, 2000.
102. LPM Lamers, SME Van Roozendaal, JGM Roelofs. Acidification of freshwater wetlands: combined effects of non-airborne sulfur pollution and desiccation. Water Air Soil Pollut 105:95–106, 1998.
103. E Brouwer, JGM Roelofs. Groundwater as an alternative for the supply of eutrophied surface water in nutrient poor, acid-sensitive softwater pools. Mitt Ag Geobot 57:121–127, 1999.
104. D Boeye, RF Verheyen. The relation between vegetation and soil chemistry gradients in a ground water discharge fen. J Veg Sci 5:553–560, 1994.
105. B Beltman, T van den Broek, S Bloemen, C Witsel. Effects of restoration measures on nutrient availability in a formerly nutrient-poor floating fen after acidification and eutrophication. Biol Conserv 78:271–277, 1996.
106. RSEW Leuven, G Van der Velde, HLM Kersten. Interrelations between pH and other physico-chemical factors of Dutch soft waters. Arch Hydrobiol 126:27–51, 1992.
107. JWM Rudd, CA Kelly, DW Schindler, MA Turner. Disruption of the nitrogen cycle in acidified lakes. Science 140:1515–1517, 1988.

108.　GHP Arts, RSEW Leuven. Floristic changes in shallow soft waters in relation to underlying environmental factors. Freshwater Biol 20:97–111, 1988.

109.　M Maessen, JGM Roelofs, MJS Bellemakers, GM Verheggen. The effects of aluminum, aluminum/calcium ratios and pH on aquatic plants from poorly buffered environments. Aquat Bot 43:115–127, 1992.

110.　MM Schoof-van Pelt. Littorelletea, a study of the vegetation of some amphiphytic communities of western Europe. PhD dissertation, University of Nijmegen, The Netherlands, 1973.

111.　H Van Dam, H Kooyman-van Blokland. Man-made changes in some Dutch moorland pools, as reflected by historical and recent data about diatoms and macrophytes. Int Rev Ges Hydrobiol 63:425–459, 1978.

112.　E Brouwer, R Bobbink, F Meeuwsen, JGM Roelofs. Recovery from acidification in aquatic mesocosms after reducing ammonium and sulphate deposition. Aquat Bot 56:119–130, 1997.

113.　H Hultberg, O Grahn. Effects of acid precipitation on macrophytes in oligotrophic Swedish lakes. Int Assoc Great Lakes Res 1:208–217, 1975.

114.　O Grahn. Macrophyte succession in Swedish lakes caused by deposition of airborne substances. Water Air Soil Pollut 7:295–305, 1977.

115.　JAAR Schuurkes, MA Elbers, JJF Gudden, JGM Roelofs. Effects of simulated ammonium sulphate and sulphuric acid rain on acidification, water quality and flora of small-scale soft water systems. Aquat Bot 28:199–225, 1987.

116.　JAAR Schuurkes, CJ Kok, C Den Hartog. Ammonium and nitrate uptake by aquatic plants from poorly buffered and acidified waters. Aquat Bot 24:131–146, 1986.

Index

Abscisic acid, 85–88, 95, 96, 125, 141, 256, 272–274
   CAM metabolism, 244
Acacia mangium, 395
Acanthamoeba castellanii, 200
Acaulospora, 394, 395
Acetabularia acetabulum, 30
acetate, 391
acetic acid, 353
acetoxymethyl esters, 78
acetyl-CoA, 237, 238
Achromobacter, 367
acid forming processes, 352
acid mine drainage, 339, 340
acid neutralizing capacity, 406
acid phosphatase, 26
acid rain, 363, 357
acid sulfate soils, 337
acid tolerance, 384, 385, 386–389, 391, 395
acid waters, 413
acid-base chemistry, 263, 264
acid-base relations
   transmembrane ion transport, 110
acid-growth theory, 125, 132
acidification of terrestrial ecosystems, 414
acidification of the rhizosphere, 262, 271
acidification of vacuoles, 241
acid-sensitive species, 329
acid-shock proteins, 387, 389
aconitase, 220

actinomycete, 392
actinorhizal species, 393
actinorhizal symbiosis, 392, 393
adenine nucleotide carrier, 196, 198
ADP^{3-}/ATP^{4-} exchange, 196, 198
Aequorea victoria, 75
Agapanthus pollen tubes, 79
Agapanthus umbellatus, 91
Agavaceae, 228
Agrostis canina, 414
airborne sulfur (SO_x), 414
Aizoaceae, 228, 230, 231, 242
Al:Ca ratios, 415
aleurone, 96
   -amylase, 87, 88, 96
alfalfa, 389, 390
algae, 2
algal bloom, 409, 410
algal rhizoids, 307
Alismataceae, 232
alkaline phosphatase, 363
alkaliphilic bacteria, 354
alkaliphilic organisms, 354
alkaloid biosynthesis, 90, 94
alkaloids, 25
alleviation of aluminum toxicity, 334
Alnus rubra, 392, 393
alternative oxidase (AOX), 189, 190, 205, 206
alternative oxidase pathway, 60
alternative ubiquinol oxidase, 175

aluminosilicate clays, 315, 334
aluminum, 260, 261, 271, 309, 325,
    326, 358, 405, 409
  availability, 329
  complexation, 335
  ionic speciation, 334, 335
  solubility, 332
  sensitivity, 329, 415
  tolerance, 315
  toxicity, 327, 333, 334, 338, 355,
    364, 391, 411
  mobilization of, 407
amelioration of metal toxicity, 333
amide-containing fertilizers, 327
amino acid permease, 223
amino acid transport, 223
amino acids, 261
ammonia, 274, 276, 277, 331, 353, 355,
    363, 366, 387, 414
  volatilization, 331
ammonia toxicity, 331
ammonia-containing fertilizers, 327
ammonium, 407–409
  availability, 331
  assimilation, 65
  flux, 306
  influx, 217
  nutrition, 262, 332
  sulfate, 416, 418
  transporters, 219
  transport, 118, 124
  uptake, 408
ammonium-fed plants, 332
ammonium nitrate additions, 416
amphibious, 233
anaerobic metabolism, 2
anaplerotic pathway, 236
anion channels, 96, 118
anion efflux, 271
anion exchanger, 256
anoxia, 264
anoxic switch, 120
antennae pigments, 161–163
Antennaria dioica, 412
anthocyanins, 25, 41
antibody, 240
antimicrobial defenses, 88
antimony, 338
antimycin, 182, 190
antiporters, 313

apoplast, 123
  acidification, 6, 7, 144
  alkalinization, 6
  conductivity, 257
  definition, 255
  ion transport, 222
  pH, 112, 117, 118, 121, 124, 132,
    136, 271
  pH determination, 280–282
  pH gradient, 228
  physical and chemical properties, 257
  structure, 255
apoplast, leaf, 124
  pH homeostasis, 274–277
apoplast, root
  pH homeostasis, 270, 271
apoplastic fluid, 124, 281, 282
apoplastic pathway, 360
apoplastic phloem loading, 227
apoplastic sap, 137
apoptosis, 198
apparent free space, 258, 260, 267
aquaporins, 35
  vacuolar type, 36
aquatic ecosystems, 409
aquatic fern, 393
Arabidopsis, 2, 3, 29
  Al-tolerant mutant, 316
  amino acid permease, 223
  AOX, 189
  columella cells of roots, 88
  det3 mutant, 36
  expansin genes, 135
  expansins, 141
  fluorophores, 78
  genes *AVA-P3* and *AVA-P4*, 37
  guard cells, 86
  H$^+$-ATPase AHA3, 5, 9
  mitochondrial carrier family, 197
  nitrate transporters, 218
  uncoupling proteins, 199
  V-PPase, 34
arbuscules, 363
archaebacterium, 2
arctic tundra, 412
Arnica montana, 412, 414, 415
arsenic, 335, 339
Asclepiadaceae, 242
ascomycin, 244
Asparagales, 242

Asparagus mesophyll cells, 80
aspartate, 3
aspartate/glutamate carrier (AGC), 196
aspartic endopeptidase, 26
Aspergillus, 218
astomatous, 233
atmospheric deposition, 414, 416
atmospheric deposition of sulfur, 416
ATP, 1
ATP hydrolysis, 4
ATP synthase, 150, 159, 160, 165, 191,
    192, 195, 197
    H^+ channel, 191
ATP synthesis, 111, 119, 122, 149, 153,
    192, 195, 201, 238, 353
$ATP^{4-}_{in}/ADP^{3-}_{out}$ exchange, 197
ATPase
    mutant (eg. D730), 4
ATPases, 60
ATP-synthase, 113, 114, 116
autophagosomes, 26
auxin, 8, 85–87, 95, 125, 136, 139,
    141, 142
available ions, 323, 324
Avena coleoptiles, 136, 139
Avena sativa, 335
Avenella flexuosa, 329
azolectin liposomes, 165
Azolla, 384
Azolla-Anabaena symbiosis, 392, 393
Azospirillum, 353, 360, 365

Bacillus, 354
Bacillus alcalophilus, 354
bafilomycin, 88, 241
bafilomycin A_1, 32, 36
Banksia attenuatta, 337
barley, 35, 51, 215, 238, 240, 241, 274,
    329, 360
base cations, 327, 328, 334, 357, 412,
    415, 417
basiphilous pioneer plant communities,
    412
bc_1 complex, 180
BCECF, 78, 83–88, 91
bean, 143, 391
bean rhizobia, 385
beans, 329
beech, 356

beetroot, 188
benzene tricarboxylic acid affinity chro-
    matography, 240
benzohydroxamate, BHAM, 190, 201,
    205
benzophenanthridine alkaloids, 89
Berberis repens, 339
berverine, 25
β-carotene, 163
β-D-glucanase, 136
beta-glucanase, 89
betaine, 25
β-(1,3)-D-glucanases, 135
β-(1,4)-D-glucanase, 136
β-(1–3,1–4)-glucans, 140
β-oxidation of fatty acids, 35
bicarbonate, 216, 265, 276, 331, 365,
    406, 410, 411, 413
    excretion, 332
bioassays, 324
bioenergetics, 192
biofertilizer, 393
bioimaging techniques, 14
bleeding technique, 273
blue light, 10, 11, 141
bogs, 413
borate, 256, 264, 329, 365
Bradyrhizobium, 365, 384, 385
Brassica napus, 274, 276
brassinosteroids, 141, 142
Bromeliaceae, 228, 242
bromocresol purple, 280, 304, 363
Bromus erectus, 124, 274, 275, 277
Bryophyllum (Kalanchoë) fedtschenkoi,
    229, 243
buffer capacity, 406, 407
buffer power, ion, 324
bulk soil, 332
bundle-sheath cells, 67, 263
butyric acid, 86, 353

C_3 pathway, 228
C_3 photosynthesis, 230, 231, 233, 234,
    237, 238, 242, 246
C_3 plants, 221, 229
C_4 photosynthesis, 228
C_4 plants, 58, 156, 220, 221, 229,
    263
Ca^{2+}/calmodulin-dependent kinase II,
    13

Ca^{2+} efflux. *See* mitochondria, cation transport
Ca^{2+} gradients, 96
Ca^{2+} homeostasis, 260
Ca^{2+} transport, 126
Ca^{2+} uptake, 166. *See* mitochondria, cation transport.
Ca^{2+}, as a secondary messenger, 124
Ca^{2+}, intracellular, 61, 95
Ca^{2+}/H^+ antiport, 165, 166
Ca^{2+}/H^+ antiporter, 40
Ca^{2+}/Pi cotransport, 205
Ca^{2+}-ATPase, 126
Ca^{2+}-induced Ca^{2+} release, 87
Ca^{2+}-ionophore, 244
Cactaceae, 228, 242
cadmium, 325, 326, 330, 338, 339, 389, 405
  flux, 307
  speciation, 330
cADPR. *See* cyclic adenosine diphosphate ribose
Caenorhabditis elegans (nematode), 29, 30, 197
calcareous grasslands, 416
calcareous soils, 337, 406
calcicole species, 329
calcifuge species, 329
calcifuge-calcicole gradient, 411
calcium, 329, 365, 387, 390, 393. *See also* Ca^{2+}
  availability, 329
  soil solution, 325
  carbonate, 406
  deficiency, 388, 411
  fluxes, 306, 317
  phosphates, 368
  supply, 394
  uptake, 408
calcium-modulating proteins, 391
California poppy. *See Eschscholzia californica*
Calluna vulgaris, 329, 408, 411, 416
calmodulin, 244, 391
CAM, 228
  biochemistry, 234, 236–238
  citric acid synthesis, 237, 238
  diurnal pH changes, 233
  ecophysiology, 228, 229
  facultative, 229, 230, 237, 244
  gene expression, 242, 243, 244

CAM [Continued]
  in aquatic plants, 232, 233
  metabolism, 61
  morphology, 229
  nocturnal CO_2 fixation, 227
  pH stat, 245
  phases, 234
  subcellular pH changes, 234
  vacuolar storage of organic acids, 238
CAM plants, 37, 58, 221, 228
Candida parapsilosis, 200
canonical correspondence analysis, 369
capsicum, 160
Capsicum annuum, 273
carbon dioxide, 410, 411, 417
carbon skeletons, 220
carbonate, 358
carbonic acid, 328
carbonic anhydrase, 246
carbonylcyanide *m*-chlorophenylhydrazone. *See* CCCP
carboxyfluorescein, 78
carboxylic acids, 57, 58, 59
carboxySNAFL-1, 82
cardiolipin, 354, 388
Carex nigra, 411, 417
Carex panicea, 411
Carex pilulifera, 411
Caricion davallianae, 413
Caricion nigrae, 413
carnitine carrier, 197
carotenoids, 163
carrot, 395
Caryophyllales, 242
Casparian band, 259, 260
cassava, 394
castor bean, 26
Casuarina, 393
Casuarina glauca, 271
Catharanthus, 240
Catharanthus cell suspension, 49
cation adsorption, 333
cation exchange, 7
  co-precipitation, 327
  non-specific adsorption, 326
  organic complexation, 326
  specific adsorption, 326
cation exchange capacity, 258, 261, 268, 276, 325, 331, 334, 406
  roots, 326

cation replacing power, 327
cation/anion ratio, 332
cation/anion uptake ratio, 331
CCCP, 83, 91, 93, 116, 118
cell elongation, 309, 312, 313
cell expansion, 36
cell polarity, 95
cell wall, 7, 121, 122, 256, 259, 263,
    387
    extensibility, 131, 141, 256
    load-bearing bonds, 135
    load-bearing crosslink, 132
    rigidity, 388
cellulase, 136
cellulose, 135, 136, 257
cellulose microfibrils, 132
Cephalosporium graminearum, 376
Ceratophyllus submersum, 414
Chara, 62, 80
    cytoplasmic pH, 51
Chara corallina, 30
chelation, 358
chemical extractants, 324
chemiosmotic energy-transduction con-
    cept, 173
chemiosmotic theory, 153, 191, 196
chemolithotrophs, 353, 366
chemolithotrophy, 352
Chenopodium rubrum
    acidification of the cytosol, 10
chilling
    V-ATPase, 37, 38
chitinase, 89
Chlamydomonas, 219
Chlamydomonas mutants, 156
chlorate, 218
chloride, 32, 36, 61
    import pump, 107
chloride fluxes, 317
chloride transport, 62
chloride, exchange with gluconate,
    117
chloro-Cd complexes, 330
Chlorococcum littorale, 38
chloromethylfluorescein diacetate
    (CMFDA), 79
chloromethylSNARF-1 acetate, 79
chlorophyll, 84, 158, 161–163
chlorophyll *a/b*-protein complex
    (LHCIIb), 162

chloroplast
    D1 subunit, 160
    Donnan equilibrium, 156, 160
    Donnan phase, 158
    efflux of H^+, 157
    lumenal pH, 157, 158, 161–163
    pH gradients, 265
    stromal, 165
    stromal pH, 156, 159, 160, 166
chloroplast-encoded protein synthesis,
    160
chlorosis, 276
chlorothalonil, 374
choline chloride, 32
chromate, 329, 338
chromium, 335
circadian rhythm, 236
circumnutations, 313
Cirsio dissecti-Molinietum, 415
Cirsium dissectum, 409, 411, 415
citrate, 271, 333
    carrier (CIC), 197, 198
    decarboxylation, 238
    synthase, 237
    transport, 240
citric acid, 233, 236, 237
Cl^-/H^+ co-transport, 314
clay particles, 325
CL-NERF, 88
Clusia, 230, 233, 237, 242
Clusiaceae, 228
cluster roots, 271
CN-resistant respiration, 189, 205
CO_2 fixation, 159, 165
coated vesicles, 26
coenzyme A, 198
cold stress, 243
    sugar beet, 10
Commelina, 87
competitive advantage, 416
complex bc_1, 183
complex I, 177, 189
complex III, 180, 182
complex IV, 183
complex V. *See* ATPsynthase
Comptonia, 393
concanamycin A, 32
confocal fluorescence microscopy,
    79
confocal imaging, 88, 93

confocal laser scanning microscopy, 78, 81, 82, 138, 281
coniferous tree species, 408
copper, 183, 326, 330, 335, 339, 357, 389
copper center, 183
copper toxicity, 339
corn, 271, 309, 310, 394
   Al toxicity, 315
   autolysis of isolated coleoptile walls, 140
   auxin application to coleoptiles, 10
   coleoptiles, 125, 205
   ion mapes in coleoptile, 85
   protein kinase, 12
   vacuolar pH, 84
   yields, 328
co-transporters, 313
cotton, 29
cowpea, 369
Crassula aquatica, 233
Crassula helmsii, 232
Crassulaceae, 228, 230, 242
cryptogein, 61
cSNARF-1, 78–80, 83, 84, 87, 89–91
C-terminus, 57, 180
Cu_B, 185
cucumber, 141
Cupressus pygmaea, 412
current-voltage characteristics (I-V curves), 114, 115
cutin, 260, 263
Cu-transporting ATPase, 389
cyanide, 120, 142, 190
cyanide-resistant respiration, 189, 205
cyanobacteria, 218
cyanogenic glycosides, 25
cyclic adenosine diphosphate ribose, 87
cyclic AMP, 124
cyclic electron flow, 156
cyclopropane, 388
cyclosporin A, 244
Cyperaceae, 413
cysteine, 58
cytochrome *a*, 183
cytochrome and alternative oxidases, 190
cytochrome *b*, 180, 182
cytochrome *bf* complex, 150, 156, 166
cytochrome b_H, 154

cytochrome *c*, 185
cytochrome *c* oxidase, 183, 185
cytochrome *c* oxidase (complex IV), 175
cytochrome c_1, 180, 182
cytochrome oxidase, 190
cytokinin, 244
cytoplasmic acidification, 89, 94
   alkalization, 125
   Ca^{2+}, 86, 87, 95, 261
   H^+, 85, 94
   male sterility, 198
cytoplasmic pH, 8, 60, 62, 86, 112, 125, 126, 234, 238, 245, 246, 265, 386
   Chara, 51
   decrease due to low oxygen, 9
   decrease in acidic soils, 9
   measurement, 50
   origin of changes, 51, 53
   pH gradients, 309
   regulation, 63–65
   streaming, 260
cytoskeleton, 97
cytosol
   buffering capacity, 80
cytosolic acidification, 94, 96, 119, 125, 126
cytosolic H^+, 96
cytosolic pH, 90, 118, 125, 236

Dactylorhiza maculata, 411
damped squared voltage-wave, 301
Datsica, 393
deciduous trees, 408, 409
deciduous forests, 416
de-epoxidation, 158, 163
defense reactions, 125, 126
dehydrogenase, 58
denaturing gradient gel electrophoresis (PCR-DGGE) of 16S rDNA, 374
denitrification, 366, 367
denitrification potential, 367
Deschampsia flexuosa, 408, 409, 411, 415, 416
desiccation, 415, 416, 418
dextran conjugates, 79
dextrane-coupled pH probes, 86
diacylglycerol, 124
diazotrophs, 368

dicarboxylate carrier (DIC), 197, 198
diffuse layer, 328
diffusion, 257, 260, 263, 306, 326
  coefficient, 303
  potential, 267, 268
Digitalis purpurea, 329
Digitaria, 78
Digitaria sanguinalis, 222
diphosphate. *See* inorganic pyrophos-
      phate
dissociation of -OH and -COOH groups,
    325
diurnal organic acid fluctuations, 230
diurnal rhythm, 273, 277
DNA profiling, 373
DNP, 118
Donnan Free Space (DFS), 139, 258
Donnan potential, 256, 267, 268
Drosera, 411
drought stress, 230, 231, 238, 242, 271
dry environments, 229
Dryopteris affinis, 91
DTPA extract, 337
Dunaliella acidophila, 12
dwarfism, 134

E. coli, 32, 189, 194, 387. *See also*
      Escherichia coli
ectomycorrhizae, 363, 364
ectomycorrhizal fungi, 384, 409
ectomycorrhizal symbioisis, 394
effective CEC (ECEC), 325, 326
efflux of H^+, 64
EGTA, 244, 311
electro neutrality, 261
electrochemical gradient, 111
electrochemical gradient of H^+, 4
electrochromic shift, 156
electrode puller, 302
electron paramagnetic resonance, 154
electron transport, 159
electroneutrality, 327
elicitor molecules, 88
Elodea canadensis, 232, 411
Elodea leaves, 122
Empetrum nigrum, 416
endo-1, 4-beta-glucanase (Cel7), 8
endodermis, 260
endoglucanases, 7, 140, 141

endoglycosidases, 135
endoglycosylases, 136, 137
endomycorrhizal symbiosis, 384, 394
endoplasmic reticulum, 26, 29, 61, 78
epidermal bladder cells, 231, 232
epifluorescence, 82
epiphytes, 228, 230
epitope-tagged H^+-ATPase, 2
Eremosphera viridis, 124
Erica tetralix, 411
Ericaceae, 408, 413
ericoid mycorrhizal fungi, 409, 416
ericoid species, 416
Escherichia coli, 32, 189, 194, 387. *See*
      also E. coli
Eschscholzia californica, 78, 83, 84,
      94
esterases, extracellular, 77
esterases, intracellular, 77, 78
ethanol, 55
ethylene, 140, 141, 143
Eubacteria, 2
eucalypts, 394
Euphorbiaceae, 242
eutrophication, 417, 418
evapotranspiration, 328
evolutionary origin of proton pumps, 2
excess cation uptake, 64, 65
excitation energy, 162, 164
exciton-coupled dimer, 162
exodermis, 260
exopolysaccharide, 388
expansins, 7, 8, 135–137, 140–143
extensor, 275
extracellular polysaccharides, 388
extractable cations, 325
extractable ions, 324
extractants, 324, 330
extramatrical mycelium, 364
extremophiles, 353

F_1F_0ATPase, 187
Fagus sylvatica, 356
fatty acid analyses, 356
fatty acid methyl ester (FAME) profiles,
      388
FCCP, 83
Fd-GOGAT, 214
Fe deficiency, 60, 271, 276

Fe reeduction
  redox systems, 61
Fe/S center, 180, 182
fens, 413, 417
ferns, 228, 409
ferredoxin, 213
ferredoxin (Fd), 156
ferric uptake regulatory (*fur*) gene, 389
ferrichrome, 361
ferricyanide, 182
ferrihydrite, 359
ferrioxamine B, 361
Festuca elatior, 337
Fick's first law of diffusion, 303
Filicopsida, 228
FITC-dextran, 138, 143, 281
5,5-dimethyl-oxazolidine dione (DMO), 50, 51
flavin mononucleotide (FMN), 177
Flavobacterium, 360
flavonoid *nod*-gene inducers, 391
flavonoids, 389, 391, 392
flavonols, 25
flax, 337
floating base-rich fens, 417
flooding, 337, 338
Fluo-3, 85
fluorescein boronic acid, 281
fluorescence dyes, 6, 124
  pH sensitive, 138, 139, 281
fluorescence imaging, 81–83
fluorescence imaging analysis, 74, 75
fluorescence microscopy, 74, 80
fluorescence ratio imaging, 61
fluorescent pH indicators, 246
fluorescent pseudomonads, 376
fluoride, 335
fluorochrome, 79, 138
fluorophore, 78
FMN-dependent rotenone-insensitive
  NADH dehydrogenation, 177
14–3-3 protein, 10, 11, 12
  fusicoccin receptor, 11
Frankia, 392, 393
  spore-producing, 392
fructose bisphosphatase, 159
F-type H^+-ATPases, 2, 27, 29
Fucus zygotes, 79
fulvic acids, 257
functional redundancy, 374

fungal hyphae, 307
fungi, 2
fungi-to-bacteria ratios, 355, 356
Fura-2, 86
Fusarium oxysporum, 376
fusicoccin, 6, 7, 9–11, 13, 86, 90, 120, 141, 309
  increase in turgor, 11
  increased proton pumping, 11
  potassium uptake, 11
Fusicoccum amygdali, 11

Galium harcynicum, 329
Geaumannomyces graminis, 376
gene duplication, 228
Genista anglica, 412
Genista pilosa, 411
genotypes, 329
Gentiana pneumonanthe, 411
GEOCHEM, 324
Geum urbanum, 329
gibberellic acid (GA), 87, 88, 143
gibberellin, 141
Gibbs free energy, 36, 113, 174
Gigaspora, 394, 395
Gigaspora margarita, 394
Glomalean (endomycorrhizal) fungi, 394
Glomus, 394, 395
Glomus etunicatum, 362
Glomus manihotis, 394
glucans, 242
gluconate, exchange with chloride, 117
glucose, 54
glucose-6-phosphate, 236, 244
glutamate, 162, 183, 185, 264, 386, 387
  carrier, 196
  synthase (GOGAT), 213
glutamic acid, 116
glutamine, 264
  carrier, 197
  synthetase (GS), 213, 222
glutathion, 79, 387
glyceraldehyde 3-phosphate dehydrogenase, 243
glycine decarboxylase, 212
glycolysis, 188, 206, 233, 237, 246
glycosidases, 26
Gnetopsida, 228
gold, 335

Goldman equation, 119
Golgi bodies, 26
Gossypium hirsutum, 29, 93, 229
gramicidin, 159
granite, 412
grasses, 329, 409
grassland ecosystem, 419
gravitropism, 88, 270, 310, 311
green algae, 61
green fluorescence protein (GFP), 75, 282
green manure, 393
groundwater, 336
GS/GOGAT cycle, 219
guard cells, 85–87, 275, 276
  anion channels, 118
  potassium uptake, 6
  water uptake, 6
gum arabic, 392
*Gus*A-fusions, 389

H⁺
  activity, 328, 333
  antiport, 112
  ATPase, 386
  carrier, 199
  channels, 116, 117, 196
  co-transport, 115
  currents, 303, 309, 310, 312
  efflux, 165
  electrochemical gradient, 173, 197
  exchangers, 122
  excretion, 353
  extrusion, 86, 91, 120, 312, 313, 354
  exudation from roots, 332
  fluxes, 307, 315, 316
  gradients, 1, 174, 175, 190, 191
  pumping, 4, 185, 186
  pyrophosphatase, 61, 240
  sucrose uptake, 5
  symport, 112, 121
  toxicity, 338
  transhydrogenase, 191
  uptake, 192
H⁺/amino acid co-transporters, 314
H⁺/ATP ratio, 197
H⁺/ATP stoichiometry, 195
H⁺/Ca²⁺ antiport, 126
H⁺/Ca²⁺ exchanger, 204
H⁺/cation antiport, 332

H⁺/cation exchange, 276
H⁺/K⁺ symport, 114, 314
H⁺/sugar co-transporters, 314
H⁺-ATPase, 1, 3, 54, 60, 61, 95, 113
  AHA3, 9
  C-terminus, 4, 11
  gene knock-outs, 14
  in root epidermis, 4
  phloem companion cells, 5
  posttranslational regulation, 11
  potassium uptake, 6
  promoter-GUS fusion, 5
  role in nutrient uptake, 4, 7
  transmembrane domains, 3
Hagen-Poisseuille law, 262
hairy type mycorrhizae, 364
halophyte, 231
heat stress, 237
heathlands, 411, 418
heavy metal loading rates, 359
heavy metal stress, 255
heavy metal toxicity, 338, 360
heavy metalloides, 338
heavy metals, 93, 326, 338, 357, 388
heme *a*, 183, 185
heme *a* and *a₃*, 185
heme *a₃*-Cu_B binuclear active center, 183
heme *b*, 183
heme *b_L*, 182
hemes *a* and *a₃*, 185
hemicellulose, 132, 257
hemi-epiphytes, 230
hemoglobin, 393
herbs, 409
heterocyst, 393
heterologous expression
  plant H⁺-ATPase in yeast, 11
heterotrophic microorganisms, 352
high affinity nitrate transporters
    (HATS), 218
histidine, 58
Holcus lanatus, 409
homogalacturonans, 257
homospermidine, 387
Hordeum vulgare, 79
Hottonia palustris, 414
Hoya carnosa, 242
humic acids, 326
humus, 365

hydration, 327
hydrocarbons, 374, 388
hydrocharitaceae, 232
hydrogen peroxide, 241
hydrophobicity, 3
hydroxyapatite, 331
hydroxyapatite chromatography, 240
hydroxyl ions, 358
hypersensitive reaction, 89
hyphae, 364, 376, 394
hypodermis, 260
hysteresis switch, 239

IAA, 86, 309
ice plant, 29, 231
infection threads, 393
inorganic P, 332
inorganic pyrophosphate, 23, 33
inositol trisphosphate (IP₃), 87, 124
internal cuticle, 263
intraradical mycelial growth, 394
ion availability, 323, 324, 328, 329,
    333, 358
ion channels, 165
ion current surrounding roots
    direction, 304–307, 309–311, 317
ion gradients, steady-state, 306
ion supply (intensity/capacity), 324
ionic activities, 324
ionic currents surrounding roots
    measurement, 300–304
ionomycin, 244
ionophores, 159
ion-selective liquid membrane, 302
iontophoresis, 77
Ipomoea tricolor, 25
iron, 299, 326, 337, 358, 405
    availability, 329, 332, 335
    chelates, 361
    deficiency, 60, 361
    oxide, 327
    phosphates, 362
    redox systems, 61
    toxicity, 333, 335, 338
iron-oxidizing microorganisms, 353
iron-sulfur cluster, 178, 183
iron-sulphur (Fe/S) centers, 177
isocitrate dehydrogenase (IDH), 219
Isoetaceae, 232

Isoëtes howellii, 229
Isoëtes lacustris, 413
Isoëtes macrospora, 232
isoetid plant species, 410, 418
isomorphous substitution, 325
isoprenoid phytoalexin pathway, 89

Jerusalem artichoke, 205
Juncus acutiflorus, 411
Juncus balticus, 412
Juncus bulbosus, 413, 418
Juncus squarrosus, 329, 408
Juncus subnodulosus, 417

K⁺ channel, 218, 263. *See* mitochon-
    dria, cation transport
K⁺ channel blocker, 165
K⁺ channel, outwardly rectifying, 85,
    86, 96
K⁺ channels, inward rectifying, 85, 86,
    96, 117
    control by pH, 117
K⁺ efflux, 125, 271
K⁺ influx, 125
K⁺ uptake
    high affinity, 314
    low affinity, 314
K⁺/H⁺ antiport, 386
K⁺/H⁺ antiporter (KHE), 202
K⁺/H⁺ exchange, 58, 165
K⁺/Na⁺ co-transporter, 314
K⁺-unselective/H⁺ exchanger (KHE),
    201
Kalanchoë, 234, 238
Kalanchoë blossfeldiana, 230, 243
Kalanchoë daigremontiana, 230, 239
Kalanchoë fedtschenkoi, 236
Kalanchoë tubiflora, 246
kaolinitic soils, 334
kinetin, 86
Kluyvera ascorbata, 360
Krebs cycle, 188, 198, 206, 220

lactic acid, 55, 353
lanthanum, 91, 96
lateral root, 310
lavendustin, 244
leaching, 325, 327, 328, 330, 333, 339,
    364, 406

lead, 330, 335, 338, 339, 358
leaf frond, 393
legume/rhizobial partnership, 126
legumes, 391–395
Leishmania donovani, 2
Lemna minor, 418
lemon
  vacuole, 73
  V-ATPase, 32, 33, 36
lettuce, 141
LHCIIb, 163
light-harvesting antennae, 164
light-induced damage, 161
lignin, 257, 326
Liliopsida, 228
Lilium longiflorum, 78, 91
lime, 240
lime pelleting of the seed, 391
liming, 325, 327, 340, 355–357, 364,
  376, 391, 394, 395
Limnobium stoloniferum, 315
lincomycin, 160
linoleic acid, 201, 205
lipochitooligosaccharides, 126
lipooligosaccharides, 390
lipopolysaccharides, 388
Littorella, 232
Littorella uniflora, 233, 413, 418
Littorelletea plant communities, 410
Lotus japonicus, 391
lucerne, 329
lupins, 329
Lupinus albus, 272. *See* also white lupin
Luronium natans, 413
Lycopersicon esculentum, 335
Lycopodiopsida, 228
Lycopodium clavatum, 411
Lycopodium inundatum, 411
lysophosphatidylcholine, 10
lysophospholipid, 388
LysoSensor dyes, 75

magnesium, 32, 33, 86, 165, 187, 202,
  236, 325
  availability, 329
  bicarbonate, 413
  carbonate, 406
  soil solution, 325
  uptake, 333

Magnoliopsida, 228
maize. *See* corn
malate, 58–60, 239, 240, 244–246, 271
  transport, 238–240
malate^{2-}, 238, 239
malic acid, 227–229, 233, 234, 237,
  238
malic enzyme, 58, 242, 245, 246. *See*
  also NADH malic enzyme
manganese, 236, 325, 330, 376
  availability, 332, 335, 337
  deficiency, 361
  oxidation, 337
  oxide, 327
  toxicity, 333, 336, 337, 391
mangrove swamps, 337
mannitol dehydrogenase, 363
mantle, 394
mass flow, 261, 262
Medicago, 390, 391
  murex, 391
  trunculata, 391
  sativa, 389
    plasma membrane potential, 9
Melica uniflora, 416
membrane permeability, 118, 119
membrane potential, 37, 54, 111, 216
  depolarization, 391
membrane-permeant acid, 83
Menyanthes trifoliate, 417
mercury, 335, 389
mersalyl, 205
Mesembryanthemum crystallinum, 29,
  37, 230, 231, 237, 240, 242–244
mesophyll cells, 213
mesophyll vacuoles, 228
Mesorhizobium, 365
Mesorhizobium loti, 384
metal bioavailability, 358
metal toxicity, 333
metalliferous ore, 338
metallothionein, 26
methaloenzymes, 213
Methanococus janashii, 2
methylamine, 90
methylation, 257, 358, 360
Mg^{2+} uptake. *See* mitochondria, cation
  transport
microbes, 327
microbial acidification processes, 353

microbial biomass, 355
microbial communities, 352, 357
microbial ecology, 352
Micrococcus, 367
microelectrodes
    double-barreled, 6, 50
    ion sensitive, 301, 315
    ion-selective, 317
    K$^+$ current, 85
    liquid-membrane ion-selective, 302
    pH sensitive, 51, 117, 124, 125, 137,
        280, 307, 315
    potassium sensitive, 314
    static, 301–303
    vibrating, 301–303, 307
microinjection, 79
micronutations, 313
microsymbiont, 384, 389, 390, 393, 395
middle lamella, 257
Milium effusum, 411
millets, 329
mimosa, 35
Mimosa pudica, 35
mine spoils, 339
mineral interlayer spaces, 326
mineralization, 405
mineroptrophic mires, 412
mires, 412
Mitchell's chemiosmotic hypothesis,
    153
mitochondria, 78
    alternative NAD(P)H oxidations, 188
    ATP synthase, 173, 174, 191, 192,
        195
    catalytic cycle, 185
    glutamate/aspartate exchange, 198
    inner membrane, 1, 173, 174, 201
    kinetics of ATP synthesis, 192
    pH gradients, 265
    pyrophosphatases, 187
    transport of cations, 201
    uncoupling protein, 199
mitochondrial carrier family (MCF),
    197, 199
Mn cluster, 151, 152
molecular rotary motor. See F-type AT-
    Pase
Molinia caerulea, 411, 415
molybdate, 329
molybdenum deficiency, 391

monoclonal antibodies, 388
Monte Carlo analysis, 371
morning glory, 25
morphogens, 390
mosses, 410, 413
motor cells, 275–277
mRNA
    phenylalanine ammonia-lyase, 9, 94
    splicing, 11
    V-ATPase, 39
mull type, 409
mung bean, 29, 32
    chilling injury, 38
    V-ATPase, 37, 39
    V-PPase, 39
    V-PPase/V-ATPase, 34, 35
mutagenesis analysis of various P-type
    ATPases, 3
mutants, 386, 389
mutation, 4
Mycelis muralis, 329
mycorrhizal fungi, 352, 363, 394, 409,
    416
myo-inositol o-methyltransferase (Imt1),
    232, 244
Myrica, 393
Myriophyllum alterniflorum, 414, 418

N,N'-dicyclohexylcarbodiimide (DCCD),
    32
N$_2$ fixation, 384, 387, 391, 392
Na$^+$ channel. See mitochondria, cation
    transport
Na$^+$ export pumps, 107
Na$^+$/Ca^{2+} antiporter, 204
Na$^+$/H$^+$ antiport, 354
Na$^+$/H$^+$ antiporter (NHE), 40, 201, 202,
    232
Na$^+$/H$^+$ exchanger, 204
Na$^+$/K$^+$-ATPases, 32
NAD malic enzyme, 242
NAD(P)H dehydrogenases, 175
NAD(P)H dehydrogenases, alternative,
    189
NAD(P)H dehydrogenases, alternative
    non-H$^+$ pumping, 188
NAD(P)H dehydrogenases, rotenone-
    insensitive, 201
NAD$^+$, 59, 198, 220
NAD$^+$-IDH, mitochondrial, 220

NADH, 59, 189, 191, 195, 213, 238
NADH and NAD(P)H dehydrogenases,
      188
NADH malate dehydrogenase, 234
NADH oxidation rotenone-insensitive,
      188
NADH oxidation, rotenone-sensitive,
      189
NADH/NAD$^+$ ratio, 188
NADH:ubiquinone oxidoreductase
      (complex I), 175
NADH-GOGAT, 214
NADP malate dehydrogenase, 243
NADP malic enzyme, 233, 242
NADP$^+$, 220
NADPH, 61, 149, 156, 189, 242
NADPH-oxidase, 89
nanopaths, 262
Nardus stricta, 408, 412
Narthecium ossifragum, 411
native rhizobia, 385
natural ecosystems, 356
naturally acidified waters, 410
n-butyl malonate, 201
net surface charge, 354
N-ethylmaleimide, 36
Neurospora crassa, 3, 63, 78, 91, 177,
      314
   V-ATPase, 32
   hyphae, 79
NH$_4^+$ assimilation, 274, 277
nickel, 326, 338, 357, 358
Nicotiana plumbaginifolia, 2
   H$^+$-ATPase, 6
nicotinamide dinucleotide phosphate
      (NADP$^+$), 149
nigericin, 83, 159
nitrate, 32, 36, 54, 57, 353, 355, 366,
      407, 408
   assimilation, 65, 66, 227
   concentration in xylem, 273
   efflux, 215
   flux, 307
   influx, 215
      high affinity transport system
         (HATS), 215
      low-affinity nitrate transporter
         (LATS), 215, 218
   reductase, 213, 363, 408
   transport, 62, 63

nitrate [Continued]
   transporters, 218
   uptake, 408
nitrate-fed plants, 332
nitrification, 327, 355, 366, 405, 408,
      414, 417, 418
nitrifying bacteria, 353, 407, 414
nitrite, 366
nitrite reductase, 213
nitrite-oxidizing bacteria, 366
Nitrobacter, 366
nitrogen
   assimilation, 65, 211, 219
   availability, 356, 405
   deposition, 414
   fixation, 327, 364
   form, 270, 272, 275
   inorganic pool, 407, 408
   mineralization, 355, 356
   oxides (NO$_x$), 414
   transformation, 366
   uptake, 408
nitrogen-fixing symbioses, 362
Nitrosomonas, 366
NO$_1^-$ reductase, 275
NO$_3^-$/H$^+$ co-transport, 314
Nocardioides, 374
nocturnal accumulation of organic acids,
      238, 239
nocturnal acidification, 231
   CO$_2$ fixation, 227, 229, 234
nod factors, 390, 391
nod gene, 391
nod gene expression, 390, 391
nod-gene inducers, 391
nodulation, 389, 391–393, 395
   capability, 365
   genes, 390
nodule, 387, 389, 390, 392
nodulins, 390, 393
nutrient dynamics, 357
nutrient mineralization, 355
Nympha alba, 417
nymphaeid plant species, 417

O$_2$ evolution, 159, 161
O$_2$ reduction, 185
oats, 329
o-chain lipopolysaccharide, 388
OH$^-$, 216

okadaic acid, 244
oligogalacturonide elicitors, 89
oligomycin, 157, 191
oligotrophic waters, 410, 411
ombrotrophic mires, 412
ononitol, 231, 232
Opuntia inermis, 229
Orchidaceae, 228, 242, 243
organic acids, 57, 59, 60, 64, 65, 118,
    261, 359, 361
  accumulation in vacuole, 246
  extrusion, 271
  release from vacuole, 241
organic compounds, 374
organic matter, 325, 326, 328, 353, 355,
    356, 359, 366
organic matter decomposition, 356, 377,
    405
organic phosphates, 368
osmolarity, 312
osmotic potential, 276
osmotic pressure, 25, 35
outer membrane proteins (OMPs), 388
oxalate, 65, 268, 333
oxaloacetate, 234, 237, 243, 244
oxidation of reduced S compounds, 327
oxidative burst, 89
oxidative phosphorylation, 198, 206
oxides, 333
Oxisols, 325
oxoglutarate carrier (OGC), 197, 198
Oxycocco-Ericion, 413
oxygen evolution, 159, 161
oxygen-evolving complex (OEC), 151–
    153, 165, 166

P680, 152, 163
palmitoyl coenzyme A, 203
Panama, 230
Panicum virgatum, 395
Paphiopedilum tonsum, 85
Paracoccus denitrificans, 180, 183
Parasponia, 392
parental material, 412
Parnassia palustris, 411
parsley, 85
patch microelectrode, 77
patch-clamping, 37, 165
P-ATPase, 93
pE = -log of free-electron activity, 336

pea, 29, 35, 329, 364, 390
  leaves, 143
  V-ATPase, 37
pea stems, 205
peach, 332
pear
  V-PPase/V-ATPase, 35
peat, 412, 413
peatland, 417
pectin, 257, 260, 267, 268
Pelvetia, 90
Pelvetia embryos, 96
pentose phosphate pathway, 62
PEP carboxykinase, 236, 242, 244
PEP carboxylase, 58, 64, 65, 67, 234,
    237, 242–246, 251
PEP carboxylase kinase, 244
PEP synthesis, 228
Peperomia camtotricha, 230, 243
Peperomia scandens, 230
peptidoglycan, 388
permanent aquatic ecosystems, 413, 417
peroxysome, 220
pH
  as a messenger, 124
  buffering, 406
  buffers, 57, 58, 143
  changes, ion induced, 110
  domains, 122, 123
  dye indicators, 304, 306
  gradients, 88, 91, 92, 158, 256
  homeostasis, 2, 66, 73, 93, 95, 96,
      354, 358, 386
  homeostasis, root, 312
  indicator dye, 74, 311, 363
  mapping, 86, 88
  microelectrodes, 50
  probes, 79, 80, 84, 86
    loading and compartmentation, 77–
      79
  shifts, 95
  stat, 9, 58
    CAM plants, 245
Phaseolus pulvini, 275
Phaseolus vulgaris, 391
phenylalanine-ammonia lyase, 89, 94,
    213
phenylpropanoid pathway, 89
phenylpropanoids, 211, 213
pheophytin, 227

phloem, 66, 222, 228, 262
phosphatase, acid and alkaline, 368
phosphatases, 86, 96, 363
phosphate, 329, 358
    availability, 368, 394
    carrier (PiC), 196, 198
    solubilization by microorganisms, 367
    transport, 62
phosphate-solubilizing bacteria, 368
phosphatidylcholine, 365
phosphatidylethanolamine, 354, 365, 388
phosphatidylglycerol, 354, 365, 388
phosphatidyl-N-methylethanolamine, 388
phosphodiesterases, 26
phosphoenolpyruvate, 233
phosphoenolpyruvate carboxykinase, 242
phosphoenolpyruvate carboxylase (PEPC), 220
phosphoinositides, 94
phospholipase A$_2$, 10
phospholipid fatty acid (PLFA) indicators, 355, 356
phospholipids, 354
phosphorus, 299, 366
    adsorption, 353
    availability, 330, 332, 405
    concentration, 395
    deficiency, 391
    fertilizers, 339
    uptake, 409, 411
phosphorylation potential, 159
photoinactivation, 164
photoinactivation of PS II, 160
photoinhibition, 238
photomorphogenesis, 36
photophosphorylation, 149, 153
photoprotective mechanism(s), 161, 164
photoproteins, 75
photorespiration, 211, 212
photosynthetic electron transport, 275
pH-sensitive fluoroprobes
    cSNARF-1, 75
pH-stat, 8, 57, 59–61
    biochemical, 59, 64
    biophysical, 60, 63
phylloplane, 255
phylogenesis, 228

phytase, 26
phytic acid, 26
phytoalexins, 89, 125
Phytophthora megasperma, 94
Phytophthora zoospores, 317
phytotoxicity, 329
Picea abies, 356
piezoelectric pusher element, 301
pine forest ecosystem, 412
pines, 394
pinitol, 231, 232
Pinus aristata, 412
Pinus contorta, 412
Pinus massoniana, 363
Pinus muricata, 412
Pinus sylvestris, 408, 414, 416
Pinus taeda, 8
Piperaceae, 230
Pisolithus tinctorius, 363
Pisum sativum, 29
Plantaginaceae, 232
Plantago major
    sucrose transporter, 5
plasma membrane, 1, 245
    depolarization, 60, 87, 113, 119, 216, 223, 314
    H$^+$-ATPase, 1–3, 141, 303, 307, 332
    hyperpolarization, 6, 61, 113, 120
    membrane potential, 111
    NADPH-oxidase, 89
    proton efflux, 54
    redox systems, 61
plasma membrane H$^+$-ATPase, 1–3, 141, 303, 307, 332
    acid-growth theory, 7
    asymmetric localization, 4
    auxin, 8
    complex with 14-3-3 protein, 13
    C-terminus, 12
    Dunaliella acidophila, 12
    H$^+$ secretion, 7
    number of molecules per cell, 8
    pH dependence, 8, 9
    pulvinar cells, 6
    rate of proton pumping, 8
    regulation, 10
    regulation by phosphatases, 12
    regulation by protein kinases, 12
    role in intracellular pH, 10
    role in regulation of cell expansion, 7

plasma membrane H$^+$-ATPase [Continued]
  serine and threonine residues, 12
  site-directed mutagenesis, 14
  spinach, 12
  stomatal guard cells, 6
  Vicia faba, 12
plasmodesmata, 222, 223, 263
  phloem sieve elements, 5
plastocyanin (PC), 150, 159
plastoquinol, 154, 159, 166
plastoquinone, 156
plastoquinonol (PQH$_2$), 150, 154, 156
Poa nemoralis, 411
pollen tubes, 307
polyamines, 258, 387
polyclonal antibodies, 187
Polygala serpyllifolia, 415
polygalacturonic acids, 257
polymerase chain reaction, 374
polyphosphate, 67
polysaccharide endoglycosylases, 136
polysaccharides, 388
Populus, 264
porins, 388
Portulacaceae, 244
Portulacaria afra, 244
Portulaceae, 242
Potamogeton, 414
Potamogeton natans, 414
Potamogeton polygonifolius, 413
potassium, 61
  availability, 329, 330, 405
  fluxes, 306, 317
  leaf apoplast, 276
  uptake, 408
  V-PPAse, 34
potato, 177, 180, 183, 191, 199, 329, 330
Potentilla palustris, 413
PPi. *See* inorganic pyrophosphate
precipitation, 328
predators, defense against, 25
pressure balance method, 282
primary cell wall, 257
primary silicates, 406
principal components analysis, 356
procaine, 60
proline, 231, 360
promoter-tagging program, 391

PaseProteaceae, 271, 337
proteases, 26
protein
  bodies, 25
  kinase, 96, 222, 244
  kinase C, 13
  phosphatase 1 and 2A, 244
  phosphatase 2B, 244
  phosphatase inhibitors, 89
  phosphatases, 86, 96
  phosphorylation, 236
proton co-transport, 62
proton gradient, 73
proton motive force (pmf), 111, 116, 126, 153, 156, 159, 216, 246, 313
  photosynthetic control, 159
proton pumping
  cytoplasm alkalinization, 8
proton pumps, 1
protonophores, 83
proton-protein interactions, 109
provacuoles, 26
Prunus persica, 332
PS I, 150, 154
PS II, 150, 152–154, 156, 163
  D1 subunit, 153
  photoinactivation, 160, 164
Pseudomonas denitrificans, 367
Pseudomonas fluorescens, 361
Pseudomonas putida, 360
P-type ATPases, 2
pulvini, 275
pulvini motor. *See Mimosa pudica*
pumpkin
  V-PPase/V-ATPase, 35
pygmy forests, 412
pyoverdin, 361
pyridine nucleotide transhydrogenase, 191
pyrite, 339, 353
pyritic materials, 337
Pyrobaculum aerophilum, 30
pyrophosphatase, 232
pyrophosphatase, proton pumping, 187
pyruvate, 55, 59, 60, 189, 233, 236, 238, 242
pyruvate carrier (PYC), 196, 198
pyruvate dehydrogenase, 237
Pythium stem rot, 376

Q cycle, 154, 156, 181, 182
Q-cycling, 156
QH$_2$, 178, 180, 182
Quercus petraea, 364
Quercus robur, 364
quinol oxidase, 189
quinone acceptor Q$_A$, 153
quinone acceptor, Q$_A$, 150

radish
 V-PPase/V-ATPase, 35
rain forest, 230
raised bogs, 410, 413
Ranunculus baudottii, 414
Ranunculus ololeucos, 413
reactive oxygen species, 89
red alder, 393
red clover, 329
red light, 141
redox equilibria, 335
redox potential, 336
regulatory system (*act*R), 389
reporter genes, 2
respiration, 352
respiratory electron transport, 238
reversal potential of the H$^+$ pump, 114
reverse electron transport, 354
rhamnogalacturonans, 257
rhizobia, 364, 365, 387, 389–393, 395
Rhizobium etli, 385, 387, 388, 391
 galegae, 388
 leguminosarum, 359, 387, 389, 390
 leguminosarum bv viciae, 384, 389
 leguminosarum bv phaseoli, 388
 leguminosarum biovar trifolii, 388
 meliloti, 387, 388, 390, 391
 trifolii, 386, 388
 tropici, 384–388, 391
rhizoid growth, 307
rhizoplane, 255
rhizosphere, 123, 272, 304, 331–333,
 337, 353, 414
 acidification, 353, 363, 414
 ecology, 352
 microbial communities, 377
 microflora, 369
 pH, 299, 309, 315, 317, 358
Rhodospirillum rubrum, 30
Rhynchospora, 411
ribonucleases, 26

ribulose-1,5-diphosphate carboxylase, 67
Riccia fluitans, 418
Riccia rhizoid cells, 60
rice, 140, 142, 217, 338, 393
 phenylalanine ammonia-lyase, 89
 V-PPase, 38
Ricinus communis, 264
Rieske Fe/S center, 182
Rieske iron-sulphur protein, 180
rock phosphates, 368, 392
 apex, 304
 cap, 304, 305, 310, 316
 curvature, 311
 distal elongation zone, 142
 epidermis, 310
 exudates, 331, 390
 organic acids, 352
 growth, 329
 hairs, 7, 307, 309, 315, 389, 390,
 392, 393
 surface, pH at, 117
 tip, 307
rotenone, 201
rubber tree, 240
Rubisco, 212, 234, 236
Rumex palustris, 143
ruthenium, 203
ruthenium red, 204, 205
rye, 329
Rynchospora alba, 413

S states, 151–153
Saccharomyces cerevisiae, 2, 4
 H$^+$-ATPases, 3
 regulatory mutants of plant plasma
 membrane H$^+$-ATPase, 10
Sagina nodosa, 412
Sagittaria, 232
saline swamps, 337
salinity, 230, 237, 243, 260, 365
 CEC, 258
Salmonella, 389
salt stress, 231, 244
 V-ATPase, 37
salt-tolerant plant species, 260
Samanea pulvini, 275
Samolus valerandi, 412
Sansevieria hahnii, 242
Saprolegnia ferax, 91
sapropelium, 417

saprophytic competence, 394
Sauromatum guttatum, 189
Scabiosa columbaria, 408
Schoenus nigricans, 412
Scirpus cyperinus, 339
Scutellospora, 394
seawater, 337
secondary cell wall, 257
secondary silicates, 406
sedoheptulose bisphosphatase, 159
Sedum telephium, 230, 234
seed storage-proteins, 25, 26
selenate, 329, 340
seleniferous soils, 340
selenite, 329, 340
selenium, 340
semi-aquatic wetland ecosystems, 412,
    416
semi-aquatic ecosystems, 405, 410, 419
seminaphthofluoresceins. *See* SNAFL
    probes
seminaphthorhodafluor dye, 75
semiquinone, 178, 179, 182
serine protease, 393
Sesleria albicans, 408
sewage sludge, 338, 359
shepherd's crooks, 389
siderophores, 361
    microbial siderophores, 361
silicon, 256
silver, 335
Sinapis root hairs, 80
Sinorhizobium fredii, 387
Sinorhizobium meliloti, 384–391
site-directed mutagenesis, 189
16S rDNA, 356
SNAFL-probes, 75
sodium, 231, 232
sodium chloride, 365
soil acidification, 327, 328, 332, 353,
    384, 414, 415, 418
soil acidity, 329, 351, 355, 395, 405
soil colloids, 327
soil microflora, 328, 333, 337, 339
soil pH, 351, 357
soil respiration, 355
soil solid phase, 328
soil solution, 323–325, 327, 328, 330
    pH measurement, 328
soluble ions, 324

sorghum, 329, 369
soybean, 273, 329, 385, 391
species biodiversity, 394
Sphagnum, 410, 413
Sphagnum cuspidatum, 418
Sphagnum magellanicum, 413
spinach, 165
    ion channels, 165
spruce, 355, 356, 364
starch, 227, 228, 233, 242
staurosporine, 97
stigmatellin, 182
stomata, 234, 277
stomatal closure, 85, 86, 125
    guard cells, 11
    movement, 227
stomates, 234, 277
Streptomyces scabies, 376
stromal pH, 156
strong ion difference, 263, 272
suberin, 257, 260, 263
subsoil, 327
subsoil acidity, 327
substomatal cavity, 274, 275, 280
succinate, 175, 201
succinate dehydrogenase, 187
succulents, 228
sudan grass, 329
sugar beet, 329
    cold stress, 10
sulfate, 337, 358
sulfide oxidation, 417
sulfur, 366
sulfur deposition, 414
sulfur oxidation, 339
sulfur-oxidizing microorganisms, 353
superoxide anion, 203
sweet potatoes, 329
symbiosis, 392, 393, 395
symbiosome, 387, 391
symplast
    ion transport, 222
symporters, 313

tailings, 339
teichuronic acid, 354
teichuronopeptides, 354
terpenoids, 25
terrestrial ecology, 352
terrestrial ecosystems, 377, 405, 418

tctraethylammonium, 165
tetrahydrofolate, 198
thermodynamic stability constant, 330
thermogenic tissues, 206
thermophilic bacterium, 194
thiamine pyrophosphate, 198
Thiobacillus ferroxidans, 353
Thiobacillus thiooxidans, 353
thiolate, 241
^{31}P-NMR, 50, 51, 74
3-hydroxy-3-methylglutaryl-coenzyme A
    reductase, 89
threonine, 2
thylakoid, 158
  acidification of the lumen, 161, 163,
    166
  efflux of Mg^{2+}, 166
  membrane, 1, 111
    electron transfer components, 149,
      150
thymidine incorporation, 356
Thymus serpyllum, 415
tidal marshes, 337
tip-growing cells, 307
  organs and cells, 309
titratable acidity, 227, 231, 233
tobacco, 2, 8, 61, 310
  benzoic acid, 89
  cultured cells, 80
  cytosolic acidification, 10, 94, 96
  expansins, 135
tomato, 8, 240, 272, 273
  dephosphorylation of H^+-ATPase, 13
  expansins, 135
  mitochondria, 205
  mobile wound signal, 90
  V-ATPase, 37
tonoplast, 60, 61, 111, 112, 238, 239,
    240, 246
topsoil, 328
transfer cells, 261, 277
transmembrane electrical potential. *See*
    membrane potential
transmembrane ion transport
  acid-base relations, 110
  pH gradient, 114, 119, 120
transpiration, 229
trehalase, 363
Trifolium, 390
Trifolium repens, 337

triosephosphate, 236
tropics
  arid, 383
  semi-humid, 383
tropism, 313
trypsin, 11
turgor, 85, 131, 313
turgor pressure, 25
two-component sensor (*actS*), 389
2,4,5,6-tetrachloroisophthalonitrile, 374
$2H^+/Cl^-$ transport, 62
2-oxoglutarate, 219

ubiquinol (QH_2), 176
ubiquinol oxidase, alternative. *See* alter-
    native ubiquinol oxidase
ubiquinol:cytochrome *c* oxidoreductase
    (complex III), 175
ubiquinone, 175, 178, 189
  protonation and deprotonation, 178,
    179
ubiquinone (Q), 176
uncoupling of mitochondria, 198
uncoupling protein (UCP), 196, 200,
    205, 206
unstirred layer, 122, 256, 259, 268,
    301–303
urea, 331
urease, 331
Urticularia minor, 410, 413
Urticularia vulgaris, 414

Vaccinium myrtillus, 412, 416
Vaccinium reichenbachiana, 416
vacuolar H^+-ATPase, 1, 220, 240, 243,
    245, 246
  membrane potential, 24
  pH, 24, 60, 80, 84, 89, 112, 238, 241
  PP_iase (V-PP_iase), 187
  pyrophosphatase. *See* V-PPase
vacuole, 23, 60
  accumulation of Ca^{2+}, 126
  accumulation of metal ions, 25
  ammonium, 217
  analogy with lysosomes, 26
  aqueous vacuole, 35
  buffering capacity, 80
  Ca-oxalate, 65
  carboxylic acids, 25, 39
  H^+-ATPase, 220

vacuole [Continued]
  ion channels, 87, 95
  iron accumulation, 38
  lemon, 73
  lytic vacuole, 35
  organic acids, 220
  pH, 41, 73, 112
  physiological roles, 25, 26
  regulation of lumenal pH, 38, 39
  tannin vacuole, 35
  volume, 39
valency, 327
Vallisneria spiralis, 232, 233
vanadate, 2, 6, 36, 64, 93, 120
Vanilla planifolia, 243
V-ATPase, 232, 240, 241, 244. *See* also
    vacuolar H⁺-ATPase
  chilling, 38
  enzymatic properties, 29, 32
  expression under stress, 37
  lemon, 36
  mRNA, 39
  organ specificity, 36
  photomorphogenesis, 36
  regulation of proton transport, 37
  structure, 26, 27, 29
vetches, 329
vibration amplitude, 301
Vicia faba, 124
  H⁺-ATPase, 6
  leaves, 143
  V-PPAse, 34
Vicia guard cells, 80
videodensitometric analysis, 280
video-microscopy, 194
Vigna radiata, 29
Viola canina, 412, 414
Viola reichenbachiana, 411
violaxanthin, 158, 163
violaxanthin de-epoxidase, 163
Vitis vinifera, 272
voltage clamp, 86
V-PPase
  enzymatic properties, 33
  expression under stress, 37
  lemon, 36
  regulation of proton transport, 37
  rice, 38
  young cells, 34
V-PPase/V-ATPase, 35, 240

wall-loosening proteins (WLPs), 132–
    137, 139–143
Water Free Space (WFS), 139, 256,
    258, 268, 275, 277
water splitting, 153
water stress, 231
water use efficiency, 229
weak acid distribution, 50
weathering, 334, 351
wheat, 271, 272, 329
  genotypic differences in Al tolerance,
    315, 316
white clover, 329
white lupin, 271
woolgrass, 339
wounding, 317

xanthophyll, 163
xenobiotics, 374
Xenopus oocytes, 216, 218
xylans, 257
xylem, 66, 222, 261, 262
  homeostasis, 273
  pH homeostasis, 271, 272
xylem parenchyma, 261
xylem pressure probe, 282
xylem sap, 262, 264, 266
  collection, 277
  pH, 271–273, 282
xyloglucans, 132, 135, 136, 257
xyloglucan endotransglycosylase (XET),
    7, 8, 135, 137, 142, 143

yeast, 2, 3, 26, 197, 218
  absence of V-PPase, 30
  ATP synthase, 192
  bc1 complex, 180
  V-ATPase, 29
yeast mutants affected in uptake of am-
    monium ions, 219

Zannichellia pedunculata, 414
Zea mays, 86, 229, *362*. *See* also corn
  H⁺-ATPase, 6
zinc, 299, 326, 330, 333, 338, 339, 389,
    405
  deficiency, 361
  toxicity, 339
zoospores, 317
zygote, 307

Milton Keynes UK
Ingram Content Group UK Ltd.
UKHW020009071024
449327UK00031B/2716